Sustainable Urbanism in Developing Countries

Sustainable Urbanism in Developing Countries

Edited by

Uday Chatterjee

Arindam Biswas

Jenia Mukherjee

Dinabandhu Mahata

CRC Press
Taylor & Francis Group
Boca Raton London New York

CRC Press is an imprint of the
Taylor & Francis Group, an **informa** business

First edition published 2022
by CRC Press
6000 Broken Sound Parkway NW, Suite 300, Boca Raton, FL 33487-2742

and by CRC Press
4 Park Square, Milton Park, Abingdon, Oxon, OX14 4RN

CRC Press is an imprint of Taylor & Francis Group, LLC

Library of Congress Cataloging-in-Publication Data
Names: Chatterjee, Uday, editor. | Biswas, Arindam, editor. | Mukherjee, Jenia, editor.
Title: Sustainable urbanism in developing countries / edited by Uday Chatterjee, Arindam Biswas, Jenia Mukherjee, and Dinabandhu Mahata.
Description: 1 Edition. | Boca Raton, FL : CRC Press, 2022. | Includes bibliographical references and index.
Identifiers: LCCN 2021052133 (print) | LCCN 2021052134 (ebook) | ISBN 9780367649340 (hardback) | ISBN 9780367675912 (paperback) | ISBN 9781003131922 (ebook)
Subjects: LCSH: Sustainable urban development--Developing countries. | City planning--Developing countries--Environmental aspects. | Urban policy--Developing countries. | Urbanization--Developing countries.
Classification: LCC HT169.E8 S87 2022 (print) | LCC HT169.E8 (ebook) | DDC 307.7609172/4--dc23/eng/20211217
LC record available at https://lccn.loc.gov/2021052133
LC ebook record available at https://lccn.loc.gov/2021052134

ISBN: 978-0-367-64934-0 (hbk)
ISBN: 978-0-367-67591-2 (pbk)
ISBN: 978-1-003-13192-2 (ebk)

DOI: 10.1201/9781003131922

Typeset in Times
by SPi Technologies India Pvt Ltd (Straive)

Dedicated to young scholars in the field of urban studies,
urban planners and urban policy makers

Contents

PART II Urban Planning, Innovation and Urban Resilience for Sustainable Development

PART III Environmental Management and Ecosystem Services

Foreword

It is indeed a pleasure to write a Foreword for *Sustainable Urbanism in Developing Countries*, which covers topics of contemporary relevance. The book is edited by a team of young scholars, Dr Uday Chatterjee, Dr Arindam Biswas, Dr Jenia Mukherjee and Dinabandhu Mahata from the fields of humanities and social sciences, geography, architecture and planning. The book is published by CRC Press on behalf of Taylor & Francis and is dedicated to urban planners and policy makers.

Urban study is an interdisciplinary understanding of towns and cities, and the process of urbanization, urban life and society. People are choosing towns and cities because of their ease of access to healthcare, education, opportunities to make a living, and efficient infrastructure, which together provide them with a better quality of life. In view of this, urban areas have developed enormous advantages in comparison to the rural areas as centers of human habitation. Since maximum GNP growth is recorded from the cities, urbanization equips countries to compete with other countries in the global arena on the economic front. Concentration and density of people in urban areas facilitates agglomeration of various socio-economic services such as health, hygiene, sanitation, education, water, energy and other essential services at a much cheaper price.

The vast expansion of urban areas and their exploding population in developing countries are starting to reveal their global implications. Cities are emerging as the focal points of transformation in the areas of environment, economy and demography. Now the world has reached an invisible and momentous milestone, where more than 50% of its population are urban. It is expected that by the end of 2030, the urban population may grow up to 5 billion, with most of them from the poorer sections of society. The future of these cities and towns in the developing world will depend on how the challenges of planning and development are envisioned in a sustainable manner. With growing urbanization, cities are emerging as melting points of the new global order challenged by political, economic and technological transformations. In view of the growing industrialization in cities with increasing rural–urban migration and the growth of squatter settlements, urban environments are posing a serious challenge in terms of their sustainability. Needless to say, it is the towns and cities of the less developed countries that are going to configure the economic growth, environmental sustainability, population stabilization and alleviation of poverty for a peaceful life.

The present volume is a collection of 24 research papers from across the academic communities of the globe covering topics ranging from changing land use and land cover to urban industrialization and pollution, urban migration and refugees, geopolitical urban networking, resilience to climate change, urban environmental planning and sustainable development. The themes identified and captured in the book are vital components of the urban environment and the issues addressed are of utmost importance in view of the growing urbanization of the 21st century. I would like to convey my thanks to the editors and congratulate them on venturing to bring out this valuable volume for readers of diverse academic disciplines in the areas of urban planning and sustainability. The case-study approach used in the chapters, with examples from across the globe covering diverse economies, brings the insights of the authors and editors to bear upon the regional pattern of urbanization and related issues of the new millenium. I hope the book will be widely acclaimed by urban planners, policy makers and scholars of allied disciplines on the emerging issues of urbanism of the new millennium.

Dr Gopal Krishna Panda
Professor Emeritus
Department of Geography
KISS Deemed University, Bhubaneswar
Odisha, India

Preface

We are in the 'Urbanocene', with more than 50% of the global population residing in urban areas, according to the United Nations' *World Urbanization Prospects*. World Bank data suggest that by 2045, the urban population of the world will increase 1.5 times to 6 billion. It is within this context that contemporary urbanization has attracted wide attention from academia, civil society and policy circles, seeking to address issues of sustainability through inclusive and resilient theories and action plans. What is remarkable and atypical of our times is the rate and scale of shifting geographies of urbanization with the urban locus anchored in the Global South. The numerous challenges associated with the nature and pattern of urbanization of the southern peripheries have been analyzed by stalwart urban scholars from David Harvey to Mike Davis to the more recent theorizations on 'peri-urbanization' by Adriana Allen, the President of the Habitat International Coalition (HIC) and Vice-Dean International at the Bartlett Development Planning Unit, University College London. However, due to the complexities and non-linear feedback loops of urban processes and dynamics, and the spatiocultural and political heterogeneity of the Global South, more and more empirical case studies offering 'place-based' narratives remain imperative. Moreover, urban research awaits the deployment of innovative methodologies situated at the crossroads of quantitative-qualitative, physical-social, historical-contemporary and micro-meso-macroscale analytics to unravel the diversity, complexity and dynamism imbricated in the 'urban'. *Sustainable Urbanism in Developing Countries* accomplishes this larger agenda by addressing various facets of urbanization from integrated perspectives across a wide and diverse spatial and temporal range.

The book comprises four Parts: (I) Local, Regional and Global Perspectives on Sustainable Urbanism, (II) Urban Planning, Innovation and Urban Resilience for Sustainable Development, (III) Environmental Management and Ecosystem Services and (IV) A Multidimensional Approach to Urban Ecology, Space, Society and Sustainability. The 24 chapters capturing the diverse and complex trends of urbanisms, informed by case studies and discussions from different pockets of the developing world, from Middle Eastern countries like Syria and Iraq (Chapter 7) to urban areas of island countries in the (south) Pacific region (Chapter 12) to Africa (Chapters 13, 14 and 20). There are chapters on South Asian cities from megalopolitan centers like New Delhi (Chapters 11 and 16) to less explored small towns like Berhampore Municipality, West Bengal (Chapter 18) and the peri-urban interface of Kolkata (Chapters 1 and 15). The novelty of this edited volume lies not only in its spatio-temporal breadth but also in the wide coverage of chapters showcasing innovative methodological techniques: cross-fertilization of geospatial data with qualitative field findings (Chapters 15 and 19); explorations of urban temporal morphology by studying racial, socio-economic and sectarian elements determining spatial expansion (Chapter 6); and formulation and application of the geopolitical urbanism network (GUN) framework to explore urban armed conflicts finally suggesting how the notion of 'sustainable peace' can be facilitated and eliciting the generative capabilities of SDGs (Chapter 7). Some chapters offer a detailed review of urban policy designs by mapping their effectiveness and efficiencies and how citizens navigate through needs-based informal mechanisms against the lack of policy-based interventions in urban fringes. Finally, moving beyond declensionist portrayals of contemporary urbanization in developing countries, various chapters suggest radical recommendations for urban planning and governance based on comprehensive understandings of specific urban scenarios across the application of robust multi-modal methodologies. Issues such as the Rural-Urban Continuum (RUC) analysis that combines census data with ArcGIS software for Kerala (Chapter 8), discussions of land value capture (LVC) finance as an alternative fiscal arrangement to generate revenues from private lands (Chapter 9), Sacred Land Capture (SLC) through pictographic models and its socio-ecological implications for urbanities (Chapter 23), and so on have crafted the radical transition from knowledge generation to action in the urban frontier. This volume can be considered a breakthrough work in terms of its epistemological, methodological and axiological rigor in facilitating integrative sciences towards just, sustainable and resilient urban trajectories.

Uday Chatterjee
Arindam Biswas
Jenia Mukherjee
Dinabandhu Mahata

Acknowledgments

First and foremost, we are thankful to the contributors from different parts of the world who kindly and enthusiastically shared their hard-earned research findings with us. The formulation of the book has been carried out under the experienced supervision of various urban planners, city planners and urban policy makers. We owe a great debt to these experts for providing meticulous and insightful feedback on the chapters in this book. We would also like to express our gratitude to the anonymous reviewers whose scholarly suggestions refined and enriched the quality of the book. The preparation of this book was a time-consuming job for the editors. Our genuine acknowledgment goes out to our families for their matchless care and encouragement. This work would not have become what it is without relentless encouragement from our students, teachers who kept us going, generous appreciation from our colleagues and collaborators, and unflagging assistance from our families, particularly during the difficult times of the Covid-19 pandemic. Professor Virendra Kumar Tewari, Director of the Indian Institute of Technology Kharagpur, provided the zeal and motivation to young faculty members to pursue cutting-edge urban environmental research that has the potential to be translated into social action. We are sincerely grateful to Professor Ajit Okay Chaturvedi, Director of the Indian Institute of Technology, Roorkee, and Dr Pabitra Kumar Mishra, Principal of Bhatter College, Dantan for their constant academic mentorship and encouragement. Finally, we also thank our publisher, CRC Press, Taylor & Francis Group, senior editor Irma Shagla Britton and editorial assistant Shannon Welch for their continuous support and vision in the publication of this book.

Editors

Uday Chatterjee, PhD, is an Assistant Professor at the Department of Geography, Bhatter College, Dantan, Paschim Medinipur, West Bengal, India and an applied geographer at Utkal University. He holds a doctoral degree in Applied Geography at Ravenshaw University, Cuttack, Odisha, India. He has contributed various research papers to reputed national and international journals and edited book volumes. He has jointly edited *Harmony with nature: Illusions and elusions from geographer's perspective in the 21st Century*. He has also convened a Faculty Development Program on 'Modern methods of teaching and advanced research methods' sponsored by the Indian Council of Social Science Research (ICSSR), Government of India. His areas of research interest cover urban planning, social and human geography, applied geomorphology, hazards and disasters, environmental issues, land use and rural development. His research work has been funded by the West Bengal Pollution Control Board (WBPCB), Government of West Bengal, India. He has served as a reviewer for many international journals. Currently Dr Uday Chatterjee is the lead editor of the Special Issue of *Geojournal*, on Urbanism, Smart Cities and Modelling (Springer).

Arindam Biswas, PhD, is an Associate Professor at the Department of Architecture and Planning, Indian Institute of Technology (IIT) Roorkee, India. He holds a doctorate in urban planning from the University of Tokyo. He is a recipient of the Institute's research fellowship for outstanding young faculty, ASEM Duo-India Fellowship, DAAD Fellowship (Germany) and MEXT Scholarship (Japan). He has also undertaken a Certificate Program on International Environmental Leadership (Asian Program for Incubation of Environmental Leadership, APIEL) from the University of Tokyo. He is presently pursuing research sponsored by different research foundations of the Government of India, including the Department of Science and Technology (DST), Ministry of Education, Indian Council of Social Science Research, Ministry of Environment, Forest and Climate Change and National Building Construction Company. He is also working on research sponsored by the Royal Academy of Engineering and GCRF-UKRI, United Kingdom. He has edited two books and contributed to four book chapters. He is currently working on three books to be published next year. He has also extensively published articles on urban infrastructure, inclusive growth, resilience, housing and informality, innovation and regional development in peer-reviewed journals.

Jenia Mukherjee, PhD, is an Assistant Professor at the Department of Humanities and Social Sciences, Indian Institute of Technology Kharagpur. Her research interests span across urban studies and environmental history. She received the Australian Leadership Awards Fellowship in 2010 and 2015. She completed her book, *Blue infrastructures: Natural Ecology, Political History and Urban Development in Kolkata* (Singapore: Springer Nature, 2020), has published extensively in peer-reviewed impact factor journals including *Environment and Planning E, International Journal of Urban Sustainable Development, Urban Research and Practice*, and has contributed book chapters. She is a member of the Editorial Advisory Board for Routledge's Equity, Justice and the Sustainable City series edited by Julian Agyeman.

Dinabandhu Mahata is a researcher at the Department of Geography, Central University of Tamil Nadu and a former research officer at the Department of Development Studies in the International Institute for Population Sciences (IIPS), Mumbai, India. He received a Government of India Fellowship while studying for his Master of Population Studies (MPS) and Master of Philosophy (MPhil) at the Department of Development Studies in IIPS, Mumbai, India. His work mainly focuses on urban geography and its link with the environment. He has written journal articles and book chapters. He received a research grant from the International Union for the Scientific Study of Population (IUSSP) for the International Population Conference in Cape Town, South Africa.

Contributors

Mohammad Afsar Alam, PhD, is an Assistant Professor of Geography in the Department of Social Science, College of Humanities and Education, Fiji National University, Fiji. His areas of research interest cover urban planning, social and human geography, hazards and disasters, tourism geography, climate change and environmental issues. He has contributed many research papers in various reputed national and international journals and edited book volumes. He has served as a reviewer for many international journals.

Mumtaz Alam, PhD, is an Assistant Professor in the Department of Social Science, College of Humanities and Education, Fiji National University, Fiji. His areas of research interest cover the history of urbanization, medical history, environmental humanities, ecotourism and ICT education with an interdisciplinary approach. He has contributed research papers in various reputed national and international journals. He is an Associate Editor for *Heliyon*, is on the editorial and advisory boards and reviews for many international journals.

Nestor Asiamah is an international public health researcher whose research network extends to Africa, Europe, North America, and Oceania. His research interests focus on physical activity, health research methodology, healthy aging (social gerontology), and health management. He leads an international research team and serves as the Executive Director at the Africa Centre for Epidemiology (ACE).

Haimanti Banerji, PhD, is an Associate Professor in the Department of Architecture and Regional Planning, Indian Institute of Technology Kharagpur, West Bengal, India. Her research interests cover sustainable urban planning, urban design, land-use planning and universal design. She has been the Principal Investigator for several prestigious projects in India such as Inclusive City under BASIIC, Declaration of Cooch Behar as a Heritage Town, and the redevelopment of the SAIL township at Burnpur.

Parama Bannerj, PhD, is Assistant Professor and Head of the Department of Geography, Nababarrackpore Prafulla Chandra Mahavidyalaya, West Bengal, India. Her areas of research interest cover urban planning, social and human geography, environmental issues, land use and rural development. She had been a Commonwealth Scholar (UK) and ICSSR Doctoral Fellow. She has contributed various research papers published in reputed national and international journals and edited book volumes. She has served as a reviewer for many international journals.

Jwngma Basumatary is an environmental planner who has worked as a project associate/consultant for the Department of Environmental Planning, School of Planning and Architecture, Delhi on various projects. He is well versed in various software tools and applications. He also holds a Google certification in data analytics.

Scott Berson is a regional and community planner at the Western Piedmont Council of Governments in Hickory, North Carolina. He was awarded a master's degree in community planning at Auburn University in 2020 and is an active member of the American Planning Association.

Aakanksha S. Bhargava is a PhD research scholar in the Department of Geography, Cotton University. She has research interests in regional, social and urban geography and is working under the guidance of Dr Barnali Gogoi (Associate Professor, Department of Geography). Currently her area of study is aging and its different aspects.

Srabanti Bhattacharya, PhD, is Principal, Saheed Anurup Chandra Mahavidyalaya, Burul, West Bengal. Her areas of research interest cover urban water supply, urban planning, hazards and disasters, and environmental issues. She has contributed several research papers published in various reputed national and international journals and in edited book volumes. She was awarded the CSSSC-NRTT Post-Doctoral Fellowship in 2009 and completed her post-doctoral research at the Centre for Studies in Social Sciences, Calcutta (CSSSC) in 2010. She also received research grants from UGC and Centre for the Study of Developing Societies (CSDS), New Delhi.

Rubia Biswas is an Assistant Professor in the Department of Geography of Raja Narendra Lal Khan Women's College, Medinipur, West Bengal, India. Her research interests are hazard and disaster studies, political economy, environmental development, and urban development.

Nilendu Chatterjee, PhD, is an Assistant Professor in the Department of Economics, Bankim Sardar College, West Bengal, India. Among his research interests are resource economics, general equilibrium, development economics, etc. He has published research papers in various national and internationally renowned journals, including the *International Journal of Sustainable Economies Management*, *Economic Affairs*, and *Foreign Trade Review*.

Uday Chatterjee, PhD, is Assistant Professor in the Department of Geography, Bhatter College, Dantan, Paschim Medinipur, West Bengal, India. His areas of research interest cover urban planning, social and human geography, applied geomorphology, hazards and disasters, environmental issues, land use and rural development. He has contributed various research papers to reputed national and international journals and edited book volumes. He has served as a reviewer for many international journals.

Shalini Chaudhary is a PhD research scholar at the Rekhi Centre of Excellence for the Science of Happiness, IIT Kharagpur. She graduated with honors in political science from Lady Sri Ram College for Women, University of Delhi and thereafter earned a master's in political science at MDS University, Ajmer. Her research interests include peace and conflict resolution, technology in conflict management, international relations, and international security.

Roy Alexander Chileshe is an Associate Professor in Land Management, Department of Urban and Regional Planning, at the Copperbelt University, Witwe, Zambia. He earned a PhD at the University of Western Cape, an MSc at ITC the Netherlands, and a BSc (Hons) at East London University. He is a member of the Surveyors Institute of Zambia, Zambia Institute of Estate Agents, and Zambia Institute of Planners.

Anuradha Choudry, PhD, is a joint faculty at the Department of Humanities and Social Sciences and the Centre of Excellence for Indian Knowledge Systems, IIT Kharagpur. A multilingual Sanskritist, she works in the areas of Indian psychology, philosophy and culture and is a resource person for national and international organizations.

Susan Cyriac is currently a research scholar at the National Institute of Technology, Calicut. Prior to her academic career, she worked as an assistant manager in the strategic consultancy wing of Jones Lang LaSalle India Private Limited. Her areas of interest, apart from property consultancy, include regional planning and interdisciplinary studies.

Pratik Dash, PhD, is currently working as Assistant Professor at Khejuri College. He has more than 11 years of research experience, including three years' teaching experience. He is an expert in remote sensing and GIS, hydrological modeling, flood modeling, disaster management, etc. He has published several articles in Scopus-indexed journals.

Nandita Deb is currently working as Assistant Professor in Adamas University, Kolkata, India. She earned a master's degree at Adamas University in 2017 with a specialization in environmental geography. Her major research interests are urban environmental problems, environmental change and food security, etc. She received two Best Paper awards in 2018 and 2020 and a Junior Scientist award in 2020.

Meenakshi Dhote, PhD, is currently the Head of the Environmental Planning Department at the School of Planning and Architecture, New Delhi. She has 30 years of professional, research and teaching experience in the fields of architecture, landscape architecture and environmental planning. Her areas of interest involve conservation of biodiversity in the urbanized environment and hill area development. She has been engaged in many national projects addressing aspects of planning in eco-sensitive zones, environmental impact assessments of townships and so on.

Megersa Olumana Dinka, PhD, is an Associate Professor in the Civil Engineering Science Department of the University of Johannesburg, South Africa. His areas of research cover water management, hydraulics and hydrology. He has more than 15 years' experience as an academic and researcher and has supervised several post-graduate students. He has published and reviewed several articles in peer-reviewed journals.

Mohammed Firoz C, PhD, is an architect and urban planner by profession. He graduated in urban and regional planning from CEPT University, earned a PhD at IIT Kharagpur and a BArch at NIT Calicut, for which he was a university rank holder. He has been involved in teaching, researching and consulting at NIT Calicut since July 2004. He was also engaged as a visiting teacher at the Architectural Association London (AA London) for the term May–June 2015. His field of interest includes rural–urban interface studies, sustainable urban design and planning, regional development and planning, etc.

Raktima Ghosh is a lecturer in geography at the Bhawanipur Education Society College, Kolkata, India. She is a research assistant at the Indian Institute of Technology, Kharagpur (India), for the Dried Fish Matters project, undertaken by the University of Manitoba and funded by SSHRC (Canada). Her research interests include political ecology, human geography, small-scale fisheries and environmental governance.

Sudeshna Ghosh, PhD, is an Associate Professor in the Department of Geography and Regional Planning, Indiana University of Pennsylvania, United States. She is a scholar in the areas of urban and regional planning, community development planning, land-use planning, spatial modeling, and planning in the developing world. She teaches planning history, global cities, planning methods, seminar and study abroad courses. She received several small grants for her research work in the areas of small-town economic development in the US and issues of urban poor and slum settlements in the developing world.

Barnali Gogoi is an Associate Professor and former Head of Department of the Department of Geography, Cotton University, Guwahati (Assam). Her research interests are fluvial geomorphology, urban and social geography.

Somenath Halder, PhD, is Assistant Professor in the Department of Geography, Kaliachak College, Sultanganj, Malda, West Bengal, India. His areas of interest cover social and cultural geography, regional studies and regional planning, human ecology, urban ecology, political ecology, environmental sustainability, and development issues. He has a number of publications in international reputed journals. He has also contributed as peer reviewer and board member to several international journals.

Md. Asif Iqubal, PhD is currently a guest faculty and GIS expert in the Department of IDRS and GIS application at Aligarh Muslim University, Aligarh. His research interests include urban planning, agriculture, livestock rearing, social, rural and economic geography, climate change and environmental issues. Dr Iqubal received the Young Geographer's Award at an international conference held in the Department of Geography, Aligarh Muslim University, Aligarh, 2009.

Sajed Bahrami Jaf is a PhD student of geopolitics at Ferdowsi University of Mashhad, Mashhad, Iran. His areas of interest are geopolitical relations, the political arrangement of space, and borders. He has collaborated in Persian and English research articles.

Niraj Jain is a Senior Lecturer in the School of the Built Environment at the Copperbelt University (Zambia)He earned a BSc and MPhil in land economy at the Copperbelt University and University of Cambridge, respectively. He has 20 years' experience in academia and real-estate consultancy. His works are mainly inclined towards land governance with emphasis on customary land.

Mohsen Janparvar, PhD, is Assistant Professor in the Department of Geography, Ferdowsi University of Mashhad, Iran. His areas of research interest cover border studies, political geography, geopolitical issues and environmental problems. He has contributed various research papers and books published in Persian and English, and he has undertaken some analysis and attitude about politics and different issue in Persian and English. He has served as a reviewer for many international journals.

Bappaditya Koley, PhD, is head of the Department of Geography, Bankim Sardar College, West Bengal, India. He is involved in investigating in the field of applied geomorphology and landslide disaster mitigation studies. His areas of research interest cover landslide hazard, vulnerability and risk assessment, landslide early warning systems, coastal risk assessment, coastal planning and management, climate change, environmental issues, and so on. He has published research papers in various national and international reputed journals.

Sukanya Koner is a project assistant in the Department of Architecture and Planning, Indian Institute of Technology, Roorkee, Uttarakhand, India. She holds a bachelor's degree in planning. Her research interests include regional economics, urban and regional development, urban clusters, environmental issues and sociological challenges. She earned an Erasmus Mundus Scholarship for 4CITIES program for the academic year 2021–2023.

Sudeshna Kumar is a doctoral student in the Department of Architecture and Regional Planning, Indian Institute of Technology Kharagpur, West Bengal, India. Her areas of research interest cover neighborhood sustainability, sustainability science, sustainability metrics and urban resilience. Currently she has been selected as the EPP Fellow 2021 at Eawag-ETH, Zurich. She has served as a reviewer for the *International Review for Spatial Planning and Sustainable Development.*

Sakul Kundra, PhD, is Assistant Professor of History in the Department of Social Science, School of Arts and Humanities at the Fiji National University. He earned a PhD in history at Jawaharlal Nehru University (India) and is the recipient of the Gold Medal for PGD in Education, USP. He has many high-ranked publications, research projects, conference presentations and op-eds to his credit. His expertise includes urbanization, historiography, Pacific history, tourism, and European travelers.

P.L.C.B. Liyanage is a town planner in the Urban Development Authority, Sri Lanka. She graduated from the University of Moratuwa, Sri Lanka and is currently following a master's degree in geographic information systems and remote sensing at the University of Sri Jayewardenepura Sri Lanka. Her areas of research interest cover urban planning, social and physical planning, disaster management and use of geospatial technology.

Jay Mittal, PhD, is an Associate Professor in the Master of Community Planning Program (MCP) at Auburn University, Alabama, USA. He has over 22 years of professional experience in private consulting, research and academic settings. His research and teaching interests are local economic development, land value capture, land markets, real-estate valuation, market analysis, and plan making both in the United States and in India.

Hassan Momin is a Junior Research Fellow at the Institute of Development Studies Kolkata at the University of Calcutta, West Bengal, India. His areas of research interest cover hazard and risk analysis, environment and development, river basin management, and urban climate. He has participated and presented research articles at national/international level and has contributed research articles to several reputed national and international journals and edited book volumes.

Kasturi Mukherjee, PhD, is an Associate Professor and Head of the Department of Geography, School of Basic and Applied Sciences, Adamas University, and a Visiting Researcher at Stockholm Resilience Centre, Stockholm University. She earned a PhD in urban climate at the University of Calcutta. She completed a post-graduate degree in geography with a specialization in environmental geography and a post-graduate diploma in geoinformatics. She is a gold medalist, securing 1st rank in the university in post-graduation and has twice received the Best Paper award, in 2010 and 2018. Her research interest focuses on urban climate, sustainable development and peri-urban development.

Ephraim Kabunda Munshifwa, PhD, is an Associate Professor and Dean of the School of the Built Environment at the Copperbelt University in Zambia. He earned a BSc and MPhil in land economy from Copperbelt University and the University of Cambridge, respectively. He also holds a PhD from the University of Cape Town.

Anthony Mushinge, PhD, is a senior lecturer in the Department of Real Estate Studies, Copperbelt University. He holds a BSc in Land Economy from the Copperbelt University in Zambia as well as MSc and PhD in Land Management from the Technical University of Munich in Germany. His research interest is customary and statutory land governance and corruption in the land sector. He has published journal articles and book chapters in various reputed international journals and books.

Anindita Nath earned an MSc in geography and presently works as a state-aided college teacher in the Department of Geography, Bankim Sardar College, West Bengal, India. She researches coastal geomorphology and hazards studies. Her areas of research interest include coastal hazard, vulnerability and risk assessment, coastal planning and management, landside hazards studies and environmental issues. She has published research papers in various national and international reputed journals.

Joan Nyika, PhD, is Assistant Lecturer in the Geoscience and the Environment Department of Technical University of Kenya, Nairobi. Her areas of research interest cover various environmental science topics including hydro-biogeochemistry, environmental chemistry, bioremediation and biochemical techniques and instrumentation. She has published several articles in peer-reviewed journals and various book chapters in edited books. She is also involved in mentoring under- and post-graduate students.

Abdul Odud is a Junior Research Fellow in the Department of Geography, Faculty of Science, Aligarh Muslim University, Aligarh, India. His research interests include urban geography, social geography, and population geography.

Himanshhu Panwar is an architect and environmental planner whose expertise and experience is in the analysis and projection of various issues related to the structure and function of the urban environment at different local, urban and regional scales. He has worked as an external consultant at the National Institute of Urban Affairs (NIUA) and as a consultant at the School of Planning and Architecture, New Delhi, undertaking various projects in the field of environmental planning.

Ravinesh Rohit Prasad, PhD, is an Assistant Professor in the Department of Social Science, College of Humanities and Education, Fiji National University, Fiji. His research interests include urban and rural geography, biogeography, human geography, climate change and education. He has contributed various research papers to reputed national and international journals.

H.M.R. Premasiri, PhD, is a Professor in Department of Earth Resources Engineering, University of Moratuwa, Sri Lanka. He has been in academia and research for over 25 years. His main research areas are remote sensing, GIS, geology, geophysics and disaster management. He has published many research papers in national and international journals.

D.K.D.A. Ranaweera is a lecturer in the Department of Geography, Faculty of Humanities and Social Sciences, University of Sri Jayewardenepura, Sri Lanka. Her areas of research interest cover urban geography, settlement geography, geographic information systems, and remote sensing. She has contributed various research papers published in various reputed national and international journals.

R.M.K. Ratnayake, PhD, is a Professor of Geography, Department of Geography, University of Sri Jayewardenepura, Nugegoda, Sri Lanka. His areas of research interest cover urban planning, settlement geography, regional development, medical geography and applications of geospatial technology for many areas. He has published many books on geography, various research papers in reputed national and international journals and contributed as an editor for government publications such as books, articles and chapters. He has served as a reviewer for many national and international journals.

N.M. Rijesh completed his post-graduate studies in traffic and transportation planning at NIT Calicut, Kerala, India. He is currently doing further post-graduate work in urban planning at NIT Calicut, on secondment from the Local Self Government Department (Planning), Government of Kerala, India. His areas of interest cover urban planning, transportation planning, environmental issues, and urban rural interface studies. He has been associated with the preparation of master plans for major urban centers in the state of Kerala, such as Kozhikode and Kannur.

Abira Dutta Roy, PhD, is an Assistant Professor at Bankura Zilla Saradamani Mahila Mahavidyapith. With over six years of teaching experience and 15 years of research experience she has a considerable number of research publications to her credit. She specializes in the application of remote sensing and informatics in the environment.

Reyhaneh Salehabadi, PhD, earned a PhD in political geography and geopolitics at Tarbiat Modares University. She has written several articles and books in Persian and English in the fields of national power, borders, and hydropolitics. She is a researcher at the Center for African Studies at Tarbiat Modares University.

T. Athulya Satheesh is a research associate in the Department of Architecture and Planning, Indian Institute of Technology, Roorkee, Uttarakhand, India. Her research interests are innovations in urban planning, use of remote sensing data for spatial investigations, localization of planning techniques, equity and social welfare.

Puspita Sengupta, PhD, is an Assistant Professor in the Department of Geography, Purash Kanpur Haridas Nandi Mahavidyalaya, Howrah, West Bengal, India. Her areas of research interest cover urban planning, social and human geography, land use and rural development. She has contributed various research papers published in various reputed journals.

Mahboubeh Shahbazi earned a master's degree in geography and urban planning and is interested in such topics as urban–rural relations, political arrangement of space, borders, environmental geopolitics, development of border cities, and has compiled articles on some of these issues in Persian.

Anugya Shandilya is a master's student at the School of Planning and Architecture, Bhopal, Madhya Pradesh, India. She earned a bachelor's degree in planning at the National Institute of Technology, Bhopal, Madhya Pradesh, India. Her research interest is focused on geographic information systems in context with environmental issues.

Kalyan Nath Somavarapu is a Research Fellow in the Department of Earthquake Engineering at the Indian Institute of Technology Roorkee, Uttarakhand, India. He earned a Master's in technology in infrastructure systems at the Centre for Transportation Systems at IIT Roorkee. His research is focused on geographic information systems and urban growth studies.

Sruthi Krishnan V completed post-graduate studies in energy and environmental engineering at VIT University, India. She is pursuing a PhD at the Department of Architecture and Planning at the National Institute of Technology Calicut, India. Her research interests include environmental planning, regional planning, spatial analysis, and satellite image analysis.

Chandrakala Tamang is a Junior Research Fellow at Kazi Nazrul University in West Bengal. She earned an MPhil at the Institute of Development Studies Kolkata under the University of Calcutta. Her diverse research interests include urban planning and development, regional development, and environmental hazards.

Anil S. Thakur, PhD, from the University of Melbourne (Australia), Architecture Building and Planning, is a freelance researcher, planner, architect, and visiting professor to many institutes. He worked in Kuwait before setting up a practice in Chandigarh in 1981. In his 21 years of practice, he has authored more than 150 projects of different scales and types. Presently working from Melbourne, he is writing a book on Chandigarh, the subject of his thesis. His research interests cover the politics of sustainability; smart technologies centered-urbanism(s); district urbanism; resilient, regenerative architecture; and assemblage theory.

Part I

Local, Regional and Global Perspectives on Sustainable Urbanism

1 Alternative Rehabilitation Program against Agricultural Land Acquisition for New Development Projects

A Case Study of New Town, Kolkata, West Bengal, India

Puspita Sengupta
Purash Kanpur Haridas Nandi Mahavidyalaya (University of Calcutta), Howrah, India

CONTENTS

1.1 INTRODUCTION

Land is a vital natural resource which is limited and immovable. Land plays a dynamic role in determining the economic, societal and cultural progression of mankind. Since the dawn of human civilization, men have been domesticating land in their need for food, fodder and shelter. However,

DOI: 10.1201/9781003131922-2

such land conversion has severely impacted the landscape over the last 300 years through the period of post-industrial revolution. Land conversion is 'the process by which land is converted from one use to another like agricultural to urban use and from one urban use to another' and 'is indeed a normal outcome of the urban development process' (Firman 1997). With the increasing dependency of the majority of the global population on land for their livelihood, agricultural land conversion is not merely a matter of environmental concern but also significantly related to social well-being. According to Grigg, in England and Wales in the 1930s, about 240,000 hectares/year of farmland was lost, while about half a million hectares of farmland was converted to non-farm uses during 1940–1950 in California (Grigg 1995). Taiwan has also witnessed significant reduction of cultivated land area by approximately 17,000 hectares, from about 907,000 hectares in 1980 to about 890,000 hectares in 1990. 'A large portion of the area lost to farming has been converted to industrial, urban and other development uses' (Tsai 1993). Estimates show that the annual increase in built-up areas was around 3.6% in developing countries between 1990 and 2000 (Angel et al. 2005). Among world regions, East Asia, including the Pacific, and Southeast Asia witnessed the most intensive land consumption, with growth rates of 7.2% and 6.4%, respectively (Siedentop and Nuissl 2021). In the year 2000, there was more than 650,000 sq. km of urbanized land globally, while another 550,000 sq km of fringe area is waiting to be urbanized by 2030 (Berg 2012). In respect of the European Union, Kuemmerle et al. (2016) made this observation in relation to land conversion:

> The most widespread changes in the extent of land-use categories in the EU between 1990 and 2006 were cropland decline (~136,660 km^2), followed by expansion of grazing land (~75,670 km^2), and expansion of forest areas (~70,630 km^2). The least common conversion among broad land-use categories was urban expansion (~16,820 km^2). … At the European scale, these area changes translate into moderate land-conversion rates in the agricultural sector between 1990 and 2006, ranging from −13.4% for permanent crops to +6.5% for meadows and pastures, while urban areas expanded by approximately 21%.
>
> (**Kuemmerle** et al. 2016)

The various reasons behind the development of the built-up areas include rapid and uncontrolled population growth, industrial development and economic growth which have continuously necessitated transformation of land-use patterns (Dutta et al. 2019). The changes in land use in the suburbs are mainly due to the expansion of urban areas due to the urban sprawl (Bhat et al. 2017) or the development of new towns and the growth of the public transport system which permits a significant amount of the labor force to commute daily (Karasian et al. 2017). The urban areas expand outwards naturally, leading to conversion of agricultural and other natural land cover types into built-up area (Dutta et al. 2020). This land- use change causes global environmental changes (Turner et al. 1994) and the livelihood of the primary sectors are also at stake.

India has witnessed speedy growth in urbanization as the urban population rose from 25.72% (1991) to 27.86% (2001) and in 2011, it was 31.16%. Another data reveals that about 42% of the total urban population (311.7 million in 2011) lives in metropolitan cities resulting in overcrowding and congestion. Average population density of the top seven metropolitan areas in India remains as high as 3,768 persons/km^2 (2011). To cope with these problems, new towns are being developed in the vicinity of major metropolises, such as Navi Mumbai near Mumbai, New Town Kolkata near Kolkata and Gurgaon near Delhi. Under the regime of neo-liberalization, these new towns are generally developed by converting agricultural land to built-up areas as they are established purely on a 'plantation' basis (Beresford 1967). In this conversion process generally three parties with unequal bargaining power are involved, namely developers, government and the land owners, with the environment, identified as the mute fourth party to this conversion process, mostly remaining unattended and ignored. This land conversion has required huge investments which are generally carried by private investors, which in turn may lead to job creation and a rise in the income of local people. Moreover, the creation of infrastructure speeds up economic development as well as facilitating foreign direct investments, leading to an increase in export revenues and helping maintain the national trade balance (Lohr 2010). On the other hand, additional revenues in the form of different taxes are earned by the local government, and these can be spent on welfare programs. Hence, land conversion and planning for new developments usually occur under the nexus of interest of private investors and the government. Land owners from primary occupations, such as farmers, fishermen and forest dwellers, are obviously weaker parties in the negotiation process, with limited ability to resist land conversion. Also, land conversion is often preceded by land acquisition and eviction of the original inhabitants (both people and other creatures). Though there is land legislation to protect individuals' property rights in most countries, in the case of India, the state enjoys the legitimate power to acquire private land for defined 'public purposes' or is indirectly legitimized to displace people from their land for the sake of 'public interest'. Estimates show that 60 million Indians were displaced between independence and 2004, and over 25 million hectares of land, including 7 million hectares of forest land, have been converted for various purposes (Fernandes 2008). 1.5 million hectares of land with a capacity to produce 1 million tons of food grain has been converted to special economic zones (SEZs) in 2005 and at least 1.4 million households lost their land and occupation. Another 8.2 million agricultural workers have lost their earnings in this process (SEZs and Land Acquisition Factsheet for An Unconstitutional Economic Policy). Therefore, displacement brings with it the risk of 'landlessness, joblessness, homelessness, marginalization, increased morbidity, food insecurity, loss of access to the common property and social dis-articulation' (Cerena 1995). State-induced displacement stands against the democratic process of participatory development. Acceptability of abrupt non-sequential changes in land use through massive land conversion and mass displacement for the 'public interest' is now questionable. With this dilemma, a serious research work is needed to understand the holistic implications of agricultural land acquisition, so as to develop a set of land policies with an eye to minimizing the negative effects of acquisition by suggesting alternative rehabilitation policies.

1.2 STUDY AREA

The study has been conducted in New Town Kolkata, located adjacent to the city of Kolkata (10 km away from central Kolkata). The New Town is encircled by Rajarhat Road in the north and the Krishnapur Kata Khal (canal) in the south. The eastern boundary of Calcutta Metropolitan Area is on its western side. Salt Lake Township is located to the south-west of the site. Netaji Subhas Chandra Bose International Airport area is situated in the north-west of the northern boundary (Figure 1.1). The East Calcutta Wetland, including a Waste Recycling Zone, is located in the south.

1.3 PLANNING AND POLICIES BEHIND THE DEVELOPMENT OF NEW TOWN KOLKATA

1.3.1 Planning Stages

The Additional Director of Planning – II, Calcutta Metropolitan Development Authority (CMDA) in collaboration with the Directorate of Housing, Government of West Bengal, prepared the initial project report on 'New Township DumDum-Rajarhat' in 1982. The fundamental purpose of this new area development project was to encourage economic growth by creating employment opportunities across all strata of the population and provide basic residential facilities for all, with a specific focus on the economically weaker section (New Township DumDum-Rajarhat Preliminary Project Report 1982). Work began on the plan in 1993 when the Government of West Bengal established a Department of Housing for the progress of a new township. A master land-use plan for the township was prepared by the Department of Architecture and Regional Planning of I.I.T., Kharagpur. The plan (Figure 1.2) emphasized the 'environment friendly' township, by allotting almost half of the project area to open space (New Town at Rajarhat, Project Report 1995). Later in 2012, WBHIDCO laid more emphasis on the residential (43%), industrial (11%) and transportation (13%) categories than large open space (reduced from 47.6% to 10%) (Figure 1.3) (Table 1.1).

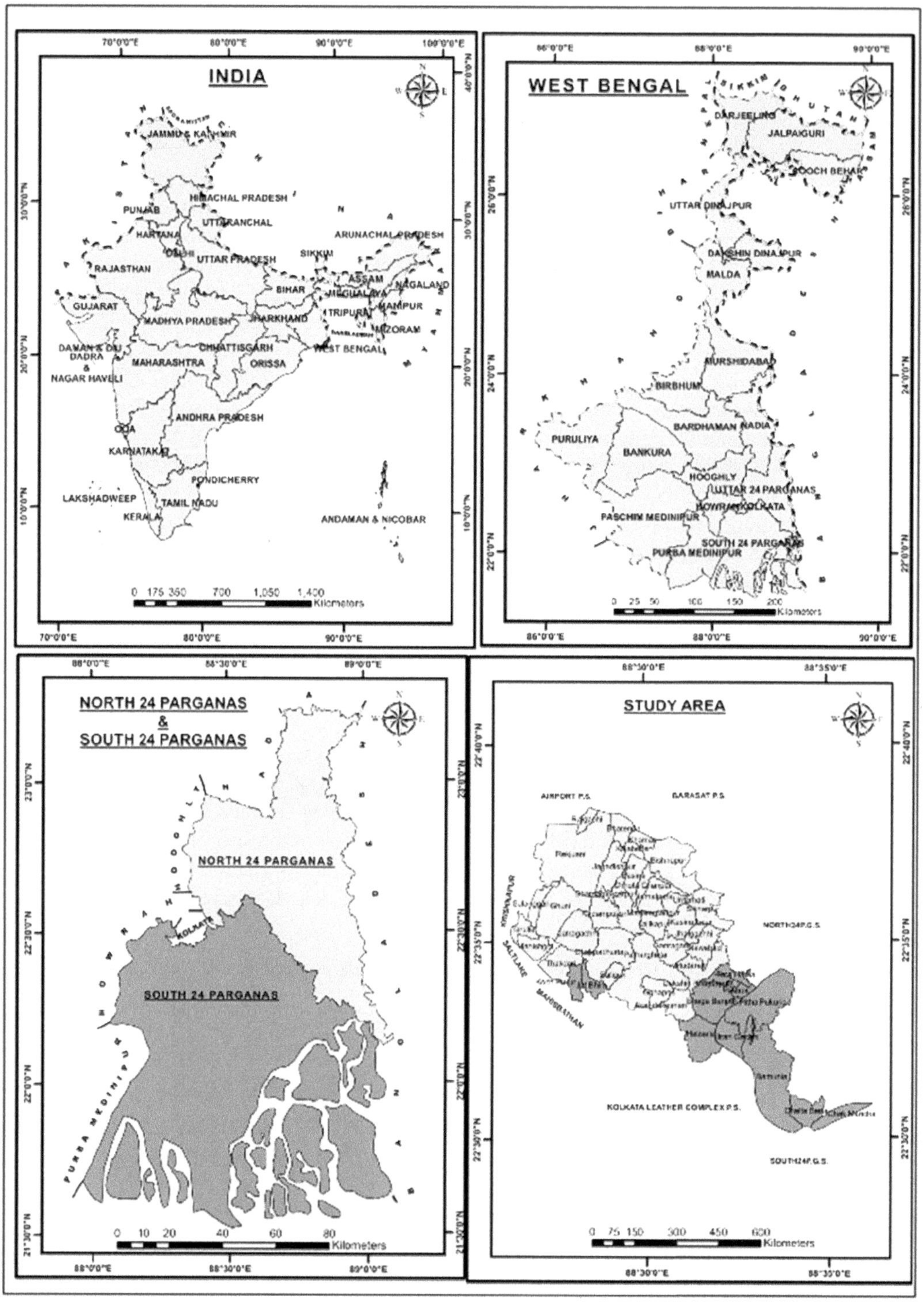

FIGURE 1.1 Study area.

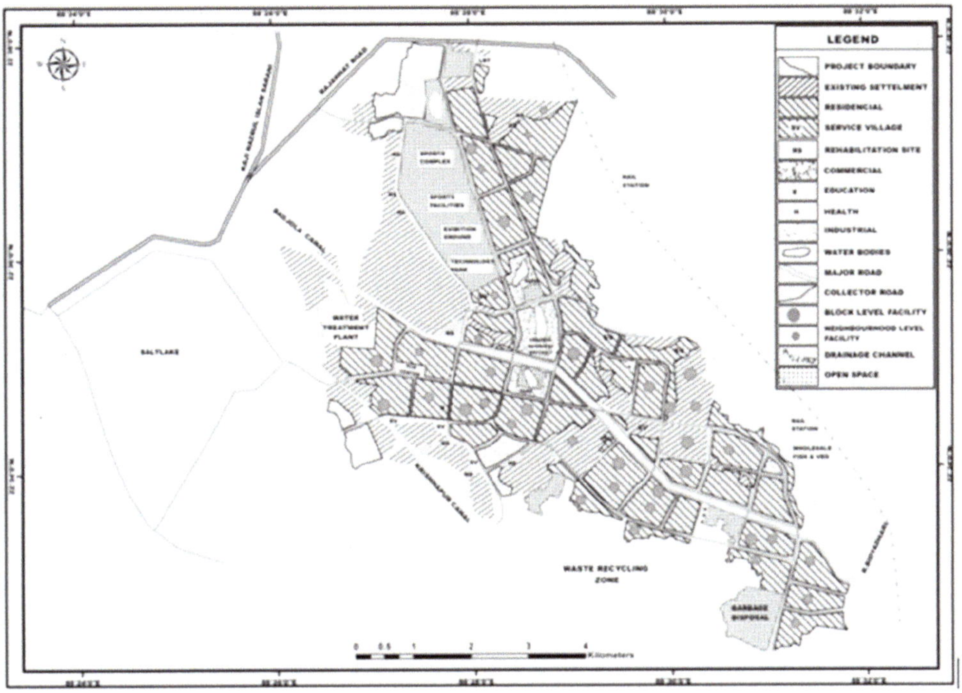

FIGURE 1.2 Master land-use plan.

(Source: Project Report, New Town at Rajarhat, Task Force, Department of Housing, 1995.)

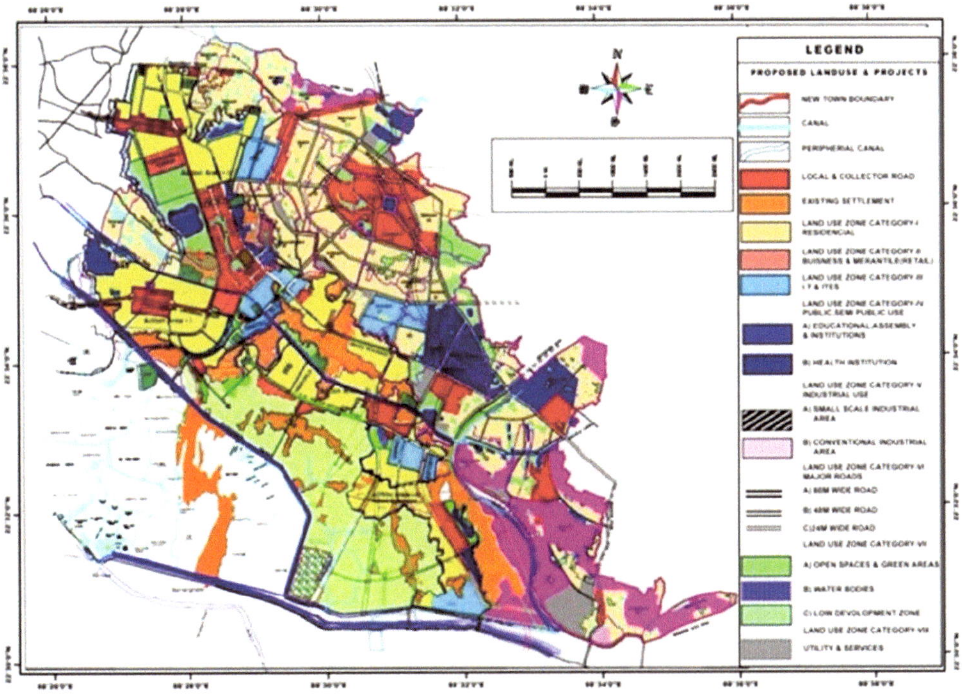

FIGURE 1.3 Land-use and development control plan.

(Source: Land use and development control plan for New Town Planning Area, WBHIDCO, 2012.)

TABLE 1.1
Stages of Development of New Town, Kolkata

Year	Function	Organization
1982	Preparation of preliminary project report on 'New Township DumDum-Rajarhat'	Additional Director of Planning – II, Calcutta Metropolitan Development Authority (CMDA) in collaboration with the Directorate of Housing, Government of West Bengal
1993	Constitution of a Technical Committee for preparation of a preliminary report on New Town	Department of Housing, GoWB
1994	Formulation of concept plan	Technical Committee, KMDA
	Establishment of Task Force	Department of Housing, GoWB
1995	Submission of suggested structure plan	Indian Institute of Technology, Kharagpur
	Draft project report	Task Force
1996	Submission of report entitled 'Master Land Use Plan for New Town at Rajarhat'	Indian Institute of Technology, Kharagpur
1999	West Bengal Housing Infrastructure Development Corporation, wholly owned by GoWB, set up under the Companies Act 1956	Department of Housing, GoWB
	Final review and publication of Master Land Use Plan	Development Consultants Ltd (DCL)
	Rapid Environmental Impact Assessment and Environmental Management Plan Report submitted	West Bengal Housing Board in consultation with Ghosh Bose & Associates Pvt. Ltd
2005	Bhangar Rajarhat Area Development Authority (BRADA) constituted	Department of Housing, GoWB
2008	New Town Kolkata Development Authority constituted under The New Town Kolkata Development Authority Act, 2007 (The West Bengal Act XXX of 2007)	
2011	BRADA area merged with WBHIDCO, thereafter termed New Town Planning Area	Department of Housing, GoWB
2012	Preparation of Land Use & Development Control Plan	WBHIDCO

Prepared by the Author.

In 1999, the West Bengal Housing Infrastructure Development Corporation Limited (HIDCO) was formed as a Special Purpose Vehicle (SPV) to develop the area in a township style, along the same lines as CIDCO for Navi Mumbai. Later, HIDCO was declared a Planning Authority under the West Bengal Town Planning and Country (Planning and Development) Act 1979. The land-use distribution varies in different plans of the area (see Figure 1.4).

Initially, the New Town plan covers an area of about 9,000 acres of land comprising four Action Areas – Action Area-I (AA-I), Action Area-II (AA-II), Action Area-III (AA-III), Action Area-IV (AA-IV) – and one vast Central Business District (CBD) (Annual Report for WBHIDCO 2017–2018). In 1994–1995, land procurement for the development of the township began via two processes: (i) direct purchase from willing land owners and (ii) acquisition through the Land Acquisition Act, 1894. According to HIDCO's Annual Report 2017–2018, the total acquired land is about 7051.92 acres, of which 212.33 acres were purchased directly and 6839.59 acres were acquisitioned (Annual Report for WBHIDCO 2017–2018).

1.3.2 Assessment of Policies

The Department of Housing, Govt. of West Bengal started the planning of New Town Kolkata and the West Bengal Housing Infrastructure Development Corporation was given responsibility for procuring the land as well as the infrastructure.

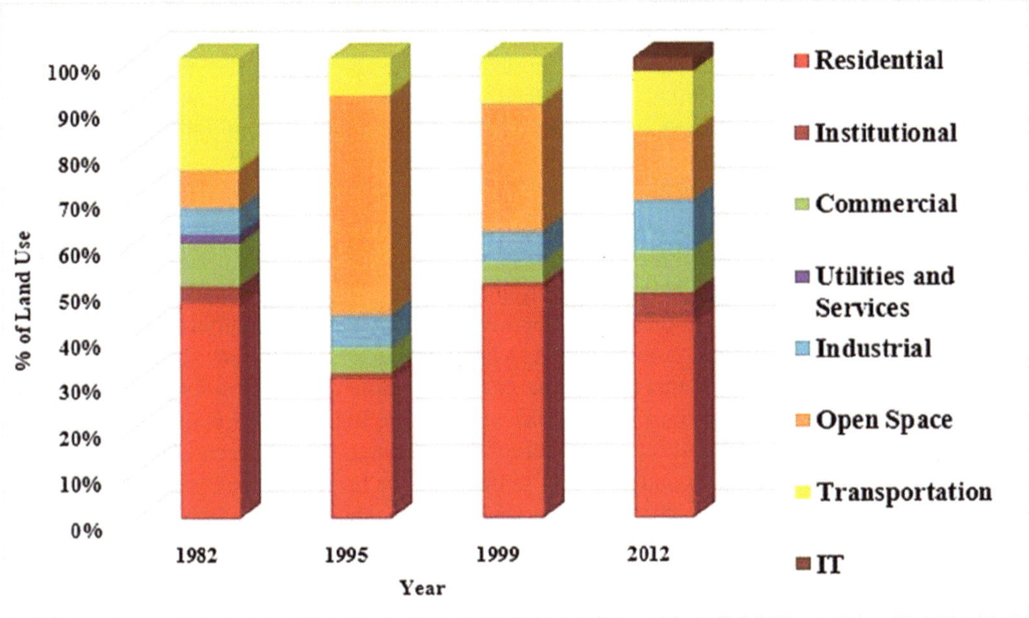

FIGURE 1.4 Comparative analysis of land-use distribution in different plans.

(Source: Prepared by the author.)

The land for New Town Kolkata was acquired by the state through the enforcement of the Land Acquisition Act (LAA), 1894. In 1998, a preliminary notification was issued by order of the Governor, West Bengal Government, demarcating the required land (Section 4(1)) for the project. Following Section 4(2), a survey including mapping of contours and identification of existing settlements was carried out by DoH (New Town Calcutta Project Report 1999), and in accordance with Section 8 of this Act, formal Gazette Notification was made to demarcate the boundary of the Project Area. In 1999, land was acquired under LAA 1894 (Section 12(2)). Following an award made by the Collector as per Section 11, possession may be taken of the acquired land, free from all encumbrances (Section 16). Selling and purchasing of the land, under the Registration Act, 1908 (Section 16 of 1908), was stopped (Section 11(4)) to prevent land price hikes. The people of Mahishgot and Thakdari villages refused to accept the preliminary notice of land acquisition on the grounds of lower compensation rate. Petitions against forcible land acquisition by the government (Chakraborty 2015) were filed in the High Court under Section 18(1).

The acquisition process violated certain existing land laws, since 100% of the land was rural and was mostly agricultural, thus was administered under the Land Reforms Laws. As per Section 4C of the West Bengal Land Reforms Act, 1955, land conversion and changes in the character of land are restricted. Yet 3,075 ha of rural land had been converted to urban land simply by serving a notice. Moreover, conversion of water bodies is restricted by Section 4C(2) of WBLRA, 1955. However, the authorities acquired 33 large water bodies for development of this town (Dey 2013). The filling up of water bodies as well as changing the character of wetlands are prohibited by Section 17A(9) of the West Bengal Inland Fisheries (Amendment) Act, 1993. The Eco-tourism Board must advise the state government on the framing of rules towards ensuring such protection as stated in Section 8C, West Bengal Town and Country (Planning and Development) Act, 1979.

In 1999, the government defined the Planning Area, with modifications and expansions, as per Section 9(1) of the West Bengal Town and Country (Planning and Development) Act, 1979. Under the same section of this Act, BRADA (Bhangar Rajarhat Development Authority) was formed.

Eventually, in 2011, the two planning areas were merged and the New Town Planning Area was officially formed. WBHIDCO was formed as a Development Authority (Section 11(1)) in 1999 and in 2012 WBHIDCO became the Planning Authority (Section 11(1)), also with retrospective effect from 1999. The powers and functions of the WBHIDCO are stated under Section 13(1) of this Act. WBHIDCO did not prepare the present LUMR within one year after constitution of the Planning Area, as stipulated under Section 28.

Another notable violation is that the LUDCP (2012) was published by WBHIDCO before the LUMR. It is definitely a case of timeline violation, due to which ordinary people were deprived of their legitimate right to raise objections (Section 29(1)) over inclusion of their lands within the ambit of Planning Area. This LUDCP was submitted to the state government for approval. The Planning or Development Authority should then, after approval from the state government, publish a public notice in the Official Gazette (Section 38(1)). Any persons can raise questions regarding the validity of the LUDCP in the High Court within one month from the date of operation of the LUDCP (Section 39(1)). The High Court may issue a stay order, if the LUDCP affects any property of the applicant (Section 39(2)). Every 10 years the authority concerned should prepare the LUDCP after required alterations and additions and submit it to the state for further actions (Section 40(1)). The preparation of the first LUDCP in 2012, thirteen years after the declaration of the Planning Area, in itself constitutes a gross violation, as the law stipulates a period of only three years, as mentioned above. Any land reserved in the LUDCP implies land designated for a public purpose within the meaning of the LAA, 1894 and may be acquired accordingly (Section 43). Restrictions on wetland conversion were also mentioned in the West Bengal Town and Country (Development and Planning) Act, 1979 (Section 8C) to advocate the protection of the entire wetlands.

Subsequently, in compliance with Sections 13 and 14 of the Environmental Protection Act, 1986, consultant Ghosh Bose & Associates Pvt. Ltd was appointed to prepare an environmental impact assessment, the 'Rapid Environmental Impact Assessment and Environmental Management Plan Report'. In 2008, the New Town Kolkata Development Authority Act, 2007, established to look after the infrastructural development of New Town Kolkata was published in the Gazette.

Wetlands were acquired without maintaining the buffer zone necessary for saving the fragile ecosystem from the urban sphere. Such activities were in violation of various sections included in the West Bengal Town and Country (Planning and Development) Act, 1979 (Section 8C), the West Bengal Inland Fisheries (Amendment) Act, 1993 (Section 17A(a)) and provisions under the Environmental Protection Act, 1986.

As well as the legislative aspects, the following observations are made:

i) The assigned land for New Town Kolkata was under the local self-government (rural) jurisdiction, where people's participation in all aspects of planning of the area (i.e., planning, implementation of plan and its monitoring) through the *Gram Sansad* (village council, i.e., forum of all electors of the ward/constituency) and *Gram Sabha* (village assembly, i.e., all electors of the area under its jurisdiction) is a constitutional right. However, in this case, land acquisition and land conversion were done by the state without any consultation with village councils.

ii) Besides, in the absence of transparent and participatory land policy, the land losers did not get due compensation and rehabilitation packages. No compensation was allocated to those who were not the owners of the land property but whose livelihood was affected due to this development (De 2012).

iii) The Department of Housing, Government of West Bengal formed a committee to examine the rehabilitation and resettlement issues of people affected or to be affected by the New Town Project, which was partially modified by replacing the words "The persons who have lost their dwelling houses" with "The persons who have lost their homesteads" in 1999. This may be viewed as narrowing down the scope of rights of the 'persons'. The idea of the formation of the Rehabilitation and Resettlement Committee may be viewed as a pro-people measure taken by the then government. No scope was provided to rectify the impoverishment that was caused by involuntary displacement.

TABLE 1.2
Distribution of Workers, Rajarhat Block

Year	Main Worker (%)	Marginal Worker (%)
1991	96.50	3.50
2001	88.08	11.92
2011	92.72	7.28

Source: District Census Handbook, Part–B, North 24 Parganas, Census of India, 1991, 2001, 2011.

TABLE 1.3
Decadal Changes of Main Workers, Rajarhat Block

Type of Main Workers	Percentage of Workers (1991)	Percentage of Workers (2001)	Percentage of Workers (2011)
Cultivators	10.38	10.14	4.84
Agricultural Workers	16.96	14.66	4.28
Household industry	1.91	5.22	2.86
Others	14.41	69.27	88.03

Source: District Census Handbook, Part–B, North 24 Parganas, Census of India, 1991, 2001, 2011.

1.4 ISSUES AND CONCERNS FROM LAND-USE CHANGE

The conversion of rural arable land to urban development has become a countrywide phenomenon. This land-use change due to urbanization has some benefits as well as adverse impacts. The positive side is the development of accommodation, commercial spaces and transportation, resulting in economic growth. But the loss of arable lands, displacement of local people and their loss of livelihood has negative implications for the economy, particularly the agro-economy of the country (Table 1.2).

Table 1.2 reflects that the marginal workers of the Rajarhat Block increased in 2001 due to land acquisition. The failure of agricultural activities might have forced them to become marginal workers. The 2011 census data shows that the situation has changed, with a minor increase in main workers. This may be due to the fact that a new set of residents are now occupying the area.

Table 1.3 shows that the internal composition of the main workers group has also drastically changed in terms of cultivators and agricultural laborers due to agricultural land acquisition.

Therefore, the acquisition of agricultural land and its conversion for non-agricultural purposes has changed the pattern of livelihood and income of the land losers. It has caused a decline in the agrarian status of the land owners. Their economic strength has been weakened. With the intrusion of an urban class, the social environment is gradually changing. Social cohesion and the identity of the agrarian society have been shattered. Land acquisition in this case needed additional or alternative rehabilitation programs to reduce the economic displacement and gross dissatisfaction among the land losers.

1.5 PROVISION OF REHABILITATION UNDER LAND LEGISLATION

1.5.1 Land Acquisition Act (LAA), 1894

LAA, 1894 was implemented over the whole of colonial India. The Act provided for acquisition of land for defined "public purposes" (Section 3f) against "monetary compensation" at "market value"

(Section 23). While calculating the compensation amount, other than market value of the land, the Court was to consider the damages, as given under Section 23(1), that happen due to the acquisition process (Section 23). The market value is also considered for the calculation of compensation, as directed by the Court. This calculation commences either on or from the date of the publication of the notice or the date of taking possession of the land (Section 23(1A), as per Amendment under Land Acquisition Act 1984).

LAA 1894 persisted for 120 years in India with nominal amendments but no noteworthy changes in its basic principles.

It was in 1998 that the Ministry of Rural Development took the initiative to change the LAA for the first time. Several bills were proposed, including the National Policy for Resettlement and Rehabilitation of Project Affected Families, 2003, the National Rehabilitation and Resettlement Policy, 2007 and the Land Acquisition, Rehabilitation and Resettlement Bill, 2011, but these bills were not passed in the parliament. Not until September 2013 was the Right to Fair Compensation and Transparency in Land Acquisition, Rehabilitation and Resettlement Bill approved and the subsequent legislative proceedings published in the Gazette.

1.5.2 The Right to Fair Compensation and Transparency in Land Acquisition, Rehabilitation and Resettlement (Henceforth RFCTLARR) Act, 2013

This Act is radically different from the Land Acquisition Act, 1894. The Act devolves power to the local government as well as its people's forum, thereby involving them in the acquisition decision-making process. Thus, it is much more participatory and transparent in nature. Besides, it makes provision for a compulsory rehabilitation package and much more justified compensation to the affected families. This Act contains a number of remarkable features. The rate of compensation is fixed as the market value of the land multiplied by a factor of 1.00 or 2.00 (The Second Schedule) as per the distance from the urban area center (Section 26) plus value of assets attached to the land plus 100% *solatium* (Section 30(2), The First Schedule). The details of entitlement to Rehabilitation and Resettlement Awards for each affected family are given under Section 31(1), such as provisions for housing units for the displaced families, land, reservation of 20% of the developed land for the people affected by the project in the case of an urban development and suitable training for employment. A Rehabilitation and Resettlement Committee is to be constituted (when at least 100 acres of land is acquired) to monitor the progress of the Rehabilitation and Resettlement Scheme and to carry out a post-implementation social audit in consultation with local government (Section 45(1)). There is provision for returning land which remains unutilized for a period of five years from the date of taking possession, the land returned to the original owner or owners or their legal heir or to the Land Bank of the appropriate government in a prescribed manner (Section 101).

1.5.3 The Right to Fair Compensation and Transparency in Land Acquisition, Rehabilitation and Resettlement (Amendment) Ordinance, 2014

On 31 December 2014, an ordinance was promulgated with a view to speedy land acquisition. It provides for amendment of certain sections of the Act (RFCTLARR 2013). The ordinance makes provision for inclusion of the words "compulsory employment to at least one member of such affected family of a farm labourer" after "affected family".

1.5.4 The Right to Fair Compensation and Transparency in Land Acquisition, Rehabilitation and Resettlement (Amendment) Ordinance, 2015

In February 2015 the RFCTLARR (Amendment) Bill was passed in the Lower House of the parliament but was strongly opposed in the Upper House for its elimination of certain clauses. The ordinance was ultimately rolled back in August 2016.

1.6 EXISTING REHABILITATION PROGRAM

The state enjoys the dominant power to acquire land for public purposes and the landowners receive monetary compensation against the land right. But persons who are indirectly dependent on the land, like agricultural laborers, were displaced without any form of compensation. The land acquisition process was not smooth at all. The *Jami Bachao* (Save the Land) movement was initiated and a *Rajarhaat Krishi Raksha* (Save Agriculture) committee was also formed to protest against the land acquisition. But this movement was not sustained because there was no powerful organization to stand by these helpless people (A History of Brutal Rajarhat Land Acquisition Bengal's New IT hub). 160,000 families (Block Land and Land Reforms Office, N & S 24 Pgs, 2013) lost mainly their agricultural land. Individuals who had lost land received cash compensation at the market value of the land. The compensation was fixed according to the then land value, i.e., at the rate of (i) INR 3000 per 1.6 decimal for fishery-cum-agricultural land, (ii) INR 6000 per 1.6 decimal land for mono-cropped low land and (iii) INR 8000 per 1.6 decimal for a multi-cropped area. Monetary compensation never creates significant benefits for the land losers. Hence, in recent times, land losers have been demanding return of land which has hitherto lain vacant (Figure 1.5) (Rahjarhat land losers demand higher compensation, Kolkata 2016), under section 101 of The Right to Fair Compensation and Transparency In Land Acquisition, Rehabilitation and Resettlement Act, 2013.

The WBHIDCO has already provided some rehabilitation for the land losers. These are as follows:

- *Fishing:* Eco Park is used for fishing with approved permits provided by West Bengal Housing Infrastructure Development Corporation (WBHIDCO) (West Bengal Housing Infrastructure Development Corporation 2017).

FIGURE 1.5 Vacant land and open space.

(Source: Satellite Image (Google Earth, February, 2017).)

- *Syndicate:* Syndicates were formed as co-operatives (15–20 members) to supply the building materials. In the New Town area, there is a huge number of syndicates.
- *Resettlement Program*: A good number of displaced people (683 people, WBHIDCO) from Narikel Bagan area have resettled on the bank of Bagjola Khal near Nandannagar.

1.7 PROPOSAL FOR ALTERNATIVE REHABILITATION PROGRAM

Development of New Town Kolkata commenced in 1999 when maximum agricultural land was acquired causing loss of occupation/livelihood of cultivators and agricultural laborers. A proposal to rehabilitate some of the victims in the agricultural sectors is given below.

1.7.1 Mango Orchard in Vacant Land

Land acquisitions on a mass scale have usually involved either agricultural land or forest land. Displaced people are adept at planting and growing food. Each planned city needs to develop green space as standards prescribed by the Town and Country Planning Organization (Urban Greening Guidelined 2014). In recent times, land losers of New Town Kolkata are have also demanded the return of land that has hitherto lain vacant (Rajarhat Land Losers demand higher compensation 2016).

It is therefore suggested that responsibility for greening the open spaces as well as land remaining vacant be handed over to willing displaced people. A modest estimate proves that a good amount of profit would be generated through developing mango orchards (Figure 1.6) in the existing vacant lands (2,977 ha, 2017) if the land could be handed over to the land losers by means of leasing out, contractual farming and *patta* distribution. At least 8,014 persons (Table 1.4) may get back their livelihood in the agricultural sector (considering income per annum of INR 18,000 per person as per MGNREGA, 2017 wage rates, http://nrega.nic.in/netnrega/writereaddata/Circulars/2058Notification_wage_rate_2017-2018.pdf).

The income can be enhanced by inter-cropping cultivation (Figure 1.7) between the mango trees if irrigation is provided (http://nhb.gov.in/report_files/mango/MANGO.htm). This will also increase the number of rehabilitated people.

FIGURE 1.6 Mango orchard in New Town Kolkata.

(Source: Primary Survey, 2017.)

TABLE 1.4

Cost and Projected Income Estimates from Proposed Mango Orchard (INR per ha) in New Town Kolkata

Spacing: 8m × 8m

Plant Population: 100

A. Estimated Cost INR/Ha (base year 2016)

		Year					
Sl. No.	**Particulars**	**1**	**2**	**3**	**4**	**5**	**Total**
1	Planting material	6000	0	0	0	0	6000
2	Manures & fertilizers	3000	1100	1100	1400	1400	8000
3	Sprayer & implements	1500	0	0	0	0	1500
4	Labor	4800	1200	1200	1500	1500	10200
5	Fencing	6200	0	0	0	0	6200
6	Irrigation	0	0	0	0	0	0
7	Miscellaneous including leasing cost, loan* interest	600	500	500	500	500	2600
	Total	22100	2800	2800	3400	3400	**34500**

B. Projected Income (in INR)/Ha

	Year										
	1 to 5	**6**	**7**	**8**	**9**	**10**	**11**	**12**	**13**	**14**	**15th onwards**
Yield (q/ha)	0	25	35	50	60	70	80	90	100	110	120
Gross value @ Rs45/kg	0	11250	15750	22500	27000	3100	3600	40500	45000	49500	5400
Income (Rs.)	0	11250	15750	22500	27000	3100	3600	40500	45000	49500	54000
Maintenance (Rs/ha)	0	3400	4500	4500	5000	5000	5000	5000	5000	5000	5000
Net Income (Rs/ha)	0	7850	11250	18000	22000	26500	31000	35500	40000	44500	49000

Source: Primary Survey, 2017.

* Subsidized financial assistance can be arranged from the following sources to meet the preliminary costs, like PMEGP.

FIGURE 1.7 Grass cultivation within the mango orchard as inter-cropping cultivation, New Town Kolkata.
(Source: Primary Survey, 2017.)

1.7.2 Floriculture in Open Space

The Forest Department of West Bengal sanctioned tenders for beautification with cherry blossom flowers along the arterial roads and adjacent service lanes. The process of beautification can be done with flowers which are in great demand in local market. Moreover, inside the housing premises floriculture can be done by the land losers. But in this case, enthusiasm from the housing society is needed.

1.8 CONCLUSION

Land legislation is intended to protect rights to land as well as conservation of land as natural resources. To serve this purpose, land laws are amended regularly and new laws are enacted. Under the discourse of the land laws, good arable land is being converted to urban uses. Mass change of the traditional livelihood of land losers for the sake of new development is never justified. Mere monetary compensation cannot generate sufficient benefit for the affected people. In recent years, sincere efforts have been made at national and state level, even at the private organizational level, to justify rehabilitation policies, though without a great deal of success. This study has proved that better rehabilitation is possible for a proportion of the victim families by utilizing their basic agricultural knowledge.

ACKNOWLEDGMENTS

I wish to acknowledge my deep sense of gratitude to Dr Subhra Chattopadhyay, Associate Professor, Department of Geography, Lady Brabourne College, University of Calcutta for the ideas and suggestions, in spite of her serious involvements with other important assignments.

REFERENCES

A History of Brutal Rajarhat Land Acquisition Bengal's New IT Hub. 29 August. Accessed February 6, 2019. http://sanhati.com/excerpted/945, 2008

S. Angel, S.C. Sheppard, & D.L. Civco. 2005. *The Dynamics of Global Urban Expansion*. Washington, DC: Transport and Urban Development Department, The World Bank.

Annual Report for WBHIDCO. Annual, Kolkata: West Bengal Housing and Infrastucture Development Corporation, 2017–2018.

M. Beresford. 1967. *New Towns of the Middle Ages*. London: Town Plantation in England, Wales and Grascony.

N. Berg. 2012. "Where Urban land growth is about to explode." *Citylab*. 19 September. Accessed April 28, 2019. https://www.citylab.com/equity/2012/09/where-urban-land-growth-about-explode/3318/

S. Bhat, M. Shafiq, A.A. Mir, & P. Ahmed. 2017. "Urban sprawl and its impact on landuse/landcover dynamics of Dehradun City, India." *International Journal of Sustainanble Built Environment* 6(2), pp. 513–521.

M.M. Cerena. 1995. "Understanding and preventing impoverishment from displacement: Reflections on the state of knowledge." *Journal of Refugee Studies* 8(3), pp. 245–264.

K. Chakraborty. 2015. *Land People and Power, An Antropological Study of Emerging City of New Town Rajarhat*. Kolkata: Gyan Publishing House.

P. De. 2012 August 1. "Rajarhat: When the government loots." Sanhati. http://sanhati.com/articles/5354

I. Dey, R. Sammadar, & S.K. Sen. 2013. *Beyond Kolkata: Rajarhat and the Dyastopia of the Urban Imagination*. New Delhi: Routhledge.

D. Dutta, A. Rahman, S.K. Paul, & A. Kundu. 2019. "Changing pattern of urban landscape and its effect on land surface temperature in and around Delhi." *Environmental Monitoring and Assessment* 191, p. 551.

D. Dutta, A. Rahman, S.K. Paul, & A. Kundu. 2020. "Estimationg urban growth in peri-urban areas and its interrelationships with built-updensity using earth observation datasets." *The Annals of Regional Science* 21–25.

W. Fernandes. 2008. "Sixty years of development induced displacement in India: Scale, impacts and search for alternatives." In *India Social Development Report*, by H.M. Mathur, 89–102. New Delhi: Oxford University Press.

T. Firman. 1997. "Land conversion and urban development in the northern region of West Java, Indonesia, Urban Studies." *Urban Studies Journal Limited* 1027–1046.

D. Grigg. 1995. *An Introduction to Agricultural Geography*. New York: Routledge.

"Guidelines on Prime Minister's Employment Generation Programme." New Delhi, 2017.

D. Karasian, K. Maat, & B. Van Wee. 2017. "The impact of urban proximity, transport accessibility and policy on urban growth: A logitudinal analysis over five decades." *Environment and Planning: Urban Analytics and City Science* 91–95.

T. Kuemmerle, C. Levers, K. Erb, S. Estel, M.R. Jepsen, D. Muller, C. Plutzar, J. Sturck, P.J. Verkerk, P.H. Verburg, & A. Reenberg. 2016. "Hotspots of land use change in Europe." *Environmental Research Letters* 11(6), pp. 1–14. IOP Publishing, Bristol.

D. Lohr. 2010. "The driving forces of land conversion. Towards a financial framework for better land use policy." *Land Tenure Journal* 61–89.

New Town at Rajarhat, Project Report. Project Report, Kolkata: Department of Housing, Government of West Bengal, 1995.

New Town Calcutta Project Report. Project Report, Kolkata: West Bengal Housing amd Infrastructure Development Corporation, 1999.

New Township DumDum- Rajarhat Preliminary Project Report. Pleliminary Project Report, Kolkata: Calcutta Metropolitan Development Authority, 1982.

Rajarhat Land Losers Demand Higher Compensation, Kolkata. Newpaper, Kolkata: Times of India, 2016.

S. Siedentop, & H. Nuissl. 2021. "Urbanization and land use change." In *Sustainable Land Management in a European Context*, by Tim Barkmann, Nadin Gaasch, Sebastian Rogga, Christian Straub, Jana Zscheischler, & Thomas Weith, 87. Bloomington: Spinger.

H.C. Tsai. 1993. "Socio economic dimension of rural land use changes in Taiwan." In *Land Polici Problems in East Asia: Towards New Choices, East-West Centre, Honololu, Hawaii and Korea*, by B. Koppel, & D.Y. Kim, 413–432. Korea: Korea Research Institute for Human Settlement.

B.L. Turner, W.B. Meyer, & D.L. Skole. 1994. "Global land-use lamd-cover change-towards an integretaed study." *Ambio* 91–95.

Urban Greening Guidelined. New Delhi: Town and Country Planning Organization, Ministry of Urban Development, Government of India, 2014.

West Bengal Housing Infrastructure Development Corporation. 5 May. Accessed March 10, 2018. http://www.wbhidcoltd.com, 2017

2 Contextualizing Urban Sustainability

Limitations, Tensions in Indian Sustainable-Smart Urbanism Perceived through Intranational, International Comparisons, and District City Ambala Study

Anil S. Thakur
Independent Researcher, Melbourne, Australia

CONTENTS

DOI: 10.1201/9781003131922-3

2.1 INTRODUCTION

'Change' is a cosmological reality and plays out variously in space and time. Conscious of this reality, we recall immense degradation and irreversible damage caused to planet Earth for ages by anthropogenic activities and environmental catastrophes, and remind ourselves of the urgency to protect it. Concerns over sustainability, including quality of the environment (natural and man-made), cleaner production and consumption processes, exploitative uses of non-renewable resources for energy needs, climate change, scarcity of clean water and energy, shrinking biodiversity, green cover and the depletion of the earth's precious topsoil, have motivated humanity to think ecologically and stop depleting non-renewable natural resources and escalating exploitation of the host planet. Our existence is inextricably linked with the well-being of the earth's ecosystems. Interestingly, humans occupy 10% of the land, and only 3% of the world's land surface area constitutes urban areas. Surprisingly, humans with their minuscule footprint on earth exercise tremendous influence on the planet. Humans make for only 0.01% of the total average share of life on earth.

Compounding adverse impacts of human activities on the planet are worrying, and nations worldwide, irrespective of their ideologies, political economies, physical and developmental distinctions, are nervously drawn to the concept of sustainability.

2.1.1 Simplicity to Complexity

In the long-drawn apocalyptic visions of ecologists and dystopic scenarios advanced by socialists/writer-thinkers such as Orwell, we locate 'environment', 'economics', and 'society and culture', the three imperatives of sustainability discourses of the late 1980s. We then relate those to the developing country-centred eight Millennium Development Goals (MDGs) and the 17 people-centred Sustainable Development Goals (SDGs) launched under the aegis of the United Nations Economic Commission (UNEC) (Brundtland 1987; Kumar et al. 2016). Whereas the developed nations aided MDGs until 2015, the SDGs (2016 onwards) are self-owned and financed by the stakeholding, emerging and developing countries (EDCS) and the developed nations. SDG 11 nurtures the ambitious agenda of raising sustainable communities across the globe with the help of the private sector. The oft-quoted Brundtland Report (1987) was the first to pronounce sustainable development as that which "meets the needs of the present without compromising the ability of future generations to meet their own needs". Since then, public perception of sustainability has changed under the SDG-led perspectives and environment-conscious and environment-friendly technologies that can potentially fulfill new generations' expectations, needs, and desires. Crucially, global agendas for sustainable urban development informed by the Brundtland Commission's idealism (under the patronage of the UN) have guided today's smart technologies-led sustainable-smart urbanism. The combination of sustainable and techno-centric urbanisms in developing countries can best be described as 'farrago urbanism' (a paradox; myth–reality; unreal–actual), a self-devised term that the author often resorts to in the following text.

Urban change occurring at global, continental, national, regional, district, and local scales, and across varying developmental levels and altering conditions of various societies is what interests us. Urbanization, an agent of planetary demographic change, is one of the critical substratum processes that sustain the concept of sustainable urbanism and ideas of 'smart urbanism' (an emerging field) (Hebe and Karin 2019). Urban spaces at different spatial scales transform variously. They even mutate in the fast-changing demographic socio-economic and socio-cultural milieu and rapidly evolving new criteria of urban spaces, a phenomenon we address later. Whereas sustainable urbanism, eco-cities, and eco-logics generically are offshoots of enviro-centric thought,[1] the lexicon of intelligent, smart, responsive cities, and smart urbanism represents the ever-changing and unsettled technocratic paradigms. Furthermore, urbanization and adoption of new technologies boost gross domestic product (GDP), enhance the economies of places, and influence the cultures of their societies. Linked to urbanization are migration, the economy and economic growth, governance and citizenry, and socio-spatial hierarchies, all of which we touch upon later.

2.1.2 Dilemmas: Poverty and Smart Sustainability

Sustainable development in India and other developing countries is beset by dilemmas related to environmental, economic, and socio-cultural imperatives, namely accelerating economic and human development with minimum per-capita emissions (low costs vis-a-vis sustainable-smart solutions). Moreover, tensions mount as we attempt to include urban poverty and informality, the two interrelated and inseparable realities of our times within the ambit of sustainable urbanism (Cobbinah et al. 2015, 26). Reconciling the clearly opposed logic of traditional urbanism of locales (economic development disregarding the environmental costs) and ambitious new big data-driven sustainable and smart urbanisms is problematic for urbanists. Juxtaposing the above two invariably results in conflicts between citizens over their clashing interests and between citizens and governments over their controversial decision-making. Additionally, enforcement mechanisms, policies, and public resistance to the top-down, lofty and costly propositions do not agree with the developing nations' overall socio-cultural milieu and urban contexts.

Divided interests give rise to well-spread out and networked social activism at various social-spatial scales and political platforms. Ananya Roy's succinct comment, "There is an urgency for urban studies and planning to move beyond the dichotomy of First World 'models' and Third World 'problems'. One possible route is through policy approaches that seek to learn from Third World cities" (Roy 2005, 147) is timely. Sanyal (1990) also supported the argument.

2.1.3 Traditional vs Institutionalized Sustainability

These dilemmas necessitate the mediation of contradictions (over developmental context) between traditional urbanism (agriculture, industry, manufacturing-based) and the intended institutionalized urbanism driven primarily by sustainability concerns and an array of ever-evolving smart, clean, and lean technologies. Economic growth is the critical underlying agenda for any mode of production, notwithstanding that today's developed nations also once pursued industrial modes of production and unsustainable practices. Another concern is fast-spreading consumerism that hunts for incremental profits and introduces social unevenness and tensions in developing countries' societies. In the atmosphere of opportunism-led betrayals, where peculiarities of developing societies are overlooked, quite unknowingly, we get sucked into a deceptive world that runs on farragos of half-truths and capitalist projections (as mentioned earlier) or on Immanuel Kant's duality of 'noumenal' and 'phenomenal' worlds.

Smart technologies that are both compulsive and emancipative are invariably ridden with the hidden costs and money-making (incentive-generating) agendas of their promotors. For example, the collection of unstructured data, its storage, and its retrieval and delivery for prescriptive analyses are attributes of big-data, OT, and cloud industries whose profits flow to the owner multinationals. Such services as may be optional today in developing nations will, over time, become essential and pave the way for new industries, markets, and a new workforce.

Irrespective of emerging new modes of production, sciences and technologies, India's post-independence settled constitution and administrative divisions and practices are likely to retain their status quos more or less. Supposedly, the ingrained flexibility and tolerances in the traditional urban development systems will absorb and bear with the impacts of transitional economies and their concomitant developments with likely increasing sustainable and smart urbanism contents. The benefits of data-based decision-making, improving professionalism, and accountable governance will trickle down to the roots in the long run. The situation will introduce contextually relevant urbanization and sustainable-smart urbanism at district and local levels.

2.1.4 Sustainability for the Developed, Developing Economies

New paradigms of sustainable and smart urbanisms in the developed world prefer new building stocks raised wholly from scratch. Key examples are the contemporary, history-denying (without inherited identities), model, sustainable and smart cities of Songdo and Masdar built in South Korea and Abu Dhabi, respectively. New model towns that were conceptualized and built in China under three categories, Industrial parks (e.g., 'Suzhou' established in 1994); eco and ecological towns (Dongan, Binhai), and thematic towns, characterized after European, Scandanavian, American, Canadian and Chinese cities (Germantown, Linghai, Pujiang, Thames and Zhujiajiao, etc.) are also exemplary. The last two categories were rolled out under the 'eco-city movement' and the 'one city, nine towns mission' in 2010 and 2004, respectively. Renowned international architects, planners and global practices were engaged to execute these sustainable and smart new projects. The presence of a small group of global practices of architects and urban planners (multinationals and global intelligence corps (GICs)) specializing in applying sustainable and smart technologies solutions to new economic and urban developments is growing in international markets (Lin 2012).

The masterplan for Dongtan in Shanghai (initiated in 2004) was prepared by multinational consultancy Arup, specializing in sustainable developments and solutions. Dongtan, a lofty exercise in planned sustainable and smart urbanism, remained unbuilt for political reasons (cancelled in 2009). It was to be the first carbon-neutral ecological city in the world. Of the many measures taken for the city's sustainability, one was to run it on renewable energy and recycled water. Fossil fuel-run vehicles were to be banned in the city. All city buildings in Dongtan were to be insulated and double-glazed; Dongtan aimed to cut car journeys by 90% with the induction of a light rail; 60% of the city's waste was to be recycled (ibid.). The second eco-city project in Tianjin, Binhai, less lofty than Dongtan, had a pragmatic plan.

Over time, such experimental sustainable eco and smart towns and nouveau planned urban developments, including Songdo and Masdar, seem to be political rhetoric. At best, they became just 'listed commodities' for sale in the realtors' market. However, the symbolic and cosmetic values of these developments escalated because of their planned sustainability agendas, but they hardly served any of their intended social and economic objectives. Incorporating ecological and sustainable development ideas, smart technologies, and smart solutions added a promotional value to these developments. They also fostered 'farrago urbanism' in China to serve international agendas as SDGs and gain global appreciation on the sustainability front. Koolhaas aptly described the new sustainable towns making and attempting to shape China's symbolic sustainable urbanism as "cities of exacerbated differences" (Koolhaas 2001).

In India, the idea of sustainability was initially limited to buildings. It slowly upscaled to clusters of human habitats, neighborhoods, towns and cities, resulting in a bricolage of disparate regional urbanisms with varying sustainability. There are geographic variations, disparate developmental levels, socio-economic and socio-cultural diversity that are peculiar to the Indian states and UTs. The underlying complexities of the country's federal structure, and center-state and interstate politics also make India distinct. India's major challenges include managing urban change, economic and infrastructural developments and attaining SDG targets. In this chapter we address multiple schemes floated in India to meet these objectives and improve the lives of citizens at all levels. India's top-down approach to economic and urban sustainable development is based on enviro-centric and techno-centric paradigms. New and associative urbanisms, smart city treatises, sustainable and smart urbanism, smart ideas, and big-data technologies altering our everyday living experiences and perceptions of the perpetually transforming world interest us. We also refer to the underlying conditions, mechanisms and traditional structures that have long regulated Indian urbanization and urbanisms. They constitute the substrata conditions on which we proceed to rest our ambitious modern edifice of sustainable urbanism. Traditional structures include the Indian Constitution, legislation, governance processes, concepts of urbanization, public participation, altering modes of productions, and allied instruments. The arguments advanced here are driven by Indian urban peculiarities, contexts, and conditions. Arguably, sustainable urbanisms (eco, regenerative, biophilic, green cities) are inescapably linked with smart urbanism in today's big data-driven world.

Indian urbanism has its peculiarities. It is conflicted by several influences including ethnic cultures, traditional, modern, modernizing and global cities, and resembles a series of regional metamorphic gneisses. To become smart, advanced, socially and economically responsive and sustainable, India needs to have plans to retrofit its cities with green infrastructure, incremental assertions of sustainable practices, developments, and innovative bespoke interventions. Also, an essential agenda for India is to link its urban growth with social and human development. Constitutional and legislative amendments that streamline the country's progress and sustainable development must be supported. A pragmatic, scientifically balanced and analytical attitude runs through the chapter.

We engage with urban change that shapes our urbanisms at global, national, regional, city, town, and local levels. We are mindful that speed, technology, and organizational capacities have become quintessential to global change facilitated by the fast mobility of goods, instantaneous transfer of capital and information from one location to another, and delinking production and consumption spaces. Furthermore, we deal with spatial-hierarchical urbanism, mainly but not limited to district urbanism. The relationships of urbanization to the economy, economic growth, urban governance, planning, citizenship, constitutional power, funds and investments concern us. So do the nexuses between spatial management, statutory planning, politics, and the constitution. We treat these factors as pre-existing or given conditions that we need to grapple with in the new ecological phase. In short, we rest the structure of our contemporary smart urban sustenance on the traditional central-federal administrative and governance systems of the post-independence era (substrata), and we must bring traditional structures within the purview of smart-urban sustainability discourses.

We view cities both as independent entities (inert stand-alone formations or physical agglomerations) and as parts of interrelated networks where a city's growth affects other settlements through the phenomenon of associative urbanism, and as ever-changing entities that, inextricably linked to life, keep shedding their identities and correspondingly emerging with new ones almost kaleidoscopically. Cities' constant transformations, ongoing metamorphoses, and fluidity are understood effectively through global-local 'assemblage thinking'. In our analysis of urban developments, we refer to the politics of inclusion (community engagement and citizen participation); co-authorship; teleocratic and nomocratic practices of governance; private entrepreneurship; top-down, bottom-up planning practices; urban agglomerations (UAs); and district urbanism.

2.1.5 Roadmap

The chapter has two distinct parts. The first comprises deliberations on sustainability in the broad context of developed and developing nations, SDGs, prevalent smart technologies that inhere immense potential to improve our livability, communicability, urban governance, and decision-making. Interestingly, enviro-centric sustainable urbanism, functioning hitherto independently, has over time become associated with fast-developing techno- or smart urbanism and vice versa (upscaling of smart cities by association with ecology). Smart technologies that notably operate and exercise over pre-existing socio-spatial networks, economic activities, and constitutional provisions (ground conditions) also necessitate new infrastructure as buildings and hardware related to their application. If we call these operational essentials for future technologies a 'superstructure', then we need to understand our substrata conditions, including a brief historicity of urban change and current urban transformations at national and regional scales. To make headway with smart-ecological urbanism, we trace the past and the present trends of Indian urbanization, draw some international and national comparisons, and delve into altering definitions and criteria of urban and rural areas. Concepts of spatial hierarchies and continuums, UAs, and administrative controls (power structures), are deployed in the upcoming discussions and analyses.

The constitution of a country is vital for its sustained progress and development. The Indian Constitution of 1950, projected by nationalists as most labored, has strengths (stability, reliability, representative government) and weaknesses (lack of agility to cope with accelerating compression of time and space, and rapidity of change and globalization). A good constitution promotes pragmatic federalism and inheres mechanisms that introduce autonomy (self-rule), equity, and flexibility in the functioning of the states and their Urban Local Bodies (ULBs). A moral constitution must include efficacious mechanisms to affect its own circumstantially necessitated amendment(s).

Importantly, in the second part of this chapter, we apply our conceptually and theoretically driven analytical framework to a live situation, the mid-tier district capital city of Ambala in Haryana, India. We develop valuable insights that inform grassroots knowledge gaps and attempt to bridge those later in this enterprise. Conclusions and recommendations figure last.

2.1.6 Struggling Sustainability of District City Ambala

Following a holistic approach, we examine the diversity of India's district urbanism and its potential role in promoting sustainable urban growth at the level of Indian towns that, over time, are graduating to cities. We do not only read the district city of Ambala from the sustainability angle but also view it as a site where traditions persist and smartness is asserted. The model town and the cantonment of Ambala keep us abreast with the 'new sustainable urbanism' of colonial times. The empirical study of Ambala also helps us understand the impediments and incentives that Indian cities meet on their paths to becoming sustainable and smart, how changing atmospheres and circumstances shape their future, and how cities plan for themselves.

2.2 FROM UNPLANNED TO SUSTAINABLE URBANIZATION AND PROJECTIONS

The average growth rate of the urban population in the last decade was 4%. India's urban population (40%) is expected to reach 600 million by 2031, and the urban sector's share in the country's GDP is expected to increase from its current 66% to 75% by 2031. According to TNN (2019), in 1950, 30% of the world's population was urban, and by 2035 and 2050, the world will urbanize to the extents of 52.5% and 68%, respectively. In a significant fast-paced demographic shift, India is going to be a major player, and it will house 17% of the global population (416 million), followed by China (10%; 255 million) and Nigeria (8%; 189 million). The three will house 35% of the global population (Ahluwalia et al. 2014, 2). Future projections also imply massive internal transformations for the nations mentioned, especially India, which will become an epicenter of the world's urbanization in the coming decades. The ensuing structural transformations will pose a big challenge of meeting the needs of the built environment and creating opportunities for bridging the rural–urban divide. India will require 700 to 900 million m^2 of residential and commercial built area per year from 2020 to 2031 to accommodate a population of about 600 million. According to the Indian Economic Survey 2017–2018, the country will require US$4.5 trillion in investments to build its essential infrastructure by 2040.

The following important questions arise: Is India ready with appropriate plans to accommodate its burgeoning urban population and migrancy;[2] and will it sustain the fast approaching mega-urban change? Any futuristic plans would be demanding not only on the country's resources but also on the old Indian systems of urban governance that present hurdles to new plans for sustainable-smart urbanism. Signs of India's growing impatience and desperation to streamline its decision-making (by employing big-data technologies, analytics, and operational technology) and new deliveries are already visible. How will the old mechanisms of organizing post-independence Indian societies cope with the new urbanisms that aspire to lean toward sustainable and smart technologies? At the present time some significant constitutional amendments and the rescinding of outdated Acts are already being attempted.

Furthermore, unplanned migration, migrancy, and uncontained poverty would add to the already problematic informality of urban spaces. The good side of informal city areas is that they play important supportive roles in maintaining local informal economies and creating casual jobs like workers in the transportation sector, city cleaners, housekeepers, assistants, and cleaners in commercial and institutional industrial enterprises, etc. The bad side is that informality is a barrier to the sustainability of cities in terms of slums and the eyesores they generate in the cities that mar the progressive images of these cities in global circuits.

The vigorous research that is ongoing on various aspects of informality is timely; nevertheless, it needs to be accelerated to avert a perceptibly bleak future for Indian sustainable urbanism. Lefebvre's (1991, 55; 374; 380) socio-spatial analysis that societies' inherent contradictions and class struggles determine the quality of their lived environment(s) fits the Indian context. Yet, it is hard to visualize the future local environmental conditions that Indians would be living in.

The picture of widespread, rather passive, Indian urbanity springing from the medium and smaller towns is undermined and displaced by the one that emanates from the metropolises and upper-class cities. Stark socio-spatial contradictions and duality of spaces inform the uniqueness of Indian urbanism that dispels some Western propositions like "urbanity is a uniform universal experience, and it imparts sameness to cities by promoting 'world standard' real estate projects" (Marshall 2003, 5). Robinson (2002) contrasts well the urbanism of global cities (representing the first world's replicable contemporary models of informational capitalist urbanism) with the "big but not powerful" in-crisis urbanism of megacities located in the developing world.

2.2.1 Comparisons

Urbanization is a precursor of urbanism. Reading the characteristics and experiences of urbanization in different political economies, regimes, and times helps us understand the diversity of the ensuing urbanisms of these places, and how that upgrades to sustainable and smart urbanisms. Unlike in India, urbanizations are aggressively regulated in some settings, although less by the center and more by states. In India, capital to plan and facilitate urbanization of the country is largely arranged and administered by the center. Under the 74th constitutional amendment (1992), the center devolves powers and finances to states and UTs to manage their urban affairs. In turn, they are to disburse funds to cities and towns through ULBs according to their hierarchical statuses and budgetary needs.

China follows a decentralized approach to urbanization. States play a major role in collecting funds (through the sale and acquisition of land) and utilizing those in urban developments. Private investments are critical in Chinese urbanization. The Indian system works through many checks and controls; for example, since the liberalization of the economy in 1991, the center has been floating different nationwide schemes for sustainable urban and city developments on an equity basis with the states and their ULBs. To obtain funds under such schemes, cities and states must raise their shares that are pre-settled with the center. Cities fund through their ULBs, and the states from their resources based on project reports prepared by the ULBs.

Challenges and approaches to urbanization and sustainable urban development are completely different in the Indian and Chinese contexts. Indian urbanization is typically marked by logic, restraint, lack of imagination and action, undersupplies, and late deliveries. In the late twentieth century, China's urban transformations beat the logic of traditional rural–urban urbanization. China's contemporary urbanization is fast-paced and mega-scaled. The surpluses (unutilized built areas) are epigraphs of the new ghost urbanisms embedded in the Chinese naturescape (Thakur 2005). Neologisms and prefixes calling urbanization 'phantom', 'fake', 'impetuous' and 'cart before the horse' convey the Chinese sans-logic and flamboyant urbanisms. Both countries share the goals of delivering adequate infrastructure, housing, commercial and other functional built areas, sustainable and environmentally conscious cities, and planned city expansions to their citizens, adopting different approaches to meet their ends.

On the one hand, India adopts a parsimonious attitude, where resources have perpetually been scarce, and the centralized disbursals of funds for the upliftment of the states are cumbersome. On the other, unrestricted mega-urban developments that defy market forces proliferate in China's western and central regions.[3] China's upcoming, mainly private, urban development is investment friendly and lacks welfare sentiment. Moreover, in China, the land belongs primarily to the state, and practically nothing comes in speedy deliveries of readymade building stock to the realtors, consumers, and investor markets. However, in both Chinese and Indian contexts, 'economic development' has traditionally been a priority. Interestingly, China is pursuing its ambitious goal of achieving 65% urbanization by 2025, by aggressively advancing its new urban agenda of ushering in a sustainable-smart urbanism era, building new towns and cities and applying contemporary ideas of sustainability and smart technologies. Meanwhile, India struggles to sell the idea of smart sustainability to its citizens and tries to include it in the ambit of city planning and management.

2.3 URBAN AGGLOMERATIONS (UAS); RURAL–URBAN CONTINUUMS; AND SMART DISTRICT URBANISMS

As mentioned at the beginning, 'change' plays out variously in space and time. With the fast transforming rural–urban interrelationship due to rapid urbanization, perceptions of regional urban change have transformed. Consequently, the new dynamic expressions 'UAs' and 'urban continuum' have been added to today's regional lexicon.

Urbanisms operate within defined geographical limits. To grapple with the idea of India's smart sustainable urbanism, it is necessary to be abreast with issues of territoriality of the states, UTs, and their subdivisions into administrative districts. The idea of UAs in Indian urban discourse was officially introduced during the 2001 census. Since then, there has been a paradigm shift in observing and analyzing urban change (van Duijne and Nirman 2019, 1979). As we climb the spatial-hierarchical ladder up to UAs, we notice peculiarities, varying characteristics, and diversities within the districts of various states. We no longer view urban settlements as independent growths but conceptualize them as parts of the phenomenon of regional, geographic, and social outspread. Census of India 2001, and e-Geopolis adopting different criteria had reported the existence of 3,279 and 6,467 units of UAs in the country in 2001 that supported 27% and 37% of its urban population(s) respectively (Denis and Kamala 2011). Significantly, in the process of UA formation, when cities taking part in the process expand, the gap between them narrows and their mutual 'competitor' relationship is transformed into one of 'competitors and collaborators'.

Anthropologist and ethnologist Robert Redfield originated the idea of the rural–urban continuum, where cultural differences between rural and urban spaces are blurred due to rapid urbanization. On the one hand, continuums are anti-dichotomous and relate to intermediary positions between the binaries; on the other hand, they signify experiencing change within the overall ever-changing environments. In the context of urbanization, contrary to the old perception of urban growth as a swelling of independent settlements in space (hinterlands) and time, the idea of continuum promotes urban growth as a multifaceted process that deals with the hierarchies of human settlements (both existing and evolving) and their changing profiles. The rural–urban continuum relates to changes in urban conditions and human living experiences as we move from rural to urban and within urban areas. In most Indian rural–urban continuums, it is widely noted that the 'primitiveness' of villages is vanishing over time, and increasingly villages are becoming integral parts of several expanding UAs. No single discourse can describe a spatial continuum. It is more than spatiality and human geography evolving together; economic and cultural dimensions are inherent attributes. A continuum is also associated with the dynamics of cultural 'change' (unending metamorphoses) and cultural reproduction(s).

Prakasa Rao (1983) was the first to extol the roles of small, medium towns and secondary cities in Indian urbanization and urbanism. His pro-mid-tier cities arguments, however, were overshadowed by metro-centric discourses. He attributed the transformation of highly stratified rural societies into cosmopolitan working-class societies to the "urbanindustrial syndrome" (ibid., 56).

The internet, digital media, IoT (internet of things), and expanding OTT (over-the-top platforms), etc. override the erstwhile sharp differences of life in rural and urban areas. Likewise, widespread consumerism and the ubiquitous advertising associated with it are further homogenizing the lived experiences of continuum settlers. Moreover, new technologies are generally adaptable, with minor alterations, to the recipient locales, whether new, old, developed, or underdeveloped. Concerns over new technologies' 'easy applications' are well taken care of by their promoters. For example, extending internet connectivity to new locations (towns, old cities, remote or developed villages) depends less on their developmental levels and more on affordability for the prospective resident consumers, and the projects' economic viability. This is also the case with the laying of other underground services infrastructure, including domestic gas supplies. In practice, settlements can benefit from smart technologies irrespective of their morphologies (formal/informal/natural) and statuses (tenured or owned) if people's repayment capacities remain intact. In this scenario of vanishing differences along the continuums and UAs, and the universalizing experiences of living, we will locate Ambala's urbanism soon after and speculate about its future on the rural–urban, inter-urban and district continuums.

2.3.1 Appraisals and Experiences of Smart City Mission

As mentioned in the 'Roadmap' section, we now deliberate on the smart city mission. In a democracy, a strong nexus exists between the politics of votes, futuristic visions, and development. As a standard practice, India designs and launches national and state-level multi-purpose schemes supporting its urban development programs. Both UPA (Congress) and NDA (BJP) have launched various welfare-oriented schemes to uplift the nation, states, and citizenry during their tenures. In a shift, the concept of sustainable and smart development is being played in the hands of Indian politicians to muster, exercise and consolidate power. After the takeover of NDA by the Modi government on 26 May 2014, some new well-intentioned schemes were floated, but in teleocratic ways and with political overtones. We focus on the experiences of the widely publicized smart city mission.

The former union minister for Housing and Urban Affairs, Hardeep Singh Puri, stated that 896 projects worth Rs 150 billion (15,000 crore) had been completed under the smart city mission. Moreover, 1,895 projects worth Rs 750 billion (Rs 75,000 crore) were being implemented. Projects worth Rs 42.63 billion (4,263 crores) were complete, and contracts worth another Rs 610.93 billion (Rs 61,093 crore) were awarded. Under the PPP, 65 projects worth over Rs 20 billion (2,000 crore) in 26 cities were ready. 116 projects worth over Rs 100 billion (10,000 crore) in 45 cities were being implemented, and tenders for 91 projects worth over Rs 96 billion (Rs 9,600 crore) in 36 cities were floated (Puri 2019). Further, "It is for the first time that a democratically elected government, in a federal framework, in partnership with the private sector, is quite literally 'building' a nation from bottoms-up" (ibid.).

Ahluwalia et al. (2014) stated that "as of 31 March 2013, close to 700,000 units had been constructed that fell far short of even their own target of constructing 1.56 million homes by December 2012."

In the author's view, the lofty and potent idea of making cities sustainable and smart with the help of green, eco and smart technologies got mired in Indian politics at the execution stage. The idea was reduced to an exercise in sourcing funds and disbursing those to various states and cities after making them compete for their e-readiness. The underlying mechanism was to devolve central funds to states and their ULBs after entering into tripartite legal agreements involving the pooling of pre-settled state and ULB shares of capital investments, followed by endorsements of their mutual legal obligations. Most Indian ULBs are perennially in dire straits, so they find it hard to furnish their obligatory shares. The situation reminds one of the banks' standard practice of disbursing loans to parties with good repayment capacities and ignoring the needy. The smart concept in the Indian context has generated awareness of new technologies that could help India achieve SDGs, but it has delivered little in real terms. Further, leakages and the possibility of some funds being diverted for politically motivated developments cannot be ruled out – a likely reason for the unmet promises of many well-intentioned transformative schemes in India, including sustainable-smart urbanism.

2.4 AMBALA: TRADITIONS PERSIST, SUSTAINABLE-SMART URBANISM KNOCKS

We now shift from a bird's-eye view (studies of Indian urbanization and urbanism) to a worm's (grassroots) and read Ambala, the district capital of Haryana,[4] minutely. With stereoscopic vision, we see two distinct realities: (i) perceptions, visions of the center and the state; (ii) aspirations, expectations and non-compliances of the citizens at the grassroots and how they are addressed. This section constructs plausible accounts and imagery of India's fractal neo-district urbanism. We learn of experiences of district living, district urbanism and urbanity, through the readings of the medium-tier city Ambala that follow. Empirical readings of the city help us visualize its socio-spatial contradictions and the dualities of Indian urbanism. Ambala, which has evolved into a multi-specialty city, helps us to read the 'change' induced and affected by post-independence urban policies; state and local-level legislation and administration; growing private entrepreneurship; informality; metamorphizing cultures and peculiarities of the place; the city's capacity to take up 'alterities' sourced by sustainability and smart technologies; altering production modes; and adoption of new sustainable technologies in waste management and garbage disposal.

2.4.1 The Old and the Colonial

The history of Ambala is fascinatingly layered. Each reign or rule in the sequence of its rise, fall, and disappearance experienced demolitions, reconstructions, expansions, modifications, and even neglect of its lived spaces. Ambala today is a palimpsest overlaid with confusing markings of times, deferred spatial productions in different styles of architecture. The ancient history of Ambala dates to the fourteenth century and remains somewhat unmeshed, with its recent history beginning with the British era. The British established their civic headquarters of 1570 km^2, comprising Ambala city and cantonment town (Ambala Cantt.), here in the 1840s. The 'twin city Ambala' had an urban population of 406,784 in 2011 (see Figure 2.1).

Colonial-era British cantonments established under the Cantonments Act 1924 (now managed under Cantonment Act 2006 through Cantonment Boards) continue to play a vital role in urbanizing India and its distinct urbanity. To conjure an idea of their scale, Kanpur (Cawnpore), the largest (population 108,000 in 2011), was laid over approximately 17 km^2 (4,245 acres), including 1.36 km^2 (335 acres) of dedicated civil area. Ambala, rated as a Class-I cantonment, was established in 1843 over approximately 32.8 km^2 (8,100 acres), housing 55,370 people.

The British combined independent *jagirs* (territories) of local Rajahs and set up Ambala district in 1847. The district that comprised the *tehsils* of Ambala, Naraingarh, Jagadhri, Pipli, Nalagarh, Ropar, Kharar, Pinjore-Kalka (present-day Chandigarh), Manimajra, Kasauli, and Sanawar has been annexed and now includes only two *tehsils*.

As our interest lies in reading the urbanization of Ambala in colonial, post-colonial, and contemporary contexts, it would suffice to mention here that the Ambala region was once ruled by local chieftains, Moguls, Sikhs, and the British. The building styles identified with different empires are traceable in the scattered archaeological remains. Considering nothing drives Indians more than their religion(s), most surviving native edifices (pre-British) in the city and its surroundings are religious buildings or delipidated *havelis* and humble remains of outdoor shelters of grazing or hunting grounds. For instance, Rani and Rajah *ka talab* (later Company Park and then Patel Park) by Rajah Ranjit Singh of Chhachhrauli are cases in point.

After capturing most of the Indian territory, the British had started thinking about cantoning their soldiers in controlled conditions and care. Massive casualties among the soldiers owing to ailments like malaria, cholera, and STDs stood behind the decision. Importantly, in the process of building cantonments and garrisons, the foundations of Indian colonial urbanization were laid. Cantonment building activities began in the 1800s. By 1818 and 1864, 18 and 55 British cantonments, respectively,

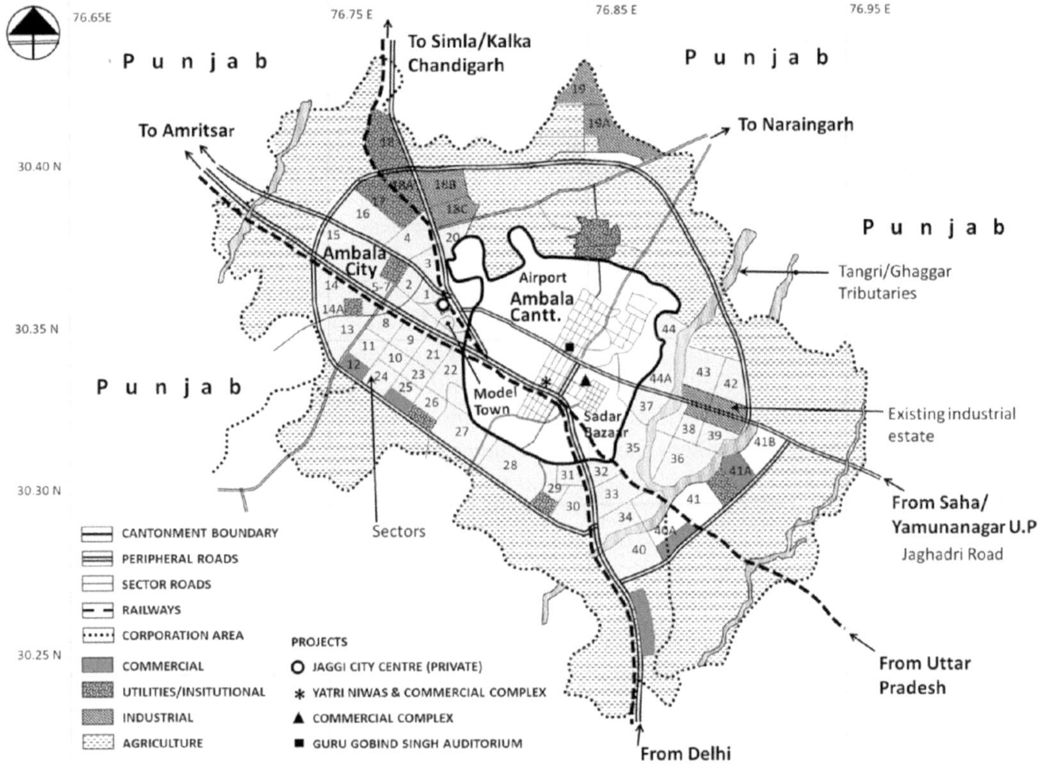

FIGURE 2.1 Ambala plan including Ambala city, Ambala cantonment, model town, Sadar bazaar, sector layout and basic land uses, location of the four projects, and the peripheral roads.

(Source: Author; developed from Town and Country Planning Department, Haryana, Drg. No. D.T.P-(A) 2923/2012 Dated: 03-09-2012.)

were operational, including Ambala cantonment (Lahiri 1990, 18–29).[5] In 1989, India had 62 cantonments, including the seven raised in the post-independence period.

Indian urbanization of the colonial period was driven by the colonizers' policy of raising four kinds of settlements in India – coastal and cantonment cities, hill stations, and railway towns (ibid.). Sheer logic (commerce and mercantile interests) governed their conception (examples include the port cities of Calcutta, Bombay, and Madras). Social exclusion and exclusiveness in civic life were at the heart of ecologically sensitive British urban planning. Cantonment sites were selected strategically, close to pre-existing towns, and followed set development patterns, controls, and norms. For details, see Kalal (1999, 18–36).

Environmental and ecological considerations also governed the development of hill stations. Railway towns did not lose their significance and followed natural growth. Ambala, a junction, can be treated as a railway-cum-cantonment town, classified since 2008 as a Tier-1 city.

2.4.2 The Post-Colonial Era

The post-independence phase was an era of distress migration and rehabilitation of the displaced. Consequently, many refugee and *mandi* towns were set up in northern India. Domestic industrialization (an outcome of India's tryst with socialism) was the *raison d'être* of the upcoming Indian industrial towns and cities. A few new towns were set up as state capitals and others to rehouse civic amenities, like universities, health, and judicial facilities that had been truncated after the partition. Inherited infrastructure like roads, bridges, canals, powerhouses, etc., played important roles in the post-independence urbanization. The early post-independence phase was marked by confusion regarding free India's urbanization policies. Some economists and planners, influenced by Gandhian ideology (small is beautiful utopia), wanted to make villages self-sustaining and promote cottage industries and healthy village living.[6] As studied earlier, India's initial slow urbanization may be attributed to the ideological conflict between Gandhi and Nehru.

On the other hand, in the region of the united Punjab, cities were becoming industrial in response to the state's drive to usher in different kinds of manufacturing. The cities were also becoming identified with specific small industries that they were engaged in. Ambala specialized in manufacturing glassware and scientific goods. Many displaced and local families that had set up and joined the city manufacturing industry for their livelihood today own flourishing businesses. Some of them have even opened export houses. Industrial development in the city has not followed a straight line. Just as the connotations of 'industry' kept changing, so did entrepreneurs' businesses and the workforce. Next, trading became a popular and lucrative occupation in Ambala (see Figure 2.2a–c).

FIGURE 2.2 (a and b) Present mixed-use development on erstwhile Nicholson Road where processions were held in the colonial era; (c) New commercial development/showrooms.

(Source: Author.)

2.4.3 The New

Keeping pace with the times, Ambala Cantt. has expanded across the river Ghaggar and beyond to Saha, a government industrial estate. This branch of Ambala Urban Agglomeration (UA) is towards the south-east. Moreover, another important interstate UA is the present-day loosely directed opportune growth along the Ambala–Chandigarh axis (Ambala's north-west spread). See Figures 2.1 and 2.3.

Today, the city is an assemblage of formal layouts, spontaneous growths that formalized iteratively, and recalcitrant informal growths (including slums) and vending areas sprinkled all over the city. The future of the city and its hitherto directionless growth is linked to the proliferating regional UAs that engulf the city. UAs and continuums set the direction for the city's future sustainable growth through gradual environmental up-grading, economic and social uplifts, and by attending to sustainability indices (SIs) (Böhringer and Jochem 2007; Pearce and Atkinson 1993). The forthcoming new highways, city-bypasses, and peripheral transportation routes will attract new growth and develop new centers (Figure 2.1).

Over time, Ambala's identity has continued to change additively from a purely administrative center of Ambala district, a cantonment town and important railway junction, to a small old industrial town, a city of food processors and blenders, an important scientific city manufacturing microscopes, allied scientific goods and sophisticated instruments (34% share of the total produced in India for domestic and overseas markets), to a regional wholesale cloth market (comprising more than 1,000 wholesale shops and showrooms) (see Figure 2.4a,b).

At present, Ambala is being projected by the BJP government in Haryana as the main center of the mutiny of 1857. In the process, Meerut's role in initiating the mutiny is allegedly being underplayed. Soon, the city's national identity will potentially morph into a famed colonial station which once sparked a mutiny.

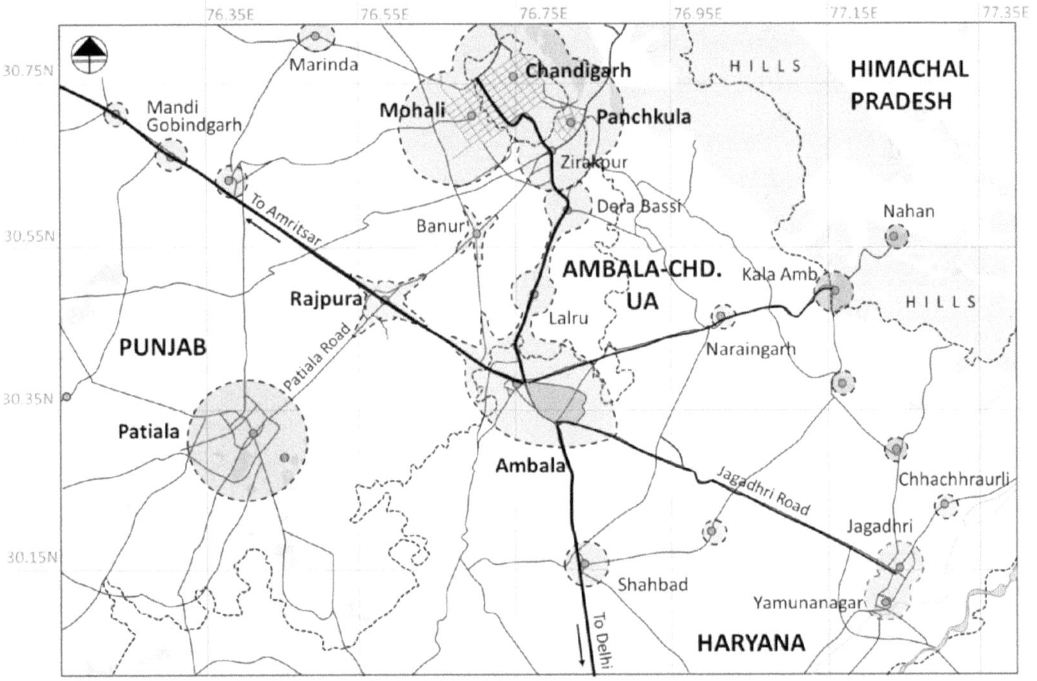

FIGURE 2.3 Regional Ambala and its emerging urban continuums and urban agglomerations.

(Source: Author.)

FIGURE 2.4 (a) New wholesale complex for wedding saris and trousseau in Ambala city; (b) Center for wedding saris and suits, Ambala city.

(Source: Author.)

Change(s) have been ushered into the city due to these shifts, reflected in the residents' prosperous lifestyles, transforming culture and social values, bustling city spaces, and enhanced prestige. Market forces, consumerism, and globalization seem to underlie the visible change. However, despite the *safai-abhiyans* (cleanliness drives), civic spaces remain neglected, and city growth is undirected despite the 2025 masterplan in force.

As described earlier, the Modi government launched a smart cities mission in India in collaboration with state and local governments. The aim was to improve the sustainability of 64 listed cities, improve their inhabitants' living conditions, make urban governance more responsive and effective, and make the cities' transportation, infrastructure, and services more efficient by employing 'e', 'k' and GIS technologies. Unfortunately, many cities, including promising Ambala, did not qualify for aid under the program.

Interestingly, the powerful BJP minister Anil Vij from Ambala Cantt. Constituency, who held many portfolios including home, health and Haryana local government, came to the city's rescue. He raised funds for constructing a new civic hospital, landscaping and lighting the cantonment bazaar streets, Netaji Subhash Chandra city park, reconstruction of the city stadium, and night shelters at the old bus station site (all completed). Other local projects under construction are Bank Square, the Martyrs of 1857 Mutiny Memorial Park on an approximately 25-acre site located on the NH-1/NH-44. This project, with an estimated cost of Rs 3 billion (Rs 300 crores), is the brainchild of Minister Vij. Interestingly, a 60 meter-high lotus tower (symbol of BJP) is the highlight of the development.

2.4.4 Public Participation

The developments mentioned above were top down and lacked public participation. Legislators and executives have a solid propensity to attain supremacy by maintaining asymmetrical power relations with the public. As a result, the commoners and inhabitants are denied access to decision-making. The challenge lies in maintaining a balance between the center at the top that holds power and the commoners at the bottom that source the power. Public participation and community engagement are essential for genuine democracy and achieving SDG 11. Unfortunately, in the cultures of scarcity and poverty, decentralization becomes challenging due to the inescapable dependence of districts and states on the center. Such crippling dependencies (inadequate internal revenues at district levels) weaken structures aspiring to decentralization, as they rely on central grants and aids. Ghana is a case in point; nearly three decades back, it implemented a decentralization program involving the devolution of financial, infrastructural resources, power, and authority to the district levels but it failed for the reasons given above (Owusu 2005, 69). This situation may not apply to India, a fast-growing and the world's sixth-largest economy (set to grow to US$3 trillion shortly). Still, there are lessons from Ghana's and other similar studies on district urbanisms and decentralization of power and resources at the state and district levels.

Vitally, the Annual Survey of India City Systems (ASICS)-17 revealed that Indian cities did not show citizen participation at all levels, including wards, committees, and area *sabhas* (Nair et al. 2018). The cities displayed little if no transparency and accountability. Circumvention of participatory processes resulted in bleak engagements between citizens and the governments. The fallacy of public participation is also common in developed nations. Australia is known for following perfunctory approaches to fixing its cities (Kelly and Donegan 2015, 154–156). The façade value of democracy is often exploited at all possible levels to shield the questionable public dealings carried out at various platforms.

Next, we will delve briefly into why public participation in urban governance has failed. In this context, democracy in planning has two faces: teleocracy (traditional directives oriented, top down, operating through elected representatives and professional central planning); and nomocracy (bottom up, direct, and inductive). Nomocracy performs well at small scales like self-organizing village communities, urban syndicates where stakeholders' engagement in decision-making and implementation of works is possible. Self-determination, or self-rule by the people, is the essence of democracy. However, the system understandably breaks down at a large scale. Teleocratic approaches suit centralized governance. Various governmental welfare schemes (mostly populist) floated intermittently are invariably teleocratic in their design and application. Nevertheless, we need to be circumspect about public participation, as it can cause delays in completing the projects. A trade-off between citizens' satisfaction and timely delivery is necessary.

Since the post-colonial period, the system of local governance and the atmosphere in which it operates have decayed almost to the point of no return, with even constitutional amendments like the 74th CAA not appearing to work. The blame for the dysfunctions and maladies of most Indian urban areas rests squarely with the inherited local self-governance structures. The outdated rules and regulations framed during the colonial period, and those not aligned to the demands of the fast-transforming world (reference globalization; new advanced sensing, cloud, and sustainable-smart technologies; the formidable speed of change; metamorphozing cultures and social values; advancing homogeneity, etc.) but still in force, are major hurdles in India's way to progress.

2.5 URBAN REGENERATIVE ARCHITECTURE; POST-COLONIAL APPROACHES; SPACES OF CONSUMPTION; PARTAKING IN GLOBALIZATION

In this section, we comb through the roots of district city urbanism and draw connections among city residents, regenerative architecture, consumerism, and changing societies. New civic projects of a medium-sized city have more than infrastructural value for its residents. Such projects can potentially recast the image of a city, the identities of a district, its constituent UAs, and residents. Retrofitted sustainable-smart city projects can even refigure the urbanscapes of a locale. For rejuvenation and renewal of cities, independent sustainable-smart projects can engage with the 'architecture of repairs' (beyond physical renewals) and rebuild images and identities of places.

Additionally, architecture of repair and 'extraordinary or iconic architecture' can make a big difference to any old city, for example, boosting its cultural life and economy. In rejuvenating cities and their renewal, smart-eco projects can engage regenerative architecture to reconstruct images and identities of old neighborhoods. Before liberalization in the early 1990s, decision-making often intersected with political interests (constituencies linked priorities); after that, while political interference continues, market forces begin to matter, and private urban entrepreneurialism and consumerism come to the fore. However, the post-colonial practice of ushering in change and improvements in small Indian cities by infusing new projects in their fabric continues as a standard approach for present-day district urbanism(s).

Re-imagining or recasting the image of an old city through public eyes (city branding with public participation) and then pursuing the pre-figured image of the city with a series of befitting civic and smart infrastructure projects through public–private and government partnerships are effective and economic strategies for invigorating decaying cities and making them eventually sustainable.

2.5.1 Jaggi Smart City Centre

Driven by the thoughts enunciated above, three projects were pursued by MCA around 2002 (see Figure 2.1): (i) the commercial complex-cum-*Yatri* Niwas (travelers' inn) at the old intercity bus terminus site on NH-1/NH-44, adjacent to the existing bus terminus; (ii) a commercial complex at the old local bus station site; (iii) Sh. Guru Gobind Singh Auditorium.

We include these three projects in our arguments here but concentrate on the first as a suitable example of expanding highway architecture in the region (see Figure 2.5a).

The building's relatively self-contained component (right) was to house rooms, dormitories, restaurants, and bars. The somewhat porous component (left) was to house food courts, the city's representative speciality eateries, exhibition areas, contact spaces desks for export and industrial houses, city tourism-related displays, stalls for networked accommodation, and miscellaneous promotional offers (a city showcase). The showrooms and public conveniences were accommodated on the ground floor. A financially assistive feature was the entry gateway with an inbuilt LED billboard displaying moving images and messages from multinational, national, and local brands on an elevated portal next to the highway.

The projects mentioned above were shelved due to lack of funds, local politics, lack of interest on the part of ULB, and inability to execute and manage them. During this period, a large private commercial project, Jaggi city center, was being developed by a local entrepreneur at the junction of GT Road (NH-1/NH-44) and the Jagadhari Road (NH-73). The project, indexed in Figure 2.1, was an instant success. The underlying causes of the pathologies and maladies of our district urbanisms cannot be truly understood without comparing the 'incompletions' of civic projects launched by

FIGURE 2.5 (a) Model of *Yatri* Niwas, Ambala Cantt. on NH-1/NH 44; (b–f) Various night-time views of Jaggi city center on NH-1/NH 44 represent emerging highway architecture.

(Source: Author.)

government agencies with the uninterrupted 'completions' of similar projects by private agencies. However, private city projects taking over and replacing equivalent government projects may generically be called 'replacement projects'. An account of Jaggi city center,[7] the replacement project for *Yatri* Niwas, follows.

First to come was a Jaggi group filling station, facing the NH-1/NH-44, with a large clientele of truck transporters. Within three years of purchasing additional land, a bizarre five-story golden hotel building and about 155 shopfronts and basements constructed by Jaggi enterprises were leased out to well-known franchise eateries, food chains and restaurants, and branded clothes shops (see Figure 2.5b–f). Travelers and locals throng the open paved courts facing the shops. Altering and giving new meanings and identity to spaces has started paying dividends to the promoters. An effective combination of strong private entrepreneurship and consumerism is producing commercial gains from the tourist facilities provided in a trendy and smart built environment.

The suspended government projects tell of how the inertia, pessimism, lack of accountability, lackadaisical attitudes, and ad-hocism of government agencies can nibble away at the city projects. We see how private replacement projects profited from the 'win-win' situation. Intensive efforts and initiatives put in by the private stakeholders stand behind their high success rates. The example shows how traditional, stable, time-consuming bureaucratic approaches to reviving old cities by instituting civic projects through district administrations are being displaced by developments floated and organized by strong private entrepreneurship.

2.6 CONCLUSIONS

1. Traditionally, India prioritized its economic growth to environmental protection. SDGs lack detailed contextual frameworks that appreciate the particularities, priorities, and inherent unevenness of the developed and developing nations. Nations' priorities and pace to accomplish SDGs vary according to their societal conditions, capacities, and political compulsions.
2. Urbanization and urban development are key to the economy, economic progress, GDP increase, and enhanced experiences of citizenship of any place. For India, which traditionally adopts a conservative approach to urbanization, three factors are important:
   i. Making governance more responsive, transparent, and inclusive at all levels, and incentivizing private entrepreneurialism,
   ii. Making ULBs self-reliant, professional, and futuristic in their outlook and working,
   iii. Genuine devolution of funds, powers, and responsibility to local levels, and relevant supportive parastatal organizations and reputable NGOs are necessary. Centralization, synonymous with teleocratic practices, affects non-inclusion, which in turn results in unsatisfactory delivery.
3. Inherited colonial bureaucratic structures and ancient Acts have proved hurdles on India's route to speedier urbanization, sustainable urbanism, economic development, and building stock deliveries. Sporadic amendments and repealing laws often stirs up serious controversies and sometimes slows down even healthy change. Considering new challenges and realities ahead of India, it is high time for the country to root out irrelevance from its urban planning and development instruments and repair if not rearticulate its outdated legislation. The time to clear heavy delivery backlogs and to leap confidently into the future has come. 'Sensible urbanization'[8] is the road to salvation. Genuine decentralization has never been a more relevant policy for India's urban development at district levels.
4. Comparisons with China come naturally to India regarding its new urbanization and emerging nondescript urbanisms. Marked differences between Indian and Chinese approaches are that the former underplays the role of its mid-tier cities and towns, and the latter underscores their role in its new policies on urbanization and sustainable-smart urbanism. The paradoxical farrago of urbanism is an expanding reality that complicates the idea of developing countries' urbanism.
5. UN HABITAT (2010)'s idea of differentiated strategies for megacities and small and medium towns (SMTs) and cities holds promise for India, particularly at the district-metro urban continuum. Sustainably spruced up mid-tier cities have tremendous potential for providing alternative 'quality living' to regional communities.
6. At the micro-scale, necessary for our analysis, Ambala figures as an assemblage of diverse city spaces: here, we have some spaces warped in time, there we find them lying unconsciously preserved in veneers of traditions. And then, we also have urban spaces frozen in time (cantonments) or an amalgamated state (civic areas, bazaars) (see Figure 2.6a–c). Products of different logics, alien *raisons d'être*, spaces co-exist in spatial neglect. Some locales are characterized by their transitive identities (shopping streets loaded with goods of fugitive brands) and others with their steadfastness (persisting specialty bazaars) (see Figure 2.6a–e). Also come trendy, aggressive, futuristic spaces (Jaggi city center; Mutiny Memorial Park) and other regressive, fatigued, and decaying spaces (slums, city informality) (see Figure 2.6f, g).
7. Complexities that stamp the history-laden Indian district urbanism remain unattended and demand attention. Autocracy seems to work here, for example, the old attractiveness of Ambala is being redeemed by the Haryana cabinet minister (Ambala Cantt. constituency), Anil Vij's teleocratic practices, and district, state, and central funds. But then, this is not a replicable solution for resolving the mounting problems of several Indian cities crying

FIGURE 2.6 (a and b) Spaces frozen in time – *Halwai* bazaar, Ambala Cantt.; (c) traditional neglected spaces; (d and e) growth of multi-brand showrooms on model town to sector-5 road; (f and g) representative of rundown city spaces and slums in Ambala.

(Source: Author.)

for attention. Developing bespoke interventions for the upliftment of district urbanisms by introducing decentralization wholeheartedly, empowering and strengthening ULBs (capacity building) to adopt data computationally, and embracing environmentally, ecologically, sensitive technologies to improve the living experiences of the residents are promising strategies.

8. Challenges lie in enhancing livability universally by following sustainable practices and adopting smart solutions for our urbanisms. The key is to maintain a balance between the roles of humans and 'sustainable-smart' technologies. Deployment of intelligent motion sensors in the street lighting systems of Singapore to support its governance substantiates the fact that as cities grow smart, the role of human agency in decision-making in their everyday lives gets relegated to IoTs (interconnected chips, sensors, software, and algorithms). These sensing technologies that are compatible with Singapore's residents may not work in the Indian context. An infringement of privacy rights may not go down well with the Indian citizenry. Therefore, India must be prudent in making choices and curb its over-enthusiasm for smart technologies.

9. If branding works well as a strategy for the future development of Ambala, then similar bespoke solutions could be replicated in other district sites. Thus, the possibility of new healthy district urbanisms flourishing in interstate regions becomes a realistic vision. Citizens may have a brighter future when investments and commitments for the upliftment of district urbanisms soar.
10. It appears that the straitjacket approach (one policy fits all situations) will not work in a diverse and extensive nation like India. A prudent choice for India seems to be to adopt multi-layered approaches like:
    i. Promoting district urbanisms (driven by local needs and priorities) and actively participating stakeholders, and focusing on decentralized (nomocratic) developments;
    ii. Attending concurrently to teleocratic developments located along metro-mega-global continuums and megapolises, and using efficient mechanisms to maintain a suitable climate for the growth and sustenance of multinationals and national institutions founded on capitalist structures.

NOTES

1 The term 'arcology' (a postmanteau of architecture and ecology) was coined by the visionary architect and urbanist Paulo Soleri in 1969. Implying sustainability in its purest form and universal context, it was applied in the Arconsi settlement, his experiment in compact biophilic living. https://www.youtube.com/watch?v=-2yZFYAPMq0.

2 Migrancy is intended to cover socio-cultural, economic, human geographic, human ecological, ethnographic, psychological, developmental, and transformative issues as related to the process of social migration. It covers all central and peripheral aspects underlying migrations including betterment of family lives and averting calamities, etc.

3 The eastern segment is nearly saturated.

4 Haryana comprises six administrative divisions and 22 districts, 163 agglomerations and outgrowths that cover 736 km^2. As per Aadhar statistics, Haryana's population is 27.39 million (2.7 crore) (2021). The state's urbanization level was 28.92.% (Census 2001); 34.88% (Census 2011); national average 31.16%; 1.7% of the state area is urban with average density 9,685/km^2. In 1951, Ambala was the most urbanized city of Haryana (37.19%) (Sangwan and Mahima 2019).

5 The mutiny of 1857 gave an impetus to building cantonments as defense forces were to be strengthened at vantage locations. In ancient times, criteria for 'urban' were whether the population lived in walled areas, citadels (urban), or open and low-lying lands (non-urban).

6 India has 641,000 inhabited villages, and 72.2 % of its population live there (Census 2011; Wikipedia; Ahluwalia et al. 2014).

7 City http://www.Jaggicitycenter.com/contact-us.php.

8 A relative term, having wide interpretations in different contexts. In the Indian context, it would be sensible to reduce primacy in its urbanization (concentration of the population in mega and metro cities) by promoting mid-tier or district urbanisms. For China to attain 65% urbanization by 2030, pursuing policies of aggressive urbanization will be sensible.

REFERENCES

Ahluwalia, I. J., Kanbur, S. M. R., and P. K. Mohanty. *Urbanisation in India: challenges, opportunities and the way forward.* New Delhi: Sage Publications (2014).

Böhringer, C., and P. E. P. Jochem. "Measuring the immeasurable: a survey of sustainability indices." *Ecological Economics* 63 (2007): 1–8.

Brundtland, G. H. *Our common future.* Oxford and New York: Oxford University Press (1987).

Cobbinah, P. B., Erdiaw-Kwasie, M. O., and P. Amoateng. "Rethinking sustainable development within the framework of poverty and urbanisation in developing countries." *Environmental Development* 13 (2015): 18–32.

Denis, E., and M. Kamala. "Toward a better appraisal of urbanization in India: a fresh look at the landscape of morphological agglomerates." *Cybergeo: European Journal of Geography*, UMR 8504 (2011): 1–32.

Coleman, M. "Immigration geopolitics beyond the Mexico-US border." *Antipode* (2007): Vol 38 54–76.

Hebe, V., and P. Karin. "Elaborating the urbanism in smart urbanism: distilling relevant dimensions for a comprehensive analysis of Smart City approach." *Information, Communication & Society* 22 (2019): 1328–1342.

Kalal, D. D. "Pune cantonments population: an analysis of its socio-economic dimensions." PhD diss., University of Pune. Available at: (http://hdl.handle.net/10603/149469) (1999).

Kelly, J.-F., and P. Donegan. *City limits: why Australia's cities are broken and how we can fix tthem*. Melbourne: Melbourne University Publishing (2015).

Koolhaas, R. "Introduction: city of exacerbated difference." In *Great leap forward, project on the city 1*, eds. J. C. Chung et al. Cologne: Taschen (2001).

Kumar, S., Kumar, N., and S. Vivekadhish. "Millennium development goals (MDGs) to sustainable development goals (SDGs): addressing unfinished agenda and strengthening sustainable development and partnership." *Indian Journal of Community Medicine* 41 (2016): 1–4.

Lahiri, M. "British influence on Indian urbanization." *Geographical Review of India* 46 (1990): 19–28.

Lefebvre, H. *The production of space*. Translated by Donald Nicholson-Smith. Oxford: Basil Blackwell Ltd (1991).

Lin, Z. "Building utopias: China's emerging new town movement." In *The emerging Asian city: concomitant urbanities and urbanisms*, ed. V. Bharne, 225–233. London: Routledge (2012).

Marshall, R. *Emerging urbanity: global urban projects in the Asia Pacific Rim*. London and New York: Spon Press (2003).

Nair, V. A., Vachana, V. R., and S. Rao. "*Five challenges India's cities must overcome to unlock their true potential*." Available at: (https://thewire.in/urban/five-challenges-indias-cities-must-overcome-to-unlock-their-true-potential) (2018, April 05).

Owusu, G. "The role of district capitals in regional development: linking small towns, rural-urban linkages and decentralisation in Ghana." *International Development Planning Review* 27 (2005): 59–89.

Pearce, D. W., and G. D. Atkinson. "Capital theory and the measurement of sustainable development: an indicator of 'weak' sustainability." *Ecological Economics* 8 (1993): 103–108.

Prakasa Rao, V. L. S. *Urbanisation in India: spatial dimensions*. New Delhi: Concept (1983).

Puri, H. S. "How India is building cities of the future?" *The Economic Times*, June 27, 2019. Available at: (https://economictimes.indiatimes.com/news/economy/infrastructure/how-india-is-building-cities-of-the-future/articleshow/69964522.cms)

Robinson, J. "Global and world cities: a view from off the map." *International Journal of Urban and Regional Research* 26 (2002): 531–554.

Roy, A. "Urban informality: toward an epistemology of planning." *Journal of the American Planning Association* 71 (2005): 147–158.

Sangwan, H., and M. S. Mahima. 2019. "Growth of urban population in Haryana: a spatio-temporal analysis." *International Journal of Research and Analytical Reviews* 6 (2019): 752z–756z. Available at: (http://ijrar.com/upload_issue/ijrar_issue_20543429.pdf)

Sanyal, B. "Knowledge transfer from poor to rich cities: a new turn of events." *Cities* 7 (1990): 31–36.

Thakur, A. "Reading 'Migrancy' and 'Buoyancy' in selected Chinese and Indian representational urban spaces Pearl river delta (China) and National capital region (India)." *Paper presented at the Fourth China India Roundtable: Population Policies and Development* (2005, September 22–23), Hong Kong University. Avaibale at: (https://www.academia.edu/3813123/Reading_Migrancy_and_Cuoyancy_in_selected_Chinese_and_Indian_representational_urban_spaces) (2005)

TNN. "7 Charts show why India is a world leader in urbanisation." *The Times of India*, July 2019. Available at: (https://timesofindia.indiatimes.com/india/7-charts-show-why-india-is-a-world-leader-in-urbanisation/articleshow/70325128.cms)

United Nations Human Settlements Programme (UN HABITAT). "*The State of Asian Cities 2010/11*." Available at: (https://www.unescap.org/sites/default/files/SoACR.pdf) (2010)

van Duijne, R. J., and J. Nirman. "India's emergent urban formations." *Annals of the American Association of Geographers* 109 (2019): 1978–1998.

3 The Geopolitics of Refugees

A Strategic Tool for Explaining the Concept of Migration in Developing Countries

Mohsen Janparvar and Sajed Bahrami Jaf
Ferdowsi University of Mashhad, Mashhad, Iran

Reyhaneh Salehabadi
Tarbiat Modares University, Tehran, Iran

Uday Chatterjee
Bhatter College, Dantan, Kharagpur, India

Mahboubeh Shahbazi
Urmia University, Urmia, Iran

CONTENTS

3.1 INTRODUCTION

In recent years, relationships between actors have become more complex than ever on various scales. One important factor influencing this complexity of relations between actors is refugees, because today an unprecedented number of people have been forced to leave their homeland due to civil wars and political tensions, causing millions of refugees to apply for asylum in different countries (Dustmann et al., 2018) and settle away from their homeland in the realm of other actors.

DOI: 10.1201/9781003131922-4

The widespread presence of these refugees in geographical areas is rising, so issues related to them are also rising. Various fields of study have studied and analyzed the refugee problem. Meanwhile, the geopolitical nature of refuge is a new concept that is increasingly being used in the global space and explored in academic books and articles. Most of these works have examined refuge from a general point of view, and its geopolitical nature has been rather less discussed and explained. The result is that work done in this area merely recognizes the problems of refuge and is unable to deal with the geopolitical issues, and achieve results and goals.

The present chapter uses a descriptive-analytical method and library data to explain the nature and geopolitical concept of refugees. First, the importance of refugees and their roles is considered. Second, the nature of refugees' geopolitics is examined. It should be noted that the geopolitical nature of refugees includes different attitudes and perspectives, providing a more nuanced and comprehensive picture, resulting in a new, important and productive discourse.

3.2 THEORETICAL BASIS OF THE RESEARCH

3.2.1 Geopolitics

Geopolitics is a concept that, due to its attractiveness and high importance, has acquired widespread applications both in the scientific field (geopolitical experts and non-geopolitical experts) and in the non-scientific field (practical users and the general public) (Figure 3.1). However, a general review of the definitions of geopolitics shows a variety of views, with the term having a number of different scientific and practical definitions and interpretations. A closer look finds some key words such as power, competition, politics, and geography in these definitions. These diverse ways of looking at the concept of geopolitics makes it more difficult to understand, with multiple users of the word each using it according to their own perception or circumstances, and not significantly related to its existential nature. This multiplicity of perceptions and views has contributed a great deal to the geopolitical knowledge literature, but has also caused ambiguity between users due to the multiplicity of applications and related concepts (Janparvar, 2017).

Ambiguity in fields of knowledge is normal, however, and different and dynamic attitudes can accelerate the forward movement of knowledge. Nevertheless, as these ambiguities increase in different ways, the foundations of knowledge can be weakened, leading to the formation and development of different and sometimes contradictory attitudes in the field (Janparvar, 2018). The following definitions of geopolitics will help to establish an initial framework.

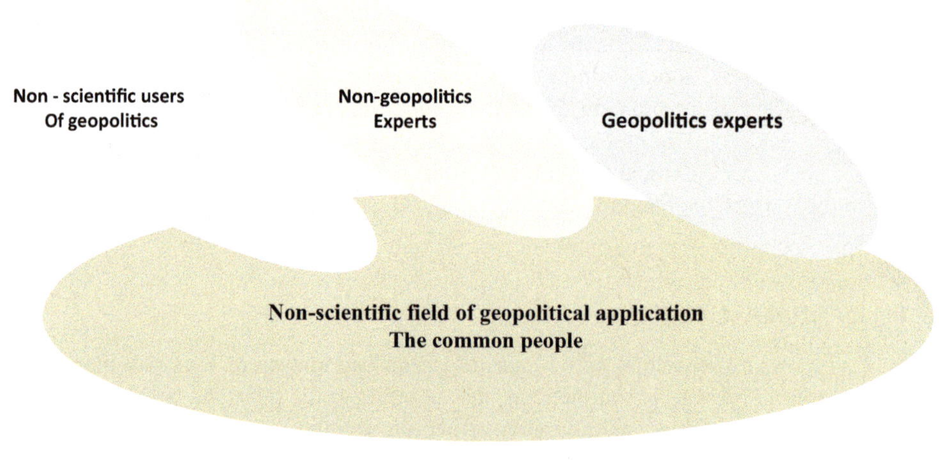

FIGURE 3.1 Scientific and non-scientific areas of application of the concept of geopolitics.

(Source: Janparvar, 2017.)

Colin S. Gray states: "Geopolitics is the unchanging relationship between geography and strategic power". Elsewhere, he equates security with geopolitics, noting the importance of the Strait of Hormuz in relation to oil supply: "Geography is energy, energy is economics, economics is security, and geopolitics is security." (Gray, 1977).

$$\text{Geography} \rightarrow \text{Energy} \rightarrow \text{Economy} \rightarrow \text{Security} \rightarrow \text{Geopolitics}$$

Peter Taylor, the English political geographer and founder of the famous journal *Political Geography*, considers that geopolitics is the study of the geographical distribution of power among the nations of the world, especially the rivalry between great and major powers. In other words, geopolitics is the knowledge of the acquisition, exploitation and preservation of geographical sources of power in transnational, national, regional and global relations. It is the study and review of the strategies of actors to achieve interests and goals based on geographical sources of power (Janparvar, 2017: 29). This definition is taken as a basis for the discussion in the remainder of this chapter.

3.2.2 Refuge

Recent events in some parts of the world have led to increasing numbers of refugees (Taylor et al., 2016). The United Nations High Commissioner for Refugees (UNHCR) has stated that thousands of people have been displaced by internal conflicts or natural disasters, which could be catastrophic for other regions (Kelley et al., 2015). In this regard, two types of displaced people, mostly civilians, can be considered. According to the definition of the United Nations, refugee status is given to people who are unable to return to their homeland due to fear of persecution, including race, ethnicity, nationality, membership in a specific social or political group (Cetin et al., 2019). The second type, displaced people, include most of the people who go abroad to seek the protection of other countries. According to the UNHCR, five million people entered European countries between 1985 and 1995 (UNHCR, 1997). Internally displaced persons are the result of spatial migration. The UNHCR describes them as people who have been forced to leave their homes and livelihoods because of persecution and armed conflict. The number of countries with refugees increased from 5% to 34% in 1996 (UNHCR, 1997).

There is ample evidence of links between internal threats and the growing trend of displaced persons (Semyenov et al., 2004; Stansfield & Stone, 2018; Verkuyten, 2009). One possible reason for the increase in refugees is a sense of threat from governments that has a significant negative impact on how people think and act (Yitmen & Maykel, 2019). Following the internal conflict in Syria, for example, a large influx of refugees from northern Syria moved to south-eastern Turkey. Many Syrians have been accepted as refugees in countries such as Turkey, Lebanon and Jordan. Prior to 2012, there were virtually no Syrian refugees in Turkey. However, by the end of 2014, the number of refugees had risen to about 1.2 million, and the trend has continued to grow. The refugees first settled in centers or camps built by the Turkish government. Many refugees also live outside the camps (Ceritoglu et al., 2015). Most Syrian refugees stay in these camps for two main reasons: (i) these places are close to Syria and they hope to return to their homeland after the internal disputes have been resolved; and (ii) as camps under government control, they have educational facilities, health care, social support and financial assistance (Balkan & Tumen, 2016). Displaced persons, therefore, impose a significant burden on their destination countries.

Figure 3.2 gives an overview of the number of displaced persons globally. Data show that since 2015, there have been approximately 15 million refugees worldwide. Only 30.7% of them have left their countries and migrated to European countries (7.8 million people) and South America (7.8 million people). This share is 25.7% and 8% for Africans and Asians displaced by conflict or violence, respectively.

Figure 3.3 shows global refugees by continent. The share of refugees for the whole period 1980–2015 was very high. However, intercontinental differences in the number of refugees are so significant that most refugees from Asia and Africa have been displaced by civil war.

	Continent of origin				
	Africa	*Asia*	*Europe*	*South America*	*World*
UNHCR population of concern (thous.)	19,124	29,423	2,672	7,765	58,991
Refugees (thous.)	5,392	9,607	507	448	15,960
of which are asylum seekers (thous.)	659	1,146	179	177	2,162
Share of refugees	0.316	0.365	0.257	0.080	0.307
Refugees distribution across destinations					
Africa	0.819	0.026	0.000		0.289
Asia	0.056	0.813	0.003	0.000	0.501
Europe	0.101	0.129	0.937	0.028	0.147
North America	0.019	0.025	0.057	0.430	0.039
Oceania	0.001	0.005	0.001	0.000	0.004
South America	0.003	0.001	0.001	0.541	0.020
Total	1	1	1	1	1

FIGURE 3.2 Origin and destination of UNHCR population of concern.

(Source: UNHCR, 2015.)

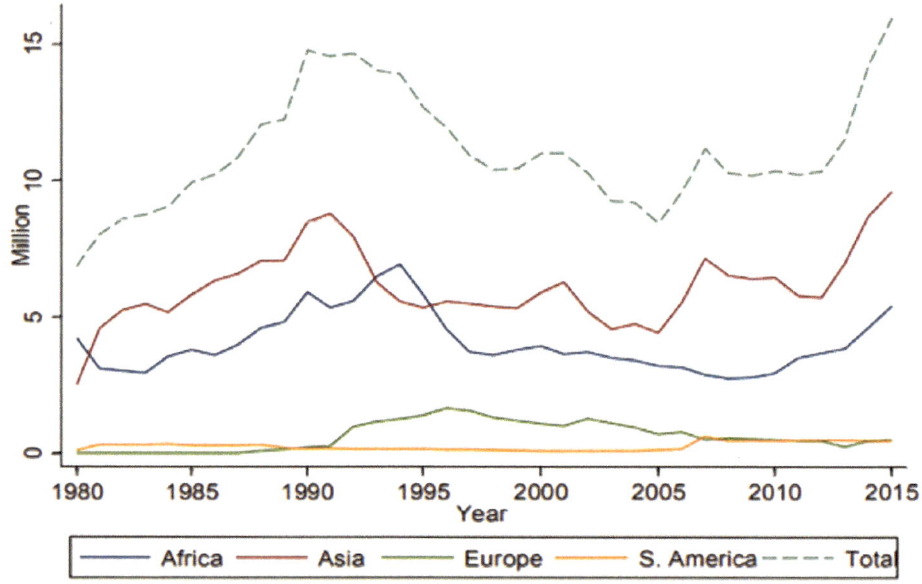

FIGURE 3.3 Evolution of refugee population by origin continent, 1980–2015.

(Source: UNHCR, 2015.)

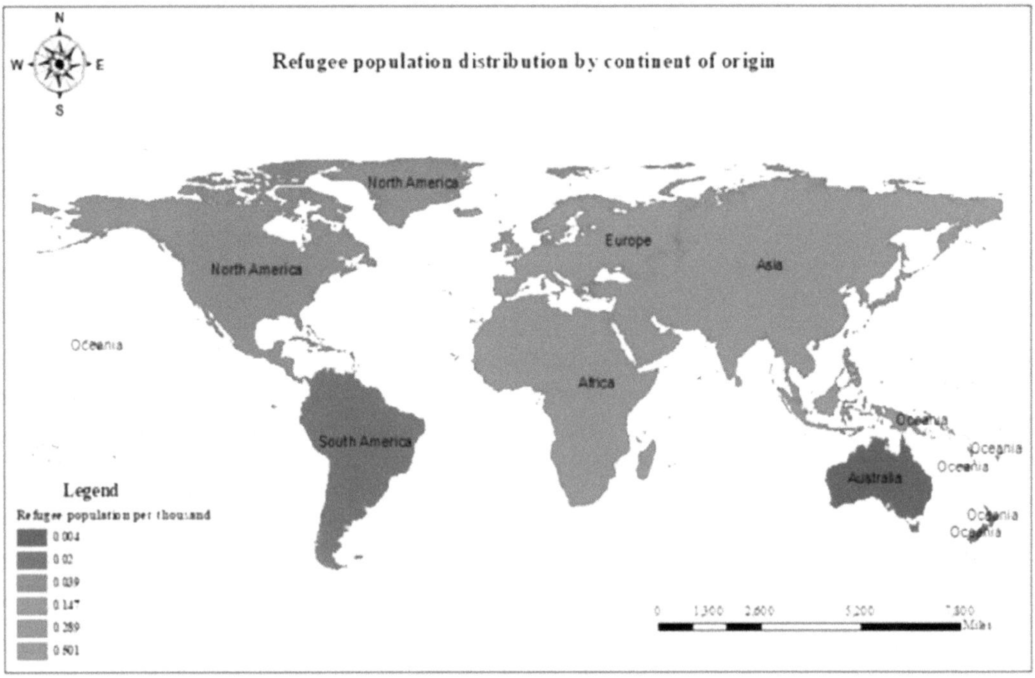

FIGURE 3.4 Population of refugees by continent.

(Source: Authors, 2020.)

Figure 3.4, based on Figure 3.2 data, shows a high number of people in Asia and Africa, shown in red, and Australia with the lowest population of refugees, marked in green. Civil wars, foreign wars, economic issues, and environmental problems on the continents of Asia and Africa have led to an increase in the number of refugees to these two continents compared to other continents.

3.3 RESEARCH METHOD

The current research is theoretical and uses the descriptive-analytical method. Due to the nature of the issue, the data collection method is library and internet. Information required has been extracted and classified from books and articles, and analyzed qualitatively.

3.4 RESEARCH FINDINGS

3.4.1 Finding the Importance of Refugees and Displaced People

The issue of refugees is one of the most dynamic issues in the national and international arena, encompassing a complex set of political, economic and social issues that can highly influence all aspects and policies both horizontally and vertically (Sasse & Thielemann, 2005). In the following sections, we examine the importance of refugees in the political, economic and socio-cultural sectors.

3.4.1.1 The Political Importance of Refugees

The contexts in which scholars have examined refugees include: (i) the involuntary movement of people in which the feeling of insecurity is the main driving force; (ii) the expectation of refugees that they will benefit from international support and assistance; and (iii) the fact that most of these refugees will return to their home countries after a period of forced relocation (Memisoglu & Ilgit, 2016). However, scientists and politicians increasingly prefer to use the concept for demographic movements of people forced to leave their homeland (Zetter, 2015). Many of these refugees are facing significant crises in global politics. For example, the United Nations has stated that the development and implementation of policies related to the protection of these refugees should be comprehensively considered by various governments (UNHCR, 2007). At the regional level, many destination countries, such as the European Union, are pursuing common strategies to reduce the influx of refugees, regardless of their mixed nature. These common strategies mainly involve strengthening the control of foreign borders as well as creating safe areas on the borders of the European Union (Del Sarto, 2010).

Many scholars claim that the situation and policies pursued by governments are always geopolitical and reflect these conditions. During the Cold War the discourse reflected the humanitarian nature and the politics of open borders. However, since the 1980s, we have seen a change in the discourse and practices associated with refugees that reflect their ideological or geopolitical value. Refugees can bring with them many problems, and since many of them have been forced to flee their homeland, they can bring security, economic and social problems to their hosts. In fact, this debate, given a technocratic approach, reflects the humanitarian and security concerns and national interests of states that are central to increasing or decreasing a country's national power. The New York Declaration for refugees and migrants is a political commitment, adopted in 2016, that contains a wide range of commitments by states to strengthen and enhance mechanisms to protect people on the move. It paved the way for the adoption of two new global compacts in 2018, one of which was the Global Compact on Refugees (GCR). The GCR response was not specifically directed at internally displaced persons (IDPs). In fact, it was criticized for "virtually" ignoring IDPs. However, it is important to note that the precursors to the GCR (notably the New York Declaration) specifically focused on refugees and migrants. As such, it is not particularly surprising that the GCR does not pay much heed to IDPs. That is not to say that IDP issues are not relevant to the GCR. One of the GCR's key objectives is to support conditions in countries of origin for return in safety and dignity. It also notes that population movements may be of a composite character and may involve IDPs. In addition, it directs states to "take into account" the situation of IDPs when contributing resources and expertise to support countries of origin (Hudson & Chrainne, 2020). Increasing numbers of people in some countries, including European countries, leads to racial mixing that can cause social and cultural crises for these countries. Therefore, it is important to pay attention to the issue of refugees.

3.4.1.2 The Economic Importance of Refugees

Refugees can bring work, skills and more to their destination countries. On the other hand, overall demand for the supply of goods and services also increases. In terms of supply and demand, refugees represent a potential economic aid. However, there are other views that consider refugees to be economically burdensome for countries (Stevenson, 2005). It is possible that refugees who have entered a country will not be able to integrate easily into the economic and commercial system of the host country and this will cause problems, because meaningful and lasting guarantees in the employment sector are considered an important factor in resolving economic tensions (MDA, 2011–2012). Experience has shown that refugees are desperately looking for work and vocational training and often accept jobs that are below their level of experience and education (MDA, 2012). For example, a survey of 227 refuges found that the majority of them were willing to do low-skilled jobs requiring no education or experience (Parsons, 2018).

There are different approaches to the economic importance of the organization in the country of destination: the first approach considers a relatively broader understanding of the value system, including the unmeasured benefits of economic aid to refugees with regard to social and cultural capital. This approach may take a more comprehensive view of reducing economic costs and will work well in the proportion of asylum seekers in the economy. This approach is more objective, independent and non-political and can work well for the maximum participation of asylum seekers in the economy. Some experts, however, believe that the economic benefits of the participation of refugees will only come to the second generation of their offspring (Hugo et al., 2006; Carrington et al., 2013). For example, according to Professor Grim Hugo's report for the Australian Department of Immigration and Citizenship (2007), the study of the contribution of first- and second-generation humanitarian entrants to the country has shown that there has been economic integration of refugees for almost all of their children.

The second approach is to have regard to the economic problems that the host country faces. Many scholars believe that refugees can play an important part in boosting economic costs, inflation, and so on, and can negatively affect the economies of nations.

3.4.1.3 Socio-Cultural Importance of Refugees

The refugee population can have a significant socio-cultural impact. For example, inequalities and access to health services through resettlement can have a significant impact on their health (Asanin Dean & Wilson, 2010). Alternatively, discriminatory behaviors (Edge & Newbold, 2013) can directly relate to socio-cultural challenges. These factors can be associated with exacerbated racism, conflict with cultural values, educational gaps, physical and mental problems, poverty and isolation (Rossiter & Rossiter, 2009). Refugees are often forcibly displaced from their country of origin and will live in dangerous conditions through displacement in refugee camps (Edge et al., 2014), and they may also be survivors of war and torture, and face the dangers of psychological problems and social injuries. For example, evidence suggests that displaced women are more vulnerable to mental health issues (Miszkurka et al., 2010). They are reported to be at higher risk of post-traumatic stress disorder, depression, and anxiety due to lack of social support, poverty, poor health, and discrimination (Le Espiritu, 2001). Thus, it can be stated that refugees, on the one hand, are mentally, socially and culturally incapacitated and, on the other, that they cause religious and racial mixing in other countries, including European countries. The influx of refugees to Europe in recent years has made them a threat to race and ethnicity due to their low growth rates. Discrimination can be seen against refugees on the basis of their religion, including Syrian refugees in European countries. So many of these countries are concerned about the influx of immigrants into their green continent and constantly seek to explain their immigration policies to suit their political circumstances so that they can make regulations and enact laws that address the scale of the continent's future problems.

3.4.2 The Nature and Concept of the Geopolitics of Refugees

Explaining the nature and meaning of new topics is a complex and difficult process; when they are ambiguous and multifaceted it becomes more complex. Geopolitical refugees are a case in point. The nature and conceptualization and the proper, systematic and scientific use of new terms is essential. Here we attempt to explain the geopolitical nature and concept of refugees using previous work by scholars in the field and based on scientific reasoning. The geopolitical nature and concept of refugees has been expressed in both traditional and critical geopolitics, and both approaches seem to be correct.

3.4.2.1 Traditional Geopolitics Frameworks

Despite different efforts made in various areas of refugee policy, governments are still the main players in this field. Governments and international organizations are posited as distinct and monolithic actors operating on scales that are taken for granted and seen as fixed, and the everyday dimensions of how refugees are governed in interaction between a range of relevant agencies is neglected. This has been the case even where crucial empirical findings point to the need for a different conceptualization. In her groundbreaking article on host country policies, for example, Karen Jacobsen discusses street-level bureaucrats' motivations and their impact on government policy, as well as power struggles between different institutions and decision makers as crucial factors shaping refugee policy. This warrants empirical analysis, but still conceives of government as a unitary and confined actor in this process, which can be squarely located on the national level (Lenner, 2020).

Discussions and frameworks in traditional geopolitics are based on actors looking at each phenomenon and geographic component only as a geopolitical component and trying to use it to achieve their interests and goals in relationships. Based on this, geopolitical actors regard refugees as one of the geopolitics components and try to use them in their relations on different levels. In this context, Hyndman (1997) states that displacement and migration have been a criterion for geopolitics for a long time and are the center of attention of experts. He argues in his book *Mobility and Geopolitics* that refugees can emerge as one of the dominant geopolitical discourses in geography (Hyndman, 1997). Spark also warns us that in the human dimension of migration, we must seriously consider existing trends, actions, and policies (Hyndman, 2012). In this approach, scientists state that the geopolitics of the land examines how power relations relate to specific human beings to monitor territory. In this regard, Huysmans (2006) refers to the expression of policies that widely create insecurity for human populations (Huysmans, 2006). Refugees are intertwined with the traditional definition of geopolitics, the initial layer of which is easily identifiable by the large population movements that result from state policies (Bobic & Jankovic, 2017). In fact, this definition states that the spatial and geographical dimension of population displacement occurs through forced and deliberate decisions. Therefore, according to this definition, two places interact with each other. Geographically, intrinsic changes must occur in the geographical location to determine the population's migration from one location to another. Based on the definition of Klein and Rutzel, who argue for a vital space in geopolitics and believe that countries must occupy more lands in order to survive, between political groups tension increases and forces the population to migrate and move, for example, forced and planned migrations to the east and southeast of the former Ottoman Empire that occurred during the last century. Traditionally based on geopolitics, the empire sought to expand its territorial space over the centuries, forcing many populations to relocate in order to gain their independence.

For example, for many years Turkey has had a fragile relationship with Syria. At the same time, we have seen that the government in Iran has enjoyed relatively friendly relations with Syria. The government of Bashar al-Assad has also tried to resolve disputes in previous periods and to avoid escalating them. It is noteworthy that Syria and Turkey have had a series of disputes over issues such as Kurdish political activity in the PKK, how the Euphrates River divides, and the province of Alexandria. On the other hand, Turkey in Syria has sought to form a government with a Sunni and fellowship approach. Hamas in Palestine, fellowship in Egypt, and the Justice and Development Party in Turkey operate as fellowship branches in these countries. Therefore, in this respect it can be stated that Turkey has about 900 km of common border with Syria. Turkey's geographical proximity to the center of the Syrian crisis is due to the influx of 36,000 Syrian refugees and the threat to its border with Syria due to the high volume of fire. On the other hand, Turkey's many years of trying to join the European Union have failed due to the crisis on the Turkish border and tension on the EU's borders. The European Union in the past has been optimistic about the security issue in Turkey; now it considers Turkey's entry into the EU as dependent on peace and security at its borders. Turkey can use refugees in order to put pressure on the European Union, and on the other hand, the refugees can be regarded as a force to expand Turkey's influence and increase its vital space.

3.4.2.2 Critical Geopolitical Approach

In contrast to approaches focusing on understanding the effects of geographical factors on the formation and shaping of foreign policy, the critical geopolitical approach emerged in the 1970s out of the ultra-structural movement formed in France (Muir, 1997). This approach expresses how geopolitical arguments enter into political issues focusing on social and political relations in the international arena. Among the theories that can be examined in this approach is the theory of Agnio and Corburij. They believe in the geopolitical concept of the present world representing specific disciplines that determine specific and disjointed patterns among global powers and emphasize geopolitical disorder. In this approach, we see that international trends in foreign policy are brought together to illustrate the global order of the world and illustrate causal factors. In relation to geography, critical geopolitics states that geography is not only a science of objects in particular places and situations, but also of objects that are at the center of social issues, so that geography is close to critical thinking (Reynolds, 1994).

In critical geopolitics, government is only one of the constituent components of geopolitics at the various local, national, transnational, and international scales. In past times, geopolitical factors were limited to governments and their actions, while in today's world these factors directly relate to individuals, groups, and organizations. So, geopolitical factors include the president or the bomber, organizations including terrorist organizations, social movements, etc., institutions including provinces, countries or multinationals, and so on. The purpose of all these is to achieve national independence, to deploy fundamentalist religious laws in countries, eliminate weapons of mass destruction, and so on (Ghalibaf & Pooyandeh, 2015). Therefore, it can be said that in this approach all the components and factors are directly related to each other and have high impact.

However, there are important and complex links between geopolitical factors and the movement of populations and their mobility that have received less attention from geographers. Massey investigated population displacement in large-scale population changes (Massey, 1993). Most urban geographers have considered developments related to demographic change in their empirical research and analytical interpretations. Geopolitical experts consider population displacement an influential factor in inter-state relations (Smith, 2001). Tesfahuney argues that the flow of migration of refugees and international refugees is one of the most important issues for geopolitical science (Tesfahuney, 1998). He cites, for example, the number of refugees in different countries who are willing to deal with the conflicts in which governments and countries are involved, such as Afghanistan, Iraq and Syria. In addition, ethnic populations in different countries can have significant effects on humanitarian action. In the 2004 tsunami, for example, Sri Lanka's ethnic groups in Canada were able to allocate funds to help the country (Hyndman, 2009). Therefore, in this approach, we can see important and diverse insights about geopolitical variables at different scales. Political geographers have considered important areas for understanding this issue and the complex relationships between geopolitics and critical geopolitical mobility. Hyndman, for example, describes transnational mobility policy and believes that the geopolitical divisions of the North and South, West and East have increased migration and attracted refugees from civil war and environmental factors to other countries. Cresswell also takes a critical view, stating that demographic movements through political and ideological factors have led to the displacement of people in different regions (Cresswell, 2006). At the same time, Coleman sought to examine geopolitical movements through the executive processes of immigration in the United States by looking at local and international population movements (Coleman, 2007: 59). Based on these insights, it is possible to understand the components and critical geopolitical elements in the relationships between populations and the extent of their displacement. In fact, nationalist discourses have legitimized the emergence of refugees at the national and regional levels, as well as legitimizing their entry into other regions (Ashutosh & Mountz, 2012).

In critical geopolitics, Sharp argues that the hierarchy of power focusing on actors' abilities can reproduce power in one country and the power of the elite suppresses ordinary people (Sharp, 2000). Sharp believes that Autotyle has stated that ordinary people inject different ideas into culture, and that diplomats and geopolitical experts only contribute to the current process as elites, and only have a supervisory role in these factors. This oversight prevents critical geopolitical experts from distinguishing between public discourse and geopolitical events and existing actors, and distinguishes between governmental and non-governmental actors. Thus, we can see a wide range of people and groups that have activities in international politics. These activities and trends affect geopolitical relations, especially human relations, and in some cases, when governments seek to gain more national and international power, they cause mobility and displacement in populations in non-populated areas. This is happening in a place where geopolitics has been the focus of bureaucratic decisions across the country as an important factor. Therefore, it can be stated that in these approaches, elements such as border and place are marginalized and the greatest emphasis is on religion, culture, humanity and so on. Meanwhile, human populations are a major factor in critical geopolitics, and the impact of political decisions on foreign and domestic policy by politicians and diplomats of any country can have a positive or negative effect on the movement of these populations. Therefore, the discourses in this approach tend to focus more on spatial rather than on space and express a shared space between countries. That is to say, the border no longer plays a colorful role, and the strategies of power in the domestic and foreign sphere are of great importance, as Turkey's policies towards Syrian refugees show. Turkey hosts 3.3 million displaced people. By opening its borders to Syrian refugees, Turkey has made significant efforts to support them. The Syrian people, despite widespread social pressure, have been able to use these services to an acceptable degree. These include health, housing, education and social assistance. In January 2016, the Turkish government allowed Syrian refugees to obtain official work permits. The aim was to help Syrian refugees in terms of economic development and social assistance, as well as to use their labor for economic development. Economic growth in Turkey has been divergent in some areas. In 2017, the unemployment rate rose to 13%, with youth unemployment averaging 19% in the country, so Turkey used the Syrian refugee population to generate economic growth in its own country. Syrian refugees have become an economic player in Turkey, which has led to the supply of labor and entrepreneurship skills, and by 2015, 157 Syrian companies were established in the country. In 2015, the share of Syrian companies in total foreign participation reached 26%. The increase in Syrian refugees has also had a significant impact on the age structure of the Turkish population, providing a good opportunity to turn them into a working population. By using Syrian refugees, Turkey has been able to reduce its demographic window (population decline by 2030). Based on these issues, in general, we can say that the geopolitical concept of refugee is defined based on a critical approach. The geopolitics of the displaced is the knowledge, acquisition, exploitation and retention of the displaced as a geographical source of power in transnational, national, regional and global relations. In other words, the geopolitics of the displaced is the knowledge of designing and reviewing actors' strategies to achieve the benefits and goals based on the displaced as a geographical source of power.

3.5 ANALYSIS AND CONCLUSION

The world arena has accommodated many actors; these actors have become more diverse than before and have created a variety of games to play. In addition to these actors, the geographical sources of power have become more diverse than in the past, and this has presented actors with different strategies to achieve their goals and interests. Strategies in various forms have shaped relationships from cooperation to conflict and the global stage. In the meantime, the world power play scene and the actors' strategies can be better described from a geopolitical vantage point. While the displaced have become more prominent as a geographical source of power in recent decades, actors are trying to use this resource to achieve their goals. The issue of refugees along the Turkish–Greek border, for example, has become a major global issue these days. There are human rights issues and humanitarian aspects of refugees along the Greek border, and on the other hand, there are behind-the-scenes actors shaping the game.

This chapter has attempted to conceptualize the geopolitical nature of the displaced. The present study shows the increasing importance of refugees raises issues of their attention and exploitation. Refugees have become a geopolitical component, to which there are two approaches (Figure 3.5).

A traditional approach to refugees is as a geographical source of power to achieve goals and interests in relationships. The geopolitics of the refugee is the knowledge, acquisition, exploitation and retention of the displaced as a geographical source of power in transnational, national, regional and global relations. In other words, the geopolitics of refugees is the design and review of actors' strategies to achieve refugee-based benefits and goals as a geographical source of power. In this approach, the geopolitical nature of the refugee is based on the actors' one-sided view of the refugee, their exploitation and captivity, their disregard for their living conditions, and, perhaps, their hostage taking. The critical approach, which seeks to exploit the geographical sources of power to

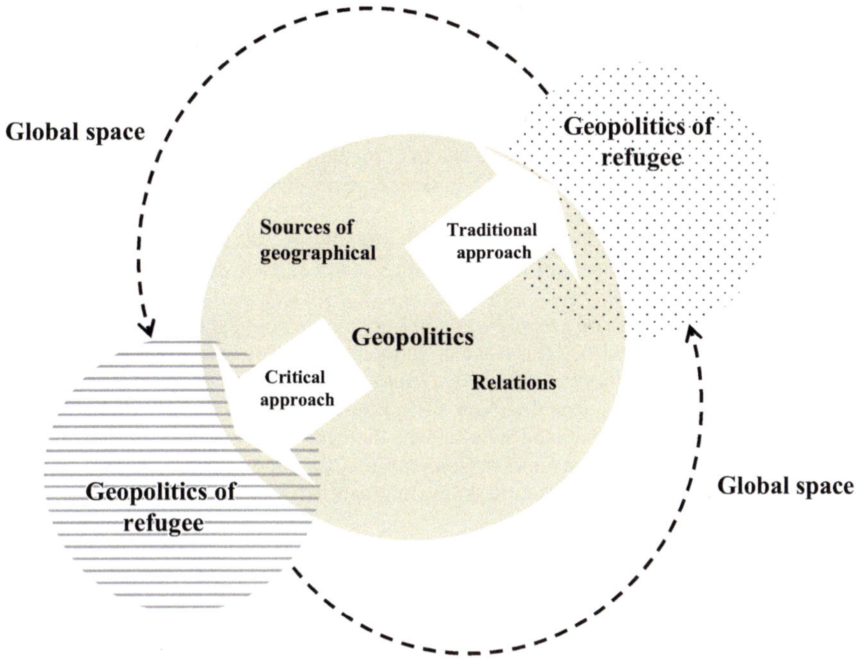

FIGURE 3.5 The nature and geopolitical concept of refugee.

create a better environment and resolve the problems of refugees, forms its strategies around the displaced in the global space to reduce their suffering. On this basis, the geopolitics of the refugee is about understanding, acquiring, exploiting and preserving the geographical sources of power in local, national, regional and global relations to resolve the problems and issues of the displaced and create a better living environment for them. In this approach, the geopolitical nature of the refugee is based on the actors' comprehensive view of the displaced, maintaining security, increasing the quality of life of the displaced, and reducing their suffering, creating better living conditions in the origin and destination countries of the displaced.

REFERENCES

Asanin Dean, J., and K. Wilson. "My health has improved because I always have everything I need here. A qualitative exploration of health improvement and decline among immigrants." *Social Science Medicine* (2010): Vol 70 (8) 1219–1228.

Ashutosh, I., and A. Mountz. "The geopolitics of migrant mobility: Tracing state relations through refugee claims, boats, and discourses." *Geopolitics* (2012): Vol 17 (12) 335–354.

Balkan, B., and S. Tumen. "Immigration and prices: Quasi-experimental evidence from Syrian refugees in Turkey." *Journal of Population Economics* (2016): Vol 29 (3) 657–686.

Bobic, M., and S. Jankovic. *Towards Understanding of Contemporary Migration Causes, Consequences, Policies, Reflections*. First edition. Belgrade: Institute for Sociological Research (2017).

Carrington, K., A. McIntosh, and J. Walmsley. *The Social Costs and Benefits of Migration into Australia*. Canberra: Commonwealth of Australia (2013, January). Available at: http://www.immi.gov.au/media/publications/research/social-costs-benefits

Ceritoglu, E., H. B. Gurcihan Yunculer, H. Torun, and S. Tumen. "The impact of Syrian refugees on natives' labor market outcomes in Turkey: Evidence from a quasi-experimental design." Unpublished manuscript, Central Bank of the Republic of Turkey (2015).

Cetin, G., A. Levent, and A. Sevinc. "International mobility." *Migration and Wellbeing Conference & Workshop* (2019). Istanbul, Turkey.

Cresswell, T. *On the Move: Mobility in the Modern Western World*. New York and London: Routledge (2006).

Del Sarto, R. "Borderlands: The Middle East and North Africa as the EU's Southern Buffer Zone," in: Dimitar Bechev & Kalypso Nicolaidis (Eds.) *Mediterranean Frontiers: Borders, Conflicts and Memory in a Transnational World*. London: I.B. Tauris (2010).
Dustmann, C., K. Vasiljeva, and A. P. Damm. "Refugee migration and electoral outcome." *The Review of Economic Studies* (2018): Vol 86 (5) 2035–2091.
Edge, S., and B. Newbold. "Discrimination and the health of immigrants and refugees: Exploring Canada's evidence base and directions for future research in newcomer receiving countries." *Journal of Immigration Minority Health* (2013): Vol 15 (1) 141–148.
Edge, S., K. B. Newbold, and M. McKeary. "Exploring socio-cultural factors that mediate, facilitate, & constrain the health and empowerment of refugee youth." *Social Science and Medicine* (2014): Vol 117 (1) 34–41.
Geddes, A. *The Politics of Migration and Immigration in Europe*. London: Sage Publications (2003).
Ghalibaf, M. B., and M. H. Pooyandeh. "The place of geopolitical factors in geographical levels (case study: Al-Ghaedeh)." *Journal of Military and Security Geography* (2015): Vol 2 (2) 130–140.(In Persian).
Gray, C. S. *The Geopolitics of the Nuclear Era*. New York: Russak & Company Inc. (1977).
Hudson, B., and B. N. Chrainne. "Enhancing state-to-state dialogue on internal displacement: Current global fora and future prospects." *Refugee Survery Quarterly* (2020): Vol 39 (4) 425–443.
Hugo, G., S. Khoo, and P. McDonald. "Attracting skilled migrants to regional areas: What does it take?" *People and Place* (2006): Vol 14 (3) 25–36.
Huysmans, A. *The Politics of Insecurity: Fear, Migration and Asylum in the EU*. London: Routledge (2006).
Hyndman, J. "Acts of aid: Neoliberalism in a war zone." *Antipode* (2009): Vol 41 (5) 867–889.
Hyndman, J. "Border crossings." *Antipode* (1997): Vol 29 (2) 149–176.
Hyndman, J. "The geopolitics of migration and mobility." *Geopolitics* (2012): Vol 16 (2) 243–255.
Janparvar, M. *Explaining the Concept and Nature of Environmental Geopolitics*. Chapter 2 Environmental security, complied by Kavianirad. Tehran: Center for Strategic Studies Publications (2018). (in Persian).
Janparvar, M. *Geopolitics for Everyone*. Tehran: Entekhab Publication (2017). (in Persian).
Kelley, C. P., S. Mohtadi, M. A. Cane, R. Seager, and Y. Kushnir. "Climate change in the Fertile Crescent and implications of the recent Syrian drought." *Proceeding National Academic Science of USA* (2015): Vol 112 (11) 3241–3246.
Le Espiritu, Y. "We don't sleep around like white girls do: Family, culture, and gender in Filipina American lives." *Signs* (2001): Vol 26 (2) 415–440.
Lenner, K. "Biting our tongues: Policy legacies and memories in the making of the syrian refugee response in Jordan." *Refugee Survey Quarterly* (2020): Vol 39 (3) 273–298.
Massey, D. *Power-Geometry and a Progressive Sense of Place, in Mapping the Futures: Local Cultures, Global Change*. London and New York: Routledge (1993).
Memisoglu, F., and A. Ilgit. "Syrian refugees in Turkey: Multifaceted challenges, diverse players and ambiguous policies." *Mediterranean Politics* (2016): Vol 22 (3) 317–338.
Miszkurka, M., L. Goulet, and M. V. Zunzunegui. "Contributions of immigration to depressive symptoms among pregnant women in Canada." *Canadian Journal Public Health* (2010): Vol 101 (5) 358–364.
Muir, R. *Political Geography*. New York: Bloomsbury (1997).
Multicultural Development Association Inc. (MDA). (2011–2012). Available at: http://www.mdainc.org.au/sites/default/files/mda_annual_report_2012.pdf, accessed 16 January 2013.
Multicultural Development Association Inc. (MDA). "*Settlement works snapshot: Employment in early settlement*." (2012). Available at: http://www.mdainc.org.au/sites/default/files/MDA_snapshots.pdf, accessed 16 January 2013.
Parsons, R. "*Assessing the economic contribution of refugees in Australia*." (2018). Available at: https://www.researchgate.net/publication/297324238
Reynolds, David R. *Political Geography*. London (1994).
Rossiter, M. J., and K. R. Rossiter "Diamonds in the rough: Bridging gaps in supports for at-risk immigrant and refugee youth." *International Migration Integrated* (2009): Vol 10 (4) 409–429.
Sasse, G., and E. Thielemann. "A research agenda for the study of migrants and minorities in Europe." *Journal of Common Market Studies* (2005): Vol 43 (4) 655–671.
Semyenov, M., R. Raijman, A. Yom Tov, and P. Schmidt. "Population size, perceived threat, and exclusion: A multiple indicators analysis of attitudes toward foreigners in Germany." *Social Science Research* (2004): Vol 33 (4) 681–701.
Sharp, J. "Remasculinizing geo(-)politics? Comments on Gearóid Ó Tuathail's critical geopolitics." *Political Geography* (2000): Vol 19 (3) 361–364.

Smith, M. P. *Transnational Urbanism: Locating Globalization*. (Malden and Oxford: Blackwell 2001); M. Davis, Magical Urbanism: Latinos Reinvent the US Big City. London: Verso (2001).
Stansfield, R., and B. Stone. "Threat perceptions of migrants in britain and support for policy." *Sociological Perspectives* (2018): Vol 61 (4) 592–609.
Stevenson, R. "Refugees and economic contributions, Hopes fulfilled, or dreams shattered? From resettlement to settlement." *Responding to the Needs of New and Emerging Refugee Communities Conference*, University of New South Wales, Sydney (2005), 23–28.
Taylor, J. E., M. J. Filipski, M. Alloush, A. Gupta, R. I. Rojas Valdes, and E. Gonzalez-Estrada. "Economic impact of refugees." *Proceedings of the National Academy of Sciences* (2016): Vol 113 (27) 7449–7453.
Tesfahuney, M. "Mobility, racism and geopolitics." *Political Geography* (1998): Vol 17 (5) 499–515.
UNHCR. "*Refugee protection and mixed migration: A 10 point plan of action*." (2007 January). Available at: http://www.unhcr.org/4742a30b4.html, accessed 7 April 2016.
UNHCR. *The State of the World's Refugees 1997–98 A Humanitarian Agenda*. Oxford: Oxford University Press (1997).
Verkuyten, M. "Emotional reactions to and support for immigrant policies: Attributed responsibilities to categories of asylum seekers." *Social Justice Research* (2009): Vol 22 (1) 293–314.
Yitmen, S., and V. Maykel. "Support to Syrian refugees in Turkey: The roles of descriptive and injunctive norms, threat, and negative emotions." *Asian Journal of Social Psychology* (2019): Vol 23 (3) 293–301.
Zetter, R. *Protection in Crisis: Forced Migration and Protection in a Global Era*. Washington, DC: Migration Policy Institute (2015).

4 Economic Development, Environmental Degradation and Sustainability

A Panel-Data Analysis of the Major Economies from 1990–2015

Nilendu Chatterjee, Bappaditya Koley, and Anindita Nath
Bankim Sardar College, South 24 Parganas, India

Uday Chatterjee
Bhatter College, Dantan, Kharagpur, India

CONTENTS

4.1 INTRODUCTION

At the heart of economic development there exists the words 'economic growth and high levels of sustained growth in income' or GDP. Without these two concepts, sustained growth in income and thus economic development is not achievable. These two concepts are very much connected with one another and certainly with the growth of manufacturing activities or industrial development. But, we all know that environmental degradation is caused by environmental problems and the nature as well as the magnitude of these problems are multidisciplinary and varying (Sankar, 2009). Some problems are global in nature, whereas others are regional; both need attention from the respective controlling bodies (United Nations, 2009, 2010).

Industrial or manufacturing activities have always occupied an important position in the developmental process of any nation. Apart from a few exceptions, it is a well-known fact that a nation cannot prosper without industrialization. But it's a fact that all the developed industrialized nations of the first world have prospered through industrialization, and the growth of national income has played an important part in this regard.

The third-world developing nations, in the era of globalization, have also been relying on industrial development for the sake of achieving higher growth rates and becoming developed. One can easily name China, India, Brazil or any other third-world developing nation in this regard. These nations, whose economies are based on service-related activities, have emerged as the world's dominant economic powers. But the manufacturing sectors in these nations have been lagging far

DOI: 10.1201/9781003131922-5

behind than expected, not only regarding their disappointing growth rates but also regarding the generation of employment, given the fact that hugely populous third-world nations expect industrial sectors to generate employment, and it is a fact that no other sector can generate employment as much as industrial activities can. So, becoming developed, and for this purpose relying on industries, has been a general phenomenon from the very beginning. There is a strong relationship between poverty or backwardness and sustainable development through the exploitation of environmental resources (Heady, 2000; Anijah-Obi, 2001). Hence, it can be rightly said that the world is under immense pressure and sustainable development is the need of the hour (Barrow, 1995).

The importance of environment started taking shape from the late 1970s and early 1980s and it has taken center stage in development economics ever since the inception of the concept 'sustainable development'. We all know that the discussion regarding growth and development is immensely important, but it cannot be completed without considering the environmental aspect. No nation can assert that they have become developed without doing any harm to the environment; neither can the developing nations in their way of becoming developed. But, to save the world and its resources is hugely important for the sake of future generations; without it all our developmental efforts along with their fruits will only be enjoyed by the present generation and will thus go on in vain, since future generations will not even exist without environmental resources.

In today's globalized world, the words 'global warming' are known as well as very much felt by people residing in every part of the world (Abe et al., 2014). No one can deny that the concept of global warming is very much connected to economic activities, mainly manufacturing activities, at the heart of which lies the everlasting thirst to achieve rapid growth and development.

We have reached such a stage today, that the words 'development' and 'rapid economic growth' have become almost synonymous with environmental degradation and the destruction of non-renewable resources without proper replenishment. This serious deterioration in environmental quality and resources has happened in the last three decades. We all know how important it is to safeguard the environment and yet achieve a commendable growth rate; striking a balance between the two has been the need of the hour for the last three decades. Several international agreements have been reached between nations, but we all know that what has been achieved has been "far below" the standards that need to be achieved. Again, it is a well-known fact that the lack of willingness on part of nations, whether developed or developing, has been the most concerning factor in not achieving the desired environmental standard, at the heart of which again arises the question of achieving economic development.

If we look towards India, we can see that the country has undergone a complete change of its trade policies since July 1991, and this shift has been motivated by the importance of trade policies for reaching the sustainable development goals. But we can see that there are shortages of forest stock in India. India possesses almost one-sixth of the world's population, 15% of the world's livestock, but only less than 2.5% of the world's land area and even less than 2% of forest stock. Hence, it is obvious that the amount of land and forest resources do not go hand in hand with the growth of India's population and livestock. Among the major reasons for the diminishing forest resources have been over-use, unsustainable use of forestry and the irrationality of developmental schemes (Government of India, 1999). From 1950 to 2005, over a period of more than 50 years, India's CO_2 emissions were 5.8% per annum on average, which made India the fourth highest nation in terms of fossil fuel CO_2. From 2006 onwards this growth rate started passing the 6% mark. Figure 4.1 and Table 4.1 represent the total pollution content ratio and the relative pollution content ratio.

An unbelievable increase in the pollution content ratio from the last decade of last century and for the first two decades of this century has been a real cause of worry.

Nations have involved themselves so much in the race to achieve high GDP growth rates that they have almost ignored the concept of sustainable development and safeguarding environmental resources. But what is good is that all nations have realized the necessity of using green, safe technology for the sake of saving this world. Several initiatives have been taken by both developed as

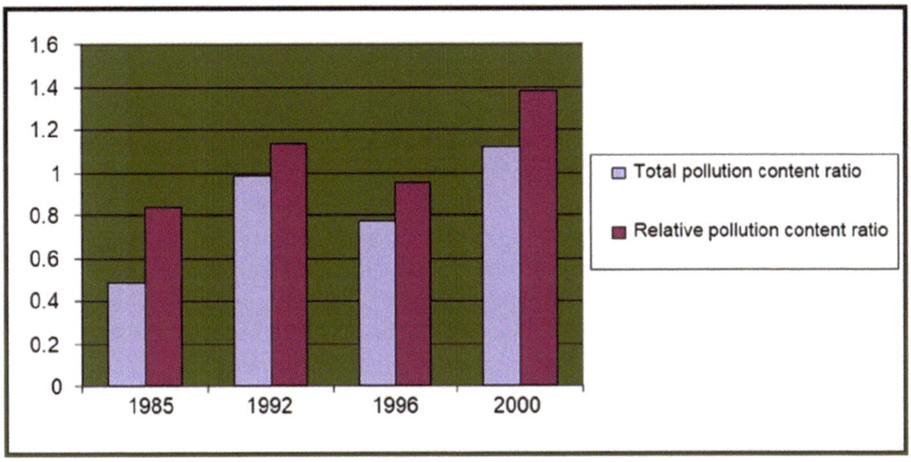

FIGURE 4.1 Total pollution content ratio and relative pollution content ratio.

TABLE 4.1
Pollution Content of India's Trade (in Rs '000)

Category	Pollution Content	1985	1992	1996	2000
(A)	Pollution content of exports	13,196,218	63,585,419	120,099,178.5	246,533,726.2
(B)	Pollution content of imports	27,484,349	65,004,807	156,264,425.1	219,352,036.2
(A/B)	Total pollution content ratio	0.480	0.978	0.769	1.12
(A/exports)/(B/imports)	Relative pollution content ratio	0.835	1.13	0.948	1.38

Source: Ministry of Environment and Forests, Government of India.

well as developing nations. Recycling is, perhaps, the most widely applied technique worldwide to reduce environmental degradation. Every type of solid waste can be recycled for further preparation of a commodity. For example, agricultural wastes can be recycled for making of various types of boards, wall panels, etc. Similarly, industrial waste can be re-used for making cement-related products. Other types of waste such as mining waste, hazardous and non-hazardous waste also have their re-uses when recycled (Pappu et al., 2007).

But we all know these initiatives are not enough. We need to take many more steps for the sake of achieving economic growth along with saving the environment. In India, the government felt that it is necessary to categorize industries according to their pollution potential to make it easy for ordinary people to understand. These categories are Red, Orange and Green in decreasing order of severity of pollution content. It means that highly polluting industries are represented in the Red category, followed by moderately polluting ones as Orange and marginally polluting or non-polluting ones in the Green category. In the Red category there are 17 industries, in the Orange division there are 34 industries and in the Green segment there are 54 industries listed at present.[1] The Central Pollution Control Board (CPCB) classified 2,744 businesses from 17 categories as highly polluting ones, till the end of 2012. 2351 of those industries have upgraded their technologies to comply with the guidelines, but 371 have failed to do so and as a consequence they have been closed down. Governments, both at state and central levels, have been taking action against the defaulting units and as a result of court cases these units have either upgraded themselves or have remained defaulting ones. Preventive measures like installation of abatement technologies, proper auditing of environment-related statements or green measures, etc. [2] have been followed but we all know they have fallen short of expectations as well as requirements. It is not that the severity of environmental degradation has hit India or developing nations only. It has hit all parts of the world and every nation has started taking initiatives to control it but failing to achieve significant improvement in the quality of nature, that is, reduction in emission.

We know that the need to achieve higher GDP has been the prime cause of environmental degradation. This way of achieving high GDP has been accompanied, mainly, by higher production of manufacturing goods. This paper seeks to throw light on all these aspects. In an era of globalization, GDP or production, environmental standards, human development, educational standards, institutional framework, environmental policies, openness of an economy, etc. – all these aspects are very much interrelated. Whether GDP growth is responsible for environmental degradation or growth of industrial activities – or the other way round – all these issues are highlighted in this paper with the help of panel-data analysis from 25 major economies across all continents. It may be that such an attempt has not been made before, and this is the prime motivation behind our study.

The rest of the paper is organized in the following manner. Section 4.2 discusses existing literature related to this field. Section 4.3 describes the data and methodology used in our study. Section 4.4 analyzes the findings of our study, and Section 4.5 concludes the paper with policy recommendations.

4.2 A REVIEW OF EXISTING LITERATURE

There are plenty of works regarding sustainable development and its environmental impact (Jansson et al., 1994). Here we briefly consider a few of those works. Meadows et al. (1972) show that higher levels of consumption and production are associated with higher emission levels as well as waste and if this continues, it will exceed the carrying capacity of the environment and ultimately cause degradation of environmental quality, as pointed out by Daly in 1991. Such an outcome may place a question mark before the functioning of economic activities and put the survival of civilization at risk, which would take us towards our steady-state values of growth (Ekins, 2000). Wu et al. (2020) have done a dynamic analysis of environment and economic growth, involving expenditure on healthcare in Taiwan, by applying wavelet analysis. They analyzed the co-movement, causal linkages between environmental pollution and the healthcare movement, by taking economic growth as the control variable. They found that there exists co-movement and causality between these two variables at different frequencies and times. They have also observed changes in the relationship during the periods of economic growth, environmental pollution, etc. Alvarado and Toledo (2017) have done analysis on Ecuador covering a time period from 1971 to 2010. They have shown that for a developing economy like Ecuador that relies heavily on exploitation of natural resources, there exists an opposite association between the two variables, indicating environmental degradation at the cost of economic growth. The researches have also found the existence of the 'pollution haven hypothesis', that is, polluting industries in the developed nations have been falling owing to tough norms, but in the developing world poor nations have been increasing, allowing these poor economies to enjoy a comparative advantage over developed economies, but in the production of "pollution-intensive" goods (Low & Yeats, 1992; Mani & Wheeler, 1998). Lucas et al. (1992) found the opposite result, with the growth of polluting products high in closed economies which are also poor. They found the environmental Kuznets curve (EKC) pattern of relationship between the pollution intensities of industries and income (Kuznets, 1995). So it is not always the case that only international business forces developing small economies to support production of pollution-oriented goods. Galeotti (2003) has reconsidered and analyzed the explanations that have been put forward by various authors regarding the existence or non-existence of the EKC. Chaudhuri and Gupta (2004), with the help of the general equilibrium framework, examined the impact of foreign capital inflow on the transfer of pollution-free technology to developing nations. They found that inflow of foreign capital positively impacts national welfare but it can also increase pollution levels. This result holds good irrespective of the shifting of green technology. International business can also harm the environmental quality of a nation in a number of ways, with rising production levels again causing depletion of resources (Siwar et al., 2008). Trade policies also influence migration. Chichilnisky and Di Matteo (1998) show, in their general equilibrium analysis, a positive impact on environmental policies, as well as discouraging migration of labor across nations with different pollution intensities. How trade policies are associated with environmental policies and how such linkages change along with national stance and export or import in international business has been neatly shown by Copeland and Taylor (2003) and Chichilnisky (1994). Phimphanthavong (2013) conducted a time-series analysis on Laos from 1980 to 2010 and found the existence of an inverted U relationship between economic growth and environmental degradation. Again, very interestingly, this analysis observed that the degradation of the environment has a strong association with the openness of a nation, its industrial activities, etc.

It is evident by now that most of these works studied one nation or a few nations with the help of time-series analysis, general equilibrium or dynamic analysis. Again, most of the work has tested the existence of the environmental Kuznets curve. But, almost no panel-data analysis has been done covering major economies from all over the world, considering various variables that affect environmental degradation and economic growth. This analysis seeks to fill that lacuna, and this has been the prime motivation behind our work.

4.3 DATA AND METHODOLOGY

Here, we have used annual data for 25 nations across the world – both developed and developing – during the period 1990–2015.[3] Our main data source is the World Development Indicators (hereafter, WDI). Our main intention is to check (i) whether the growth of GDP (Y) is influenced by the area of forest cover of a nation (FC) and the output of its manufacturing sector (M), (ii) whether the area of forest cover (FC) is influenced by the output of the manufacturing sector (M) and the growth of GDP (Y), and finally (iii) whether M is influenced by the other two variables, that is, Y and FC. In summary, we want to examine the relationship which exists among FC, M and Y for major economies of the world. We do this with the help of the following three equations in order to form a simultaneous equation structure or model. The simultaneous equation model is investigated for individual cross-section as well as for overall panel.

$$g\left(\mathrm{M}\right)_{it} = \alpha_0 + \alpha_{1i} g\left(\mathrm{Y}\right)_{it} + \alpha_{2i} g\left(\mathrm{FC}\right)_{it} + \alpha_{3i} g\left(\mathrm{FDI}\right)_{it} + \alpha_{4i} g\left(\mathrm{CO_2}\right)_{it} + \alpha_{5i} g\left(\mathrm{GHG}\right)_{it} + \varepsilon_{it} \quad (4.1)$$

$$g\left(\mathrm{Y}\right)_{it} = \beta_0 + \beta_{1i} g\left(\mathrm{M}\right)_{it} + \beta_{2i} g\left(\mathrm{FC}\right)_{it} + \beta_{3i} g\left(\mathrm{OPN}\right)_{it} + \beta_{4i} g\left(\mathrm{EI}\right)_{it} + \beta_{5i} g\left(\mathrm{EC}\right)_{it} + \varepsilon_{it} \quad (4.2)$$

$$g\left(\mathrm{FC}\right)_{it} = \lambda_0 + \lambda_{1i} g\left(\mathrm{Y}\right)_{it} + \lambda_{2i} g\left(\mathrm{M}\right)_{it} + \lambda_{3i} g\left(\mathrm{HS}\right)_{it} + \lambda_{4i} g\left(\mathrm{EE}\right)_{it} + \lambda_{5i} g\left(\mathrm{UP}\right)_{it} + \varepsilon_{it} \quad (4.3)$$

Here, $i = 1, \ldots\ldots, N$ denotes the number of cross-sections which are used (in this case $N = 25$) and $t = 1, \ldots\ldots, T$ illustrates the time period on which this work is based on. In Equation (4.1), α_1, α_2, α_3, α_4, α_5 show the effect of growth of GDP, forest cover area (FC), foreign direct investment (FDI), carbon dioxide (CO_2) emissions (metric ton per capita), and total greenhouse gas emissions (kt of CO_2). Equation (4.2) β_1, β_2, β_3, β_4, β_5 reveals the impact of growth of manufacturing output, forest cover area (FC), openness (OPN) and energy consumption (EC)[4], expenditure on infrastructure (EI)[5] on GDP. Equation (4.3) λ_1, λ_2, λ_3, λ_4, λ_5 shows that the growth of forest cover can be influenced by the growth of GDP, manufacturing output, health status (HS),[6] education expenditure (EE)[7] and urbanization or proportion of urban population (UP).

TABLE 4.2
Results of Panel Unit Root Tests

	LLC Test	IPS Test	ADF Test	PP Test
g(M)	−2.65* (0.00)	−3.289* (0.00)	34.15* (0.00)	43.31* (0.00)
g(Y)	−1.71** (0.04)	−1.42*** (0.08)	36.40* (0.00)	51.19* (0.00)
g(FC)	−5.62* (0.00)	−4.76* (0.00)	35.57* (0.00)	51.11* (0.00)
g(FDI)	−4.87* (0.00)	−3.41* (0.00)	30.23* (0.00)	38.60* (0.00)
g(CO_2)	−7.87* (0.00)	−6.36* (0.00)	47.87* (0.00)	47.65* (0.00)
g(GHG)	−7.41* (0.00)	−7.43* (0.00)	50.23* (0.00)	43.93* (0.00)
g(OPN)	−6.09* (0.00)	−5.19* (0.00)	43.39* (0.00)	47.53* (0.00)
g(EC)	−3.98* (0.00)	−2.97* (0.00)	33.17* (0.00)	34.07* (0.00)
g(EI)	−14.87* (0.00)	−7.28* (0.00)	41.71* (0.00)	41.32* (0.00)
g(HS)	−6.41* (0.00)	−8.43* (0.00)	55.97* (0.00)	45.93* (0.00)
g(EE)	−4.99* (0.00)	−7.81* (0.00)	43.39* (0.00)	43.53* (0.00)
g(UP)	−0.95** (0.01)	−0.33** (0.04)	25.32* (0.00)	57.11* (0.00)

Note: Values in parenthesis are the estimated p-values. *, **, and *** indicates significance level at the 1%, 5% and 10%, respectively.

Here we have followed Newey (1985) and Smith and Blundell (1986) and used the Durbin–Wu–Hausman test and the Hansen test for estimation of the simultaneous equation system. We know that the endogeneity issue for all the three equations of our study is suggested by the Durbin–Wu–Hausman test. This test also helps us to understand if instrumental variables techniques are needed or not because a rejection of the null hypothesis implies the significant as well as meaningful impact of endogenous variables on the parameters. Then, we have used the Hansen test for finding the proper instruments to determine the over-identifying restrictions in order to test the instruments' validity. The validity of instrument implies that a null hypothesis of over-identifying restrictions cannot be rejected; it also implies that the instruments are appropriate.

We also have applied panel unit root tests for smooth supervision or conduction of estimation. The generalized method of moments (GMM)[8] has been used to estimate the relationships with the help of simultaneous equations. Again, panel unit root tests have been applied for the overall panel. The following panel unit root tests have been applied: LLC test (Levin et al., 2002), IPS test (Im et al., 2003), PP-Fisher chi-square (Maddala & Wu, 1999) and Fisher-type augmented Dicky Fuller (ADF) test.

Apart from the LLC test, the other three tests are applied with the assumption of heterogeneous coefficients across different cross-sections. Only the LLC test is performed with the assumption of homogeneity. Here, we have found that all the variable of our study follow I (0). The results are given in Table 4.2.

4.4 RESULT AND DISCUSSION

From Equation (4.1) we have obtained the following result as depicted in Table 4.3. It shows that there are nations like USA, UK, etc. in which forest cover has no significant impact on the output of the manufacturing sector. Perhaps, the capital–intensive nature of production or replenishment of forests are the reasons behind this. But, GDP does have a significant impact on M which suggests that growth of GDP encourages the nations to go for further industrialization. FDI has a positive impact on M for most of the nations. FD has a significant impact on M in most of the nations and CO_2 and Green House Gas emission (GHGs) bear significant impact in all nations. Interestingly, the result of Panel GMM estimation illustrates significant impact of all the variables on M.

TABLE 4.3
Results of GMM Estimation for Equation (4.1)

	Dependent Variable: Growth of Manufacturing (M)					
Country	Intercept	Y	FC	FDI	CO_2	GHG
1	0.14 (0.95)	1.06* (0.00)	−0.08 (0.41)	−0.17** (0.08)	0.20** (0.02)	1.36*** (0.061)
2	1.78* (0.00)	1.41* (0.00)	0.007 (0.70)	0.08** (0.04)	0.10** (0.04)	0.81*** (0.054)
3	1.53* (0.00)	0.96* (0.00)	0.009** (0.02)	−0.089* (0.00)	0.09** (0.01)	0.02** (0.03)
4	5.94* (0.00)	0.34* (0.00)	0.008 (0.167)	0.24* (0.00)	0.42* (0.00)	1.20** (0.012)
5	6.412* (0.000)	0.051*** (0.06)	0.01** (0.02)	0.14 (0.35)	0.20* (0.00)	0.36** (0.02)
6	8.30* (0.00)	0.045* (0.00)	0.003*** (0.051)	−0.10 (0.66)	0.51* (0.00)	1.01* (0.00)
7	8.59* (0.00)	0.14** (0.03)	0.05*** (0.09)	−0.32 (0.76)	0.16** (0.014)	0.08** (0.03)
8	6.57* (0.00)	0.27** (0.01)	−0.01 (0.88)	0.51* (0.00)	0.27* (0.00)	0.91** (0.012)
9	5.82* (0.00)	0.01** (0.04)	0.08** (0.01)	0.30* (0.00)	0.24** (0.01)	1.15*** (0.51)
10	4.85* (0.00)	0.39* (0.00)	0.06* (0.00)	1.30** (0.03)	0.36* (0.00)	0.77* (0.00)
11	4.59* (0.00)	0.12** (0.02)	−0.09 (0.43)	0.19* (0.00)	1.09* (0.00)	0.48* (0.02)
12	6.89 (0.11)	0.95* (0.00)	0.08 (0.84)	−1.08* (0.03)	0.76** (0.02)	1.09** (0.02)
13	5.30* (0.00)	0.07* (0.02)	0.18*** (0.05)	0.04** (0.06)	0.91** (0.03)	0.36* (0.00)
14	4.64* (0.00)	0.27** (0.04)	1.19*** (0.051)	−0.59** (0.03)	−0.03*** (0.07)	0.23* (0.00)
15	1.66*** (0.09)	0.07*** (0.09)	0.002** (0.04)	−2.78* (0.00)	0.37* (0.00)	−0.54* (0.01)
16	5.26* (0.00)	0.02** (0.03)	0.19** (0.03)	0.12* (0.00)	0.73* (0.00)	0.11* (0.00)
17	−1.07 (0.28)	1.16* (0.00)	0.06*** (0.06)	0.39* (0.00)	0.15** (0.04)	−0.17** (0.04)
18	1.53* (0.00)	0.96* (0.00)	0.009** (0.02)	−0.089* (0.00)	0.09** (0.01)	0.02** (0.03)
19	5.94* (0.00)	0.34* (0.00)	0.008 (0.167)	0.24* (0.00)	0.42* (0.00)	1.20** (0.012)
20	6.412* (0.000)	0.051*** (0.06)	0.01** (0.02)	0.14* (0.01)	0.20* (0.00)	0.36** (0.02)
21	3.11* (0.00)	0.19* (0.00)	0.05 (0.38)	0.21** (0.04)	0.26* (0.00)	0.76** (0.013)
22	9.04* (0.00)	−0.011 (0.219)	0.51*** (0.07)	0.58* (0.00)	0.93* (0.00)	0.07*** (0.06)
23	1.92 (0.38)	0.56* (0.00)	0.70* (0.00)	0.16** (0.017)	0.11** (0.03)	−0.06*** (0.000)
24	0.91 (0.87)	−0.37** (0.04)	0.01* (0.00)	0.66*** (0.06)	0.56* (0.00)	0.72** (0.03)
25	2.98* (0.00)	0.07** (0.03)	−0.05*** (0.09)	0.41* (0.00)	0.91* (0.00)	0.36* (0.00)
Panel	4.42* (0.00)	0.07*** (0.09)	0.13* (0.00)	0.29* (0.00)	0.03*** (0.09)	0.82* (0.00)

Hansen Test (p-value): 11.25 (0.72).
Durbin–Wu–Hausman test (p-value): 10.77 (0.00).
Note: Values in parenthesis are the estimated p-values. *, **, and *** indicates significance level at 1%, 5% and 10%, respectively.

The result of Equation 4.2 is shown in Table 4.4, where M, FC and EC can be seen to have significant impacts on Y in most of the nations, but OPN and EI have a significant impact only on a few nations, not all. Rather, energy consumption (EC) has a significant impact on most of the nations' GDP growth. Here also, the overall panel depicts the significant impact of all the variables on the growth of GDP.

The result of Equation 4.3, as depicted in Table 4.5, suggests that growth of GDP (Y) as well as growth of manufacturing (M) have a significant impact in all the nations on FC, as do educational expenditure, urbanization and health status. The overall panel shows the significant impact of all the independent variables on the dependent variable. We can say that all the independent variables have a significant impact on forest cover, which establishes our argument that economic production and development hampers forest cover or environment, but education can bring a change as shown by its significant impact.

TABLE 4.4
Results of GMM Estimation for Equation (4.2)

	Dependent Variable: Growth of GDP (Y)					
Country	Intercept	M	FC	OPN	EI	EC
1	−3.97** (0.04)	1.59* (0.00)	−0.12*** (0.08)	0.49* (0.00)	0.73** (0.02)	0.09*** (0.08)
2	4.89* (0.00)	0.15*** (0.07)	−0.03 (0.66)	0.05** (0.03)	0.09*** (0.052)	0.82* (0.00)
3	0.51 (0.33)	0.25** (0.03)	0.03** (0.03)	0.45* (0.00)	0.09 (0.42)	0.03*** (0.08)
4	9.80* (0.00)	0.57*** (0.06)	0.52* (0.00)	0.19 (0.40)	0.06*** (0.08)	0.91** (0.01)
5	−32.22 (0.31)	0.06*** (0.07)	−0.10 (0.22)	−0.07 (0.64)	1.36* (0.00)	0.77** (0.02)
6	1.87 (0.49)	0.97*** (0.09)	0.14*** (0.08)	0.47*** (0.07)	−0.26 (0.65)	0.04* (0.00)
7	0.91 (0.87)	−0.37** (0.04)	0.01* (0.00)	0.66*** (0.06)	0.56* (0.00)	0.72** (0.03)
8	2.98* (0.00)	0.07** (0.03)	−0.05*** (0.09)	0.41* (0.00)	0.91* (0.00)	0.36* (0.00)
9	8.07* (0.00)	−0.32** (0.03)	−0.11 (0.62)	−0.35 (0.25)	0.13** (0.05)	0.74*** (0.08)
10	8.70* (0.00)	0.32** (0.03)	−0.11 (0.72)	−0.35* (0.00)	0.05*** (0.06)	0.13*** (0.053)
11	15.49* (0.00)	0.38 (0.27)	0.41** (0.02)	0.24** (0.012)	1.61** (0.013)	1.55* (0.00)
12	−1.33 (0.58)	0.82* (0.00)	−0.46 (0.493)	0.43** (0.02)	0.80** (0.02)	0.57*** (0.09)
13	3.11* (0.00)	0.19* (0.00)	0.05 (0.38)	0.21** (0.04)	0.26* (0.00)	0.76** (0.013)
14	9.04* (0.00)	−0.011* (0.00)	0.51*** (0.07)	0.58* (0.00)	0.93* (0.00)	0.07*** (0.06)
15	1.92 (0.38)	0.56* (0.00)	0.70* (0.00)	0.16** (0.017)	0.11** (0.03)	−0.06 (0.24)
16	1.80 (0.57)	0.61** (0.02)	0.98* (0.00)	0.18** (0.03)	0.27* (0.00)	0.53*** (0.08)
17	6.412* (0.000)	0.051*** (0.06)	0.01** (0.02)	0.14 (0.35)	0.20* (0.00)	0.36** (0.02)
18	−0.67 (0.32)	1.28* (0.00)	0.07 (0.66)	0.23** (0.01)	0.45** (0.04)	0.13** (0.03)
19	5.94* (0.00)	0.34* (0.00)	0.008 (0.167)	0.24* (0.00)	0.42* (0.00)	1.20** (0.012)
20	6.412* (0.000)	0.051*** (0.06)	0.01** (0.02)	0.14 (0.35)	0.20* (0.00)	0.36** (0.02)
21	8.30* (0.00)	0.045* (0.00)	0.003*** (0.051)	−0.10 (0.66)	0.51* (0.00)	1.01* (0.00)
22	8.59* (0.00)	0.14** (0.03)	0.05*** (0.09)	−0.32 (0.76)	0.16** (0.014)	0.08** (0.03)
23	6.89 (0.11)	0.95* (0.00)	0.08 (0.84)	−1.08 (0.37)	0.76 (0.16)	1.09** (0.02)
24	5.30* (0.00)	0.07* (0.02)	0.18*** (0.05)	0.04 (0.84)	0.91** (0.03)	0.36* (0.00)
25	4.64* (0.00)	0.27** (0.04)	1.19*** (0.051)	−0.59 (0.36)	−0.03*** (0.07)	0.23* (0.00)
Panel	1.54* (0.00)	0.51** (0.012)	0.81* (0.00)	0.65*** (0.07)	0.62* (0.00)	0.15* (0.00)

Hansen Test (p-value): 8.62 (0.20).
Durbin–Wu–Hausman test (p-value): 5.54 (0.04).
Note: Values in parenthesis are the estimated p-values. *, **, and *** indicates significance level at the 1%, 5% and 10%, respectively.

4.5 CONCLUDING REMARKS AND POLICY PRESCRIPTIONS

Here we have considered 25 major economies. A few of them are developed and the rest are developing but they represent all the continents. With the help of a simultaneous technique, we have employed GMM and estimated the impacts of mainly Y and FC on M; M and FC on Y; and Y and M on FC. We have also considered other relevant variables which should have an impact on the dependent variable in every equation such as FDI, openness, educational expenditure, urbanization, CO_2, GHGs etc. We have also done the analysis for the overall panel. We have considered data covering a period of 15 years (1990–2015). From the overall panel analysis it has been found that all our included endogenous variables have a meaningful impact on the dependent variables, that is, M, FC and Y. But from the individual GMM estimation of individual equations, it has been found that for a few nations, FC, Y have no significant impact on M, but for almost all nations M and Y have a significant impact on FC and so the impact of M and FC on Y is correct. Variables such as EE and UP have a significant impact on FC. Emissions of CO_2

TABLE 4.5
Results of GMM Estimation for Equation (4.3)

	Dependent Variable: Forest Cover Area (FC)					
Country	Intercept	Y	M	HS	EE	UP
1	10.96* (0.00)	−1.47*** (0.07)	−0.96*** (0.08)	1.16** (0.012)	0.09** (0.03)	1.42* (0.00)
2	−13.37 (0.94)	−0.14*** (0.081)	−1.34* (0.00)	3.06* (0.00)	0.59* (0.00)	1.05* (0.00)
3	−10.33 (0.58)	−2.51* (0.00)	−1.72* (0.00)	1.04** (0.03)	0.75** (0.02)	3.02* (0.00)
4	−4.24 (0.63)	−2.17* (0.00)	−0.77** (0.02)	1.08*** (0.051)	1.06** (0.03)	2.03** (0.04)
5	−29.96* (0.01)	−1.74*** (0.054)	−3.83* (0.02)	1.03** (0.04)	1.01*** (0.07)	0.81** (0.015)
6	−17.84* (0.00)	−2.36* (0.00)	−0.17* (0.05)	−0.66* (0.00)	1.07*** (0.09)	4.04* (0.00)
7	5.97 (0.90)	−9.75* (0.00)	−2.68* (0.00)	2.40** (0.016)	1.80** (0.014)	2.08** (0.02)
8	2.53 (0.93)	−0.45* (0.01)	−5.81* (0.04)	3.34** (0.01)	0.75** (0.02)	0.10*** (0.09)
9	−8.07 (0.38)	−0.46*** (0.058)	−1.59*** (0.09)	3.88** (0.02)	−3.27 (0.73)	3.98** (0.015)
10	−9.70 (0.21)	−2.60* (0.00)	−3.10** (0.03)	0.70* (0.00)	3.39* (0.00)	1.05* (0.00)
11	25.07* (0.00)	−1.41* (0.00)	−0.56** (0.02)	1.79** (0.03)	0.33*** (0.06)	0.70*** (0.08)
12	34.54 (0.57)	−0.78* (0.00)	−1.79*** (0.057)	5.59** (0.04)	1.84** (0.03)	0.07** (0.07)
13	−58.05** (0.02)	1.42*** (0.10)	−0.56** (0.03)	2.64 (0.10)	3.99** (0.02)	0.97* (0.00)
14	31.12* (0.00)	−3.08** (0.03)	−5.09*** (0.07)	2.26** (0.011)	8.07* (0.00)	1.37** (0.012)
15	18.09 (0.37)	−1.75* (0.00)	−13.06* (0.00)	4.04 (0.35)	4.03** (0.015)	1.71*** (0.052)
16	23.03*** (0.06)	−3.53** (0.04)	−2.05*** (0.08)	3.04** (0.03)	0.85** (0.04)	12.61*** (0.08)
17	−45.08* (0.00)	−0.04** (0.06)	−4.10*** (0.09)	0.35** (0.02)	6.09* (0.00)	7.15*** (0.07)
18	1.66*** (0.09)	−0.07*** (0.09)	−0.002** (0.04)	−2.78 (0.35)	0.37* (0.00)	−0.54* (0.00)
19	5.26* (0.00)	−0.02** (0.03)	−0.19** (0.03)	0.12* (0.00)	0.73* (0.00)	0.11* (0.00)
20	−1.07 (0.28)	−1.16* (0.00)	−0.06*** (0.06)	0.39* (0.00)	0.15** (0.04)	−0.17*** (0.062)
21	1.92 (0.38)	−0.56* (0.00)	−0.70* (0.00)	0.16** (0.017)	0.11** (0.03)	−0.06*** (0.08)
22	1.80 (0.57)	−0.61** (0.02)	−0.98* (0.00)	0.18** (0.03)	0.27* (0.00)	0.53*** (0.09)
23	6.412* (0.000)	−0.051*** (0.06)	−0.01** (0.02)	0.14 (0.35)	0.20* (0.00)	0.36** (0.02)
24	15.30* (0.00)	−1.17* (0.02)	−0.08*** (0.05)	−5.54 (0.84)	0.71** (0.03)	1.56* (0.00)
25	34.64* (0.00)	−0.27** (0.03)	−1.05*** (0.07)	3.59*** (0.06)	0.03*** (0.051)	2.23** (0.03)
Panel	−10.42* (0.00)	−1.40* (0.00)	−0.34** (0.02)	0.83* (0.00)	0.86* (0.00)	0.44** (0.01)

Hansen Test (p-value): 10.08 (0.25)
Durbin–Wu–Hausman Test (p-value): 21.08 (0.00)
Note: Values in parenthesis are the estimated p-values. *, **, and *** indicates significance level at the 1%, 5% and 10%, respectively.

and GHGs have a significant impact on M in most of the nations. None of our three models is over-identified. If we are looking to maintain a sustainable growth rate for GDP along with protecting the environment, we should not only look to invest in FC to enhance M and GDP but should also look to invest in HS, EE, OPN, and EI. The use of green technology, allowing FDI to bring it (as developing nations do not have enough capital to use such technologies), strengthening infrastructure to reduce dependency on non-renewable energy, improvements in education that can bring awareness regarding saving the environment for future generations, are very much both crucial and essential.

NOTES

1 Source: Ministry of Environment and Forests, Government of India.
2 See Reports of Ministry of Environment and Forests, Government of India for details.

3 The nations we have selected are-
1) USA, 2) UK, 3) France, 4) Germany, 5) Norway, 6) Italy, 7) Canada, 8) Australia, 9) New Zealand, 10) Austria, 11) Belgium, 12) Spain, 13) South Korea, 14) Japan, 15) Chile, 16) Brazil, 17) Mexico, 18) China, 19) India, 20) South Africa, 21) Indonesia, 22) Singapore, 23) Iran, 24) Pakistan, 25) Bangladesh. In the analysis section the row of the tables will be numbered as per the numbering of the nations, that is, row 1 will represent USA; row 19 will represent India, etc.
4 Energy use (kg of oil equivalent per capita).
5 Combining all those infrastructure-related data as released by WDI.
6 Measured by life expectancy of a nation.
7 Expenditure incurred by government as a percentage of GDP spent on education.
8 Though we have employed all the possible techniques, like 2SLS, 3SLS and GMM to estimate the simultaneous regression, we get more robust results in case of GMM estimation and hence we report results of this only.

REFERENCES

Abe, K., Hattori, K., and Kawagoshi, Y. "Environmental regulations on international transportation", *The Japanese Economic Review*, Vol. 65, No. 4 (2014): 468–482. https://doi.org/10.1111/jere.12044

Alvarado, R., and Toledo, E. "Environmental degradation and economic growth: Evidence for a developing country", *Environment Development and Sustainability*, Vol. 19 (2017): 1205–1218. doi: 10.1007/s10668-016-9790-y.

Anijah-Obi, F. N. *Environmental Protection and Management: Planning, Process and Strategies*. Calabar: Clear Lines Publications (2001).

Barrow, C. J. "Sustainable development: Concept, value and practice", *TWPR. UK: Liverpool University Press*, Vol. 17, No. 4 (1995): 369–386.

Chaudhuri, S., and Gupta, K. "Inflow of Foreign Capital, Environmental Pollution and Welfare in the Presence of an Informal Sector". *Third Annual Conference on Models and Methods in Economics organized by Economics Research Unit of Indian Statistical Institute at Kolkata, SSRN Electronic Journal* (January, 2004). doi: 10.2139/ssrn.459060.

Chichilnisky, G. "North south trade and the global environment", *The American Economic Review*, Vol. 84, No. 4 (1994): 851–874.

Chichilnisky, Graciela, and Di Matteo, Massimo "Trade, migration, and environment: A general equilibrium analysis", Chapter 2.3 in *Sustainability: Dynamics and Uncertainty*. Kluwer Academic Publishers (1998): 109–127. https://mpra.ub.uni-muenchen.de/8819/1/MPRA_paper_8819.pdf

Copeland, B. R., and Taylor, M. S. *Trade and the Environment: Theory and Evidence*. Princeton, NJ: Princeton University Press (2003).

Daly, H. E. *Steady-State Economics, San Francisco*. Freeman & Co., 2nd ed. Washington, DC: Island Press (1991).

Ekins, P. *Economic Growth and Environmental Sustainability*. Dordrecht: Kluwer Academic Pub (2000).

Galeotti, M. "Economic development and environmental protection". NOTA DI LAVORO 89.2003. Social Science Research Network Electronic Paper Collection (2003). http://papers.ssrn.com/abstract_id=XXXXXX

Heady, C. "Natural resource sustainability and poverty reduction", *Environment and Development Economics*, Vol. 5, No. 3 (2000): 241–258.

Im, K. S., Pesaran, M. H., and Shin, Y. "Testing for unit roots in heterogeneous panels", *Journal of Econometrics*, Vol. 115, No. 1 (2003): 53–74.

Jansson, A. M., Hammer, M., Folke, C., and Costanza, R. (eds.) *Investing in Natural Capital: The Ecological Economics Approach to Sustainability*. Covelo: ISEE Press/Island Press (1994).

Kuznets, Simon "Economic growth and income inequality", *American Economic Review*, Vol. 45 (1995): 1–28.

Levin, A., Lin, C., and James Chu, C. S. "Unit root tests in panel data: Asymptotic and finite-sample properties", *Journal of Econometrics*, Vol. 108, No. 1 (2002): 1–24. doi: 10.1016/S0304-4076(01)00098-7.

Low, P., and Yeats, A. "Do dirty industries migrate?". In: P. Low (Ed.), *International Trade and Environment*. Discussion Papers 159. Washington, DC: World Bank (1992).

Lucas, R. E. B., Wheeler, D., and Hettige, H. "Economic development, environmental regulation and the international migration of toxic industrial pollution: 1960–1988". In: P. Low (Ed.), *International Trade and the Environment*. Discussion Paper No. 159, New York. Washington, DC: World Bank (1992), 67–87.

Maddala, G., and Wu, S. "A comparative study of unit root tests with panel data and a new simple test", *Oxford Bulletin of Economics and Statistics*, Vol. 61, No. S1 (1999): 631–652.

Mani, M., and Wheeler, D. "In search of pollution havens? Dirty industry in the world economy: 1960–1995", *Journal of Environment and Development*, Vol. 7, No. 3 (1998): 215–247.

Meadows, D. H., Meadows, D. L., Randers, J., and Behrens, W. W. *The Limits to Growth*. New York: Universe Books (1972).

Newey, W. K. "Semiparametric estimation of limited dependent variable models with endogenous explanatory variables", *Annales de L'inséé*, Vol. 60 (1985): 219–237.

Pappu, A., Saxena, M., and Asolekar, S. R. "Solid wastes generation in India and their recycling potential in building materials", *Building and Environment*, Vol. 42, No. 6 (2007): 2311–2320. https://doi.org/10.1016/j.buildenv.2006.04.015

Phimphanthavong, H. "The impacts of economic growth on environmental conditions in Laos", *International Journal of Business Management and Economic Research*, Vol. 4, No. 5 (2013): 766–774.

Sankar, U. "Ecology, environment and sustainable development in Indian fiscal federalism", Development Economics Working Papers 22941, East Asian Bureau of Economic Research (2009).

Siwar, C., Al-Amin, A. Q., and Huda, N., "Globalization, poverty inequality and sustainable livelihood diversification in third world countries: An assessment". In: C. Siwar et al. (Ed.), *Linking Environment and Rural Poverty: Governance and Sustainable Development Policies*, Malaysia: Institute for Environment and Development (LESTARI), UKM (2008).

Smith, R., and Blundell, R. "An exogeneity test for a simultaneous equation tobit model with an application to labor supply", *Econometrica*, Vol. 54 (1986): 679–685.

Ministry of Environment and Forests, Government of India. https://moef.gov.in/en/resource/annual-reports/. Accessed on February 21, 2021.

Wu, C., Li, F., Hsueh, H., Wang, C., Lin, M., and Chang, T. "A dynamic relationship between environmental degradation, healthcare expenditure, and economic growth in wavelet analysis: Empirical evidence from taiwan", *International Journal of Environmental Research and Public Health*, Vol. 17 (2020): 1386. doi: 10.3390/ijerph17041386.

United Nations *The Millennium Development Goals Report 2009*. New York: United Nations (2009).

United Nations *The Millennium Development Goals Report 2010*. New York: United Nations (2010).

5 Demarcation and Spatial Distribution of Urban Centres in Sri Lanka

D.K.D.A. Ranaweera and R.M.K. Ratnayake
Faculty of Humanities and Social Sciences, University of Sri Jayewardenepura, Nugegoda, Sri Lanka

H.M.R. Premasiri
University of Moratuwa, Moratuwa, Sri Lanka

CONTENTS

5.1 INTRODUCTION

According to 2018 data, more than half of the world's population (55%) lived in urban areas (UNDP, 2018). This will undoubtedly increase over the next few decades, taking variations of population, concentrations, and developments into account. Significantly, the complex social and economic landscape has become a dynamic process.

Robert E. Dickinson (1948) points out that the term urban refers to the use of land for a variety of purposes other than cultivation and therefore describes the city as a place where specific cultural, commercial, industrial, and service-related functions are concentrated. Hence, the continuous growth of cities is called urbanization. Urban settlements have become one of the most dynamic phenomena in the world's settlement hierarchy. Urbanization occupies a special place in the criteria identified in the development of a country. Planned urbanization has become a turning-point in a country's socio-economic development and an essential process for present and future prosperity.

Researchers interpret urbanization from different perspectives. From a demographic point of view, the concentration of people in cities is regarded as urbanization (Ratnayake, 2009). Natural population growth and migration in urban areas have a direct impact on urban spatial growth or urbanization. Urbanization is the percentage of the population concentrated in cities relative to the total population of a country. A rapid increase in urbanization can be seen in the developed countries, while at present urbanization is rising at a higher rate in developing countries too. As a result, governments in developing countries are now increasingly focusing on urbanization when planning their future activities.

DOI: 10.1201/9781003131922-6

The formal definition of urbanization is the number of people living in areas of a country defined as cities. Another point to note is that not all the world's countries or regions use a single criterion for urbanization. In most countries, population is used as a primary criterion for determining city status, and criteria such as ranking in the local government hierarchy are used to determine urban areas. Two or more criteria are used in most countries to define urban areas, although in some countries there is only one criterion. Due to the variation of urbanization criteria from country to country and region to region, it is difficult to express the precise value of global urbanization, and this has made it hard to pinpoint the exact situation. A minimum number of population, administrative functions, population density, urban functions, employment in non-agricultural sectors, growth of service sectors, the concentration of people, infrastructure facilities such as highways, electricity and water supply, and services such as education and health are main among the criteria used by different countries around the world, depending on their requirements. However, Frey and Zimmer (2001) noted that three groups of criteria, based on specific characteristics, are used for urban classifications: ecological criteria, economic criteria and social criteria (Table 5.1).

TABLE 5.1
Criteria Used to Demarcate Cities in the World

Country	Criteria
North American region	
USA	A place of public concentration with population of more than 2,500 and population density more than 1,000 per square mile
Canada	Areas with more than 1,000 population and population density of 400 per km^2
Mexico	Areas with a population of 2,500 or more
South American region	
Argentina	Areas with a population of 2,000 or more
Peru	Areas with 100 or more houses
Brazil	Cities and suburbs demarcated administratively
Chile	Highly populated areas where urban and state services located
African region	
Senegal	Areas with a population of 10,000 or more
Botswana	A public center with a population of 5,000 and 75% of this population engaged in non-agricultural employments
Malawi	Townships, town planning areas and district centers
South Africa	Administrative boundaries
Asian region	
China	Areas declared by the State Council and areas with a population of 1,500 or more per km^2
India	Special areas governed by provincial government institutes and areas with a population of 5,000 and more and 75% of the male labor force engaged in non-agricultural occupations
Japan	Areas with a population of 50,000 or more
New Zealand	Cities with a population of 1,000 or more
Malaysia	Areas with a population of 10,000 or more
Sri Lanka	Cities have been demarcated based on administrative boundaries and only Municipal Councils and Urban Councils consider cities
European region	
Norway	Areas with a population of 200 or more
Switzerland	Areas with a population of 1,000 or more
France	Areas with a population of 2,000 or more

Source: Prepared by author, based on United Nations, 2019.

Table 5.1 highlights some of the features that can be found in developing countries rather than developed countries when classifying cities based on different criteria. The population is simply a single criterion used in developed countries, and the population density can be relatively lower. But the fact that not only demographic factors but also administrative boundaries and employment are used to classify these categories in the developing countries shows that there are visible fluctuations in demographics in these countries.

By 2015 there were 63 countries in the world with an urbanization rate exceeding 80%. Out of these, 12 countries recorded 100% urbanization. Five of these countries are independent states (Singapore, Nauru, Monaco, Kuwait, Vatican) while the remaining seven are security areas (islands) of other countries (Dangalla, 2019). The unique feature of all of these countries is that their land area is a minimal unit. It is also noteworthy that countries with an urbanization rate of less than 20% have expanded, mainly in Africa and Asia. Nevertheless, the point to note here is that urbanization is calculated by measuring cities on different criteria. The difficulty in applying most of these measures to African and Asian countries has resulted in recording a low urbanization rate.

Thus the level of urbanization in developing countries is low due to changes in the definition of cities from country to country and region to region. Changes in the criteria have also affected the development of a country. Data studies from the past to present show that the criteria for allocating city status have influenced the variability of urbanization in Sri Lanka as a developing country.

In the first census of Sri Lanka, conducted in 1871, the number of cities was reported to be 19. The Urban Council Ordinance of 1939 gave the relevant minister the authority to upgrade any center that indicated any urban character to a small-town council; hence most of the service centers could be upgraded to the level of cities, irrespective of population limits or land area (Gunathilaka, 2005). A further change occurred in 1987 with the administrative decision to establish provincial councils, followed by the 13th Amendment to the Constitution (Ratnayake, 2009). With this development, small-scale urban councils were no longer considered cities. Calling them provincial councils reduced the number of cities in Sri Lanka, and consequently, the urbanization percentage also fell. The percentage of urbanization stood at 21.5% in 1981, fell to 14.6% in 2001, and was 18.2% in 2012. Statistics used by Aaron O'Neill (2019) show the degree of urbanization in Sri Lanka from 2009 to 2019. In 2019, 18.59% of Sri Lanka's total population lived in urban areas and cities. The country as a whole will have to face the consequences of the lower urbanization percentage. While this does not depict a real picture of urbanization according to international standards of development, it shows a drawback in the development process. Accordingly, the present study aims to identify the success of the criteria used to classify urban centers and urban dynamics in Sri Lanka.

Overall inequality can change as the development process progresses. First, with the increase in urban share, inequality in urbanization starts to make a greater contribution towards overall inequality. The second factor is that industrialization can widen the urban–rural gap; however, the impact of industrialization can be counterbalanced by factors such as technological spillovers, changes in the terms of trade, and an increase in per-capita land. Third, the development rate can be instrumental in changing inequalities within both urban and rural areas (Cem Oyvat, 2019).

A study conducted by Cem Oyvat (2019) in some developing countries shows the impact of urbanization on the change in income distribution. Inequality can be changed through urbanization and industrialization in several ways. When urbanization continues, the urban share increases, and inequalities between and within it can be observed.

5.2 THE IMPORTANCE OF IDENTIFICATION AND DISTRIBUTION OF URBAN CENTRES

Urbanization is one of the significant factors deciding the level of development of a country or a region. Although development is a common factor for dividing regions in the world, no single criterion has been adopted to measure the level of development as well as urbanization. It varies from country to country and region to region; urbanization in one country is entirely or partially different from in another country. In this context, the question arises whether it is possible to highlight the nature of development. Census reports from 1871 to 2012 in Sri Lanka identify statistical changes in the number of cities. The other issue is whether the cities that underwent such a transformation were considered out-of-town cities and whether those centres withdrew from urban functions. It is problematic that the real urbanization of Sri Lanka is manifested in such a change of cities by the timely amendments brought there. At present, there are a large number of centres in Sri Lanka that perform non-urban functions and do not belong to the groups classified under administrative criteria.

The objective of this study is to identify the success of the criteria used to classify urban centres and their unequal distribution in Sri Lanka.

Mainly primary and secondary sources are used for data collection. Primary data are collected from questionnaires, observation and interviews (structured and informal) with selected groups and individuals from different categories. Published and unpublished sources, institutional information and other relevant data sources constitute the secondary data. Finally, the data represent and analyze the use of the geo-spatial method.

The scope of the study is to analyze the urban centers in Sri Lanka and analyze their functional diversity. Areas for future research include preparation of an index based on urban dynamics and analyzing the changing pattern of urbanization in Sri Lanka.

5.3 DEMARCATION OF CITIES IN SRI LANKA

5.3.1 Historical Development of Urban Centres

Overcoming the inter-relationships between man and nature, cities are being created as integrated hubs of settlements built to accommodate human needs. These cities expand from the centres thus created to encompass a range of extended functions and services. Road networks and population are the main factors affecting this: dense populations and buildings at the focal points of roads and the concentration of various functions in those buildings have created the artificial environment that led to the origin of cities. Centres that are regarded as mainly urban have a wide range of functions, including financial, commercial, administrative, tourism, transport, economic, countermeasures, and so on, led by the residential function. Thus, the size and the spatial expansion of service centres and their growth are determined by the increase in the size of urban functions. Agriculture is the main livelihood of Sri Lanka. According to the settlement hierarchy introduced by Walter Christalar in 1933, urban service centres of this kind provide services to several surrounding rural settlements. Accordingly, the isolation of cities, service centres, or peripheral villages results in no development taking place. The level of development in cities or rural areas is determined by the inter-relationships created there. The formation and growth of urban settlements are also important. The origin of urban settlements coincided with the growth of monarchy, and there is evidence that it originated in the kingdom of Anuradhapura. Sri Lanka was not completely inhabited at that time, and people living a traditional way of life did not need urban services. In the kingdoms where there were only a handful of small industries, urban services were often limited to the king and his entourage. Public hearings were often held at the command of the king, and the community had the opportunity to enter the city only on that day. The city was a fortress demarcated by physical boundaries in the form of open space, moat or a large wall.

Cities originating from the Anuradhapura period changed with the trajectories of the kingdoms, often with only one major (capital) city. Urban settlements changed with the periodic movement of kingdoms from different places, such as Anuradhapura, Polonnaruwa, Sigiriya, Dambadeniya, Yapahuwa, Kurunegala, Gampola, Raigama, Kotte and Kandy respectively.

Clear records of the cities of Sri Lanka date back to 1871 when the first census was conducted by the British imperialists. There were cities in Sri Lanka at the time of the census, but there is no evidence that they were officially recognized as such. During the colonial period which started in 1555 with the invasion of the Portuguese, Sri Lanka started trading through Galle, Colombo, Trincomalee, and Jaffna, which were the main ports. The centers set up for these trading activities and their administrative purposes functioned as urban centers. There is evidence that the emergence of modern urban settlements in Sri Lanka after the kingdoms began from the south-west and southern coastal belts (Ratnayake, 2009). After the conquest of the country by the British as their colony in 1796, the urban settlements which performed urban functions were expanded to the hinterland with the commencement of the plantation industry.

At the 1871 census, there were 19 cities in Sri Lanka. An urban population of approximately 260,000 was reported, and the rate of urbanization was 10.8%. A closer look at Sri Lanka's urbanization also reveals that although the process was slow, there was steady growth. By 1881, there were 20 cities, and 20,000 more were added to the urban population. However, the rate of urbanization came down to 10.2%. This is because the rural population was growing faster than the urban population. By this time, the growth rate of the total population in the country was 1.4% and the growth of the urban population was 0.8%. Because by 1891 urban population had grown at a rate faster than the country's total population ratio, the urban percentage had increased to 10.7% although the number of cities had not increased. A subsequent examination of the census data reveals a gradual increase in the number of cities as well as the rate of urbanization (Table 5.2).

The first formal census conducted in Sri Lanka in 1871shows that 19 cities were then established in the country, and their number gradually increased until 1953. The growth of these cities was slow: it took 82 years for a further 24 centres to grow into cities. The proposal introduced by the Urban Council Ordinance in 1939 to upgrade centres that perform urban functions to town councils was implemented in 1946. After this, the number of urban centres rapidly increased, and by 1963, a further 56 centres had been added to the existing number of urban centres. By 1971, another 36 cities had been added, bringing the total number of cities to 135. This implies that urban functions and service centres were working for the development of rural peripherals, even though they had not been recognized as urban until then. It was a turning-point in the urbanization as well as the

TABLE 5.2
Number of Cities in Sri Lanka, 1871 to 2019

Year of Census	No. of Cities	Percentage of Urban Population	Percentage of Population Growth	Percentage of Urban Growth
1871	19	10.8	1.4	N/A
1881	20	10.2	0.9	0.8
1891	20	10.7	1.7	1.4
1901	28	11.6	1.4	2.6
1911	37	13.2	0.9	2.7
1921	42	14.2	1.7	1.6
1931	42	13.9	1.5	1.5
1946	42	15.4	2.8	2.2
1953	43	15.3	2.6	2.8
1963	99	19.1	2.2	4.9
1971	135	22.4	1.7	4.2
1981	134	21.5	1.7	1.2
2001	55	16.5	1.3	1.9
2011	64	18.2	0.6	N/A
2019	65	18.5	0.4	N/A

Source: Prepared by author, based on Department of Census and Statistics (2018, 2019).

national development of Sri Lanka. However, in 1987 urban status was removed from urban centres which were considered town councils through the 13th Amendment to the Constitution, transforming them into *pradeshiya sabhas*. Although administratively removed from urban status, all of these service centres performed urban functions, gradually increasing their services. The abandonment of centres that once administratively determined the Urbanization Index was another major factor in the decline in the level of urbanization in Sri Lanka. Accordingly, the total number of cities in Sri Lanka fell to 64 by 2012.

5.3.2 Categorization of Urban Centres in Sri Lanka

Each country determines the minimum number of people who should live in a city relative to the population of the country, regardless of the minimum population statistics, according to various countries in the world and the United Nations. Accordingly, there is no clear definition of the minimum population that a city should have in Sri Lanka, and a minimum population has been used only to determine the size of cities. Cities with a population of 5,000–10,000 people have been recognized as small-scale urban centres, a population of 10,000–50,000 as medium-scale centres, and cities with more than 50,000 population as large-scale cities (Mendis, 1982) (Table 5.3).

According to these statistics, the number of small urban centres in Sri Lanka has been reduced to eight by 2019. The population of small centres can be divided into three main categories: less than 2000, 2000–4999 and 5000–9999. Table 5.3 shows a rapid decline in small urban centres with a population of less than 2000, and a gradual increase in the number of the other two categories until 1981. However, by 2001 the total number of small urban centers had fallen to eight from a total of 59 in 1981. Especially during the census periods of 1963, 1971 and 1981 there was a sharp increase

TABLE 5.3
Expansion of Urban Population Based on the Size of Cities, 1871 to 2001

	Cities according to Population							
	Small Cities			Medium-Sized Cities		Large Cities		
Year	Less than 2000	2000 –4999	5000 –9999	10000 –19999	20000 –49999	50000 –99999	More than 100000	Total No. of Cities
1871	4	4	2	2	2	1	N/A	19
1881	3	8	4	1	3	N/A	1	20
1891	N/A	9	4	3	3	N/A	1	20
1901	3	10	6	4	4	N/A	1	28
1911	6	6	13	6	4	N/A	1	36
1921	11	7	11	6	6	N/A	1	42
1931	9	5	12	9	6	N/A	1	42
1946	3	9	7	13	5	4	1	42
1953	3	8	4	15	6	6	1	43
1963	9	21	23	21	18	5	2	99
1971	6	32	30	34	25	5	3	135
1981	3	28	28	35	31	3	6	134
2001	8				30	17		55
2011	8				39	18		64
2019	8				39	18		64

Source: Prepared by author, based on Department of Census and Statistics (2018, 2019).

in town councils. Another special feature is that the 1946 and 1953 census years showed a decline in the overall number of small urban centres, while medium-sized urban centres showed a significant increase. As regards large urban centres with a population of 50,000 to 99,999, from 1881 to 1946 not a single center with that population was reported, but 4 and 6 such cities were reported in 1946 and 1953 respectively.

5.3.3 Distribution Pattern of the Urban Centers in Sri Lanka

There are some disparities in the expansion of urban settlements. Many urban centres, in particular, are found in the coastal plains, with large urban centre in the region. The fact that more urban centres are concentrated in the wet zone has also exacerbated the disparity (Table 5.4).

As per the classification in Table 5.4, based on the national criteria, 30% of the municipal councils and 27.5% of the urban councils are located in the dry zone, and that figure is 28.7% of the total number of officially recognized urban centres in Sri Lanka. Among the cities declared by the Urban Development Authority, which do not belong to these categories, 178 *pradeshiya sabha*s are also considered urban centres (Figure 5.1).

Very few of the total number of urban centres recognized as urban according to national criteria are located in the dry zone, but service centres in this zone provide urban services without urban status.

From the very beginning, administrative boundaries have been used to demarcate cities in Sri Lanka. All Commonwealth countries adopt this criterion, and some have adopted other additional criteria, making the task of identifying cities very successful; for example, in India, in addition to the administrative boundaries, minimum population (average 5000), population density (400 per sq. mile), and non-agricultural employment opportunities (at least 75% of the city's male population must be employed in the non-agricultural sector). However, the failure to change the criteria in Sri Lanka has led to an inability to accurately calculate the level of urbanization. The other factor was the demotion of town councils – which used to demarcate urban status until 1987 – which reduced urbanization from 21.5% in the 1981 census to 18.1% in 2011. The fact that the real development of the country is not highlighted, that urbanization appears not to be accelerating along with the economic growth of Sri Lanka and that only economic growth is taking place, is problematic and strongly affects research. According to nationally accepted criteria, urbanization was 18.9% in 2019. The total number of urban centers was 65. In addition to that, the Urban Development Authority declared 178 centres by 2019, in addition to the 65 recognized urban centres mentioned above.

Although there is a total number of 271 *pradeshiya sabha*s in Sri Lanka, there are many urban service centres throughout the country that are not officially recognized as peripheral centres, and these have not yet been declared urban centres by the Urban Development Authority. Such service centres bring together several government agencies in the settlement to provide urban services to surrounding areas. According to the survey, there are 303 more such rural centres spread across the country that provide various services to the peripherals. To identify such centres, factors such as

TABLE 5.4
Expansion of Towns in Sri Lanka according to Climate Zones

City Type	Wet Zone	Intermediate Zone	Dry Zone	Total
Municipal Councils	12	04	7	23
Urban Councils	26	05	10	41
Total	38	09	17	65

Source: Prepared by author, based on Urban Development Authority (2019).

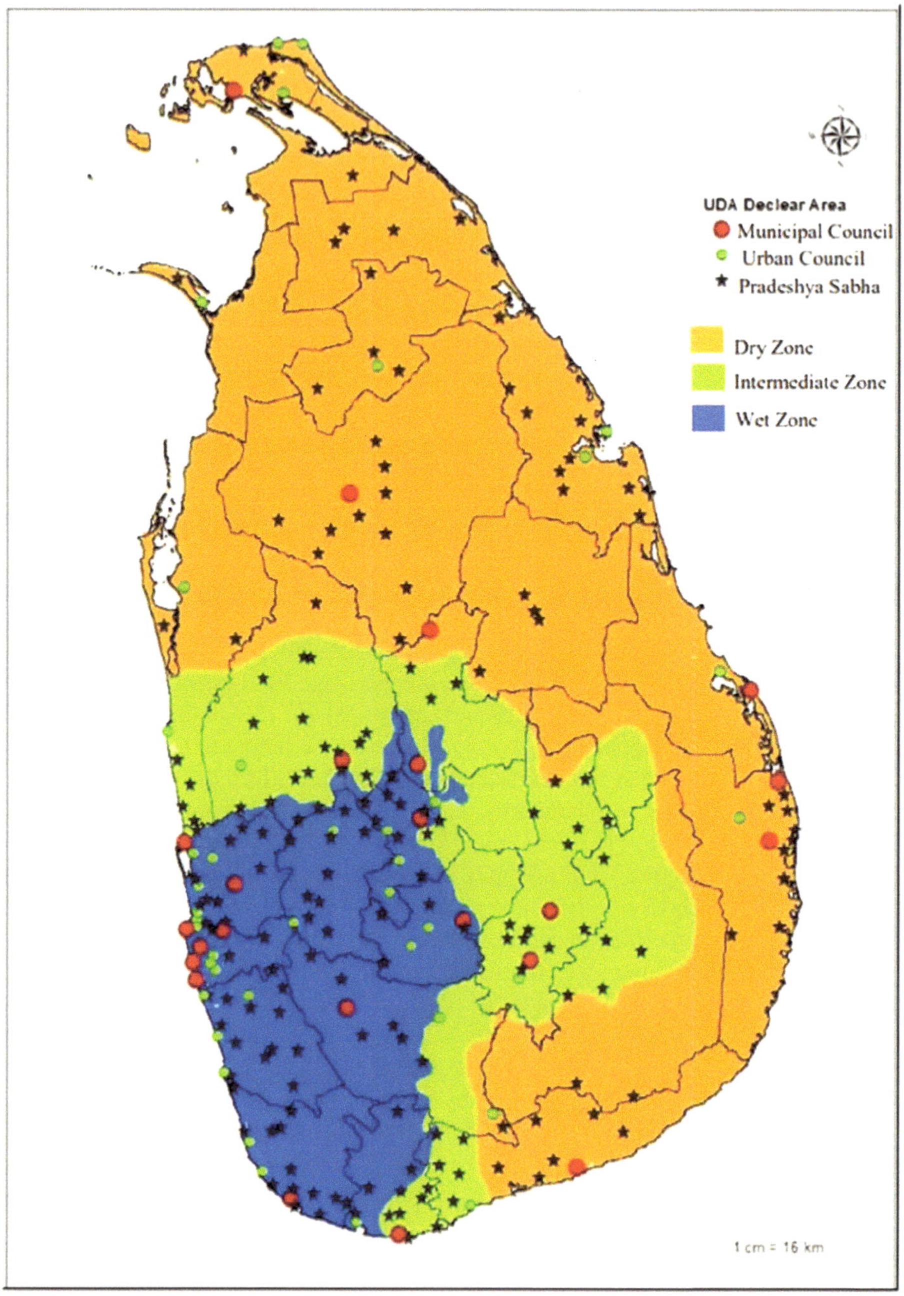

FIGURE 5.1 Expansion of cities in Sri Lanka.

(Source: Field Data, 2020.)

growth centered on a junction, density of buildings, nature of their expansion with urban fringes, and the presence of one or more government institutions such as sub-post offices, schools, banks, police stations, Divisional Secretariat Division Officials and judicial services were considered. There are around 545 service centres throughout the country, including such service centres mentioned above (Figures 5.2 and 5.3) (Table 5.5).

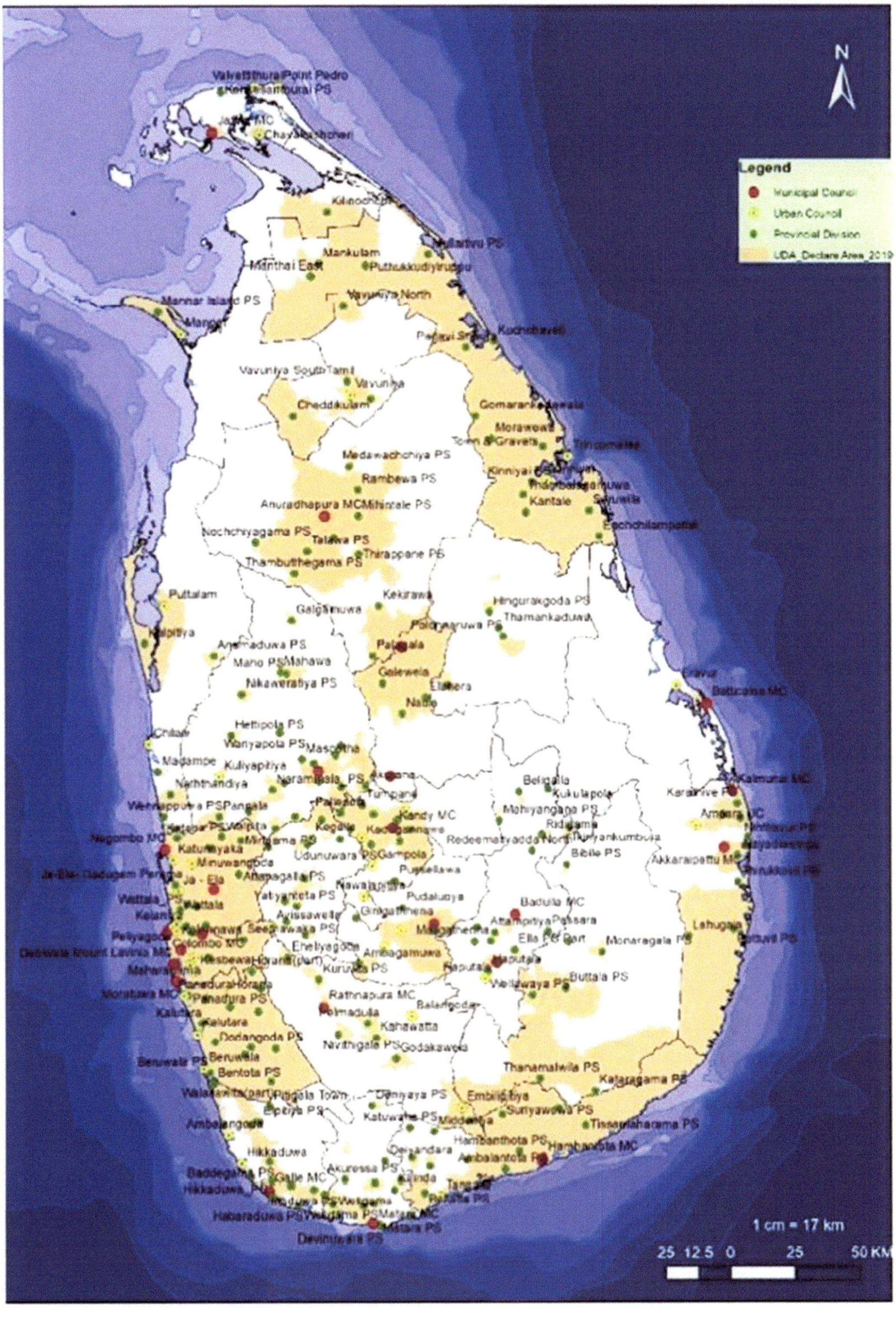

FIGURE 5.2 Expansion of cities in Sri Lanka named by Urban Development Authority.

(Source: Field Data, 2020.)

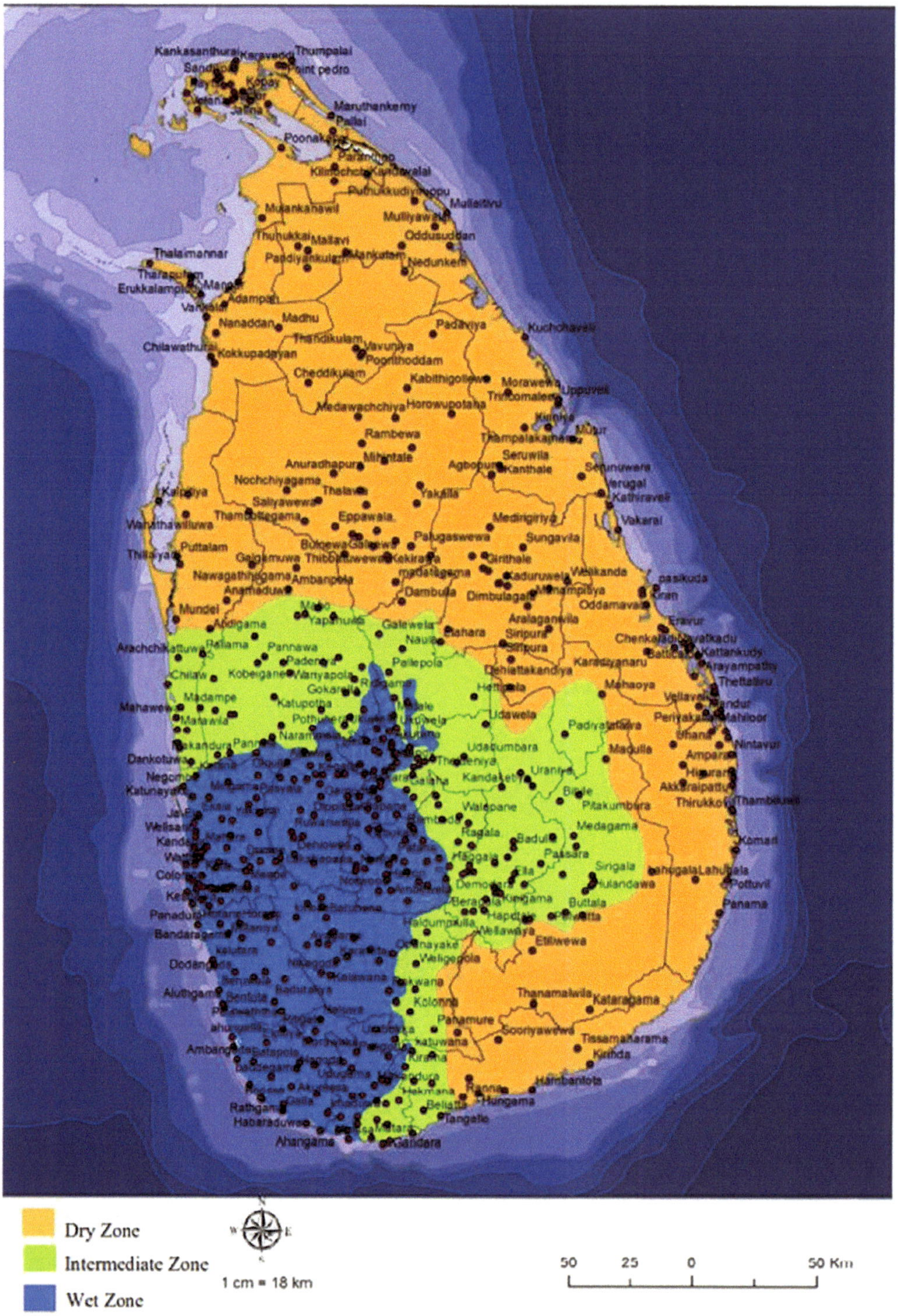

FIGURE 5.3 Expansion of urban service centres in Sri Lanka which are not declared by the Urban Development Authority.

(Source: Field Data, 2020.)

TABLE 5.5
Expansion of Service Centres in Sri Lanka That Are Not Declared Administratively by the Urban Development Authority

Type of City	Total
Urban Centres (officially announced)	65
Towns (declared by UDA)	178
Service Centres (identifying above-mentioned services)	303
Total	545

Source: Based on Urban Development Authority (2019) and field data.

TABLE 5.6
Expansion of Urban centres in Terms of Climate Zone, 2020

Climate Zone	Land Area (Square km)	Percentage (Out of Entire Land Area)	Total No of Cities (According to National Criteria)	Square km Covered by an Officially Declared City	Total of Urban centre (Officially Recognized, Mentioned by UDA)	Square km Covered by a City according to All Service centres in Sri Lanka
Wet Zone	12,696	19	38	334	189	67
Intermediate Zone	14,203	22	09	1578	142	100
Dry Zone	38,575	59	17	2030	214	180
National Level	65,474	100	64	1023	545	120

Source: Based on Urban Development Authority data and field data, 2020.

Although the overall expansion of urban centres is as above, Table 5.6 shows it in terms of land area according to climate zones.

The density of the dry zone service centres can also be calculated according to the terrain shown in Table 5.6. Accordingly, as per national criteria, there is at least one urban centre to serve 334 km^2 in the wet zone, 1578 km^2 in the intermediate zone, and 2030 km^2 in the dry zone. (When the entire country is considered, there is one urban centre to serve 1023 km^2.) Whatever the urban density expansion, according to national data, the figure can be identified as one urban centre per 120 km^2 when calculating urban expansion density in conjunction with all urban service centres in the dry zone. The wet zone is in an excellent position, and the dry zone is more than twice the national average. This means that people in the dry zone have to travel long distances for urban and higher-quality services. Accordingly, the mapping of the distances in zones to meet their needs is the special feature that covers the service centre in terms of national criteria in the dry and intermediate zones relative to the centre, but also in terms of locally connected service centre (Figure 5.4).

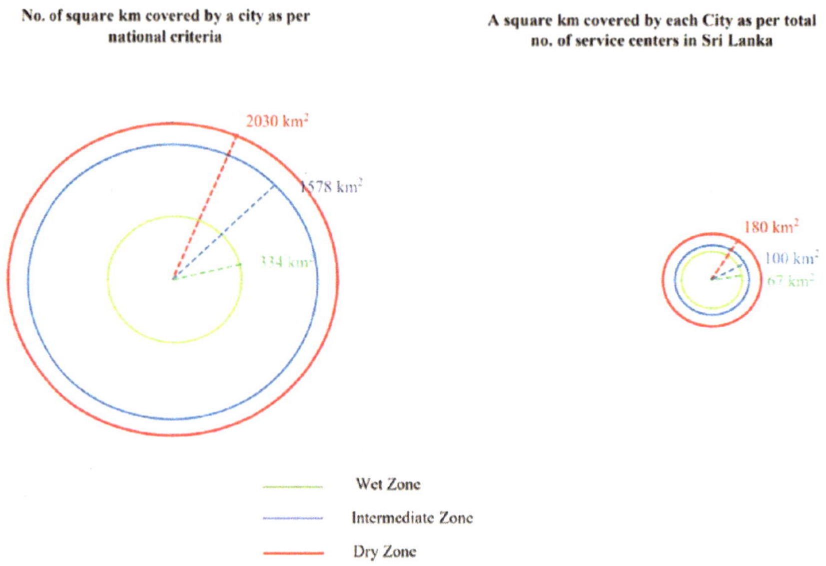

FIGURE 5.4 Density of the city and service in Sri Lanka.

(Source: Field Data, 2020.)

5.4 CONCLUSION

Accordingly, an apparent discrepancy in the density of urban centres can be identified in the current classification of towns based on national criteria. There is a maximum limit of 334 km^2 for those living in the wet zone to enter a city; it shows a large density gap of 1000 km^2–2000 km^2 in the intermediate and wet zones. When looking at urbanization in Sri Lanka based on climate zones, it is clear that the gap is due to low population density, lack of infrastructures such as roads and transport, and poverty. Those factors contribute directly to regional development disparities in Sri Lanka.

However, in Sri Lanka, although it is not administratively classified, it takes 2030 km^2 of land on national criteria to enter a city in the dry zone when measuring urban densities in conjunction with the service centres mentioned. Second, people can meet their urban function service requirements within a range of 180 km^2. This difference in the expansion of urban centres caused by the administrative criteria used to classify cities nationally has led to a numerical decline in Sri Lanka's Urbanization Index. The lack of real urbanization in a country causes its development to decline. Unfortunately, real national development conditions do not manifest themselves promptly in a so-called developing country.

Compared to absolute urbanization, Sri Lanka has not yet fully urbanized. According to the present demarcation of the urban, more than 40% of the total population lives in urbanized areas. However, according to national criteria used in 2018, the prevailing urbanization level is 18.9%. According to the pattern identified by these criteria, the highly urbanized area is the wet zone, the intermediate zone is moderately urbanized, and the least urbanized area is the dry zone. However, the trend is the highly urbanized areas extending beyond the Municipal Councils in the country. The impact of urbanization, which began in Colombo, is spreading across the country.

Accordingly, it is crucial to recognize that there are new urban service centres beyond the city of Colombo and that they are expanding to the eastern and northern parts of the country. As per this framework, no area can be subdivided into absolute urban or rural areas. This is because the creation of rural services for people living in rural areas is a trend in urbanization. In addition, the provision of urban services through rural service centres also implies that such centres play an essential role in the urbanization of Sri Lanka, albeit quietly. The level of urbanization in an area is reflected in the social and economic development of the people living there. It can be identified by evaluating the lifestyle of the people.

Most of the urban services in Sri Lanka are located around economic peripherals based on agriculture. However, the study confirmed that although the urbanization rate currently based on 64 recognized urban centres in Sri Lanka was as low as 18.9% in 2019, around 547 service centres currently contribute to accelerating urban services by providing complete urban functions. Accordingly, it can be concluded that at present, a development that exceeds more than 40% of the actual urbanization in Sri Lanka can be identified. This shows that the problematic situation that exists from the first step of identifying urbanization in Sri Lanka does not reflect the reality of urbanization in the country.

Accordingly, there is a need to update the benchmarks of urbanization in Sri Lanka, and a practical measuring system is required. If the urban economic conditions based on the functions at service centres can be calculated to illustrate this realistically, then a fixed value of urbanization in Sri Lanka, as well as in service centres that have developed as urban centres can be clearly defined, which will be a great impetus to national development.

5.4.1 Recommendations

The level of urbanization is one of the indicators for the development of a country. Sri Lanka uses a single criterion to officially demarcate urban centres. This is not suitable for a country like Sri Lanka, because the pattern of the population and resource distribution is uneven. Therefore, it is better to use many criteria to demarcate the country, which will help to distribute urban services all over the country and gather more people in one place.

It is suggested that sociological and physiological parameters are used as well as the present criteria to identify urban centers in the less populated areas as well as purely agricultural areas.

It is better to maintain an equal traveling distance for accessing urban services across the whole country, since rural poverty affects people's ability to arrange personal transport to supply input and produce market output.

The size of the area and the quantity of urban services are not important in promoting centres as urban. It depends on the economy and population density of the area. Therefore, relevant government institutions have to demarcate the new centres and maintain the urbanization level.

Settlements in Sri Lanka are far beyond international definitions of rural, because Sri Lankan rural settlements are in the peri-urban state, according to the criteria used. The Sri Lankan authorities need to revisit the definition and demarcate urban centres accordingly.

REFERENCES

Dangalla, N. K. *Study of Human Geography*, Author Publications. S & S Graphics Printers, Colombo (Sinhala) (2019).

Department of Census and Statistics. *Statistical Pocket Book*, Department of Census and Statistics, "Sankayana Mandiraya", Battaramulla, Colombo, Sri Lanka (2018).

Department of Census and Statistics. *Statistical Pocket Book*, Department of Census and Statistics, "Sankayana Mandiraya", Battaramulla, Colombo, Sri Lanka (2019).

Dickinson, Robert E. "The Scope and Status of Urban Geography: An Assessment," *Land Economics*, Vol. 24, No. 3 (August, 1948), pp. 221–238, Published by: University of Wisconsin Press, doi: 10.2307/3159689, Retrieved on 12.12.2019 (2019).

Frey, W. H. and Zimmer, Z. "Defining the city", In R. Paddison (Ed.), *Handbook of Urban Studies*, Sage Publishing, London, UK (2001), pp. 14–35.

Gunathilaka, W. L. D. P. D. A. *Patterns and Trends of Urbanization in Sri Lanka*, Department of Census and Statistics, Colombo, Sri Lanka (Sinhala) (2005).

Mendis, M. W. J. G. *Urbanization and Urban Development in Sri Lanka*, Marga Institute, Colombo, Sri Lanka (1982).

O'Neill, Aaron. "Urbanization in Sri Lanka, Economy & Politics", (International) (2019). Available at: (https://www.statista.com/statistics/728547/urbanization-in-sri-lanka/-) (2021, June)

Oyvat, Cem. "How Urbanization Affects the Inequality in Developing Countries: A Critique of Kuznets Curve", (2019). Available at: (https://www.mdpi.com) (2021, June)

Ratnayake, R. M. K. *Settlement Study*, Author publications. Hansi Printers, Nugegoda, Sri Lanka (in Sinhala) (2009).

United Nations. *World Urbanization Prospects - 2018*, Highlights, Department of Economic and Social Affairs, United Nations, New York (2019).

United Nations Development Programme. *The 2018 Revision*, United Nations, New York (2018).

Urban Development Authority. *Geographic Information System Division*, Urban Development Authority, Sethsiripaya, Battaramulla, Colombo, Sri Lanka (2019).

6 Colonial Urbanism and Morphology
The Case of the Colonial Port City of Kolkata

Parama Bannerj
Nabarrackpore Prafulla Chandra Mahavidyalaya, Kolkata, India

CONTENTS

6.1 INTRODUCTION

The growth of urban areas around the world, whether in developed or developing countries, has evoked human interest in urban spatial aspects such as location, urban form and structure. A theory of urban growth was first put forward by Burgess in 1925 while he was studying the urban morphology of the city of Chicago. Thus, intricately associated with the urban morphology of the city are its past as well as its present.

Several urban morphological models have developed over the years since 1925 with a focus on the developed world. However, urbanization in the developing world is more complex and in cities where European colonial power thrived it has developed its own peculiarities. Urban morphology in the developing world is the product of the cultural and social characteristics of the group which laid it out and its morphologies, now superfluous, were designed to serve the purpose of the colonists.

In India, for example, it is the effect of English colonialism that has impacted the urban landscape. Here the two cultures – indigenous and Western – continued side by side to produce a distinctive urban form made up of the two elements (Prakasa Rao, 1971). King (1995) has opined that colonial cities operated as important sites in the deployment of technologies of power, through

DOI: 10.1201/9781003131922-7

which indigenous populations were categorized and controlled. Urban architecture and planning in such cities mimicked the cities of their imperial rulers. An example may be the case of Kolkata, established by the British colonists, whose genesis or morphology stood in contrast to the indigenous cities of the country. Munshi (1980) also points out that Kolkata, one of the finest examples of India's colonial port cities, shows remarkable variation in urban growth across the city core (old city) and the periphery (new city).

This study follows a case-study approach. Using historical maps and narratives the study endeavors to make a comprehensive analysis of the genesis, urban transformation and morphology of Kolkata, as it appears in historical narratives or maps. Additionally, the study focuses on the process of historical evolution of maps of the selected urban area and its association with urban space. The study also focuses on the socio-cultural meanings of maps, the possibilities of the agency of mapping, and what maps reveal in spatiality.

The study looks into the physical environment within the spatio-temporal framework and the cultural environment that is conditioned and shaped by the population groups which have inhabited it over time. The study identifies the historical nuclei of growth, the various processes that have operated in the cityscape, ranging from urban agglomeration to migration or spatial segregation. The study also looks at the population groups inhabiting the city and how the dynamics of power amongst them have shaped its morphology. Spatial segregation has found expression in the form of segregated urban areas which are distinct in terms of layout, planning or density from other segregated units, as in the case of development of the Black/Native Town or White/European town. These population groups were also not homogeneous and were segmented in terms of caste, clan, occupation or class. An example of this was the development of territorial units called *para*s or neighborhoods.

The study looks into this and many more aspects of colonial urbanism.

6.2 REVIEW OF LITERATURE

Urban morphology has always captured the interest of urban social scientists and has been described by Conzen (2012) as the study of the built form, layout and spatial expression of the city in the light of the forces that have created them. However, this explanation has been subjected to further scrutiny by Kropf, who points out that urban morphology not only analyzes the relationship between built forms and the purposes of builders, but also investigates those aspects of urban form that have unintentionally developed as a byproduct during the development process.

Thus, intrinsically related to the urban morphology is the history of an urban area which has been studied by a number of scholars. According to Wolpert (1967), the development process of urban form and pattern passes through three stages: histogenesis, patternogenesis and morphogenesis. The first phase, histogenesis, refers to initiation of the nucleus around which the urban area develops. The next phase, patternogenesis, includes the phases of development of multiple nuclei of human settlement. The last phase is the morphogenesis, which can be perceived as when the settlement develops its morphological character. The morphological character is said to be associated with the functional characteristics of present-day cities.

The pattern or morphology of human settlements are determined by several factors. These determinants may be physical or cultural. The physical factors are related to the geographical location and include climate, topography and availability of local construction materials (Morris, 1974), while cultural determinants include society, the economy, the nature of political institutions and the functionality of regions.

With the growth of the urban area, organically, the city assumes certain functional characteristics which become reflected in its physical layout. Scholars like E.M. Burgess and H. Hoyt (1939) identified the central business district, the nucleus around which the zones of function of the cities developed. However, C.D. Harris and Ulman (1945) proposed that cities have not a single nucleus but multiple nuclei. However, these theories were developed before the Second World War and they were found to be applicable only in the case of Western cities. Prior to the 1960s, urban studies did not recognize or distinguish between the cities of the third world and elsewhere. Decolonization after the Second World War in Asia and Africa led to the development of many cities with a colonial inheritance (Simon, 1984).

However, the process of urban development is complex and inadequately understood in third-world countries (Bromley, 1978). It has been observed that there is a dearth of reliable statistics for most third-world countries. Hence, most researchers have confined themselves to a small number of cases.

6.2.1 The Case of Colonial Port Cities in Developing Countries

If one is focusing on the urban areas of developing countries, it is quite appropriate to speak of neo-colonial or post-colonial cities where there is distinct Western dominance, even where there is political independence of the former colonies. These cities were a product of Europe's economic history.

From the mid-1950s, European and North American social scientists started studying aspects of urbanization in the recently decolonized countries of Asia or Africa which were once settled by colonizing Europeans. They found certain peculiar spatial characteristics. According to Grunjen (1985), colonial cities were founded under the influence of a Western colonial power and had a central function in the colonial administrative or economic system. The city also had a substantial portion of Western architectural elements.

The physical layout of the colonial cities depended on the colonies of which they were a part. An example may be that of the Spanish colonizers who almost completely erased indigenous settlements of the area and culturally assimilated with the natives to become a part of it. Thus, Spanish colonial cities followed Spanish elements. Most of the cities in Latin America west of the Andes followed this pattern. In contrast to this, the French and the British colonists never mingled with the natives and carved out separate quarters. This phenomenon carved out 'dual cities' – separate European and native quarters with distinctive urban morphologies (King, 1995).

Thus, French and British colonies were segregated into the town of the colonists, called the "White Town", consisting of big spatial houses and neatly laid-out streets, and a Native Town or "Black Town" which were densely populated and housed the indigenous population. In British colonies of India or in apartheid countries of Africa, the division was very well defined.

Most colonial cities were either coastal, for the purpose of trading, or administrative. For the colonizing powers, a port was the ideal location and ensured access to the resource-rich hinterland. According to Neild (2008), in Asia, colonial cities were deliberately developed in coastal locations with port facilities, and these cities have attracted the attention of those concerned with urban morphological transformation. The colonial port cities served as links between the colonizers and the colonized. Their importance in the European colonizing system, their coastal locations, their cultural heterogeneity and emphasis on commercial rather than ritualistic activities, distinguish them from other indigenous cities of Asia.

India, too, has a number of such cities, being a former colonized state. These cities, Bombay, Calcutta and Madras, were built by the East India Company and displayed similar morphological characteristics. These cities were mostly port towns. Karmakar (2012) points out that while the port cities of pre-colonial India like Surat were essentially mercantile, colonial port cities were systematically chosen to fulfill colonizers' need. It has also been pointed out that the pre-eminence of these colonial cities came at the cost of the decline of other urban centers: Bombay rose at the expense of Surat, Calcutta at the expense of Murshidabad, and Madras at the cost of the decline of a number of ports in Coromandel. These Indian colonial port city designs were replicated in other colonies of the British empire, such as Durban over Cape Town in South Africa, or Colombo over Gale.

Colonial port cities, particularly their morphological characteristics, are a rich source of interest for academics. A number of context-specific studies have been conducted to understand the morphological character of the urban system developed due to colonial heritage in developing countries and the factors which have affected it. A study by Wan and Beisi (2016) on 'Urban Morphology of Commercial Port Cities of South East Asia' marks a similar point. The paper describes how from ancient times dating back to the emergence of the Silk Road paving the way for European colonialism up to the era of their independence from colonial rule, with the dual effects of both national and foreign elements, South-east Asian port cities have gone through a transition process. Another study has been done by Kubat (2007) who points out that after a morphological comparison between two historical towns, the patterns of settlements are shaped due to interplay of a number of cultures.

These are reflected in traditional cartographic maps. Traditional maps with landmarks of that time frame form cognitive anchors, markers, or reference points for orientation, way finding and communication. Urban morphological study also includes the concept of maps and mapping from an intellectual point of view. Such reflections are observed in the writings of Cosgrove, Edney and James Corner. According to Cosgrove (1996), mapping is a process which involves a "complex architecture of signs through which the worlds the map makers select to construct are shaped". Winichakul (1994) stressed the concept of 'imagined geography' and acknowledged that there is something imaginational about mapping and people of various cultural background, have mapped their world, differently. Pickles (2004) argues that "mapping is about the transfer of information from one form of presentation into a re-presentation of that information".

Of the various maps produced through survey, the importance of perception cannot be ignored. The book *Image of the study* by Kevin Lynch (1967) provides an empirical insight into urban planning with a study of how individuals perceive the urban area. This book, first published in 1960, also explores the presence of temporal elements in the urban environment, and how these factors affect people.

In this context, mention may be made of Dasgupta's (2009) *Mapping Calcutta: The Collection of Maps and Visual Archives*. She used historical maps to describe the evolution of Kolkata. According to Dasgupta, for colonial Kolkata, early maps published in the pre-survey phase were rudimentary in character and produced by travelers. Cartographic productions focusing on Kolkata made their appearance through a series of plans of the old fort and the area around it. She points out that cartography belongs to the terrain of the social world, and the power structure and changing relationships in the colonial city of Kolkata were reflected in the maps.

Against this backdrop, the present study attempts to understand the significant politico-economic and historical factors that have shaped the use of urban areas in general, and colonial cities in particular. The study also tries to understand the effect of visual representation of urban forms. As discussed, third-world countries have always not found a space in Western urban models and since the Second World War, after decolonization, though a number of individual city studies have been conducted in Asia or Latin America, these are mostly fragmented and empirical. Generalization and important social scientific tasks are difficult to undertake with these studies. The existing literature reveals that there is a dearth of examples as to how this sequential transformation of the colonial urban landscape has found expression in its visual representation, which may be rooted in a plethora of socio-cultural factors. The study attempts to bridge this gap and to understand the process of urbanization of the cities with a colonial heritage. It also attempts to identify the factors that have shaped urban areas and draw some generalizations about the phenomena present in them.

6.3 MATERIAL AND METHODOLOGY

Human settlement and its patterns have been hypothesized for many centuries. The objective of this study is twofold: to understand the morphological evolution of the city, including the physical and cultural factors influencing the morphology; and to examine how the morphology finds its place in graphic representation, particularly maps. Maps themselves are a contested topic that expose subjective biases. Hence, the graphic representation in maps is supported by contemporary historical narratives.

The main objectives of the study are as follows:

- to understand the evolution of the cities with colonial heritage and the subsequent socioeconomic phenomena shaping it;
- to identify the factors, both physical and cultural, influencing the morphology of the city;
- to identify the various processes of urban growth like urban agglomeration, urban expansion, migration or segregation, operating in the city;
- to understand how such cities have been historically and spatially represented in maps.

The study is qualitative in nature and uses both historical analysis and a case-study approach. Case-study methods involve systematically gathering enough information about a particular person, social setting, event, or group to permit the researcher to effectively understand how it operates or functions. The case study is not actually a data-gathering technique, but a methodological approach that incorporates a number of data-gathering measures (Hamel, Dufour, & Fortin, 1993).

The study selects the case of Kolkata, an Indian colonial port city, and attempts to understand, primarily through historical maps and literature, how political, socioeconomic and cultural processes have shaped the city and its urban forms. The methodology adopted here is a comprehensive analysis of historical maps, surveyed at various times, and investigation of Kolkata's urban form, as it has appeared in literature. The time frame selected is from Kolkata's evolution through its journey through the colonial period, up to the decolonizing period. A sequential visual representation of the urban form has been attempted.

In *The New Nature of Maps* (Harley, 2001), the authors suggest that we should look at maps as texts. They argue that historical maps can only be interpreted in the context in which they were developed. This principle is followed here.

The production of maps clearly coincided with the changing requirements of the colonial power. The study selects some representative cartographic ventures of the city from the early colonial ventures to the late. Four historical maps of Kolkata are selected for study, ranging over three centuries, typifying the socioeconomic contexts of each phase:

1. Two maps produced by Mark Woods
   - *Survey of the Country on the Banks of Hooghly River extending from the town of Kolkata to village of Otoobareah, 1780–1784*
   - 'Plan of Calcutta', 1784
2. Map produced by actual survey in the years 1847–1849, by East India Company
3. Map showing the 'Southern Black Town', 1914.

To conclude, the study uses the concept of urban morphology as an analytical framework for understanding the evolution of a city. Urban morphology deals with the knowledge of the logic of the form, in this case, urban form. It is studied in several disciplines and it involves looking at physical characteristics, structure, relations and transformations of things and their constituent elements. Narrowing down the scope, this research focuses on the dynamics of urban morphology and how various factors can influence various urban forms.

Time period of the study: The study was conducted over a two-year period, 2019 to 2021. The research involved studying the development of Kolkata for a period from the late 18th century to the early 20th century, that is, from the time of the early European settlers to the time they were leaving India.

Limitation of the study: The major limitation was availability of adequate data in terms of maps, published reports and literature. There was a limited number of reliable maps and literature published during the time frame under research.

6.4 PROFILE OF THE STUDY AREA

In order to be context specific, the study selects the case of Kolkata, whose history, urban form, etc. will be explained through maps and literature.

Kolkata is the former British capital and is presently a metropolitan city of India. It is was one of the earliest colonial settlements of the British, and was used by imperial Britain primarily as a center for organizing the linkages necessary for extracting from the vast hinterland of the Indian subcontinent, the resources required for developing Britain's internal economy. The nature of these linkages influenced the city's pattern of growth.

6.4.1 Geographical Setting of the City

Kolkata rests on a clay bed deposited by rivers forming the lower plain of the Gangetic Delta. The clay overlies a thick pile of alluvial sediments deposited in the recent geologic era (Dasgupta, 1973). Situated on the eastern bank of the river Hugli, Kolkata was surrounded on the north, east and south by low-lying swamps and marshes and numerous tidal creeks, while on the west was the Hugli, a large tidal river draining down to the sea some 130 kms to the south.

Kolkata has a subtropical climate. The maximum temperature in summer is around 41 Celsius while the minimum temperature during winter is as low as 9 Celsius. Kolkata receives rainfall during the monsoon months of June to September.

6.4.2 Evolution of the Urban Form

According to Deb Roy (1988), in the sixteenth century, the port of Saptagram went into decline. During the late seventeenth century, the British East India Company acquired the three villages of Sutanuti, Kalikata and Gobindapur on the eastern bank of the Hugli river, which became the original British holdings forming the city of Kolkata, formerly Calcutta.

Deb Roy (1988) also points out that Kolkata's land use was laid out during the colonial rule. Although the boundaries of Kolkata were defined in 1794, they underwent revisions till 1889.

Kolkata city was defined in the early days as the area under the "original jurisdiction" of Kolkata High Court, bounded by the Hugli river on the west, the Bagbazar canal on the north, the Maratha Ditch (Circular Road) to the east, and to the south the Lower Circular Road (now known as Acharya Prafulla Chandra Road) to Khidderpore Bridge and Adiganga (Tolly Nulla) to the river Hugli including the Fort and Colly Bazar (Hastings). According to Ganguly (1967), in 1889, the boundaries were redefined, leading to an extension of the area from 2,022 hectares to 4,836 hectares, adding parts of Entally, Ballygunge, Bhawanipur and Tollygunge to the city in the process. The next phase of growth occurred in 1953, when the Tollygunge municipal area was added to Kolkata, and thus Kolkata became 9,560 hectares excluding the fort, port and canals.

Broadly, three phases can be identified in the growth of pre-independence Kolkata: before 1793, 1793–1870 and after 1870. The first cited map in the present thesis is Mark Wood's map of 1784 (Figure 6.1(b)), where the city is shown to be growing along the major transportation spine of the river Hugli. The important settlements of the eastern bank included Barrackpur, Baranagar, Chitpur, Cossipore, Dakshineshwar, and Panihati. On the other side, Bali, Haora, and Baidyabati were important ones. The first phase is illustrated in Figure 6.1(a).

A survey of the river Hugli was conducted by the British East India Company and Captain Mark Wood in the aftermath of significant territorial acquisitions by the East India Company in the second half of the 18th century. While the East India Company was establishing itself, surveying became important for acquiring knowledge and control over the colonized area. While Mark Wood was surveying the area, he was given the additional task of producing a map in 1784 for the police force of Calcutta (now Kolkata), which was being augmented and strengthened at that time. The survey under discussion was undertaken 'for military information' (Dias, 2015).

(a)

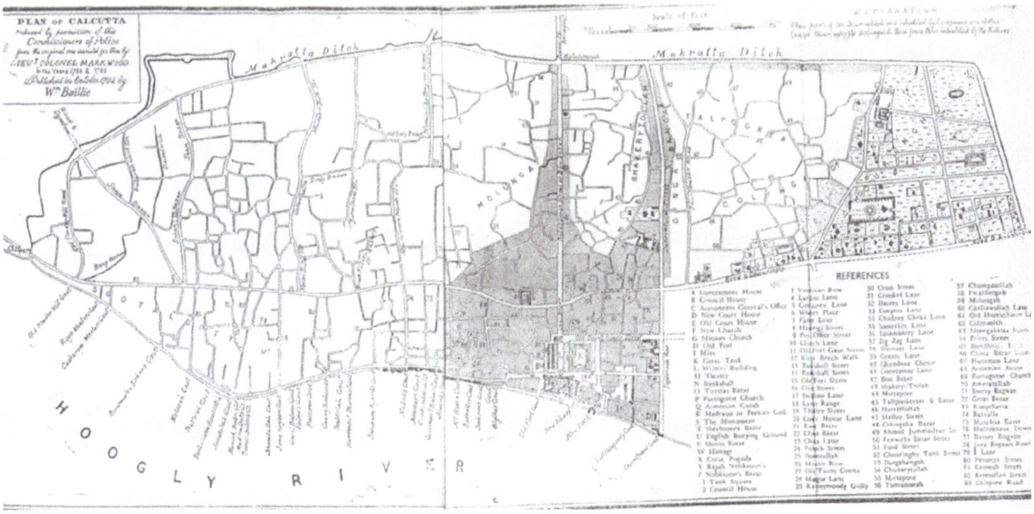

(b)

FIGURE 6.1 (a) Kolkata in 1785; (b) Kolkata in 1784.

(Source: (a) *Survey of the Country on the Banks of Hooghly River extending from the town of Kolkata to village of Otoobareah, 1780–1784;* Source: (b) Mark Wood's 'Plan of Calcutta', 1784.)

As we can see, the emphasis of Wood's map is more on the rural settlements than the urban. The map shows details of individual houses along the river Hoogly. The suburbs of Garden Reach can be located on the right bank of the river Hoogly. The map shows the extensive layouts (trees, carriage drives) that surrounded them. These individual houses, called Garden Houses or *bagan bari*, housed wealthy families. The suburban surroundings showed the spatial and social formation of Kolkata's environs.

According to Chattopadhyay (2000), Kolkata in the 18th century had remained confined within a fortification. The British East India Company responded to military threats and the reaction to these threats helped them to shape the then Calcutta's (now Kolkata's) built environment. The conflict between the British and the Mughal rulers and the threat of military attacks by them prompted several specific developments, one of which was the construction of Fort William. The new fort can be seen in the top right corner of the map. Fort William, or the original fort built in 1696, was destroyed in that year by the Nawab of Bengal's forces. In the aftermath of the British success at the Battle of Plassey the following year, a new fort was constructed and a new settlement arose extending beyond the boundaries of the fort itself.

The second map by Mark Wood discussed here was based on a survey by the Commissioner of Police and clearly reads, "Those part of the town inhabited by Europeans, are dotted…". The map became well known as the areas inhabited by Europeans were for the first time highlighted.

The next phase is the period between 1793 and 1870 (see Figure 6.2). In the next significant series of maps prepared during 1825–1832 for the use of the Lottery Committee, which was established for raising funds for the city of Kolkata, the area within the Upper and Lower Circular Roads (the erstwhile 'Town of Calcutta') looks quite densely built up. The southern part of the Lower Circular Road looks relatively unbuilt with large incidental open spaces and private gardens. The road leading to Tollygunj (the then Russapuglah) is seen as the most important link from the core towards the south. The fact that this spine would later become the North–South Metro Corridor seems an obvious and logical conclusion. A description of this has been provided by Radha Raman Mitra in his book *Kolkata Darpan* (1952), where he describes how the Circular Road system was made *pucca* and the two areas of Metiabruz and Tollygunj coalesced with the main city along the Circular Road. During this phase, parts of Ballygunj, Narkeldanga, Alipur, Kalighat, Bhawanipur and Ultadanga developed. Running parallel to the river, the Calcutta to Naihati road developed. On the western bank of the river, the development was linear and continuous from the Botanical Garden to Champdani. Baranagore by this time was already a consolidated settlement. Growth around this area was quite significant. In September 1854, the first train from Howrah to Chinsurah was introduced by the East Indian Railway, (Mitra, 1952). With Howrah Bridge across the river Hugli becoming operational, Kolkata was part of the rail network system.

The year 1844 marked the establishment of Fort William in Kolkata to protect the East India Company's trade in the city of Calcutta, the principal city of the Bengal presidency. This was done as the there was a threat of military confrontation by France in 1756. The year 1844 also marked the occurrence of the Black Hole tragedy in Kolkata

FIGURE 6.2 Map of Kolkata showing the territory of Calcutta attacked by Shiraj ul Daula's forces (published in 1880.)

FIGURE 6.3 Kolkata in 1858.

(Source: John Walker, Geographer to the Honble. East India Co., Jan. 20th, 1858.)

Another map was prepared just after the Black Hole tragedy but published in 1880 (Figure 6.2). The emphasis on marking Fort William, the entry points and the transit nodes points to its purpose as essentially military. At the top 'the position of Suraja Dowlah's Forces' in 1756 is clearly mentioned. The map was generated to mark the territory attacked by Shiraj Ud Daula in 1756.

The map (Figure 6.3) prepared under the aegis of the East India Company on the basis of a survey between 1847 and 1849 includes an index of districts. The map shows a number of new roads which were not present in the previous maps, such as Circular Road, Chitpore Road in the north and Chowringee Road in the central part of Kolkata. The esplanade area is distinctly marked while to the south-east of the map, the sparsely settled areas of Bhowanipur, Ballygunj and Chakraberia are marked. In the eastern part, one finds the emergence of trunk routes like Rowdon Street, Hungerford Street and Camac Street, neatly planned, parallel to Chowringee Road. The map clearly shows the emergence of a very dense settlement pattern in the northern part of the city, while the southern part was yet to develop. The developed areas around the Great Tank extended along Bytaconnah (*Baithak khana*) or Bowbazar Street, and along the present-day Lenin Sarani, both ending at the Mahratta Ditch. The Circular Roads, both Upper and Lower, run along the ditch, and the area between the present-day Park Street and the Lower Circular Road was known as Chowringhy. The hatched parts of the map located in the north, primarily for the native population (the Black Town) had irregular road or lane patterns. The ditch, which extended from the Bagbazar Ghat to the point where the Lower Circular Road meets the river, formed the physical boundary of the city in its earliest configuration.

This map marks the emergence of two different urban layouts with perceptual urban space inhabited by two different ethnic communities. There are clearly divided boundaries in this map, and scholars have assumed that Calcutta, like any other colonial city, worked on the basis of separation into dual cities, Black Town and White Town (Chattopadhyay, 2000). In the case of Calcutta, the idea of Black and White Towns is based on the perception that European residents of the town inhabited an area which is seemingly different from the native city in terms of layout density, architecture etc. The town was a settlement reserved exclusively for three Christian countries, English, Portuguese and Armenian.

The period after 1858 is sandwiched between the nineteenth century traditions of urbanism. It may also be pertinent to point out that the landscape of colonial Calcutta (now Kolkata) was complex and cannot be described merely in terms of the duality of Black and White Towns. Initially, the boundaries between Black and White Towns were blurred. However, the 1914 map (Figure 6.4)

(a)

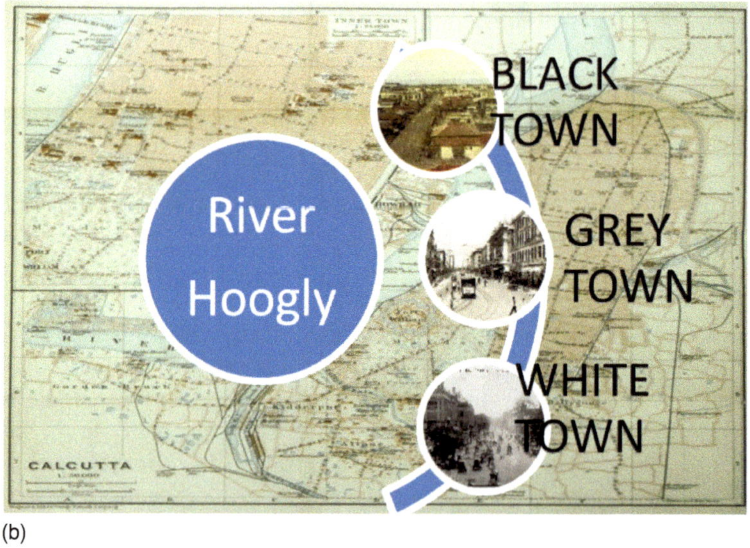

(b)

FIGURE 6.4 (a) and (b): A schematic diagram showing the original nuclei (fort) and racially seggregated urban space of Calcutta, 1862.

(Source: After Dutta, 2004.)

shows clear demarcation between Black and White Towns. This map of the city of Calcutta was prepared around 1914 by cartographers Wagner and Debes at a scale of 1: 50,000, with a smaller, inserted map showing the White and Inner Town (1:25,000). The residential segregation is prominent and due to racial exclusion, the British had forcibly moved the native population to a site in the north of the city, thus creating a White and a Black Town with the intermediate zone inhabited by Eurasians.

This phenomenon has also been described by Chattopadhyay (2000) as illustrating how the desire of the erstwhile colonizers was rooted in an 18th-century British obsession with classification, division, and separation. It gained impetus through incidents such as the Black Hole, reminding the British residents of the constant native threat to their existence in the city. In their zeal to protect islands of sociability and symbols of imperialism, the colonizers resorted to building elaborate artifices of delimitation – wrought-iron railings, masonry walls, and gates in White towns. The early township was occupied only by the White Christian populace, while the natives lived outside but came to the town as domestic helps. Natives were also engaged in respectable business profession like *Brahmins* (priest) or *munshi* (accountants).

6.5 ANALYSIS

This section will discuss the main findings and address the study's research questions. Different urban processes operate during the process of urban growth which shape the urban morphology. The chronology of events and the sequential growth of the colonial city have been discussed above. This section seeks to analyze the process of evolution and understand the process and the factors which have shaped the urban space. It is divided into two subsections, identifying the processes which have shaped the city, and the subsequent urban morphology due to the processes that have operated.

6.5.1 Processes Shaping the Colonial City

Any urban area is shaped by a number of processes like urban agglomeration, assimilation or spatial segregation. Colonial *Calcutta* also experienced it, as illustrated below.

6.5.1.1 Process of Urbanization

According to Tisdale (1942), urbanization is a process of population concentration driven by a multitude of factors. It proceeds both by multiplication of points of concentration as well as increase in individual concentration. This is exactly what happened in the case of colonial *Calcutta*.

Like any other colonial city, *Calcutta* was born out of the British need for fortification. The original historical nucleus of the city was the old fort, later replaced by Fort William. Thus, the development of this city was initially an economic phenomenon. As discussed above, the city was consolidated from three villages. Ghose (1960) observed that Hamilton discussed the condition of the town between 1710 and 1727, and in *East India Gazetteer* (1815), he gives a description of how each village had ten to twelve houses, surrounded by puddles. As the population of *Calcutta* began expanding, two factors may have played a part – push and pull. There cannot be a definite conclusion drawn as to the origin of the factors. However, the pull of a new industrial city with greater employment opportunities as well as the push of the deteriorating condition of the rural land laws due to the company's policy resulting in rural to urban migration, may have contributed to the expansion of the urban area.

However, this demographic growth of *Calcutta* is not entirely due to industrialization but also to the development of trade and commerce. Population poured into to serve as domestic servants or in the mercantile firms of the newly emerged urban aristocrats, both Europeans as well as Indians. According to Benoy Ghosh (1960), most of the native aristocrats were not industrialists but family fortune holders with titles like *Dewan, Sarkar*, etc. Examples may be found in the records of the Home, Foreign and Public Departments and assorted family papers which point out that Maharaja Nabakrishna, the founder of the Sobhabazar Raj family, was Lord Clive's *Dewan*, or the founder of the Andul Raj family was Ram Churn Race, who made his fortune under Governor Vansittart to whom he was private *Dewan*.

Thus the urbanization process took place through nucleization by fortification, spatial and demographic expansion around the original nucleus due to both push and pull factors, primarily with economic motives. Hence, as urban agglomeration occurred, primarily due to socio-economic, occupational and cultural reasons, the city expanded.

6.5.1.2 Process of Spatial Segregation

Spatial segregation of population groups is not a new phenomenon. Residential segregation is a form of segregation that sorts population groups based on certain peculiarities. Population groups may be segregated on racial or ethnic grounds as well as income inequality. In colonial *Calcutta*, too, such segregation was observed primarily on racial lines. But within each racial quarter, there was sub-segregation based on income, economic class, occupation and caste (in the Black Town).

Like other colonial towns, *Calcutta* housed a diverse population and this cultural pluralistic society was segregated into spatial units. Figure 6.5 illustrates that though this segregation, whether race, economic or caste based, was a social construct, it later led to residential segregation and took spatial forms, emerging as territorial units.

6.5.1.2.1 Race-Based Spatial Segregation

Spatial segregation between the colonizers and colonized was observed in *Calcutta* on the basis of race and ethnicity. There were distinct Black (native), White (European/British) and Grey (non-British foreigners like Greeks, Armenians) towns. This segregation is a characteristic feature of

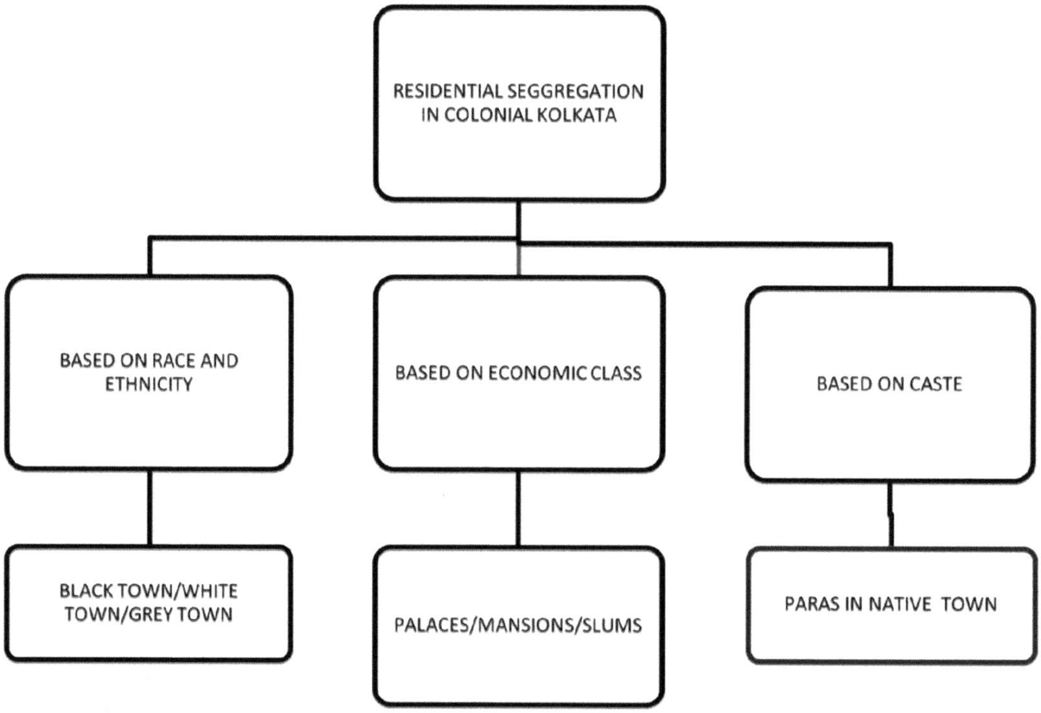

FIGURE 6.5 Residential segregation and its consequent spatial expressions in the study area.

the British and French colonial cities (Karmakar, 2012). Following this compartmentalization, Partha Chatterjee (2014) identified Calcutta as a typical colonial city of Asia, like Shanghai. Frantz Fanon (1961) also discussed the concept of colonial city, describing the colonial world as divided into compartments, just as colonial cities have experienced spatial and social segregation. Hence, *Calcutta*'s exhibition of this particular characteristic of a colonial city conforms with the existing literature.

According to the European Commission against Racism and Intolerance, racism is "the act by which a person separates other persons on the basis of one of the enumerated grounds without an objective and reasonable justification, in conformity with the proposed definition of discrimination". As discussed above, Calcutta was divided into European and native quarters. This distinction came after the emergence of the original historical nucleus, Fort William. The village of Dihi Kalikata on the eastern bank became the White Town while to the north, the cloth-weaving village of Sutanuti became the nucleus for the Black Town. In the south lay Govindapore, which became the home of Hindu trading and money-lending families, also known *as banians*.

This segregation assumed spatiality and the city was divided into sectors based on the communities residing in them, which influenced the morphology of the city. Even though the segregation may have been based on racial distinction or cultural distinctiveness, it also emerged out of hygiene and health concerns. Living in segregated neighborhoods strengthened ethnic bonds and there emerged distinct settlement patterns. In *Calcutta*, the Black and White Towns were based on the perception that European citizens inhabited an area that differed from the native inhabited area terms of density, physical layout or architecture. This segregation revealed the complex relationship of how alien power exerted control over the indigenous population.

6.5.1.2.2 Class-Based Spatial Segregation

Needless to say, as pointed out by scholars like Fanon, the spatial segregation observed in the case of Kolkata (then *Calcutta*) had its roots not only in raciality or ethnicity but also in class.

Although the central spaces of the city were not industrialized, public works attracted migration of casual laborers belonging to low socioeconomic classes, leading to the formation of slums in both the Black and White Towns. The book *Calcutta: Strange Memoirs and Foreign Perception* by Barett (1960) mentions Nathaniel Kindersley's 1768 description of the urban façade with the emergence of slums:

> Here is not as in Madras, a black town near for the servants of the English to reside in; therefore Calcutta is partly environed by their habitation, which makes the roads rather unpleasant; for the huts they live in which are built of mud and straw, are so low that they can scarcely stand upright in them.

6.5.1.2.3 Caste-Based Spatial Segregation

Other than class or race, a unique level of sub-segregation occurred within the Black Town based on caste and the consequent occupation affiliations. According to Sekhar Bandyopadhyay (2004), the most powerful status group in West Bengal was the *Bhadralok* which initially comprised three traditional higher castes – the *Brahmin, Kayastha and Baidya* – who had control over landholding, professions and education. They started carving out *para*s or neighborhoods for themselves. *Paras* were territorial units, similar to *mohalla*s in other cities, and each *para* represented a community bonding. Such *para*s are marked in the old maps of colonial *Calcutta*. These *para*s were social constructs. In other cities in the world these are called racialized ghettos. According to Sengupta (2017), one such *para* was the *sahebpara*, or the then White Town of colonial *Calcutta*.

In this context, it is important to distinguish place and space. While *para* was a place, a territorial unit, it was constructed mentally and is also a spatial construct. Mention may be made here of human geographer Yi Fu Tuan, who distinguishes space from place. Space is more abstract than place and its construct begins as undifferentiated space. Space becomes place once it is endowed with value. The description of native towns as haphazard definitely indicated the inability of the colonial rulers to know the 'space' of the natives.

These *para*s constitute social identities and emotive bonds. Chattopadhyay (2005: 88–90) provides a description of what a *para* was in colonial Calcutta. From the Bengali point of view, the city was divided into many *para*s, *tola*s *and tuli*s. These *para*s were approximately a quarter of a mile long and received their sense of identity from a particular physical feature, which might be a garden (*bagan*), pond (*pukur*) or a *bazaar* (market). The eighteenth-century maps of *Calcutta* showed the presence of such *para*s as Champatola, Sikdarbagan, Chorebagan and Jhamapukur.

Some of these names indicate and confirm that the *para*s were sometimes formed along caste or occupational status, like *Kansaripara* (neighborhood of metallurgical workers), *Kumartuli* (neighborhood of potters and sculptors*) or Coloototola.*

However, in addition to the homogeneous entities based on caste, clan or occupation, more mixed middle-class neighborhoods arose mainly from the second quarter of the nineteenth century (Chattopadhyay, 2005).

These *para*s provide the first evidence of urban agglomeration in colonial *Calcutta.*

6.5.2 Urban Morphology of *Calcutta*, as a Product of Urban Growth

Cities are visual and city morphology is a reflection of the growth of the city. According to Batty (2008), the city usually grows from bottom to top, is linked to economic and transportation behavior and is the result of the intense competition for space. Regarding urban morphology, geographers have attempted to apply Western models. But Dutta (2004) points out that South Asian city models cannot be explained using the Western urban morphological models of Burgess or Ullmann and Harris. The earliest descriptive model of colonial urbanism was given by Arthur Smailes (1969), and later modified by Dutt between 1983 and 1992.

Following Dutta's morphological model for colonial cities, a descriptive morphological model may be presented below (Figure 6.6) based on the map of 19th-century colonial Kolkata by the Church Missionary Society, 1862.

The figure illustrates the presence of the original historical nuclei and the racially segregated town – Black, White and the Grey Town comprising lower-class foreigners like Armenians and Greeks who were not the colonial powers. In addition to racial segregation, there was also caste-based segregation.

6.5.3 Presence of Other Urban Morphological Characteristics

As discussed, colonial cities showed evidence of the existence of native Black towns and European White towns, a feature also present in Kolkata.

Table 6.1 shows most of the other morphological characteristics of the colonial era demonstrated in the case of Calcutta (Figure 6.7).

Figure 6.6 compares the colonial urban morphological model of Dutta (2001) with the study area. The satellite plate of Kolkata shows an elongated development around the waterfront, which served as the main transit line to carry goods from the hinterland and transport it out of the country. Further, the fort (Fort William) had developed around the water body within the surrounding open space (*Maidan*). North Kolkata with its dense built-up area was once the Native Town while further south was the European Quarter.

The present urban form of Kolkata is represented by a satellite image. It reflects a continuous but narrow growth along the East–West dimension. As population and its activities increase in a city, the boundary of the city expands to accommodate growth along the urban fringes, leading to fragmented urban morphology, thereby impacting local ecology. This is exactly demonstrated in the satellite image (Plate 1).

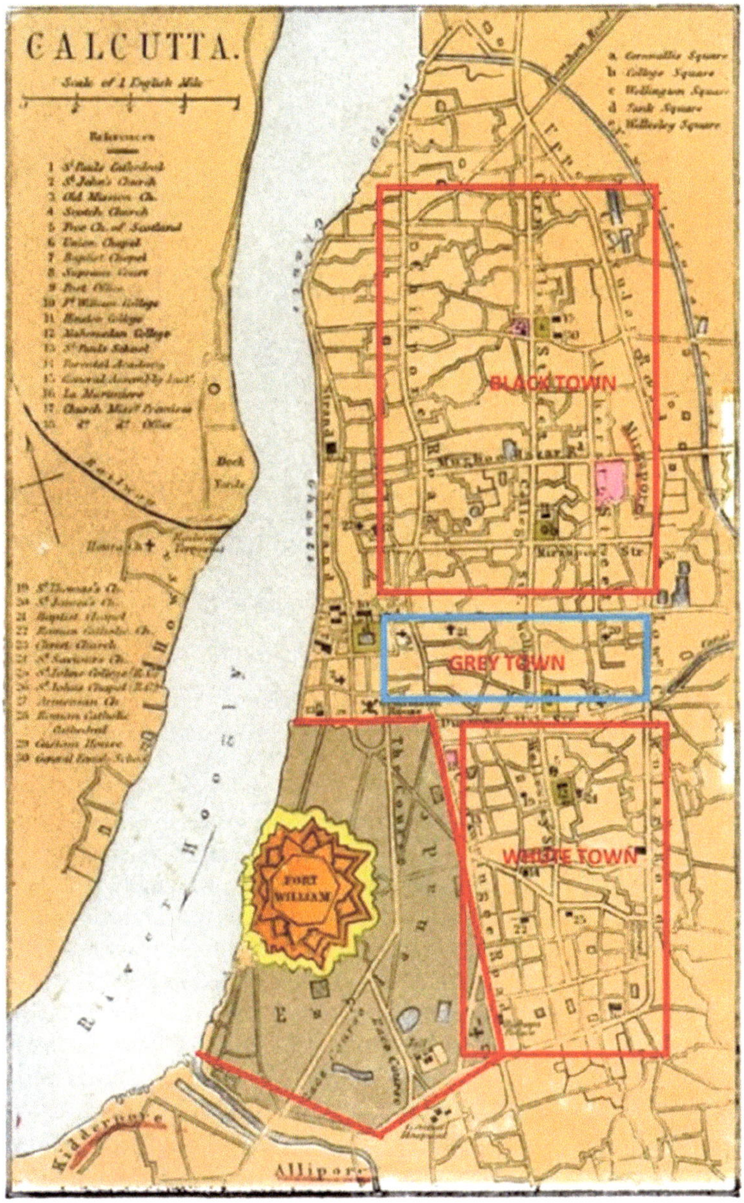

FIGURE 6.6 Kolkata in 1914, showing Southern Black Town.

(Source: Baedeker, Karl 1914.)

TABLE 6.1
Colonial Urban Morphological Characteristics of Kolkata

Typical Colonial Urban Morphological Features	Development Around Waterfront	Fortification	Open Space	Racial Segregation with Separate European and Non-European Quarters
Kolkata	*Present*	*Present*	*Present*	*Present*

Source: Author, 2021.

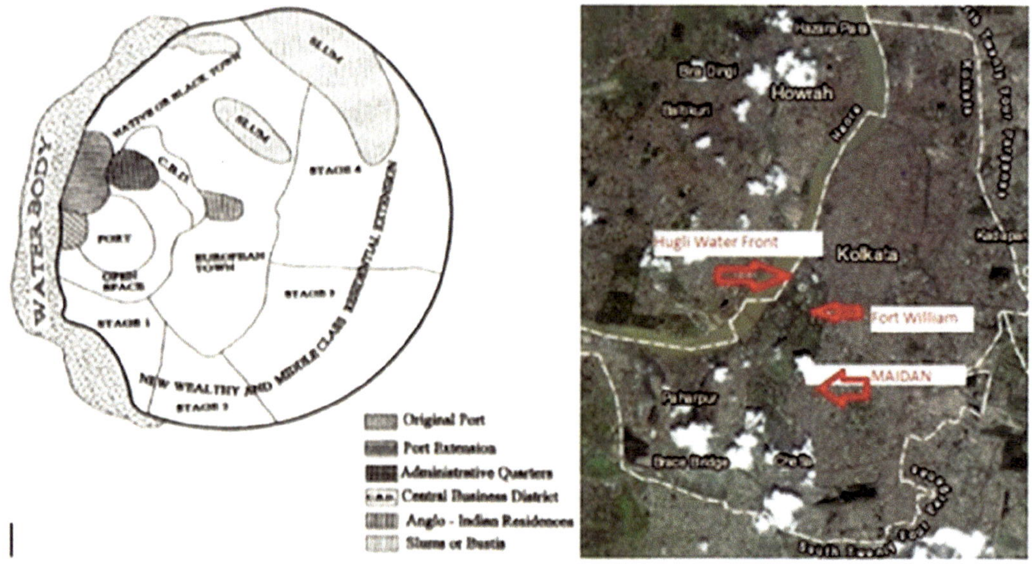

FIGURE 6.7 Comparing colonial urban model (Dutta, 2001) with Kolkata's urban pattern.

(Source: Dutta, 2001 and Google Earth, May, 2019.)

6.6 CONCLUSION

Colonial urban space finds its reflection in urban structure and form. The Europeans who founded their empire outside their own countries were primarily interested in extracting the products and hence built cities which were easily accessible to them. Hence towns emerged to serve specific function, commercial or military, or as in the case of Kolkata, where the town emerged after the fortification in the existing landscape. The fort remained the historical nucleus. However, the city of *Calcutta* reflected many other peculiarities. The growth of this city was due to push factors – the deteriorating condition of the rural agricultural base due to the exploitative policies of the colonists – and to the pull of the urban employment opportunities due to the colonists' policies. Thus, the city was created due to colonial policies. The urban expansion occurred when Indian merchants were attracted by the commercial opportunity under English protection. Side by side with this, spatial segregation developed. Broadly, this segregation was racial between Europeans and non-Europeans. However, neither community was homogeneous. The Europeans were again segregated on the basis of socioeconomic class – the higher class lived in the White Town while the lower class lived in the Grey Town, bordering the White Black towns. The Black Town was again segregated on the basis of socioeconomic class, occupational status, caste, sect, clan etc. An interesting spatial expression was the development of neighborhood or *para*, which were symbols of urban accumulation. The functional need and the topography brought changes to the urban morphology. All these were reflected in the contemporary literary and cultural elements as well as the traditional cartographic maps, which were a product of a number of surveying tasks undertaken for administrative and military purposes at that time.

REFERENCES

Barett, T (1960). *Calcutta: Strange Memoirs and Foreign Perception*. Calcutta: Deep Prakashan.

Bandyopadhyay, S (2004). *Caste, Culture and Hegemony*. New Delhi: Sage.

Babalis, D (2010). The Sustainable Urban Change. In: *Chronocity: The Scale of Sustainable Change: Heritage Value and Future*, D. Babalis (ed.) Firenze: Alinea International; Venice. 19–21.

Bromley, J (1978). Introduction to Urban Informal Sector. *World Development*, vol. 6, issue 9–10, 1033–1103.
Caquard, C and Cartwright, W (2014). Narrative Cartography: From Mapping Stories to the Narrative of Maps and Mapping. *The Cartographic Journal-The World of Mapping*, vol. 51, issue 2, 101–106.
Chattopadhyay, S (2000). Blurring Boundaries: The Limits of "White Town" in Colonial Calcutta. *Journal of the Society of Architectural Historians*, vol. 59, issue 2 (Jun., 2000), 154–179.
Chatterjee, P (2014). The Colonial City in the Postcolonial Era. *Inter Asia Cultural Studies*, vol. 15, 25–42, Taylor and Francis.
Cosgrove, D (1996). Windows on the City. *Urban Studies*, vol. 33, issue 8, 1495–1498. Sage Publications, UK.
Conzen, MP (2012). Experiments in Cross-Cultural Urban Morphology, unpublished keynote address presented at the Nineteenth International Seminar on Urban Form, Delft, The Netherlands, 2 October.
Dias, R (2015). *Old Kolkata Maps*. Accessed from https://puronokolkata.com/tag/old-calcutta-maps/on 3rd May, 2017
Dasgupta, SP (1973). The Site of Calcutta, Geology and Physiography in Calcutta. In: *Calcutta, The Living City*, S. Chaudhuri (ed.) vol 2. Kolkata: Oxford University Press.
Dasgupta, K (2009). *Mapping The Collection Of Maps And Visual Archives*. Calcutta: Centre For Studies In Social Science.
Deb Roy, R (1988). Glimpses on the History of Calcutta 1600-1800. *Annales de Démographie Historique*, 243–257.
Dutt, AK and Noble, AG (2004). Urban Development of South Asia. In: *Challenges to Asian Urbanization in the 21st Century*, AK Dutt, AG Noble, G Venugopal and S Subbiah (eds.) vol 75. The GeoJournal Library. Dordrecht: Springer.
Fanon, F (1961). *The Wretched of the Earth*. France: Grove Press.
Ganguly, M (1967). The Nature of Fringe Areas of Calcutta. *Geographical Review of India*, vol 29, issue 4, 1967.
Ghose, B (1960). The Colonial Beginnings of Calcutta Urbanisation Without Industrialisation. *The Economic Weekly*. vol 12, Issue 33, 1255–1260.
De Bruijne, GA (1985). The Colonial City and the Post-Colonial World. In: *Colonial Cities. Comparative Studies in Overseas History*, RJ Ross and GJ Telkamp (eds.) vol 5. Dordrecht: Springer.
Hamel, J, Dufour, S and Fortin, D (1993). *Case Study Methods*. Newbury Park: Sage Publications.
Harris, C and Ullman, E (1945). The Nature of Cities. *The Annals of American Academy of Political and Social Science*. vol 242, 7–17. Sage.
Han, W and Beisi, J (2016). Urban Morphology of Commercial Port Cities and Shophouses in Southeast Asia. *Procedia Engineering*, vol 142, 190–197.
Harley, JB (2001). Essays in History of Cartography. In: *The New Nature of Maps*, J Laxton (ed.) John Hopkins University Press.
Hoyt, H (1939). *The Structure and Growth of Residential Neighbourhoods in American Cities*. Washington, DC: Federal Housing Administration.
Karmakar, J (2015). Retrospect of Post-Colonial Metropolitan Planning in India: Critical Appraisal. *Journal of Geography and Regional Planning*, vol 8, issue 6, 166–173.
King, A (1976). *Colonial Urban Development*. London: Routledge & Kegan Paul.
King, A (Jul., 1995). Writing Colonial Space: A Review Article. *Comparative Studies in Society and History*, vol. 37, issue 3, 541–554.
Kubat, AS, Kaya, HS, San, F, Güler, G and Özer, Ö (2007). The effects of proposed bridges onurban macroform of Istanbul: A syntactic evaluation. *Proceedings, 6th International SpaceSyntax Symposium, 1, 003*, Istanbul.
Lynch, K (1967). *Image of the City*. Cambridge, MA: MIT Press.
Mansfield, T (2011). Calcutta, From Fort to City: A Study of a Colonial Settlement, 1690–1750. Accessed from https://lra.le.ac.uk/bitstream/2381/27917/1/2012MansfieldTPhd.pd
Mitra, RR (1952). *Kolkata Darpan*. Calcutta: Subarna Rekha.
Morris, J (1974). *History of Urban Form: Before Industrial Revolution*. Canada: John Wiley & Sons Limited.
Munshi, SK (1980). *Geography of Transportation in Eastern India Under the British Raj*. Kolkota: K.P. Begchi and Company.
Neild, S (1979). Colonial Urbanism: The Development of Madras City in Eighteenth and Nineteenth Centuries. *Modern Asian Citities*, vol 13, issue 2, 217–246.
Prakash Rao, VLS (1971). *Some Aspects of Urbanization*. Prasaranga, University of Mysore, 4.
Pickles (2004). *A History of Space. Cartographic Reason, Mapping and Geocoded World*. New York: Routledge.
Ramachandran, R (2012). *Urbanization and Urban Systems in India*. India: OUP.
Simon, S (1984). *Cities in Translation. Intersections of language and Memory*. New York: Routledge.

Sengupta, K (2017). *Community and Neighbourhood in a Colonial City: Calcutta's Para*. Sage Publication.
Smailes, A (1969). The Indian City: A Descriptive Model. *Geographical Journal*, vol 57, issue 3, 177–190. Franz Steiner Verlag.
Tisdale, J (1942). The Process of Urbanization. *Social Forces*, vol. 20, issue 3, 311–316. Oxford University Press.
Winichakul, Thongchai (1994). *Siam Mapped: A History of Geobody of a Nation*. University of Hawaii Press.
Wolpert, J (1967). Distance and Directional Bias in Inter-Urban Migratory Streams. *Annals of Association of American Geographers*, vol. 57, issue 3, 605–616.

Part II

Urban Planning, Innovation and Urban Resilience for Sustainable Development

7 Geopolitical Urbanism Network (GUN)

Building Networked Understanding of Cities in Armed Conflicts for Sustaining Peace

Shalini Chaudhary
Rekhi Centre of Excellence for the Science of Happiness, IIT Kharagpur, Kharagpur, India

Anuradha Choudry
IIT Kharagpur, Kharagpur, India

CONTENTS

7.1 INTRODUCTION

> Cities… provide much more than just the backdrop or environment for war and terror. Rather, their buildings, assets, institutions, industries, infrastructures, cultural diversities, and symbolic meanings have long actually themselves been the explicit target for a wide range of deliberate, orchestrated, attacks.
>
> (**Graham** 2004a)

Cities as primarily human spaces have the potential to be characterized equally by human achievements and human sufferings in their most absolute forms. It is no wonder that when the UN decided to materialize the concept of sustainability into a 21stcentury setting, cities inevitably became the vanguards of the SDGs. The urbanization trends and the resultant movement of population towards cities in the coming decades have been vastly documented in various sources (United Nations 2019). Additionally, with the growth of cities, both in terms of number and size, the way cities are understood and defined is also rapidly changing, with newer concepts and terms (like 'megacities', 'smart cities', 'urban agglomerations') devised to accommodate their diversity (United Nations 2018).

DOI: 10.1201/9781003131922-9

Sustaining peace in a 21st-century world marked by the 'globalized'/'interconnected'/ 'networked' world would, therefore, be a function of collective human efforts across systems, stages, and societies, with the centrality of cities reflected throughout their intersections and interstices. The systemic manifestation of peace ranges from generic natural and human systems to political, sociological, and cultural systems. The stages for securing peace incorporate the events and issues of contestation and conflict, which cover the trajectory of conflict and peace. Furthermore, societies signify the eclectic mix of cultures, identities, and motivations that may have a lasting impact on peace or the absence thereof. And cities in this schema are not accidental, nor are they simply spaces and contexts to events larger than themselves. Rather they are powered centers fueled by lines of social, political, economic, cultural, humanitarian, and technological transmission within and around them.

With this contextual understanding of a deeply interconnected contemporary world, this chapter seeks to develop a networked framework for understanding sustainable peace in cities, primarily those affected by conflicts. This attempt relies on three major propositions narrowed down from an integrated reading of resources from the disciplines of International Relations, Peace and Conflict Studies, Urban Studies, and Defense and Strategic studies.

First, urban conflicts challenge the traditional reasoning on conflicts in terms of the legitimacy of the means used and actors involved because of the indiscriminate destruction being inflicted upon the cities and their population by almost all the parties to the conflict. This view gains further traction when we examine the post-Arab Spring conflicts in Western Asia, particularly the cities affected by the Syrian civil war. These urban armed conflicts have multiple multi-level state and non-state actors with shifting allegiances, indeterminate goals, using ambiguous methods, and blurring the extent and implementation of government accountability and legitimacy because of their action–consequence similarity with other 'illegitimate' non-state parties to the conflict (like armed militias, rebel groups, insurgents, terrorist organizations). In this reference, the chapter creates a Venn diagram mapping the conflict in cities through the political positionalities of actors, strategies used, and converging goals. Thereupon, the consequences of these positions, strategies, and goals are assessed through diverse elements of the city clubbed under three heads that we will call 'the three Ps' – people, power, property. It is a requisite for two reasons – first, it helps foreground the idea of peace and conflict in cities in the three subjective yet empirically analyzable categories of people, power, and property, and second, it strengthens the case of cities to be seamlessly integrated into broader geopolitical networks that are spun around them, connecting the agenda of global and local peace.

The second proposition is the need to build a networked understanding of cities in order to dissect their role in the geopolitics of the region and also the politics of the globe, as close analysis of urban conflicts demonstrates the central role of cities as arenas for conflicts despite their strategic origin and ideological creation elsewhere. While the three Ps framework outlines the horizontal expanse the conflict can take and peacebuilding processes would have to consider, the network construction presents a mix of vertical yet non-hierarchical influences and pressures on the cities from three different directions – nation-states, regional actors, and global actors. The logic of multi-directionality, instead of multi-layeredness, is consciously applied here based on the assessment of recent urban armed conflicts that have displayed the frequency of these actors interacting directly with the city and its forces for the creation, continuation, or culmination of these conflicts. Furthermore, it also does not preclude the multiplicity of actors, institutions, goals, interests, and principles functioning from these three directions and inflicting varying degrees of pressure on the city. This attempt is rooted in the idea of geopolitical urbanism which, while it has seen some landmark studies in the field is still is in the nascent stage of its development because of the perpetual and fluid nature of the fundamental concepts underlying it – the urban, the geopolitics, and the network interweaving them. The chapter briefly explores the concept of geopolitical urbanism and develops a geopolitical urbanism network (GUN) framework to build a comprehensive understanding of urban conflicts and pathways of peace.

Third, the COVID-19 pandemic has exposed the susceptibility of cities to the mega-scale humanitarian crisis that emerges unpredictably and evolves rapidly (exploiting the channels of globalization), thereby limiting the effectiveness of peacetime control and response mechanisms that most cities are used to deploying. Moreover, a crisis of this scale has the potential to turn into a conflict or cause a relapse of cities recovering from a conflict. The GUN framework provides a readily available structure of analysis and response for such emergencies, taking into consideration the newer and updated realities of geopolitics in terms of actors, their refined goals and principles, and scope and speed of the relief and reinforcement that are plausible.

Additionally, it is impractical to try and understand a conflict only in terms of when and where it is played out. Conflict construction (strategic, ideological, or actual) and its replication in a city's perimeter can also be effectively explained and examined through the GUN framework. Furthermore, since establishing peace would require navigating similar forces and channels with similar contemplations of the compatibility of goals and principles as are postulated through the GUN framework, it can also assist in envisioning the contours for sustaining peace globally and in cities. Considering these lessons and contingencies aggregately, the final section of this chapter progresses from conflict towards the direction of peace, cautiously making observations for securing and sustaining peace in cities applicable through systems, stages, and societies laid bare through the GUN.

7.2 CITIES IN CONFLICT: THE CONTEMPORARY

The case of urban wars (or armed conflicts in cities) provides an existential challenge to modern cities. The urbanization of warfare is becoming a phenomenon of strategic and political importance for most nations in the world, within and without. For instance, terrorist groups are increasingly targeting city centers for better effect and audience, especially in the post-9/11 era, thereby 'leveling the risks' across cities of the North and South (Beall 2007); the risk perceptions of the city residents are also amplified because of their "proximity to primary terrorist targets" like cities (Woods et al. 2008); the guerrilla tactics of warfare are being superimposed on urban spaces (Spencer 2021); and the security and surveillance apparatus of most global cities are suggesting state paranoia and preparedness for a potential physical threat to their population and critical infrastructure but also as a means of maintaining global hegemony (Graham 2009). Further, the attraction of urban warfare for the parties is evident from increased media attention and access to urban centers (individualized video generation, crowd-sourcing of local news, real-time updates on social media platforms by people living in conflict zones); ease of maneuvering and hiding among the civilian population; physical security provided by concrete structures against the most advanced and heavy military operations; psychological, moral, and symbolic defeats or victories; and increased policy focus on cities as part of internal security concerns.

Armed urban conflicts since the Second World War have had many distinctive lessons for military strategists, urban planners, and policy makers alike, evolving from being fought as conventional battles, to taking the form of counterinsurgency and counterterrorism operations, to hybrid urban warfare marked by the combined application of human and technological intelligence, to a more visible yet subtle representation through "military urbanism". To chart the utility of cities in the domain of urban warfare, it is observed that while they became frequent targets by the Second World War, the Cold War facilitated their emergence as "strategic sites" (Rokem et al. 2017).

Urban warfare, for our times, was popularized by the Iraq War (2003–2011) and in some scholarly assessments will continue to hold centrality in 21st century conflicts as cities become "critical strategic objectives" (DiMarco 2012). The 'urban triad' of this warfare includes "a complex man-made physical terrain, a population of significant size and density, and a supporting infrastructure" (Urban Operations 2017, 1–3). In terms of the planning and execution of military operations in such complex city environments, it is emphasized that built environment and population density as dominant features would require not only planning of operations but consistent consolidation of gains and ultimate engagement of legitimate civilian authority for the smooth transitioning of city control in post-conflict times. It has also been argued that the 'total urban environment' – which includes city centers (the geographical extent) along with their neighboring areas (rural, semi-urban) – must also focus on the connections between these two disparate locations as they have mutual political, economic, and cultural impacts (Urban Operations 2017). This all-encompassing demand and effect of urban wars is particularly prominent in contemporary conflicts.

There are some mutations in the way threats to urban spaces are conceptualized and urban wars are characterized for contemporary times. The US military is functioning and practicing urban warfare tactics even in US cities – the "homeland" (Stephen Graham). Furthermore, they are now being used as 'military rhetoric' by politicians in the developed West in response to largely peaceful protests (like protests in response to the death of an African-American, George Floyd, in Minneapolis, US and the Yellow Vest protests in France). The increased "militarization of policing" in the city spaces is cited as the dominant reason for such a shift (Konaev and Brathwaite 2020). The case of 'new military urbanism' is also one direct consequence of diverse and evolving threats faced by modern cities. This includes not just the aspects of securitization and militarization of built environments but also the ideas and norms that bring civilian lives into the 'battlespaces' (real or perceived) of cities (Graham 2009).

7.2.1 The Conflicts on the Ground

Of the ongoing conflicts in cities, Syrian and Iraqi cities have garnered much attention because of the lethality of the tactics used and the brutality of actions of the different parties (both state and non-state) to the conflict. The Syrian crisis started as a result of the spread of Arab Spring of 2011 in the region. However, even after a decade of strikes and attacks by the parties involved in the conflict (including ground-based assaults, aerial bombardments, cluster munitions, siege warfare tactics, use of chemical weapons), that caused intentional/collateral damage to Syrian cities like Aleppo, the war continues to be messy (BBC 2021).

The war in Aleppo, the country's most populous governorate (with 23% of Syrian population as per 2011 estimates), began in 2012 with rebel/opposition forces controlling the eastern parts of the city for roughly four years before the government of Syria (GoS) forces regained control over it in 2016. In the meantime, Aleppo faced large-scale internal and external displacement of people with limited access to life-saving humanitarian assistance, destruction of its infrastructure and heritage sites, targeted bombings of buildings and medical facilities (Médecins Sans Frontières 2016), and so on.

War in Aleppo was considered the "real, strategic, greatest battle" of the larger Syrian civil war (BBC 2016). The strategic relevance of Aleppo can be gauged from the fact that it shares a border of around 221 km with Turkey, with which Syria has had turbulent relations. It hosts almost all kinds of parties to the war – the GoS forces of the Syrian regime of President Bashar al-Assad, the rebel/opposition forces fighting against the regime, the Islamic State of Iraq and al-Sham (ISIS), the al Qaeda-affiliated terrorist fronts, the Kurdish forces. Most of these parties are further backed by regional and global powers like Turkey, Iran, Saudi Arabia, Qatar, the US, and Russia, depending on the alignment of their respective geopolitical goals and interests with the forces on the ground. The free flow of money, weapons, and mercenaries by these external entities has ensured the perpetuation and continuation of the war in Aleppo despite the frequent sieges the city and its population have been subjected to.

Aleppo had 1,787,000 internally displaced people (IDPs) and 2.57 million people in need of humanitarian assistance in mid-2014 (OCHA 2014). Studies of displacement patterns and humanitarian assistance in Aleppo have also highlighted how civilians are being forced into displacement more than once because of shifting battle lines in the war (Doocy et al. 2015). Apart from the humanitarian cost of war, the infrastructural destruction caused in Aleppo is also alarming. REACH, in partnership with UNITAR-UNOSAT, conducted a satellite detected analysis of damage for Syrian cities and regions. It found that "on average, four buildings per hectare can be classified as damaged or destroyed within Aleppo city" (REACH 2019).

Mosul, an Iraqi city that was severely affected by the larger regional-geopolitical war against terrorist forces of ISIS, was of immense symbolic importance in the war. With its status as the second-largest city of Iraq and as a home to the 12th-century Grand Mosque of al-Nuri, from where the declaration of Caliphate by ISIS was made in 2014, the war in Mosul provides an insightful case study. It illustrates how all major parties to the messy urban conflict saturated with proxy actors, irrespective of their political placement in the conflict, may still inevitably cause civilian casualties, forced displacements, and sexual and gender-based violence (CIVIC 2017).

In their case study on how Raqqa (Syria) came to become the capital of ISIS, Rosenblatt and Kilcullen note "that multiple state sponsors, including Russia, Iran, Qatar, Saudi Arabia, Turkey, and the United States, lent support to armed groups" operating on the ground, and despite the changing of hands of the city between different parties in the war "many of the same conditions that made Raqqa vulnerable to an ISIS takeover… still exist in the city today" (Rosenblatt and Kilcullen 2019).

ISIS, in these city-centric conflicts, also used city structures and economy to sustain itself by taking up residence and operation among the civilian population and strategic placement of IEDs to hold city defense lines, allowing them to plan and operate in the city freely. Heavy military presence in the civilian quarters of the city (like Baghdad and Sadr, marked by concrete T-walls to restrict 'enemy' movement) changes the picture and perception of cities, turning them into so-called 'fortresses'. Attempts at locating, isolating, and targeting guerrilla fighters/insurgents by state forces cause damage to cities (population and infrastructure) that is often termed 'collateral'. Enforced sieges to uproot the militant armed groups, generally to rescue the civilian population, also leave cities in tatters, while deliberate dismantling and destruction of the city's critical infrastructure is also undertaken to prevent their misuse by rival forces. Citizens and businesses, forced into sieges, have also responded, adapting to conflict environments by engaging in economic activities with the armed groups as seen through ISIS operations in Mosul (Iraq); operating as default participants to conflict by providing human intelligence (HUMINT) to combating forces on both sides; or returning to the destroyed cities with the aim of preserving, restarting, and rebuilding.

7.2.2 Conflict Mapping: Through Political Positionalities of Actors and the Three Ps Framework

Informed by empirical journalistic and humanitarian accounts from the ground and broader academic analysis of urban conflicts, we attempt to map out urban conflicts through an inward-moving Venn diagram. For ease of analysis, we list three major 'political positionalities' (based on the assumption that the armed conflict and violence automatically bestows a political position on all the parties involved irrespective of the nature of the original contest) in an urban conflict with different strategies but some common overlapping goals (Figure 7.1).

These include:

- *Offensive*: mostly official, 'legitimate'/'sovereign', state-driven (collectively or individually) with goals of rescuing and repossessing the territory (the city) and dismantling the networks and strongholds (ideological and physical) of the defensive non-state forces. It also includes investing, (re)building, and (re)constructing the city along with the civilian population.
- *Defensive*: mostly non-state actors that are often identified as 'illegitimate'/'informal'. The broad goals are surviving the offensive onslaught by state/official forces and rescuing, possessing, and dismantling the city architecture based on perceived notions of state injustice, corruption, immorality. This may also include (re)organizing and counter-constructing narratives and politics embedded in the conflict and in the cities.
- *Incidental*: mostly the civilian population stuck in the cross-fire between offensive and defensive forces, and consequentially drafting their strategies and goals within a larger context of conflict. Their interaction with both sides is largely incidental to their physical presence in the conflict, mediated through channels like local intelligence sharing, human-shield functions, emergency economic exchanges, and humanitarian events like displacement and rehabilitation, etc.

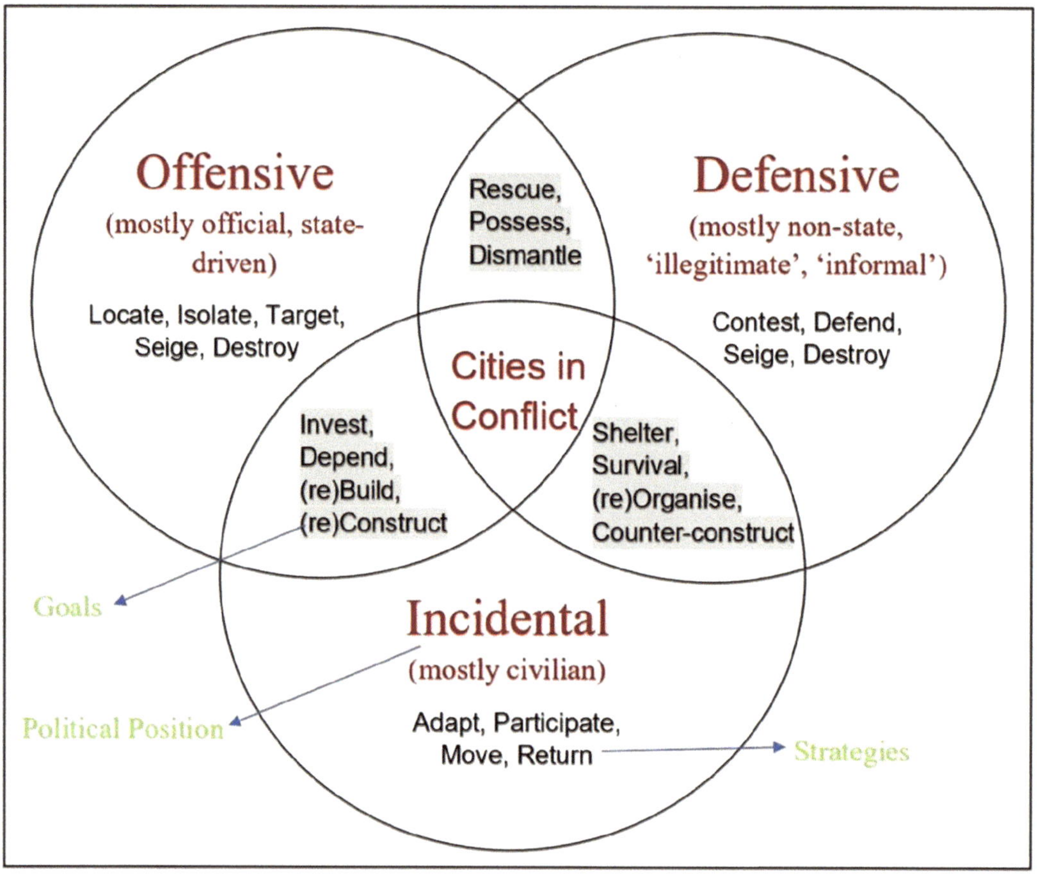

FIGURE 7.1 A Venn diagram for conflict mapping in cities in terms of political positions, strategies, goals.

We must, however, be cautious of the transientness of political positionalities in terms of actor constitution (state/non-state/civilian). A non-state force can hold an offensive or defensive position in the conflict, just like any state actor. However, we propose that despite this, the generality of goal (though guided by positionality) and the resultant consequences for the city concerned in a conflict cannot be ignored. Moving inwards, we observe that despite the different political positions in the conflict, the strategies deployed by the actors operating from different positions allow for convergence of goals in terms of city architecture that is subsequently understood through the three Ps framework.

Cities (Figure 7.2), here, are seen as composites of heterogeneous units – people, property, power – that must be preserved and protected based on the priorities set in accordance with national/global interests and the goals of the war being fought. The positions opted, strategies deployed, and goals crafted (Figure 7.1) have consequences which are cataloged through the three Ps framework (Figure 7.1):

- *People*: distinctively placed within the city, differently affected by the conflict, and possessing different capabilities to rebuild for peace
- *Power*: in forms indivisible yet destructible (resources), legitimate yet co-optable (institutions), resilient yet corruptible (people), guided yet potentially autonomic (technology)
- *Property*: outlines ownership of cities through infrastructural (both public and private), socio-cultural (tangible and intangible), and governmental (natural and built) elements.

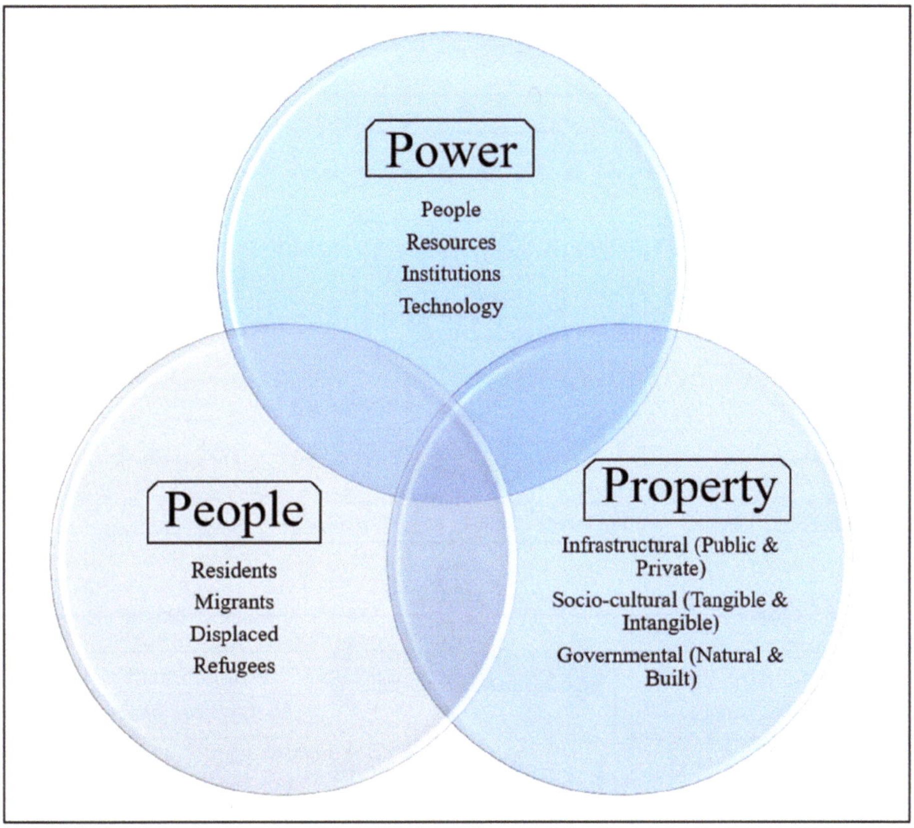

FIGURE 7.2 Cities: The three Ps framework for understanding urban conflicts.

This framework relies on the previous assessment of urban armed conflicts in this section, which helped us identify the major 'political positionalities' in terms of strategies and goals. It, in turn, provides a crucial bloc for the GUN framework (next section), hinting at the critical touchpoints (and sub-points) of cities through which geopolitics becomes cityfied.

7.3 GEOPOLITICS OF CONFLICTS IN CITIES: THE LENS

> *Some form of geopolitics is almost bound to occur, given the gross inequality of states and the weakness of the United Nations.*
>
> (**Falk** 2012)

Urban conflicts evolving with the changing visage and functions of modern cities, more so in the post-Arab Spring context, urgently require new lenses to review the consequences for cities, not only in terms of independent measures of destruction and displacement but as integrated loops and networks of all things urban. This must further take into account the dominant geopolitical links through which the city functions because these influence and create pressure on the internal dynamics (the three Ps) of the cities.

The collective design includes two components: the tri-centric assessment of urban conflicts through the three Ps – people, power, and property – and the tri-directional influence and pressure on cities from national, regional, and global levels (Figure 7.3). The consequently drafted geopolitical

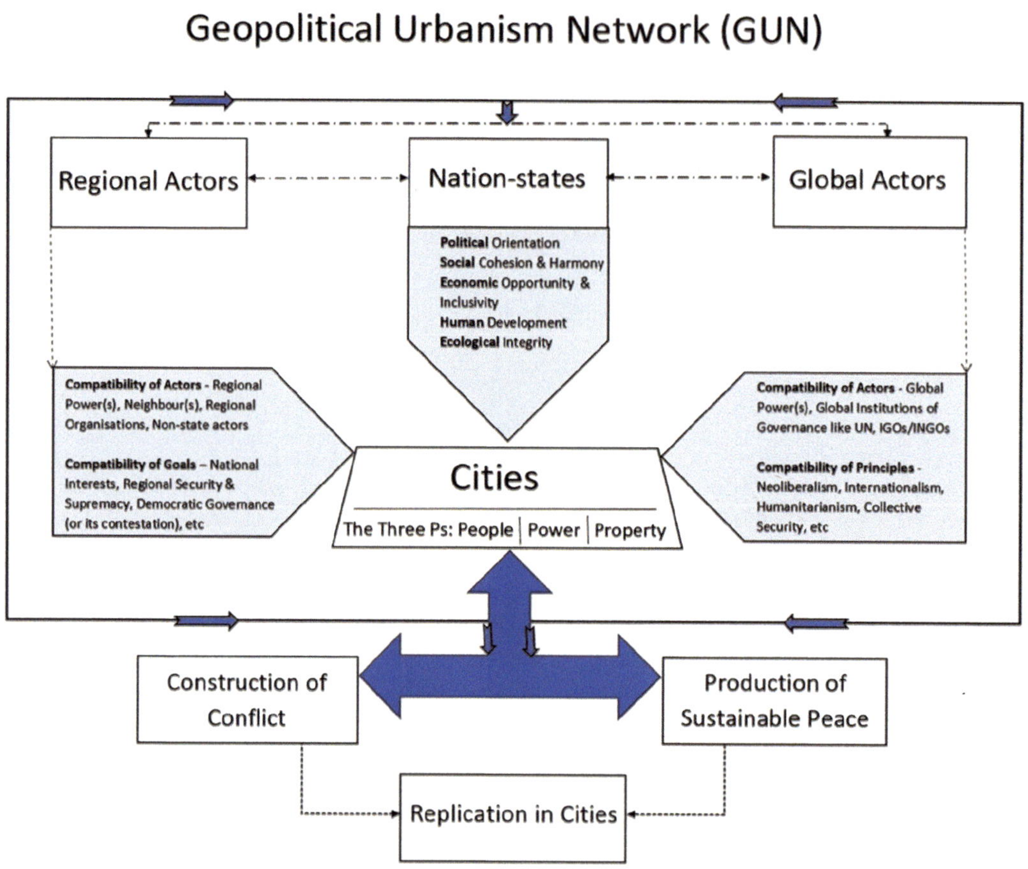

FIGURE 7.3 Geopolitical Urbanism Network (GUN).

urbanism network (GUN), therefore, incorporates the realities of geopolitics and its interface with the actualities of cities in a comprehensive frame of analysis for urban conflicts.

The *'network' dimension* of the framework seeks to highlight the intersectionality of the vernacular elements (people, power, property) of the horizontal city and the verticality of pressures and influences experienced from and through different stations beyond the city (national, regional, global). This in no way suggests that all horizontal elements are correspondingly equal or all vertical ones are essentially hierarchical vis-à-vis the city. On the contrary, the framework attempts to locate these different yet converging elements in their original stations and then outline the modes and means of interaction between them with implications for conflict and peace in the city. The term 'network' here could also be assumed to have an "abstract, metaphorical meaning" where the elements do not necessarily have a definitive and direct causal relationship with each other (Dematteis 1994). The concept also identifies the verticality of interactions operated through technological networks as understood in the sense of surveillance mechanisms and military securitization (Graham 2009). The economic exchanges fueled by advanced technology also "produce a spatiality for the urban that pivots on cross-border networks" (Sassen 2002). The generalist use of networks allows us the leverage of including diverse links between cities and geopolitical elements that may range from security, economic, political, or technological even to abstract.

Contemporary analysis of *urban geopolitics*, specifically so in the field of political geography, has reflected upon the link of global and local (city) in great detail. Basic comprehension of geopolitics involves a focus on the "role played by dominant political actors on the global stage" (Falk 2012).

But there emerge two points of inquiry: what if these dominant actors are operating through proxy actors in local spaces, and what does this interpretation entail for peace? Conventionally, and simply put, geopolitics has been linked to territories and resources, while anti-geopolitics has been associated with grassroots responses.

Urban geopolitics, on the other hand, is understood as the exploration of "complex and intimate relations between urban space and diverse and multi-scalar political geographies" (Rokem et al. 2017). In recent work, diversity and multiplicity of urban geopolitics are also evaluated in terms of "ordinary" (segregation and mobility), "domestic" (dynamics of (household) domesticity), and "refuge(es)" (displacement and refugee crisis) dimensions (Rokem et al. 2017), thereby supplying the field with updated inquiries in conflict and peace's manifestation in cities.

The urban geopolitics lens is also used to link the city and the state in a way that exposes "hybrid sovereignties involving *both* State legitimate forces *and* non-state 'illegitimate' actors" (Fregonese 2012). Furthermore, since territorial spatiality is a sine qua non for the materialization of politics and its associated institutional processes, cities provide a localized, relatively developed, and assimilatory space for it, thus binding the politics (global) with the city.

The actors – national, regional, global – also highlight an important poly-directionally linked component of the GUN framework. The cities interact with these actors differently and sometimes independently. Moreover, as the sovereignty of nation-states comes under the microscope of globalization, the effectiveness of state boundaries, regional-geographical limits, and socio-cultural 'essentialism' are all challenged by the interactions forged by the cities with supra-city elements through their three Ps.

However, that said, the same routes of interaction also expose the vulnerabilities of cities to the idiosyncrasies of geopolitical calculations in times of armed conflicts. Of the Iraq War (2003–11), academia, media, humanitarian organizations, and official state reports have long contested its 'righteousness' and 'justness'. It has been defined as an "essentially neocolonial, racist, hegemony-grabbing, killing spree" that was based on flawed evidence to secure geopolitical goals and with effective disregard for "principles of multilateral international politics" (Graham 2004b). A unique study based on the displacement data provided by the Internal Displacement Monitoring Centre (IDMC), the International Organization for Migration (IOM), United Nations High Commissioner for Refugees (UNHCR), and the UN Office for the Coordination of Humanitarian Affairs (OCHA) recently found that the US displaced 37 million people in post-9/11 wars, most of which had been fought in cities – a "conservative estimate", they said (Vine, et al. 2020). Furthermore, it is argued that the neoliberal reforms introduced by IMF and WTO and US military interventions have "added to the sense of crisis in many cities because they have resulted in the erosion of social and economic security and the further immiseration of the urban poor" (Graham 2004a). The incompatibility of goals (at the regional level) and principles (at the global level) is what accentuates the risks of conflict and risks to peace in the urban spaces.

The *cities* have been argued to possess the 'capacity for autonomous action', and the interrelationships carved out by them are seen to be beyond the state's capacity to regulate and control (Dematteis 1994). This may induce a sense of independence in the structures and processes of the cities, but their susceptibility to conflicts is as high as ever, if not more, apparently due to the very principles that attempt to liberate them from the confines of geography and politics – the "geopolitical phenomenon of mass urbanization" (Evans 2016). It is noted that "physical spaces of cities are increasingly recognized as producers of identity politics, citizenship as well as ethno-national and… socioeconomic reefs" (Fregonese 2012). The "distanciated, verticalized killing" (from outer-space to underground resources and infrastructure in cities) observed in the context of US military operations during the Iraq War (2003–11) (Graham 2004b) have effectively been reproduced for cities like Aleppo in the Syrian war. Cities, especially in the post-9/11 world, are therefore becoming "hypersecuritised and yet insecure" (Rokem et al. 2017). Moreover, as cities turn 'smart' and cyber systems become a crucial linkage in social, political, and economic processes, the overall manifestations can be seen in the form of "cyber-physical social systems" (Cassandras 2016). In the context of urban conflicts, this is portrayed through concepts like "Internet of Battle Things" (IoBT) that relies on networked communication and machine intelligence to assist the combat forces in the conflict environment (Kott, Swami and West 2016).

Conflict construction and replication in cities, therefore, operates through these channels and linkages of local, global, regional, and national; material with ideological; political with socio-economic and cultural; governments with population; state with non-state; and conflict with peace – thereby creating a network that can be used for both construction of conflict and production of peace. The GUN framework attempts to incorporate these multi-structural and multi-layered realities into one holistic framework. It possesses the following characteristics:

- *Multi-directionality of influences and pressure* on cities arising out of geopolitical networking
- *Plurality of actors and forces* driven by conflicting interests and aspirations that can be exclusive to the city concerned or be extra-territorial in nature
- *Adaptive flexibility for reading and responding to the newer crisis* with the potential of turning into conflicts (read on COVID-19 pandemic in the next section)
- *Commonality of actors, factors, and processes* responsible for conflict construction and securing long-term peace.

7.3.1 GUN for Newer Crisis-Led-Conflicts?

The Sustainable Development Goals Report 2020 noted that more than 90% of COVID-19 cases are from urban areas, and the pandemic has only increased their vulnerability given the reality of heavy population density and sprawling slums (UN 2020, 46–47). The pandemic is also exploiting the pre-existing fragility of regions due to conflict and violence (UN 2020, 56–57). Saskia Sassen, while describing the virus as the "invisible actor", notes that the situation has made "aggression in cities more visible" (Sassen 2020). In terms of response, though the supply chains for the vaccine are unavoidably global, their allocation assumes the realistic calculations of a zero-sum game and has exposed inter-state inequality in the global system or what is being increasingly called "vaccine nationalism" (Bollyky and Bown 2020).

The COVID-19 pandemic has brought the concept of 'fragility' and associated issues to the forefront of urban studies, albeit with a slight twist. First, it has not only influenced the praxis of city life by bringing to a standstill its most common operations like mobility and economic exchanges, but has also affected the idea of a city in the sense of dismantling the processes that are responsible for its social, cultural, and political transfusion and regeneration.

Second, while the political reading of the pandemic has raised questions about the legitimacy and accountability of the political dispensations (inability to secure affordable vaccines for its citizens; prolonged lockdowns; disruption of economic and political processes), it also signals towards the coordinated effort by governments, international organizations, humanitarian and economic bodies at different levels needed to fight a crisis 'forced' into a global scale by their perceived inability to pre-empt and respond to it in the first place. This also brings in international organizations with different mandates like health, poverty reduction, economic stability, humanitarianism, peace-keeping as direct stakeholders in the city-specific functions, bypassing the rigidities of national sovereignty and politically imposed institutional hierarchies, just as in the event of an armed conflict.

Additionally, the pandemic has further fed the vicious cycle of poverty, violence, and disaster in the conflict zones, forcing us to re-imagine the orientation of the life-saving humanitarian assistance and application of international humanitarian law. Humanitarian assistance in conflict areas focuses primarily on the concentrated but temporary human settlements like refugee camps where the population density could result in reversal of the gains made in conflict so far in terms of physical and mental health, creation of security structures and safety nets, quasi-social engagements – all, despite the larger contexts of war. The situation is further worsened by the conflict-driven dysfunctionality of the city administration, which is often a result of the inability of political institutions to address such a crisis and eventually translates into indifference of the local and national governments. The postponement of democratic politics (like free and fair elections, working national legislatures, corruption and accountability checks) further erodes the motivation of governments to perform and respond to emergencies like these.

The pandemic in this thread of analysis is, therefore, nothing short of a war or conflict facing human societies, just that the one party is not human but an 'invisible' virus. Moreover, it is similar to fighting an urban war because it requires continuous collection of data from the populace to inform better policy planning and implementations; targeted efforts of the city administration working in tandem with the military/police in some instances (imposition of lockdowns) to regulate the spread of the virus and consolidate the gains already made; awareness of the unavoidable presence of human populations amid counter-virus operations; and stock of the level of vulnerability of popular and most used urban spaces to become zones of infection and contamination.

This bio-infused fragility is, therefore, similar to conflict-infused fragility in urban contexts and may require a coordinated multi-agency, multi-directional response to the crisis, maneuvering and building on the GUN links.

7.4 PEACE (CITIES AND THE GLOBE): THE FUTURE

Peace is not and probably cannot be either perfect or unending...

(**Webel** 2012)

Peace is not always unilaterally, unidirectionally, or linearly established. The ideational projection may be globally mapped out, but the empirics of peace essentially demand local condensation of its principles into practices. Moreover, it is not always 'sacredly' secured in the sense of means deployed – the implicit and blatant violence of the so-called global power-chaperoned 'just wars' executed with strategic geopolitical goals being a case in point. The justness, here, is adjacently infused with legalized (United Nations' principles of 'humanitarian intervention', 'collective security', 'responsibility to protect') recourse to sanitized violence as a means of establishing peace. In the same light, cities and their populations are perceived by militaries and strategists in the Western conceptualization of urban warfare in external and distorted ways. As Stephen Graham observes, the "urban civilians are present only as real or potential collateral damage" and the cities in the South as central elements to the larger "urban turn" in the security and military doctrines of the global powers (specifically the US) (Graham 2008). Furthermore, it is worth noting that the "false or manipulated" response to the crisis may result in a "possible increase in safety but an increase in insecurity" (Marcuse 2006).

The United Nations, however, has progressed on the matter of peace in the past few decades in an attempt to conceptually contemporize peace with geopolitical realities. From its 1992 *Agenda for Peace* (Boutros-Ghali 1992), which emphasized the procedural forms to secure peace in the form of preventive diplomacy, peacemaking, peace-keeping, and post-conflict peace building, its newly adopted resolutions in 2016 on sustaining peace now focus on peace as part of building a "common vision of society... that needs to be fulfilled by the government and all other national stakeholders,... and needs sustained international attention and assistance" (UN, Dag Hammarskjöld Library n.d.).

Conflict and Peace theorist Oliver Richmond highlights the fallacy of assuming peace as an "ontologically stable" concept with a 'universal ethic'. On exploration of the idea of "one peace or many peaces", he instead argues in favor of the possibilities of "contextualising peace" in geographies, cultures, identities (Richmond 2012). It can then instead be interpreted as a "historically shifting condition" (Webel 2012). In terms of the peacebuilding process, it has been expounded that "Western epistemologies" and "centralised narratives of peace and the role of a vanguard", selected global institutions assume, have had their limitations in terms of credible ground achievements

(Mac Ginty and Richmond, 2013). In visible forms, the rise of intrastate wars (as opposed to the inter-state wars) and the corresponding emergence of a new paradigm ("New Wars"—Mary Kaldor), have focused on the "the strategy of violence" that is marked by "terror and destabilisation to displace populations and gain control of territory" (Cortright 2012). Further, total transitionality of peace in post-conflict societies may be difficult to achieve due to the possibility of "continuation of war in peace" through various means (Gusic 2020) that are often fueled by geopolitical forces that have vested interests in 'freezing'/'de-freezing'/'sustaining' the conflict.

Urban armed conflicts have also highlighted one relationship of extreme interdependence that the democratic countries cautiously and consciously circumvent. The necessity of "civil-military coordination" (DiMarco 2012, 212, 215) to win an urban conflict cannot be sufficiently emphasized. The civilian representation, in this case, is not only politically relevant but is a prerequisite for establishing legitimacy (DiMarco 2012). Newer events and crises like socio-political and economic protests, pandemic-related lockdowns and curfews, have further brought the militaries within the frontiers of the city, albeit with different objectives. The underlying conditioning of these events and crises by perceptions of social threat, economic busts, and political paranoia only betray the continuing dependence of cities on militaries, in addition to traditional law enforcement practices. This has "changed the political geographies of major cities" (Fregonese 2012).

A shift from open-space battles to city conflicts that are *asymmetric, insurgential, and audience-centric* has left the strengths of modern militaries, precision warfare, and networked communication systems redundant to some extent. But there are also indications that urbanization of terrains and modernity will further simulate the technological adaptation and vertical penetration of militaries to expand their reach from outer-space to complex undergrounds within the cities (Graham 2004b). However, as ISIS' 'assimilation' in the urban fabric of cities suggests, it would still be difficult for state forces to locate, target, uproot, or eliminate the enemy in an ongoing conflict. Moreover, even after the end of the conflict, there are chances of reorganization and relapse. Any idea for sustaining peace must take this probability of constantly changing war-space, rapid remodeling and updating of formal military doctrines informed by "technoscience" (Stephen Graham), the corresponding acclimatization and reorganization of 'informal' non-state conflict parties, and the opportunistic shifting of actor allegiances for geopolitically guided 'exploits of cities' (resources or territory) into account.

With the net vulnerability of cities overshadowing their invincibility, it is necessary to review them as potential sites of armed conflicts but with an inbuilt ability to respond to future disruptions within their larger orientation as 'complex adaptive systems' (Roggema 2017). This, contextually placed against the backdrop of the pandemic, would help us to re-assess conflict-affected cities in a new light – from passive entities in need of constant external assistance to ones with agency and strength induced by their experience in conflict. Furthermore, as the COVID-19 pandemic has exposed the fragility of even the most developed Western cities, conflict-affected cities may provide crucial lessons for cities, populations, and states that are 'new' to crisis.

As noted throughout the chapter, the projection and application of peace can 'exploit' the same channels as are being used to understand conflict and violence. Peace in this sense can be viewed "as an active and transformative process, that shapes and is shaped by urban space", further leveraging city spaces to "design peace" (Fregonese 2012). And geopolitics becomes a means to "focus on peace(s) across places, including the place of the body" (Koopman 2011). Reconciliation in post-conflict societies, it has been suggested, would involve "memory work" that would, in turn, require people to reconcile with the past loss and suffering by reinterpreting the past (Rigby 2012). Cities, too, possess memory and demand reconciliation with themselves and reinterpretation of themselves after the conflict. They do so not only through the surviving population, re-constructed and superimposed infrastructure, reworked institutional mechanisms, reimagined politics, and renewed perceptions of threat and security, but also independently as spaces used and as spaces diminished. As a response, developing "localised epistemologies and discourses" of peace (Mac Ginty and Richmond, 2013) and countering "anxiety and terror that are induced or threatened both from above and from below" (state and non-state violence) (Webel 2012), together provide a bedrock for the geopolitics of peace emanating from cities.

Feeding on the GUN loop, the larger agenda for peace must, therefore, incorporate the multidirectionality of pressures and influences experienced by cities through the network of geopolitical urbanism, recognize the plurality of actors needed to secure and sustain peace, and impart adaptive flexibilities to the concept and practice of peace to effectively face newer crises/challenges to urban spaces. In addition to this delineating of the networked existence of cities in the cradle of geopolitics, the elusiveness of peace for our times demands serious inquiry beyond just outlining the channels through which conflicts are constructed, and peace can potentially be produced. Further, since peace exists through its multiple, culturally relative, and at times politically unique, variants (like law and order, absence of violent conflict, just and equal society), one thing can be established – peace is not atomic, but networked and channelized through the very people, power, and property (the three Ps) that make up our idea and the physicality of cities. What remains for policy makers, urban planners, and strategic thinkers to explore, then, is a dilemma – a city that relies on peace for its existence and peace that can physically exist only through the city. Their coterminousity is what is sought and what is elusive. This was conceptually attempted by the GUN framework through this chapter in a limited way.

7.5 CONCLUSION

A cumulative reading highlights a few dominant themes and arguments made throughout the chapter vis-à-vis urban armed conflicts, geopolitics, and peace. The first is the presence and operation of conflicts in urban spaces. It considered the epistemology of cities as lucrative destinations for playing out violent armed conflicts and different strategies followed by the parties involved to engage and sustain themselves in these conflicts. Second, three diagrammatic representations are used to explain the shifting and complex actor politics and positioning in urban conflicts, the consequence of goals and strategies employed by these actors for cities in conflict understood in terms of the three Ps, and embedding this understanding within the geopolitics using the geopolitical urbanism network (GUN) framework. Third, the equation of the COVID-19 pandemic with a conflict-like situation was undertaken to comprehend the response mechanism a crisis of this scale should initiate. Further, the effect of the pandemic on already fragile ecosystems like conflict-ridden cities is also considered. Lastly, conceptual exploration of the concept of peace is undertaken, and a few cautious insights are made to guide our framework of *GUN for conflict and peace*.

REFERENCES

BBC News. 2021."Why has the Syrian war lasted 10 years?" *BBC*, March 12, 2021. https://www.bbc.com/news/world-middle-east-35806229

BBC News. 2016. "Aleppo: Is besieged Syrian city facing last gasp?" *BBC*, July 2016. https://www.bbc.com/news/world-middle-east-36853689

Beall, Jo. 2007. "Cities, Terrorism, and Urban Wars of the 21st Century." *Crisis States Research Centre*, February. https://www.lse.ac.uk/international-development/Assets/Documents/PDFs/csrc-working-papers-phase-two/wp9.2-cities-terrorism-and-urban-wars.pdf

Bollyky, Thomas J., and Chad P. Bown. 2020. "The Tragedy of Vaccine Nationalism: Only Cooperation can End the Pandemic." *Foreign Affairs*, September/October 2020. https://www.foreignaffairs.com/articles/united-states/2020-07-27/vaccine-nationalism-pandemic

Boutros-Ghali, Boutros. 1992. "An Agenda for Peace." *International Relations* 11, no. 3: 201–218. doi: 10.1177/004711789201100302.

Cassandras, Christos G. 2016. "Smart Cities as Cyber-Physical Social Systems." *Engineering* 2, no. 2: 156–158. doi: 10.1016/j.eng.2016.02.012.

CIVIC. 2017. *Protection of Civilians in Mosul: Identifying Lessons for Contingency Planning*. A Center for Civilians in Conflict (CIVIC); Inter Action Roundtable.

Cortright, David. 2012. "What Is Peace?" In *Peace and Conflict Studies: A Reader*, by Charles Webel and Jorgen Johansen, 118–131. New York: Routledge.
Dematteis, Giuseppe. 1994. "Global Networks, Local Cities." *Flux* 10, no. 15: 17–23. doi: 10.3406/flux.1994.973.
DiMarco, Louis. 2012. *Concrete Hell: Urban Warfare from Stalingrad to Iraq*. Oxford: Osprey.
Doocy, Shannon, Emily Lyles, Tefera D. Delbiso, and Courtland W. Robinson. 2015. "Internal Displacement and the Syrian Crisis: An Analysis of Trends from 2011–2014." *Conflict and Health* 9, no. 1. doi: 10.1186/s13031-015-0060-7.
Evans, Michael. 2016. "Future War in Cities: Urbanization's Challenge to Strategic Studies in the 21st Century." *International Review of the Red Cross* 98, no. 901: 37–51. doi: 10.1017/s1816383117000066.
Falk, Richard. 2012. "Nonviolent Geopolitics: Rationality and Resistance." In *Peace and Conflict Studies: A Reader*, by Charles Webel and Jorgen Johansen, 221–226. New York: Routledge.
Fregonese, Sara. 2012. "Urban Geopolitics 8 Years on. Hybrid Sovereignties, the Everyday, and Geographies of Peace." *Geography Compass* 6, no. 5: 290–303. doi: 10.1111/j.1749-8198.2012.00485.x.
Graham, Stephen. 2009. "Cities as Battlespace: The New Military Urbanism." *City* 13, no. 4: 383–402. doi: 10.1080/13604810903298425.
Graham, Stephen. 2008. "Imagining Urban Warfare: Urbanization and U.S. Military." In *War, Citizenship, Territory*, by Deborah Cowen and Emily Gilbert, 42. New York: Routledge.
Graham, Stephen. 2004a. "Postmortem City." *City* 8, no. 2: 165–196. doi: 10.1080/1360481042000242148.
Graham, Stephen. 2004b. "Vertical Geopolitics: Baghdad and After." *Antipode* 36, no. 1: 12–23. doi: 10.1111/j.1467-8330.2004.00379.x.
Gusic, Ivan. 2020. "The Postwar City." In *Contesting Peace in the Postwar City: Belfast, Metrovica and Mostar*, by Ivan Gusic, 1–17. Macmillan: Palgrave.
Konaev, Margarita, and Kirstin J.H. Brathwaite. 2020. "Analysis | U.S. Officials are Talking about Urban Warfare. Here's What Urban Warfare Really Involves." *The Washington Post* July 30, 2020. https://www.washingtonpost.com/politics/2020/06/04/us-officials-are-talking-about-urban-warfare-heres-what-urban-warfare-really-involves/
Koopman, Sara. 2011."Alter-Geopolitics: Other Securities are Happening." *Geoforum* 42, no. 3: 274–284. doi: 10.1016/j.geoforum.2011.01.007.
Kott, Alexander, Ananthram Swami, and Bruce J. West. 2016. "The Internet of Battle Things." *Computer* 49, no. 12: 70–75. doi: 10.1109/mc.2016.355.
Mac Ginty, Roger, and Oliver P. Richmond. 2013. "The Local Turn in Peace Building: a Critical Agenda for Peace." *Third World Quarterly* 34, no. 5: 763–783. doi: 10.1080/01436597.2013.800750.
Marcuse, Peter. 2006. "Security or Safety in Cities? The Threat of Terrorism after 9/11." *International Journal of Urban and Regional Research* 30, no. 4: 919–929. doi: 10.1111/j.1468-2427.2006.00700.x.
Médecins Sans Frontières. 2016. "Syria: Eastern Aleppo Hospitals Damaged in 23 Attacks since July: MSF." *Medecins Sans Frontieres*, October 7, 2016. https://www.msf.org/syria-eastern-aleppo-hospitals-damaged-23-attacks-july
Ocha, UN. 2014. *Syrian Arab Republic Governorates Profile*. United Nations.
REACH, UNITAR-UNOSAT. 2019. "Syrian Cities Damage Atlas: Thematic Assessment of Staellite Identified Damage." https://reliefweb.int/sites/reliefweb.int/files/resources/reach_thematic_assessment_syrian_cities_damage_atlas_march_2019_reduced_file_size_1.pdf
Richmond, Oliver P. 2012. "Peace in International Relations." In *Peace and Conflict Studies: A Reader*, by Charles P. Webel and Jorgen Johansen, 36–45. New York: Routledge.
Rigby, Andrew. 2012. "How do Post-conflict Societies deal with a Traumatic Past and Promote National Unity and Reconciliation?" In *Peace and Conflict Studies: A Reader*, by Jorgen Johansen and Charles Webel, 234–246. New York: Routledge.
Roggema, Rob. 2017."The Future of Sustainable Urbanism: Society-Based, Complexity-Led, and Landscape-Driven." *Sustainability* 9, no. 8: 1442. doi: 10.3390/su9081442.
Rokem, Jonathan, Sara Fregonese, Adam Ramadan, Elisa Pascucci, Gillad Rosen, Igal Charney, Till F. Paasche, and James D. Sidaway. 2017. "Interventions in Urban Geopolitics." *Political Geography* 61: 253–262. doi: 10.1016/j.polgeo.2017.04.004.
Rosenblatt, Nate, and David Kilcullen. 2019. *How Raqqa Became the Capital of ISIS: A Proxy Warfare Case Study*. New America; ASU Center on the Future of War. https://d1y8sb8igg2f8e.cloudfront.net/documents/How_Raqqa_Cecame_the_Capital_of_ISIS_2019-07-26_134456.pdf
Sassen, Saskia. 2002. "Locating Cities on Global Circuits." In *Global Networks, Linked Cities*, by Saskia Sassen, 1–36. New York: Routledge.

Sassen, Saskia. 2020. "The City and the Virus." *Shapes*. June, 2020. https://www.iberdrola.com/shapes-en/saskia-sassen-crisis-coronavirus

Spencer, John. 2021. "The Eight Rules of Urban Warfare and Why We Must Work to Change Them." *Modern War Institute*. mwi.usma.edu/the-eight-rules-of-urban-warfare-and-why-we-must-work-to-change-them/.

UN. n.d. *Dag Hammarskjöld Library*. Accessed May 2021. https://ask.un.org/faq/203718

UN. 2020. *The Sustainable Development Goals Report*. United Nations.

United Nations, Department of Economic and Social Affairs, Population Division. 2019. *The World Urbanization Prospects: The 2018 Revision (ST/ESA/SER.A/420)*. New York: United Nations.

United Nations, Department of Economic and Social Affairs, Population Division. 2018. *The World's Cities in 2018—Data Booklet (ST/ESA/SER.A/417)*. United Nations.

Urban Operations. 2017. Army Techniques Publication; Marine Corps Techniques Publication. https://fas.org/irp/doddir/army/atp3_06.pdf

Vine, David, Cala Coffman, Katalina Khoury, Madison Lovasz, Helen Bush, Rachael Leduc, and Jennifer Walkup. 2020. "Creating Refugees: Displacement Caused by the United States' Post-9/11 Wars." https://watson.brown.edu/costsofwar/files/cow/imce/papers/2020/Displacement_Vine%20et%20al_Costs%20of%20War%202020%2009%2008.pdf

Webel, Charles. 2012. "Thinking Peace." In *Peace and Conflict Studies: A Reader*, by Charles Webel and Jorgen Johansen, 65–74. New York: Routledge.

Woods, Joshua, Toby A. Ten Eyck, Stan A. Kaplowitz, and Vladimir Shlapentokh. 2008. "Terrorism Risk Perceptions and Proximity to Primary Terrorist Targets: How Close is Too Close?" *Human Ecology Review* 15, no. 1: 63–70. http://www.jstor.org/stable/24707484

8 Sustainable Spatial Planning for the Rural–Urban Continuum Settlements of Kerala, India

N.M. Rijesh, Mohammed Firoz C, Susan Cyriac, and Sruthi Krishnan V
National Institute of Technology Calicut, Kozhikode, India

CONTENTS

DOI: 10.1201/9781003131922-10

8.1 INTRODUCTION

Urbanization is unavoidable when there is too great a population increase, low income from primary activity, and high pressure on land. Migration to cities occurs because of the promise of economic and social stability, and availability of opportunity, even though jobs in the rural areas are present(Government of Kerala 2012). Urbanization is a natural phenomenon and it gives momentum to the development of areas. It has both positive and negative impacts on people and the environment. Urbanization enhances access to higher-order facilities to a large sector of the population and provides better employment opportunities. It improves the living condition of people both in urban and rural areas, both directly and indirectly. But this positive aspect will only be visible if the cities act as the generator of economic growth for the region in a sustainable manner(Government of Kerala 2012).

The difference between urban areas and rural areas is clear in most parts of India. But the situation is different in the state of Kerala. Large numbers of small and medium urban areas are distributed all over the state against a rural background. There are a few millionplus urban agglomerations within the small state spaced about 100km apart. It is very difficult to distinguish a rural area in the state of Kerala from an urban area or vice versa(Government of Kerala 2012). With this rural–urban continuum(RUC) type of settlement, the rate of urbanization is steadily increasing in the state. The rate increased from 26.39% in 2001 to 47.72% in 2011. The settlement pattern of the state in the context of its rapid urbanization requires analysis. While spatial plans for the sustainable development of the areas in the state are in preparation, the typology of the urban and rural dichotomy of settlements needs to be re-evaluated in the context of RUC developments.

8.1.1 SETTLEMENT PATTERN

Settlements in India are classified as urban or rural based on criteria that are defined by the census. Accordingly, rural areas are characterized in terms of "revenue villages" forming a much smaller unit than urban areas, i.e., under various district administrations. A village is a separate administrative unit and can also be composed of a group of livable areas together forming a unit. Defining urban areas, however, has proved hard for the census, and with the 1971 census, the concept of the Standard Urban Area (SUA) was developed. The requirements for a SUA are: "Population which exceeds 5,000 persons, a minimum population density of 400 persons per square km and at least 75% of the male main working population is engaged in non-agricultural pursuits." If a settlement satisfies these criteria in two consecutive decades, that area will be demarcated as a census town by the Census of India(Government of Kerala 2012).

The unique character of the settlements in the state defies the very criteria of the census and makes it difficult to define the state's RUC. This type of settlement pattern includes a mixture of both urban and rural development without a clear distinction between the two. Such patterns are also seen in other Asian countries like Sri Lanka, Indonesia, some parts of China, and in the USA(Firoz, Banerji, and Sen 2014).

8.1.2 RURAL–URBAN CONTINUUM SETTLEMENT PATTERN OF KERALA

Kerala is one of the most densely populated states in India, characterized by its unique settlement pattern with independent houses on individual plots scattered across the livable areas

(Firoz, Banerji, and Sen 2014; Kallingal and Firoz 2018). The dispersed settlement pattern owing to the preference for homestead-type development, the geographical peculiarities of the state, the availability of sub-soil water for agricultural practices,and so on, can be considered as both a prospect and a problem(Government of Kerala 2012). Settlement development in Kerala happened linearly with high densities. At the same time, these settlements are interconnected. This character and spatial distribution differentiate the settlements in Kerala from those in other parts of India (Firoz, Banerji, and Sen 2014) and has resulted in the rural–urban continuum type of settlement development all over the state.

Rural–urban continuum is the term used to describe the connections between urban and rural areas where no sharp differences can be observed in the quantity or degree of the difference between these interlinked areas. RUC development in Kerala started a long time ago. The evidence for this type of development can be seen in Ibn Batuta's records in the 14th century or in Logan's Malabar manual (1887) (Firoz, Banerji, and Sen 2014).

The co-existence of primary-sector activities with the secondary and tertiary sectors can be seen in most parts of Kerala. The difference between urban areas and rural areas is much less.This settlement pattern can be seen along the coastline of the state, stretching for 30km from north to south of the state(Firoz 2015).

The majority of urban agglomerations in Kerala with over one million population as per the 2011 census are located along the lowland region, and spaced about 100 km apart (Figure 8.1). In other words, they are spread along Kerala's coastline. Population details and their ranking within India are shown in Table 8.1.

The plan boundary of almost all statutory masterplans in Kerala is the political boundary of the municipality or municipal corporation. The conventional planning area delineation process is the delineation of areas based on the boundary of local self-government institutions. But due to the RUC developments, the urban areas in the state are not confined to the statutory boundaries. Their spread outside the political boundaries and the developments of urban–rural interfaces need

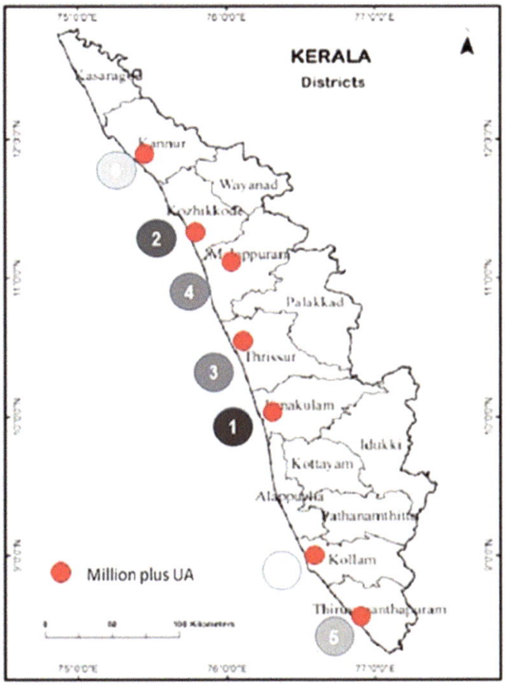

No.	Urban Agglomeration	Population		Rank
		2011	2001	
1	Kochi	21.17.990	13.55.972	17
2	Kozhikode	20.30.519	8.80.247	19
3	Thrissur	18.54.783	3.30.122	21
4	Malappuram	16.98.645	1.70.409	25
5	Thiruvananthapuram	16.87.406	8.89.635	26
6	Kannur	16.42.892	4.98.207	27
7	Kollam	11.10.005	3.80.091	49

FIGURE 8.1 Urban agglomerations in Kerala.

TABLE 8.1
Urban Agglomerations in Kerala, 2011

		Population		
No	**Urban Agglomeration**	**2011**	**2001**	**Rank (India)**
1	Kochi	2,117,990	1,355,972	17
2	Kozhikode	2,030,519	880,247	19
3	Thrissur	1,854,783	330,122	21
4	Malappuram	1,698,645	170,409	25
5	Thiruvananthapuram	1,687,406	889,635	26
6	Kannur	1,642,892	498,207	27
7	Kollam	1,110,005	380,091	49

Source: Census of India 2011a; Mundhe and Jaybhaye 2018.

to be considered when delineating planning areas. Current planning practices in Kerala need to be reviewed before further analysis of settlements.

8.1.3 Current Planning Practices in Kerala

Three main types of plans are prepared in Kerala's spatial planning framework.

8.1.3.1 Detailed Town Planning (DTP) Schemes

These are plans prepared at microlevels. Detailed plans for core town centers are prepared in this segment. Plot-level land use along with building footprints are surveyed to prepare the base map for these schemes. Site-specific infrastructure proposals and road proposals with detailed alignments are the output of these schemes. The area of DTP schemes is so small that they address only areas around a major artery in the city center or a major junction and surroundings in the city (*Gazette* 2016).

8.1.3.2 Masterplans

Masterplans are comprehensive plans for a local planning area that provide long-term policies, programs, and detailed proposals for the overall development of the area. A local planning area may be a whole area or part of a statutory town, or the town and part of its peri-urban area, or more than one statutory town along with its contiguous areas. These plans will be prepared within the framework of the district perspective plan or regional plan, if any. The local planning area is the area under the jurisdiction of a municipal corporation, a municipal council or a town *panchayat* constituted under Section 8.4 of the Kerala Municipality Act, 1994 (20 of 1994), or of a village *panchayat* constituted under Section 8.4 of the Kerala Panchayat Raj Act, 1994 (13 of 1994)(*Gazette* 2016). A detailed report containing a development vision for the area, sectoral studies, analysis results, policies, proposals and existing and proposed land-use maps are the outcomes of a masterplan. The zoning regulations within this statutory plan will control the land and infrastructure development in the planning area.

8.1.3.3 District Perspective Plan

According to the Kerala Town and Country Planning Act, 2016, a district perspective plan (DSP)is a "long-term plan approximately for twenty years providing goals, policies, strategies and general programs for the spatial development of the State, District or a Metropolitan area, as the case may be"(*Gazette* 2016). This plan is prepared for a larger area than the other two, but unlike them, it will not have statutory power. This plan only gives guidelines for developments within the district. Similarly, metropolitan perspective plans are prepared for the whole of the urban areas within a district or the contiguous urban areas which are spread over more than one district. The planning strategies and the plan outcomes of these are very similar to the DSP.

All these plans are prepared for a local body or part of a local body. Only in the case of the metropolitan perspective plan is the whole urban area in a district along with its hinterlands considered. A clear demarcation of settlements based on their urban character and morphology is absent in Kerala. Instead of this, imaginary political boundaries are used as boundaries for the preparation of perspective plans. The urban–rural interfaces (URI) developed due to the rural–urban continuum nature of settlements may not be included in the plan area because of this kind of delineation.This URI transition zone is where the expansion of urban infrastructure is happening in a planned or unplanned manner. The growth of public and semi-public facilities and transportation infrastructure, along with residential and commercial activities, are high in these areas. If the area is not properly planned, all these developments will result in uncontrolled urban sprawl. This will cause pollution of land, air, and water in those areas (*Gazette* 2016). Hence identification of these interface areas and inclusion of them in the spatial plan preparation process is necessary for the sustainable development of the state.

The spatial plans are prepared to integrate urban areas and urban–rural interfaces in the state. To achieve this goal, the following objectives are proposed:

- To study the settlement pattern in Kerala and analyze the RUC typology in the context of urban–rural interface.
- To study the issues and potentials of RUC developments in Kerala in their environmental, economic, and governance aspects, and the need for URI delineation.

This chapter is divided into four sections. The first section introduces the need for studying the RUC settlements of Kerala, and the second section discusses materials and methods, including different settlement analysis methodologies.The third section discusses the results of the settlement analysis, including different issues due to the RUC type of developments. The conclusions and recommendations are detailed in the fourth and final section.

8.2 MATERIALS AND METHODS

8.2.1 Methods of Settlement Analysis

The concept of the rural–urban continuum settlement pattern in Kerala, its issues, and the need for urban–rural interface delineation is considered in detail before preparation of an integrated spatial plan for the regions in Kerala. Growth and urbanization trends are analyzed for a better understanding of settlement characteristics. The functionality of settlements and the hierarchy of settlements are further identified for greater clarity. Population and employment data were projected to the year 2021, and the urbanization analysis was conducted using the projected data. The different methods used for the analysis of settlements are described in the following sections.

8.2.2 Urbanization Trends

There is no standard definition of an urban area and such classifications vary amongst the different urban areas of the world. Generally, settlements which serve certain administrative functions are classified as urban. Along with this, population size, density, and land use are used to delineate urban areas in most countries. Some of the classifications extend to other indicators, such as infrastructure (street lighting, road density), urban employment (non-agricultural workers), facilities (higher education institutes, hospitals), etc. (Mcgranahan and Satterthwaite 2014).

Spatio-temporal analysis of census data is carried out to understand the urbanization nature and settlement trends of Kerala. Census data for the years 2001 and 2011 were collected and analyzed using ArcGIS software. The base map used for the analysis is the village map of Kerala and the census details were added as attributes to the villages. Traditionally, urbanization patterns or any areas becoming urban depends on three census criteria: "Population which exceeds 5,000 persons, a minimum population density of 400 persons per square km and at least 75% of the male main working population is engaged in non-agricultural pursuits."(Census of India 2011b). The census urban areas were plotted in the GIS map of Kerala. Settlements that satisfy all three criteria for the last two decades were identified and compared to the urban areas in the census.

8.2.3 Present Urban Areas (2021)

Census data for the year 2021 is not published yet, so population and employment details have been projected to the current year (2021) for the urbanization analysis, using a trend-based projection method. The same three census criteria were proposed for identifying urban areas in the year 2021. When the trend of the population growth rate was analyzed, it was seen that there has been a constant decline from the year 1971 onwards (Figure 8.2).

The same declining growth trend is used for the estimation of the population for the year 2021. The estimated growth rate is 2.36% and is applied to all settlements.The same method is applied for estimating the total employment. First, the growth rate of employment from 2001 to 2011 is estimated. The same growth rate is given to total employment and projected to a future decade. From the four-fold classification of employment data of the 2001 and 2011 censuses, the total of cultivators and agricultural laborers, and household industrial workers and other workers were calculated. The shift of workers from the former to the latter is estimated for all villages.The same rate of shift is applied to the projected employment data for 2021 and total employment numbers were divided into two groups, employment in agricultural pursuits and the non-agricultural sector.

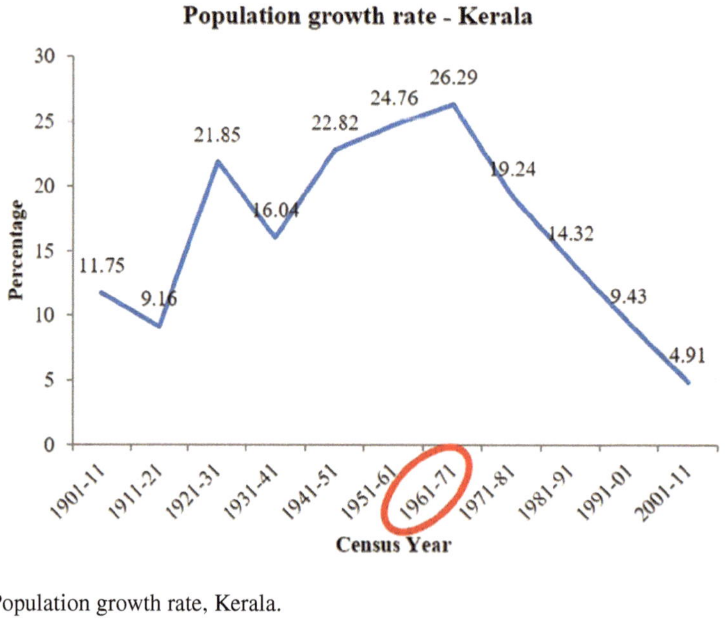

FIGURE 8.2 Population growth rate, Kerala.

8.2.4 Population Density Changes of Settlements

The population density of the core urban center will always be high compared with the surrounding rural settlements. When people migrate to the urban core or its periphery, the density of those areas will also start to increase. The changes in population density at settlements over time will give some understanding about their development or growth. Two types of population density changes are happening in city centers. Population density will stabilize or increase in growing cities. The second type is shrinking cities. Here population density in the city core will be reduced and that of the surrounding peri-urban areas will increase (Fee and Hartley 2011). The population densities of settlements in the last two decades were calculated and comparative analysis was done. Changes in population densities of core urban settlements in the state and their surrounding settlements were assessed.

8.2.5 Functional Character of Settlements

The function of a settlement is the major activity that can be brought into the settlement. It can generally be divide into three: primary with agricultural and allied, secondary with industrial, and tertiary with the service sector. Settlements generally show prominence in any or a combination of these activities. Accordingly, the functional character of the area may be purely rural or purely urban or in between the two with varying degrees of urban content(Government of Kerala 2012).

The residential population factor has a limited influence on the demographic distinction between urban and rural areas. The social and economic land uses that coincided with residential land uses are now spatially separated due to the increased opportunities for local commuting and increased distribution and availability of facilities. Now secondary and tertiary employment opportunities are concentrated in urban areas, and the types of employment determine the character of settlements(Charles Stewart 1958).

The functional character is usually analyzed using the concentration of workers across the region. The employment criterion used for delineating census towns is the percentage of non-agricultural employment. Hence for the analysis of the functional character of settlements, the same non-agricultural workers' density is used. At the village level, the workers may not live in the same village where they work. The urban/rural mixed character of Kerala, accessibility of transport, and land price are the main reasons for this change in work/home locations. So instead of analyzing individual settlements, cluster analysis has been used as an approach to defining the functional character of Kerala. The density of non-agricultural workers was used for developing the clusters and the Anselin Local Moran's I clustering method was used for the analysis.

8.2.6 Settlement Hierarchy

The attraction potential of a settlement increases with the increase in the availability of different functions within it. This determines the size of the population dependent on that settlement for their needs and thus the hierarchy of that settlement (Government of Kerala 2012). Settlement hierarchy can be used for ranking the settlements considering their size and range. Size can be determined by population and range by the availability of services and facilities. Since population size is the major factor determining the hierarchy, the simplest way of categorizing the settlement is according to its population. Another way is the categorization of settlement according to the availability of facilities or services in it. The combination of these two will give a better result. But this will fail in odd situations such as where a settlement contains most of the higher-order facilities it but has fewer people than other areas. To overcome this anomaly, other factors like employment status or road density may be used (East Cambridgeshire District Council 2017). These two basic factors, population and facilities, are related to the employment opportunities in those areas. The availability of a larger number of facilities in an area will increase the job opportunities in that area. Hence for hierarchy analysis, the population size, urban employment details, and availability of different facilities in the settlements were used. The cumulative ranking of the three factors, the population score, centrality score, and functionality score of the settlements, was used to obtain the hierarchy.

8.2.7 Population Score Estimation

Rank-size theory is used for ranking settlements according to their population. This theory says that if cities are ranked from largest to smallest populations, then the 'n' ranked city is expected to have a population equal to the top-ranked city divided by the rank of that city (n). This rank-size relationship is not present throughout India. The population sizes of the four major cities in India are almost the same. At state level, most of the states follow this rule to some extent. But states such as Kerala, Bihar, Punjab, Orissa, and Goa have never followed this rule. There were no primate cities and the population sizes of the first few cities are almost equal in these states. In general terms, India is an exception in rank-size rule (Kumari 2015). The three major cities in Kerala, Thiruvananthapuram, Kochi, and Kozhikode, have almost the same population size, thus rank-size analysis is not suitable for ranking cities in Kerala. Census of India classified urban areas into six classes. Class 1 is a settlement with a population of more than 1,00,000 and class 6 is those with a population of fewer than 5,000 persons. Other classes fall between these population values (Census of India 2001). The same ranking was used for classifying the settlements.The population score was estimated by numbering each settlement from 1 to 6. Density was also incorporated as a parameter to ascertain the weighting. Like the six population classes, six density classes were developed. The settlement with the highest population and density had a total rank of 36.

8.2.8 The Functionality-Based Ranking

Another method of determining the hierarchy of settlements is based on the number of services or facilities available within a settlement. A list of essential services is prepared for this analysis. Then a weighting is assigned to each service according to the importance of that service. In some cases, the quality of the facilities is also included in the analysis, and a particular facility's score is reduced if its quality is not up to the mark.Then the total mark obtained for each settlement is estimated based on the presence of the facility in it. This method is a subjective assessment as it considers the quality of the facility along with its number; it also depends on the precision of the information (East Cambridgeshire District Council 2017).

The weighting of a function is calculated using Equation 8.1.

$$Wi = N/Fi \tag{8.1}$$

where

Wi = Weighting of the functions/sub-functions
N = Total number of settlements
Fi = Number of settlements having the functions/sub-functions

The cumulative score of all the functionality indices is used to rank the settlements.

The functions considered for the analysis are physical infrastructures such as hospitals and schools, and social infrastructures such as stadiums and police stations. The functions are weighted against their total number in the state with the number of settlements. The functions in each settlement are multiplied with these weights and added together to estimate the total functionality score of the settlements.

8.2.9 Centrality Index-Based Ranking

Combining these two basic methods for the estimation of the settlement hierarchy will give a better result than the individual methods. In some situations, where higher-order facilities are available in an area and fewer people reside there or where fewer facilities are available for a high-density populated area, this method may show some anomaly (East Cambridgeshire District Council 2017). Due to the RUC nature of the settlements, the chances of this anomaly are very high in Kerala. So one more ranking is introduced for the hierarchy analysis – the availability of urban employment opportunities within it. Settlements that provide more urban employment opportunities will generally have more facilities and will attract more people into it.

As this third method is introduced to reduce the anomaly of the first two methods, the settlements with higher employment rankings should have more weighting than the lower ones. Hence a comparative analysis is proposed here. Location quotient (*LQ*) analysis is most suitable for comparing the employment availability in an area with that of the region. A centrality index-based ranking (given by Equation 8.2) was conducted based on the location quotient of a settlement's non-agricultural workers (Wheeler 2005).

$$LQ = \frac{\left[\dfrac{Non-agricultural\ workers\ of\ Settlement}{Total\ workers\ of\ settlement}\right]}{\left[\dfrac{Non-agricultural\ workers\ ofthe\ District}{Total\ workers\ of\ the\ District}\right]} \qquad (8.2)$$

8.2.10 Cumulative Ranking

The factors considered for the hierarchy analysis are population, employment, and availability of facilities within the settlements. All three have equal importance in the calculation and hence equal weighting is given for the three scores. To calculate the rank of a settlement, all three are added together after normalizing (given by Equation 8.3).

$$Cumulative\ Rank = Population\ Score + Centrality\ Score + Functionality\ Score. \qquad (8.3)$$

In the study, different hierarchies of settlements were identified based on their overall score. A score above 2.1 (70%) out of 3 is considered as a first-order settlement. Scores from 1.5 to 2.1 (50% to 70%) are considered as second-order and below 1.5 as lower-order settlements (Government of Kerala 2012).

8.3 ANALYSIS AND RESULTS

8.3.1 Urbanization in Kerala

The three major urban centers in the state, Thiruvananthapuram, Kochi, and Kozhikode, are positioned at almost equal distances from each other. This has helped in the uniform distribution of facilities to the people in the hinterland which in turn reduces the pressure on a single center. Along with these three major centers, the class I and II towns, mostly in the lowlands and midlands of Kerala, and a few in the highlands, form a series of urban nodes within the rural backdrop (Chatopadhyay and Sreekumar 1988; Firoz, Banerji, and Sen 2014; Chatopadhyay and Sreekumar 1995).

During the period 2001–2011, the population increase in Kerala was only 4.6%, which is one of the lowest in India, so the change in population density is correspondingly reduced. But the number of urban areas in the state increased from 159 towns to 520 towns (Figure 8.3) (Firoz, Banerji,

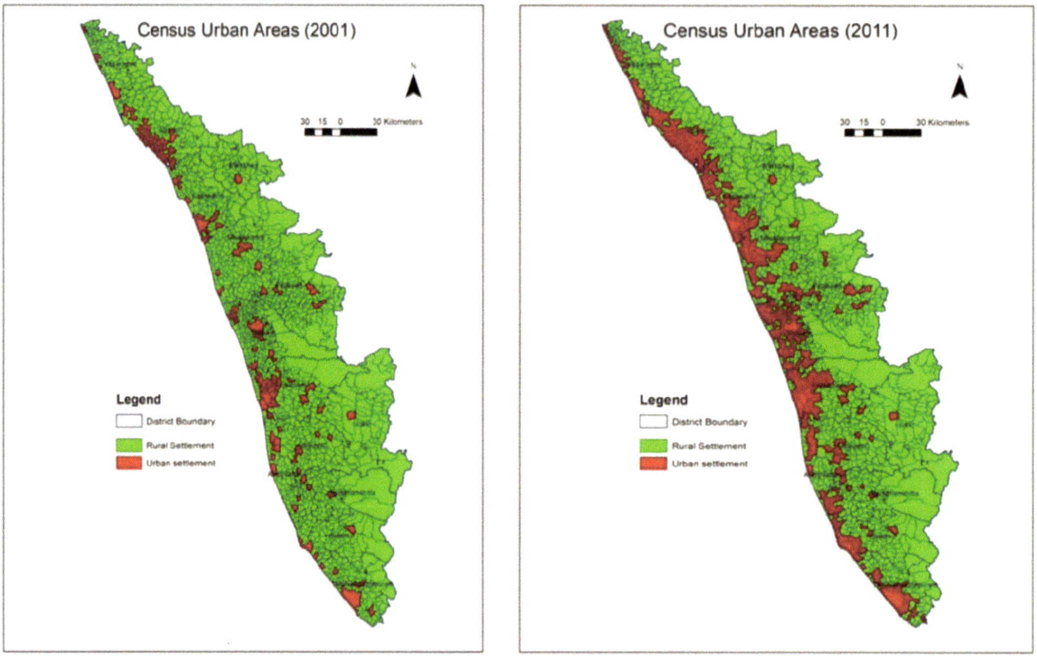

FIGURE 8.3 Census urban areas in 2001 and 2011.

and Sen 2014; Planning studio 3 2020). The urbanization rate of the state showed a sharp increase (47.72 %) in the year 2011 (Figure 8.4) (Mundhe and Jaybhaye 2018).

This growth in the number of urban centers may not be due to the increase in the population or population density of the state, as those are meager considering their values in the last decade, but it does show that a detailed analysis of Kerala's urbanization pattern and RUC is highly relevant.

For a detailed analysis, the census criteria were applied to all settlements, and the settlements which satisfy each criterion were identified. They were combined to identify the areas which are

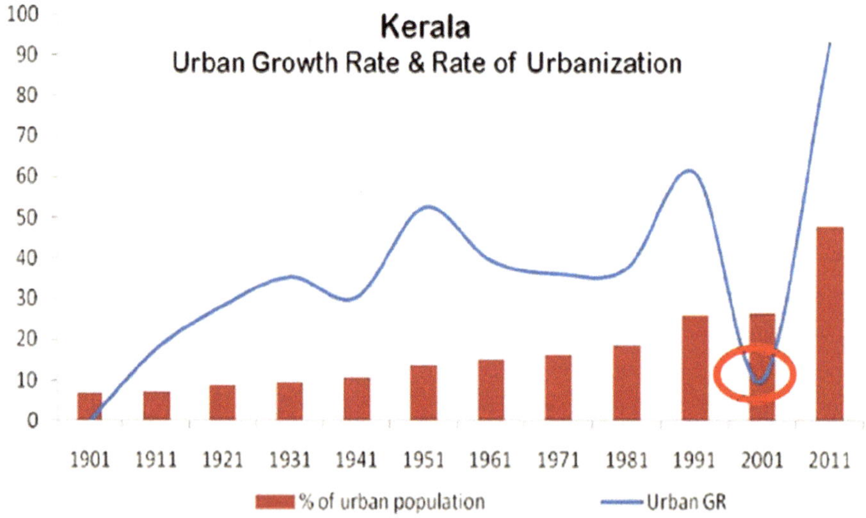

FIGURE 8.4 Kerala urban growth rate.

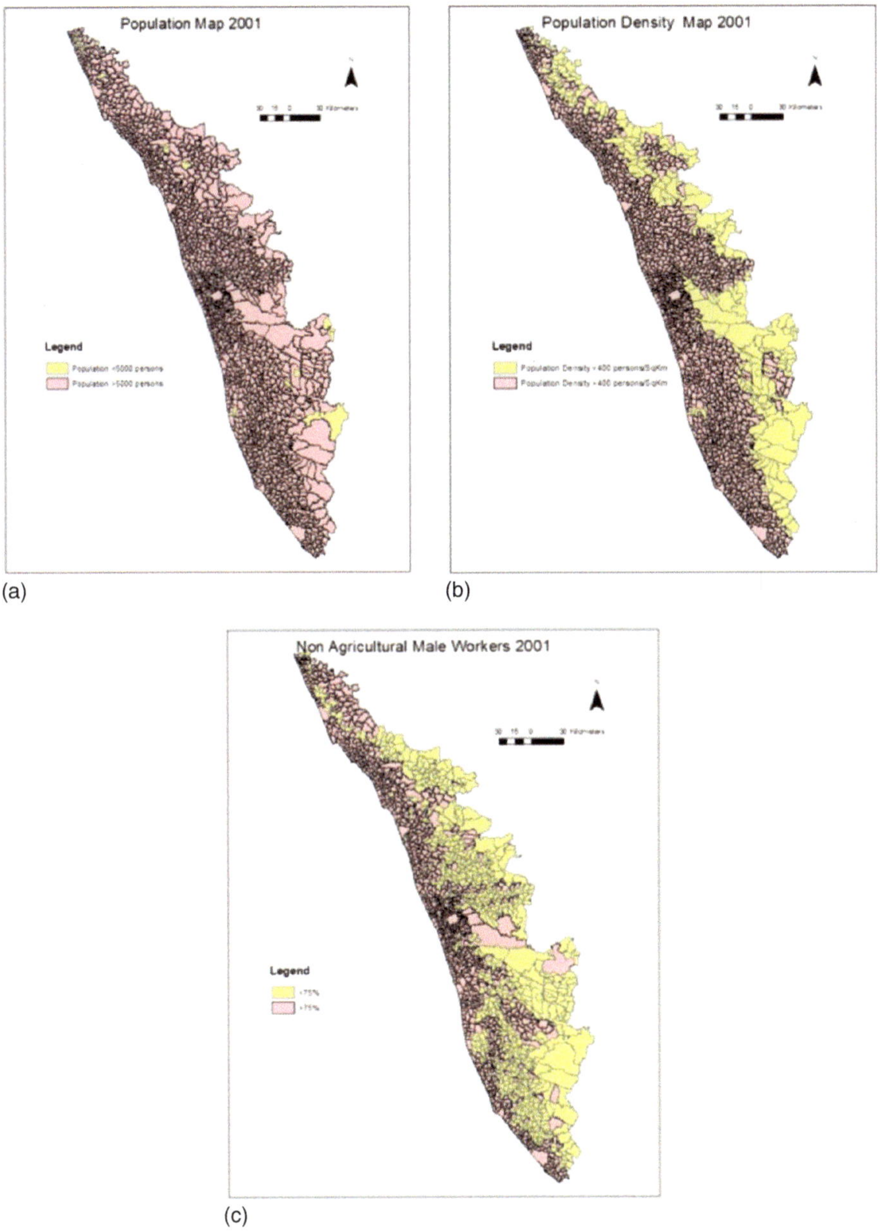

FIGURE 8.5 (a) LSGs satisfying population criteria, 2001, (b) LSGs satisfying population density criteria, 2001, (c) LSGs satisfying employment criteria, 2001.

urban according to census criteria. The settlements which satisfy the three criteria for urban (population, population density, and employment) are given in Figures 8.5a–c and Figures 8.6a–c for the years 2001 and 2011.

Analysis of 2001 census data finds that 1,447 (94%) settlements satisfy the population criterion and 1,337 (86%) satisfy the density criterion. Only 909 (59%) of settlements satisfy the employment criterion. The same pattern was found in 2011 except for the employment criterion (satisfied by 1,112 villages) with an increase of 14% from the previous decade.It can be seen that the

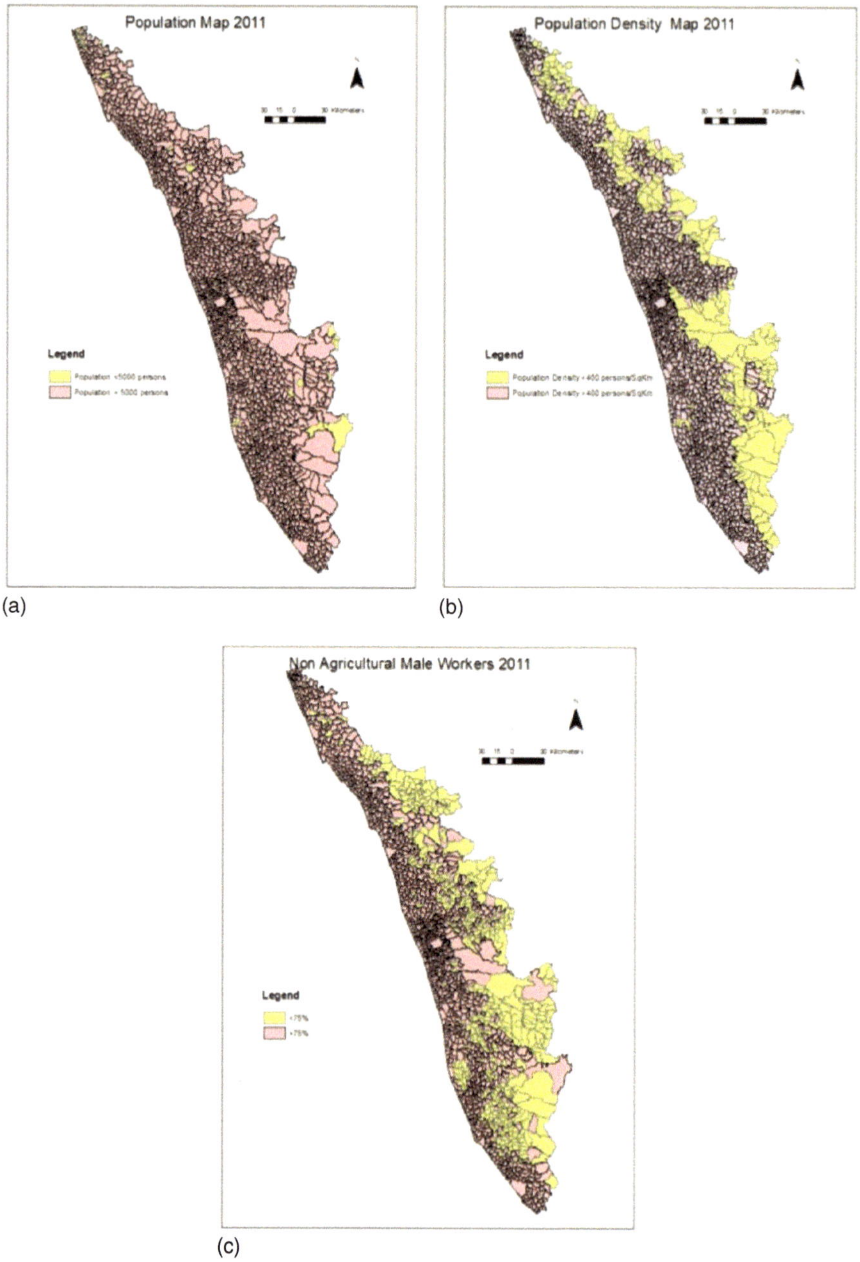

FIGURE 8.6 (a) LSGs satisfying population criteria, 2011, (b) LSGs satisfying population density criteria, 2011, (c) LSGs satisfying employment criteria, 2011.

employment criterion is the deciding factor for the identification of urban areas. The urban settlements satisfying all three criteria are shown in Figure 8.7.

The number of urban areas identified in the year 2011 according to census publication was only 520 instead of the 957 villages that satisfied all three criteria. The census considered villages urban if they satisfy the criteria for two consecutive decades. Hence all 957 settlements will not be considered for census town status until the census year 2021. Also, the state government will provide suggestions and directions for the census authority regarding the incorporation of a census village into the census town category.

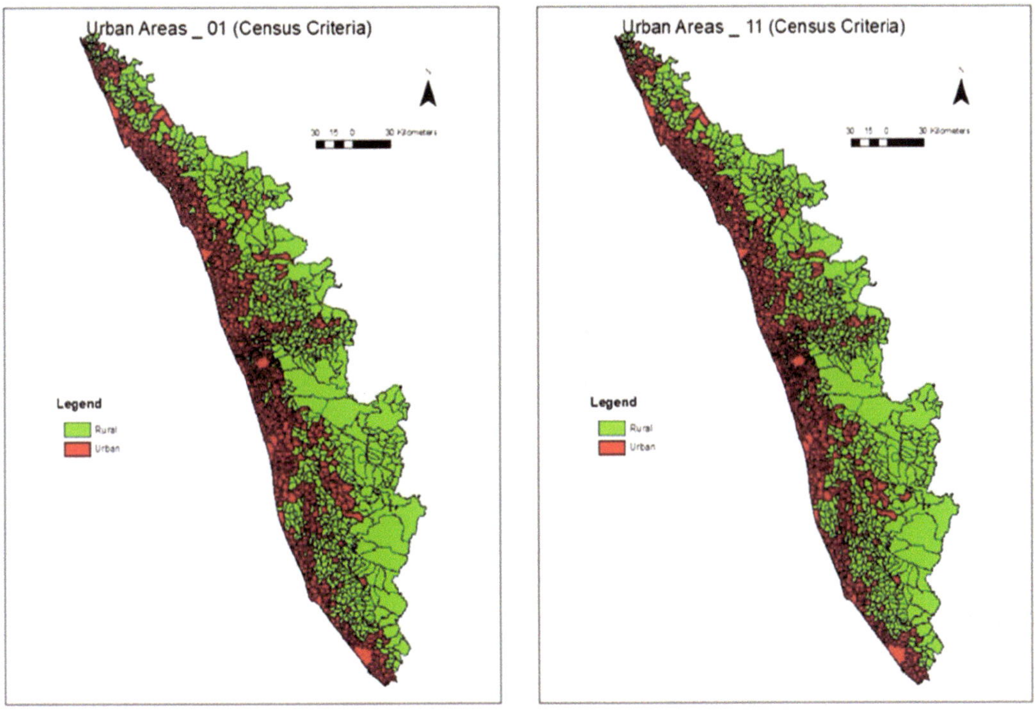

FIGURE 8.7 Urban areas as per census criteria 2001 and 2011.

8.3.2 Urban Areas in 2021

Three census criteria (population above 5000 persons, population density more than 400 persons per km^2, and more than 75% male workers in non-agricultural pursuits) for the identification of census towns were applied to the settlements using the projected population and employment values for the year 2021. Settlements satisfying all three criteria of population, density, and occupation were identified as urban areas in the year 2021. The details are shown in Table 8.2. Other than these three criteria, the census will consider settlements as census towns only if they satisfy these criteria for the two consecutive decades. The number of urban settlements as per census criteria in the year 2021 is 1098, which is more than 70% of the state. The number of settlements that satisfy census criteria in the year 2011 is already identified as 956. So if we consider the 'two consecutive decades' criterion of the census, the number of census towns in the year 2021 will be around 950, which is more than

TABLE 8.2
Possible Villages Satisfying Census Criteria in 2021

		Villages Satisfying Census Criteria Based on Projected Population and Occupation Data for 2021	
No	**Census Criteria**	**Number**	**Percentage**
1	Population >5000	1452	94.22
2	Density >400 Ppkm^2	1350	87.61
3	>75% male main workers	1244	80.73
	Total urban (1,2 & 3)	1098	71.25

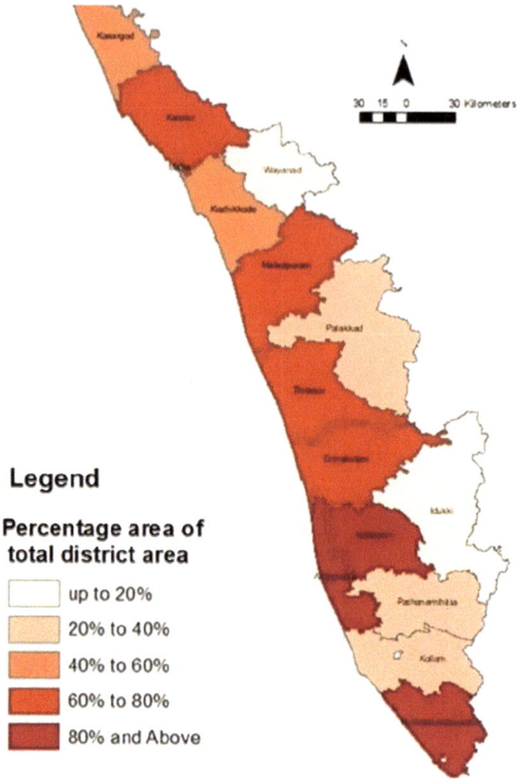

FIGURE 8.8 District-wise urban areas, 2021.

60% of villages in the state. So like past decades, in this decade also the growth of urbanization will show an increasing trend.

When district-wise analyses of these identified towns (in 2021) are conducted, it can be seen that half of the districts in the state have more than 50% urban areas within them and the Thiruvananthapuram district with more than 80% urban occupies the top position. Only the districts with a large extent of highland regions have less than 40% urbanized areas (Figure 8.8). From the last three decadal data, it can be inferred that the growth of urbanization in the state of Kerala is high. Also, initially, urbanization was mainly seen in the coastal plains of the district. In the last decade, it has spread to the midland regions and is now gradually encroaching on the highland areas of the state.

8.3.3 Population Density Variation

The population density of settlements for the last two decades was analyzed. The overall population density has not changed drastically from 2001 to 2011(Figure 8.9). A higher density is seen to be present near the coastline than in the highlands, which could be due to the presence of major urban centers in these areas. Also, the availability of major transport infrastructures such as airports, seaports, and railway lines in these areas will be an added reason for the increased population density. The population density has reduced in the major urban centers of the state. The top three urban centers in the state are Kochi, Thiruvananthapuram, and Kozhikode. Kochi city in Ernakulam district and Calicut city in Kozhikode district show a decrease in population density. But almost all surrounding local bodies show an increase (Figure 8.10). This is evidence for urban fringe development and the development of urban–rural interfaces in the state.

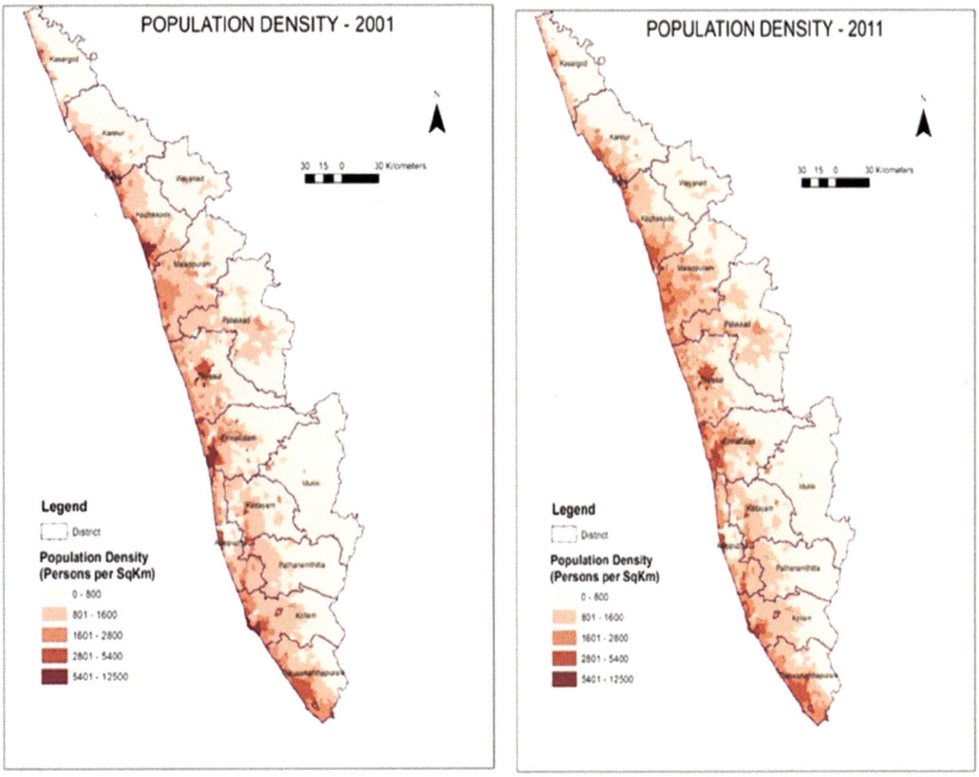

FIGURE 8.9 Population density maps, 2001 and 2011.

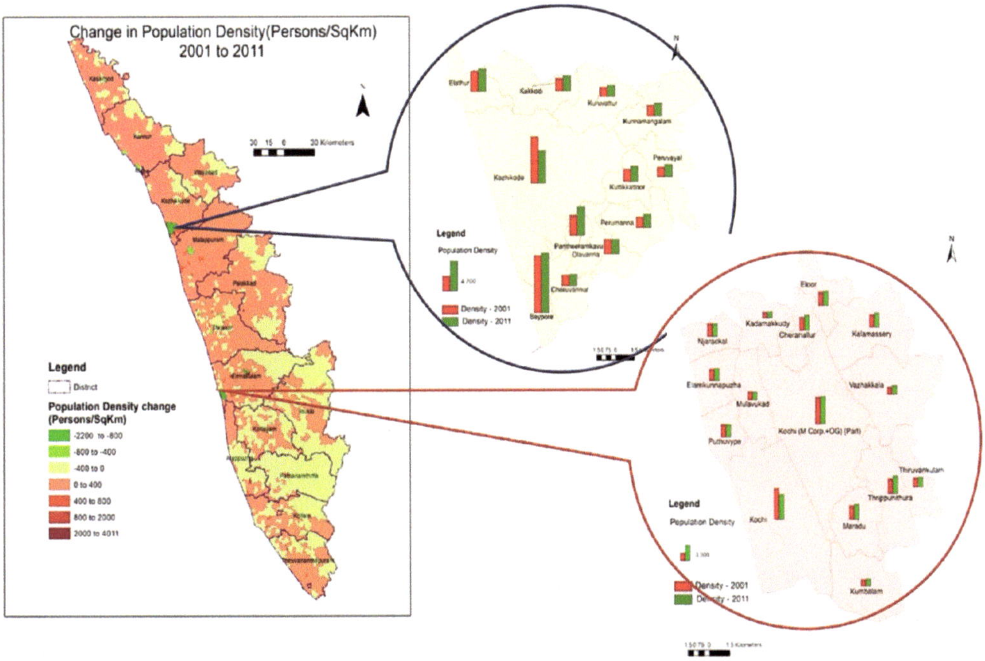

FIGURE 8.10 Changes in population density, 2001 and 2011.

8.3.4 The Functional Character of Settlements

Cluster analysis has been used to define the functional character of Kerala. From the analysis, three output classes were identified: high concentration of urban employment/activity for an urban area; low concentration of urban employment/activity for arural area; and insignificant or no concentration of urban activity/mixed employment/activities area (Figure 8.11).

The urban employment activities are mainly concentrated in the coastal region of Kerala. Some coastal regions like Alappuzha have an insignificant or no concentration of urban activity. The low concentrations of urban activities from the analysis results substantiated that the rural settlement and activities are predominant in the eastern side of Kerala. The functionality analysis

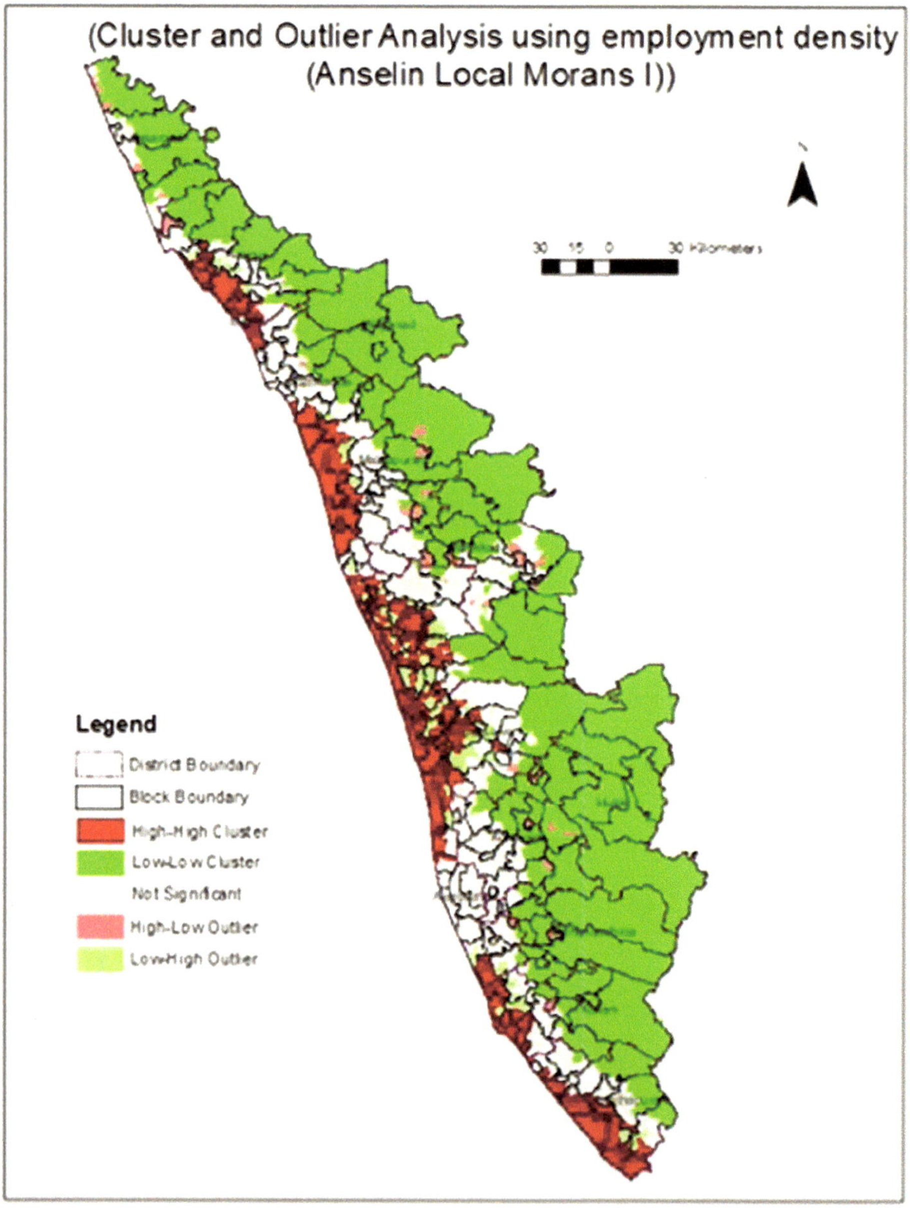

FIGURE 8.11 Functional character of settlement of Kerala (census villages 2011.)

shows five urban clusters along with the lowland which is surrounded by non-significant urban activity areas.

The high–high urban employment activity clusters were demarcated as urban areas and low–low clusters were demarcated as rural areas. The areas with insignificant urban activity are termed semi-urban areas. The settlement clusters with urban, semi-urban, and rural characters are shown in Figure 8.12. The urbanization trend analysis showed that the urban areas are rapidly spreading towards the midland regions of Kerala and then slightly towards the highland part. When the results of the urbanization trend analysis and functionality analysis are combined, it can be seen that the chances of conversion of the semi-urban areas in the midland portion to urban areas will soon be very high.

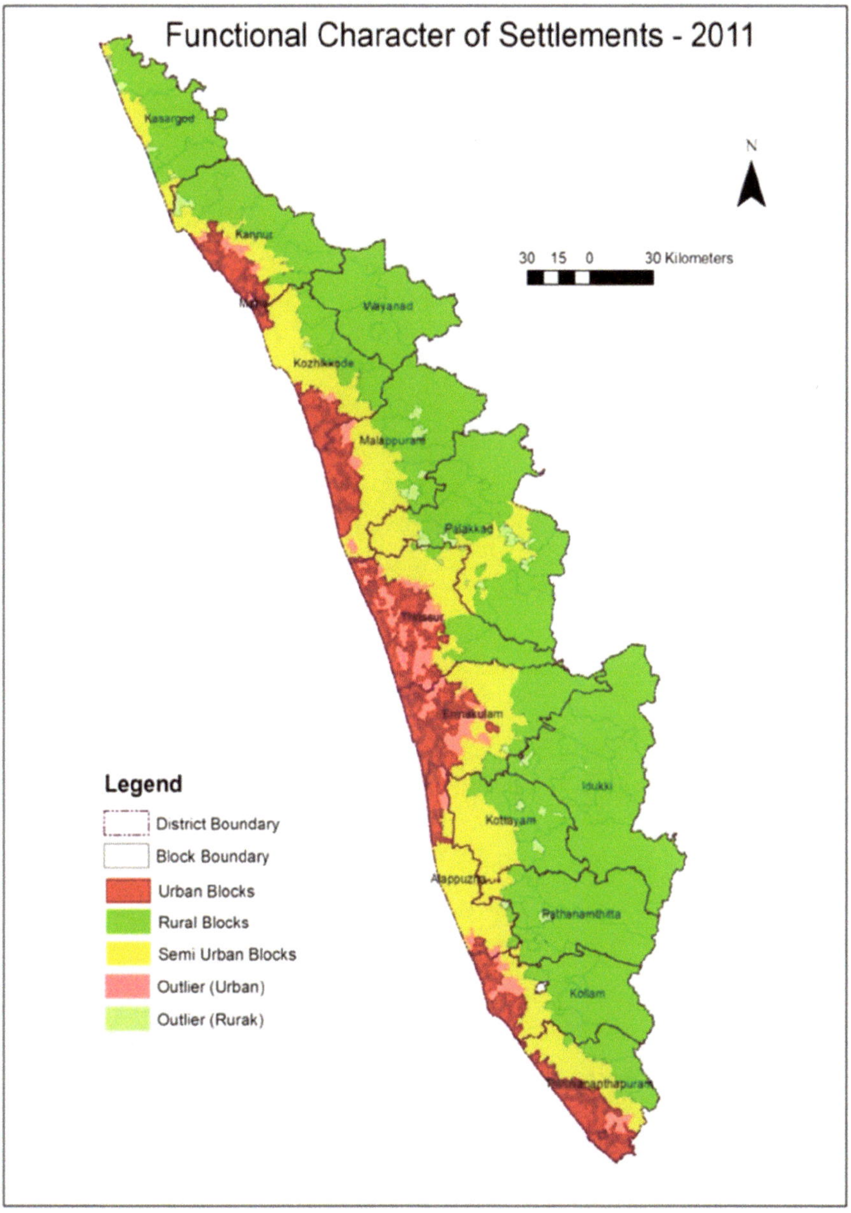

FIGURE 8.12 Functional character of settlement of Kerala, 2011.

8.3.5 Hierarchy of Settlements

Settlements are ranked based on their population size and density. Settlements with higher scores are mostly seen in the lowland area of the state (Figure 8.13).

The functionality-based rankings were obtained using the number and weighting of infrastructures in each settlement (Figure 8.14). Clustering analysis was also done to check whether any major clustering of facilities is present in the state.

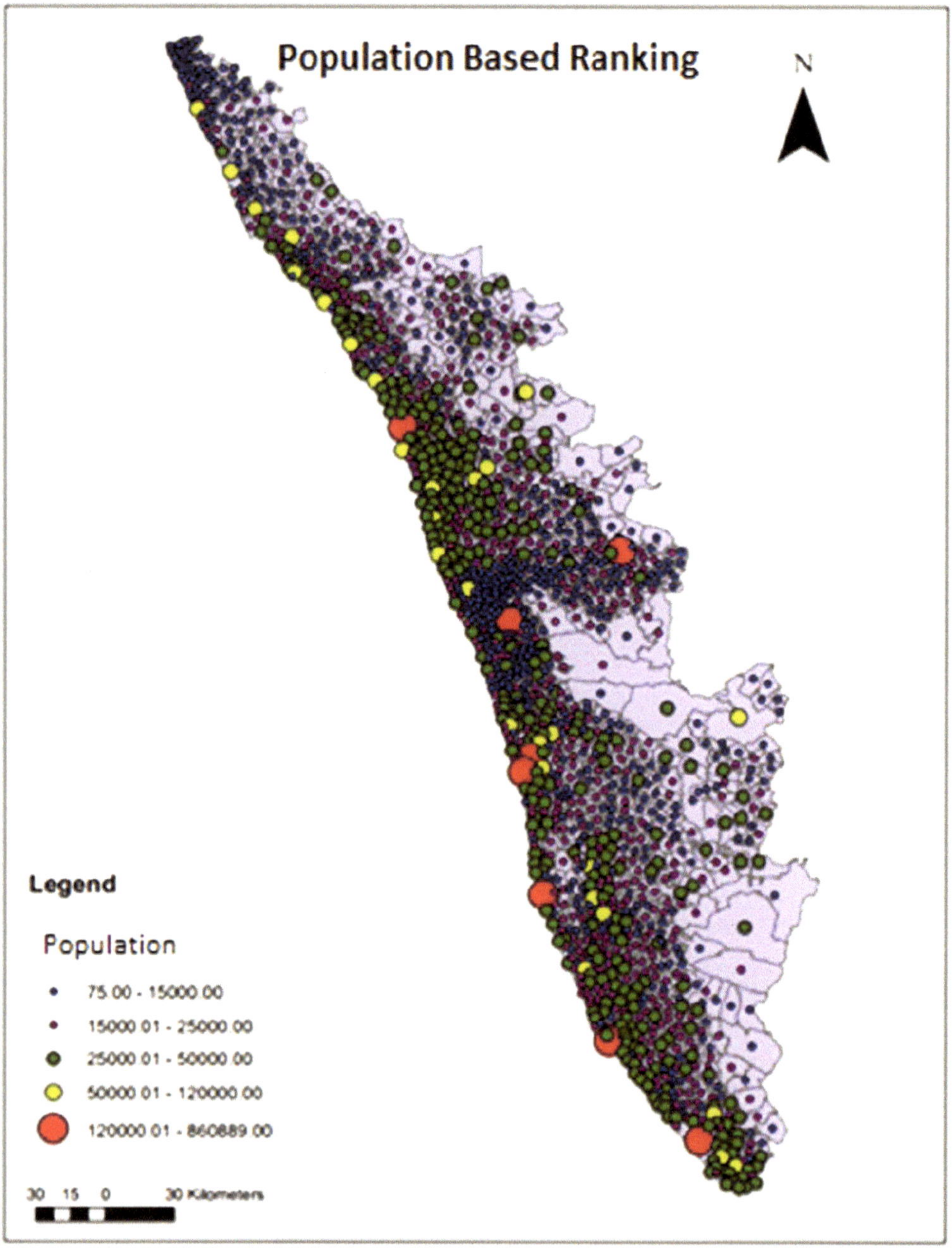

FIGURE 8.13 Population-based ranking.

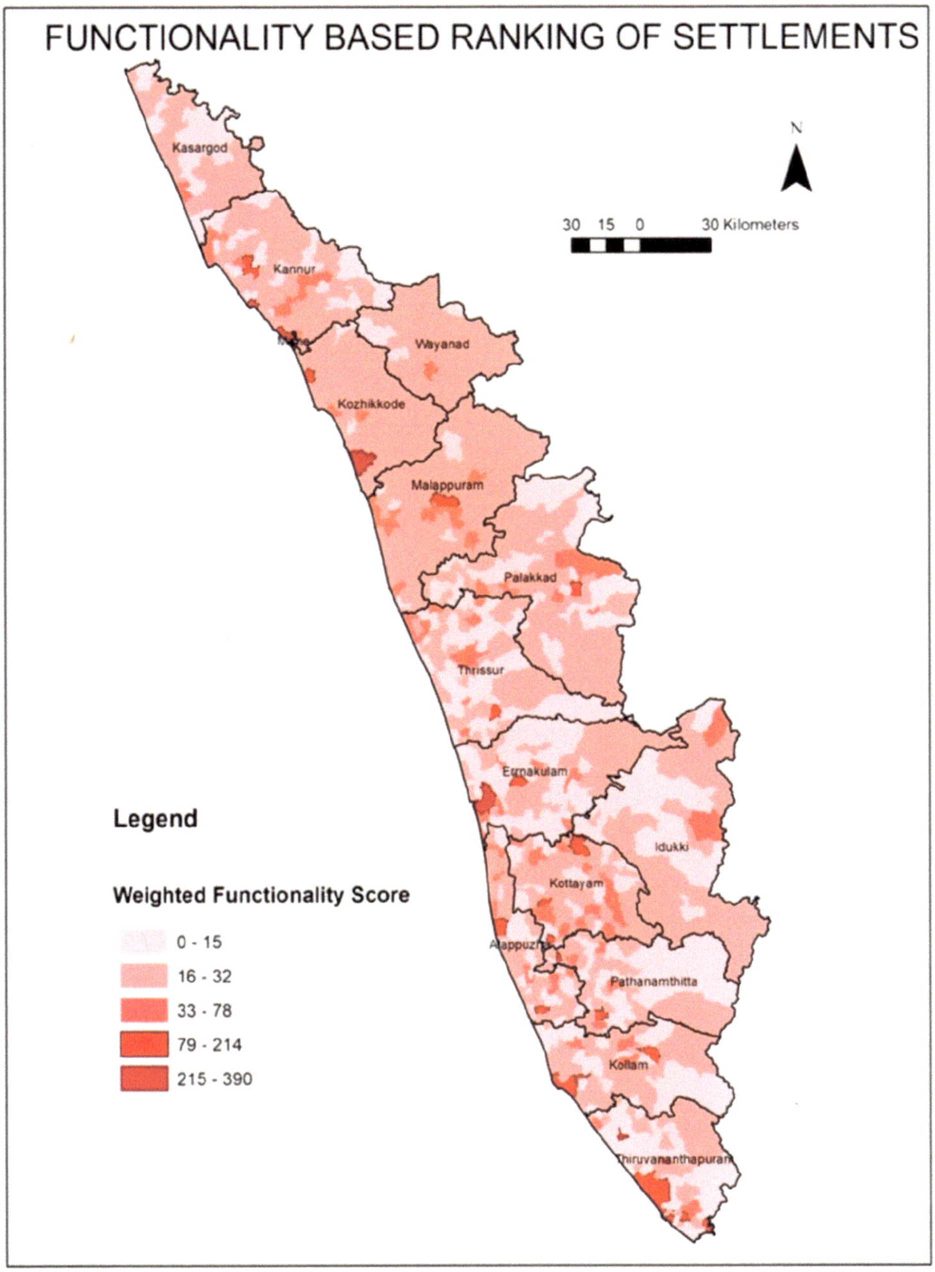

FIGURE 8.14 Functionality-based ranking of settlements.

The highland areas of Idukki and Ernakulam, having a functional score between 0 and 32, have significantly fewer facilities (Figure 8.15). But the population size of the settlements in these areas is also less. Most of the districts in Kerala have an insignificant concentration of facilities. It can be inferred that facilities are distributed evenly across the state.

Location quotient analysis of non-agricultural employment details was used for centrality analysis. Cluster analysis of LQ values was also conducted (Figure 8.16), which infers that the

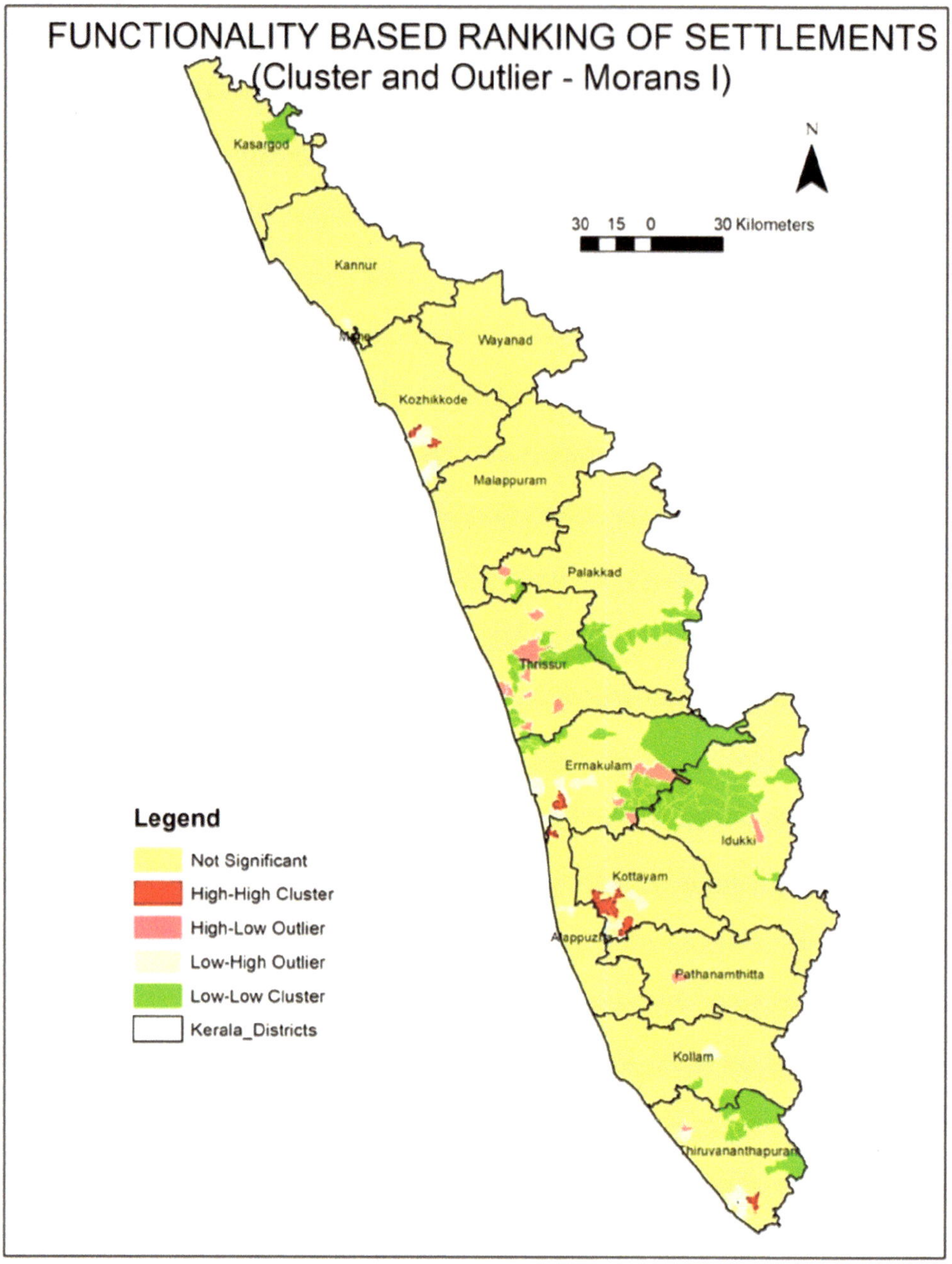

FIGURE 8.15 Moran's I clustering of functionality of settlements.

districts near the coastline contain predominantly non-agricultural workers, whereas areas like Wayanad in the eastern highland regions have a mixture of clusters.

The geographical terrain of the area determines the type of occupation. The lowland region of Kerala has predominantly non-agricultural workers,whereas, towards the midlands, concentration becomes very moderate. The midlands have an insignificant cluster type due to the mixture of the agricultural and non-agricultural workforce. Meanwhile, in the highland region, the concentrations

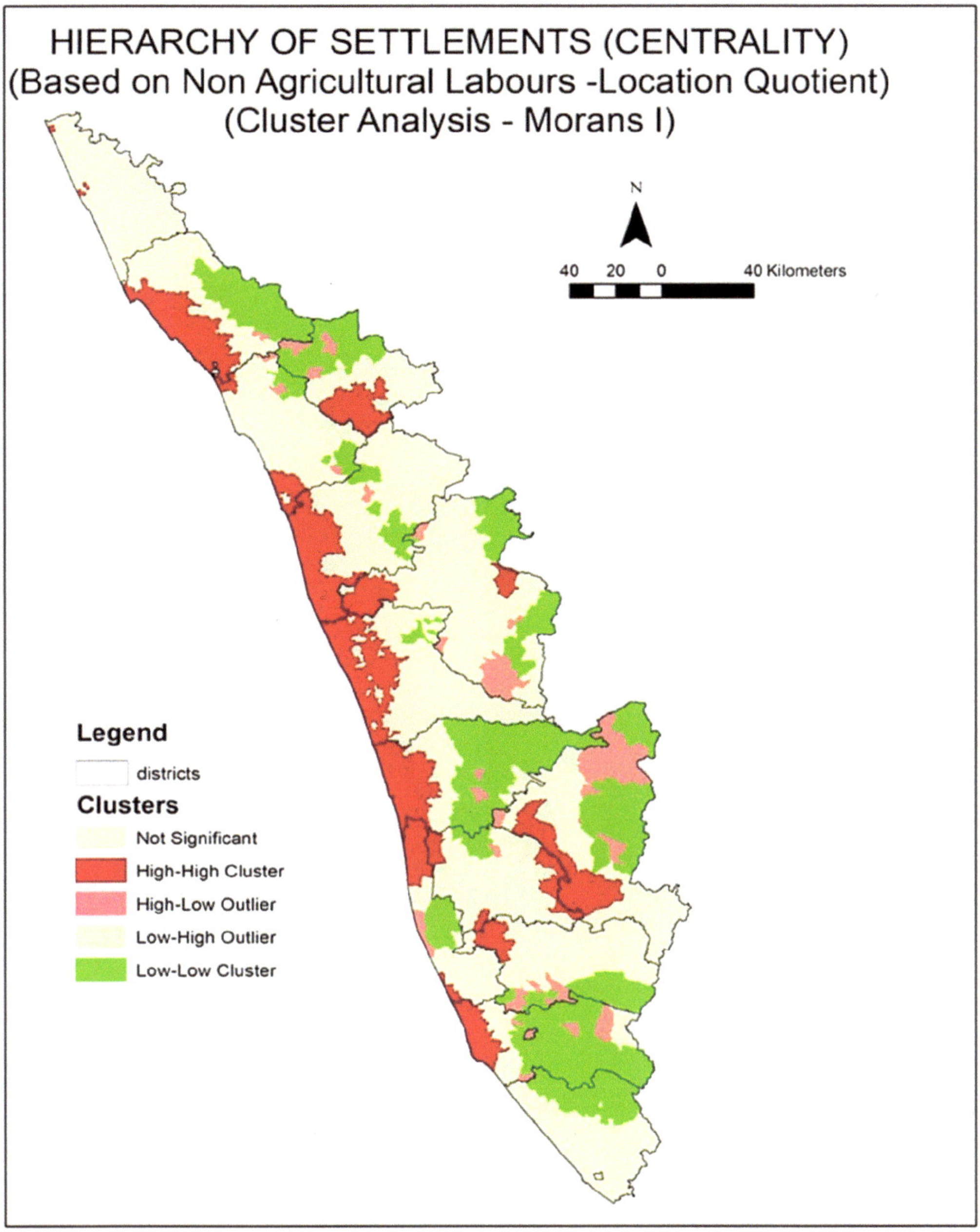

FIGURE 8.16 Hierarchy of settlements based on centrality.

of non-agricultural workers are deficient, but districts like Wayanad have a mixture of clusters, since the economy of the district depends on both agriculture and tourism.

For cumulative calculation, all the three scores of population, functionality, and centrality are normalized to 0 to 1 range and added together (Equation 8.3). A score above 2.1 (70%) out of a maximum 3 is considered as a first-order settlement and 1.5 to 2.1 (50% to 70%) as a second-order and below 1.5 as lower-order settlements. Figure 8.17 shows the hierarchy of settlements and the

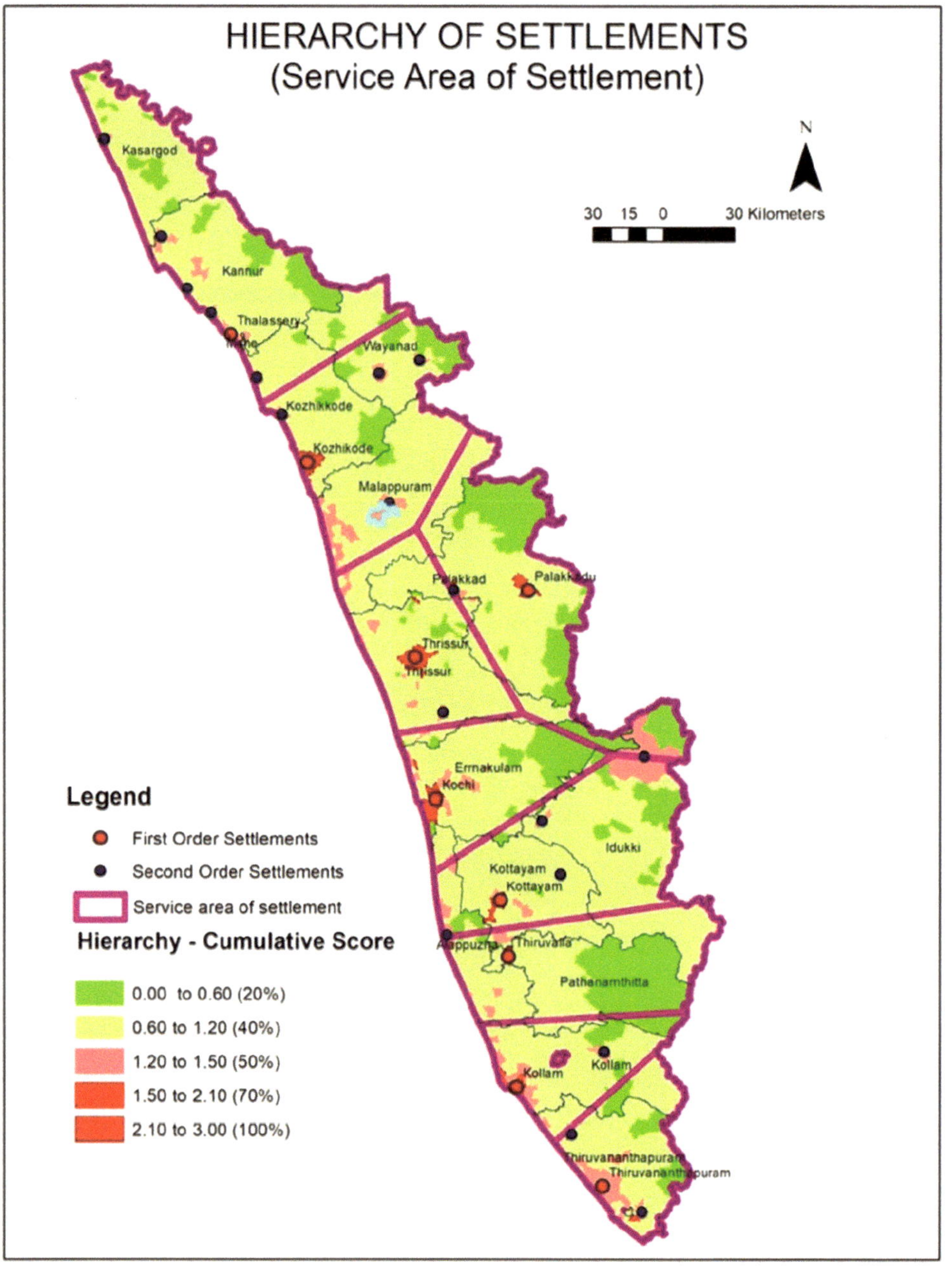

FIGURE 8.17 Hierarchy of settlements.

service areas of first-order settlements. Table 8.3 gives the service area of each first-order settlement. It can be seen that the service area of northern settlements is high compared to the settlements in the southern portion of Kerala state.

Even though the facilities are distributed evenly in the state, there are fewer first-order settlements in the northern part of Kerala than in the central region and southern part. There are more second-order settlements in the northern area. The presence of more higher-order settlements within small geographical areas also confirms the RUC nature of settlements in the state.

TABLE 8.3
Ranking of Settlements Based on Cumulative Score

No	Name of First-Order Settlement	Service Area of Settlements: Area (Km^2)	Percentage
1	Thalassery	6684.4	17.2
2	Palakkad	5246.5	13.5
3	Kozhikode	5246.5	13.5
4	Kottayam	5207.6	13.4
5	Thiruvalla	4080.6	10.5
6	Thrissur	3847.4	9.9
7	Kochi	3070.2	7.9
8	Thiruvananthapuram	2798.1	7.2
9	Kollam	2681.5	6.9
Total		**38863.0**	**100.00**

8.4 DISCUSSION

8.4.1 Settlement Characteristics of Kerala

It is understood from the settlement analysis that a linear, densely agglomerated and dispersed but interconnected spatial pattern of the settlement system is present in Kerala (Firoz 2015). A high rate of urbanization in the state is seen due to the shift of the labor force from primary to secondary and tertiary activities. The population density of major urban cores has reduced and that of their surrounding local bodies increased. Density has also reduced in the southern region compared to the northern and central areas. Even though the population density of the urban core is reducing, a high concentration of urban employment/activity is seen in five clusters around the major urban cores in lowland regions. The cluster consists of 319 local self-government institutions, which comprise an area of 5642 km^2.

The functional character of settlements cannot be limited to the binary of urban and rural. Semi-urban areas exist in most of the midland region. No major disparity is seen in the distribution of facilities among the districts. But the northern region has fewer first-order settlements than other areas. The higher-order settlements in these areas need to serve large hinterlands compared to their southern counterparts.

The analysis of urbanization, functional character and hierarchy of settlements confirms the RUC nature of settlement development in Kerala. This type of development will provide urban services in almost all areas in the state, ensuring access to state-of-the-art facilities and infrastructures for people in rural areas too. However, along with these advantages, the RUC settlement development has also caused some problems.

8.4.2 Issues of RUC Settlement Development

The major issues related to RUC developments are assessed under three categories, environmental, economic, and governance issues.

8.4.2.1 Environmental Issues

The temporal analysis of population density shows that the density of the urban fringe is gradually increasing. This will cause the development of more built space in this area and thus more pressure on the land. The exploitation of natural resources in this area will be high and the uncontrolled developments in these areas will cause loss of vegetation and natural habitat (Krishnan and Firoz 2020). High surface runoff of stormwater and lowering of the water table are additional issues. Altogether it will result in high levels of air, water, and soil pollution in these areas.

Urbanization has occurred in Kerala mainly due to the shift of employment from the primary sector to others. Due to the change in this occupational pattern, the land will be used as a real-estate commodity rather than for agriculture. This will increase the price of land in fringe and rural areas. Thus land will become unaffordable for poor people in rural areas for their housing needs.

8.4.2.2 Economic Issues

The distribution of population in all areas of the state will increase the per-capita cost of developing public infrastructures. High population density, mainly along the roadsides, may hinder infrastructure developments. This ribbon development pattern causes more demolition of houses and shops when road development or other proposals are implemented. Also, the high land values due to the urbanized nature of the area result in very high infrastructure development cost.

8.4.2.3 Governance Issues

The 73rd and 74th Constitutional Amendment Act identified and listed some functions and responsibilities related to governance for statutory urban and rural local bodies. Administrative staff are allocated based on the status of urban and rural local bodies (with more staff in urban local bodies). This has affected the administration of rural designated 'urban characteristic' settlements, whose status imposes limitations on their own fund raising. Also, there are differences in tax revenue slabs between *panchayat* and municipalities/corporations.

Another major issue caused by this RUC development in the governance sector is town planning-related activities. There are no planning measures to restrict the uncontrolled growth of areas in rural local bodies, as there are no statutory masterplans or detailed town planning schemes for controlling development. Also, Kerala *panchayat* building rules are invariably applicable to all *panchayat*s, irrespective of their development/urban nature.

8.5 CONCLUSION

From the study, it can be inferred that a rural–urban continuum type of settlement pattern exists across Kerala. The functional character of settlements cannot be restircted to the binary of urban and rural. Existing imaginary political boundaries have failed to delineate exact urban and rural areas. Also, in the case of settlements like Kerala, the exact division points of urban and rural areas cannot be identified. The urban–rural interface plays a major role in the peri-urban areas of major urban settlements. So the masterplan preparation and project formulations cannot be limited to political boundaries. Planning for rural–urban interfaces should also be included (Sruthi Krishnan and Mohammed Firoz 2021). An integrated approach to the preparation of spatial plans should be adopted, considering the urban core and urban–rural interfaces. At the microlevel, different development guidelines should be formulated for various areas, including building rules.

The basis of all these activities should be a new typology of settlements, which takes urban areas, rural areas, and urban–rural interfaces into consideration. Methods like 'degree of urbanization' may be used in Kerala for the delineation. Degree of urbanization is a delineation method based on population density, population size, and contiguity, using 1 km^2 grid cells (Statistical Commission 2020). The distortions caused by different shapes and sizes of villages can be avoided by using grid cells of identical shape and size. This is a considerable advantage over methods based on the population size and density of local administrative units.

There is a need for an alternate typological classification other than the dichotomous classification of urban and rural in Kerala. Such a classification could contribute to improved sustainable and equitable spatial planning and management in the state.

REFERENCES

Census of India. 2001. "General Population Tables." Office of the Registrar General and Census Commissioner, Government of India. https://censusindia.gov.in/2011-common/census_data_2001.html

Census of India. 2011a. Office of the Registrar General and Census Commissioner, Government of India. http://Censusindia.Gov.in/2011census

Census of India. 2011b. "Rural-Urban Classification for Census 2021." Circular No. 2, Office of the Registrar General, India. No 9/26/2018-CD, dated 04-09-2018.

Chatopadhyay, Sreekumar. 1988. "Urbanisation in Kerala." *Geographical Review of India* 50(2).

Chatopadhyay, Sreekumar. 1995. "Environmental Impacts of Urbanization." *Trans Ins Indian Geographers* 17(1).

East Cambridgeshire District Council. 2017. "Settlement Hierarchy in East Cambridgeshire."

Fee, Kyle, and Daniel A. Hartley. 2011. "Urban Growth and Decline: The Role of Population Density at the City Core." *Economic Commentary (Federal Reserve Bank of Cleveland)* (27): 1–6. doi:10.26509/frbc-ec-201127.

Firoz, Mohammed. 2015. "Reclassification of the Typology and Pattern of Composite Settlements: A Case of Kerala, India." Indian Institute of Technology, Kharagpur.

Firoz, Mohammed, Haimanti Banerji, and Joy Sen. 2014. "A Methodology to Define the Typology of Rural Urban Continuum Settelments in Kerala." *Journal of Regional Development and Planning* 3(1): 2014, no. September 2014.

Gazette, Kerala. 2016. "The Kerala Town and Country Planning Act, 2016."

Government of Kerala. 2012. "State Urbanisation Report."

Kallingal, Fathim Rashna, and C. Mohammed Firoz 2018. "Composite Index of Social Development: A Case of Kerala, India." *Proceedings of the International Conference on Working Class Districts/Urban Transformations and Qualities of Life in the Growing City*, Vienna, Austria, no. 1: 46–52. https://wiener-wissenswelt.fh-campuswien.ac.at/images/pdf/confbooks/UrbanTransformations2018A4.pdf

Krishnan, Sruthi, and C. Mohammed Firoz2020. "Regional Urban Environmental Quality Assessment and Spatial Analysis." *Journal of Urban Management* 9(2): 191–204. doi:10.1016/j.jum.2020.03.001.

Kumari, Anchal. 2015. "City Size Distributions and Hierarchy among Cities in India." *The Journal of Development Practice* 2(November): 26–34.

McGranahan, Gordon, and David Satterthwaite. 2014. *Urbanisation Concepts and Trends.* International Institute for Environment and Development. London. http://pubs.iied.org/10709IIED

Mundhe, Nitin, and Ravindra G. Jaybhaye. 2018. "A Spatio-Temporal Analysis of Urbanization in India." *IJRAR-International Journal of Research and Analytical Reviews* 5(4): 30–37.

Planning studio 3, Department of Architecture and Planning, National Institute of Technology Calicut, Kerala, India. 2020. "Planning for an Evolving Mega Region in Kerala, India."

Sruthi Krishnan, V., and C. Mohammed Firoz. 2021. "Impact of Land Use and Land Cover Change on the Environmental Quality of a Region: A Case of Ernakulam District in Kerala, India." *Regional Statistics* 11(2): 1–34. doi:10.15196/RS110205.

Statistical Commission. 2020. "A Recommendation on the Method to Delineate Cities, Urban and Rural Areas for International Statistical Comparisons." 3. https://unstats.un.org/unsd/statcom/51st-session/documents

Stewart Jr, Charles T. 1958. "The Urban-Rural Dichotomy: Concepts and Uses." *American Journal of Sociology* 64(2): 152–158.

Wheeler, James O. 2005. "Geography." *Encyclopedia of Social Measurement* 2: 115–123.

9 Self-Financing Urbanism in Developing Countries

A Ready Reckoner for Land Value Capture (LVC) Instruments

Jay Mittal and Scott Berson
Auburn University, Auburn, Alabama, USA

CONTENTS

DOI: 10.1201/9781003131922-11

9.1 URBANIZATION AND CHALLENGES

As rapid urbanization puts increasing pressure on developing-world cities, governments are struggling to secure adequate funds for capital-intensive infrastructure investments to meet the needs of their urban populations (Arfanuzzaman and Dahiya, 2019). The enormous urban infrastructure deficit and high-pressure urban growth have resulted in chaotic urban development and unsustainable sprawl, which threatens to erase traditional economic gains from growth while harming quality of life. Urban growth in south-east Asian cities is also highly unplanned, haphazard, and formed in a leapfrogged manner with "little concern for nature, life-supporting ecological systems, and the environment" (Arfanuzzaman and Dahiya, 2019). Access to affordable housing, easy access to work, efficient and equitable transportation networks are major challenges cities face to remain competitive as they grow (Ameen and Mourshed, 2017). Ensuring affordable housing for current and new residents is increasingly challenging due to a lack of serviced urban land, highly restricted land-use zoning and development regulations creating artificial scarcity and impacting affordability. In major urban centers, it also results in a high price-to-income (PIR) ratio – a ratio of median housing price and median income (Rizvi, 2016; Zhong and Li, 2016).

One of the biggest challenges for developing countries is the lack of financial resources for covering deficit infrastructure in water, sanitation, roads, affordable housing, and other utilities. The estimated cost required for urban infrastructure by 2050 in India alone is enormous. A government report estimated the urban infrastructure investment deficit at US$827 billion for the period 2012–2031 at 2009–2010 prices. This estimate was just to cover the minimum standards of basic infrastructure (Ahluwalia, 2019). Another report by McKinsey Institute (2010) estimated US$1.2 trillion to finance India's urban infrastructure deficit by 2030. The infrastructure deficit is enormous in India and several other developing-country cities and needs special attention, given the high degree of urbanization. For example, a UN World Urbanization Prospects: 2018: Highlights, United Nations, Department of Economic and Social Affairs, and Population Division (2019) report projected that 416 million people will be added to Indian cities in the next 30 years, while the rural population will decline by 111 million. This kind of urban growth will create enormous demand for serviced land and housing and will stress urban infrastructure and municipal budgets.

9.2 THE NEED AND RATIONALE FOR LAND VALUE CAPTURE (LVC)

The value of a particular piece of land is innately tied to its locational context. The more convenient and connected a piece of land is to the surrounding metro, and the more high-quality services it is provided with, the more likely it is that people will bid up the price. A robust network of quality urban infrastructure and efficient public services can enhance economic productivity, improve accessibility, and thus increase locational desirability for commercial and residential uses (FHWA, 2019). These desirability factors are reflected in higher demand for those locations and higher value for the land. When public agencies invest in a neighborhood through infrastructure improvements and redevelopment, or make zoning and land-use changes that increase the investment potential, the economic value of land rises. These "unearned value" gains are a windfall for private owners, who see an increase in the value of their property even though they did not make any specific investments (Germán and Bernstein, 2018; Peterson, 2009). Land value capture policies aim to "recover and reinvest" part of those private value gains for public benefit, rather than allowing them to fully accrue to private developers. LVC instruments promote equity and economic efficiency by following the "beneficiary-pays" principle (FHWA, 2019).

Land value capture seeks to allow governments to at least partially self-finance infrastructure improvements by leveraging the incremental land values which those public investments or public actions have added to the surrounding developments. Several cities in India and other parts of the developing world have started focusing on innovative self-financing instruments (Mittal, 2014; Mittal and Kashyap, 2015; MOUD, 2017; Peterson, 2009, 2013; Sharma and Newman, 2018). There are several LVC financing instruments, including tax increment financing (TIF), land readjustment (LR) or land pooling (LP), transferable development rights (TDR), development exactions and impact fees (DE&IF), and others as discussed in the chapter.

9.3 LAND VALUE CAPTURE INSTRUMENTS

While there are many permutations of LVC instruments, and innovative new value capture instruments are developed regularly, this toolkit focuses on seven important instruments which have a relatively stable body of research and implementation.

For each LVC instrument, a basic explanation details the essential concepts of the intervention. Subsequent sections detail the legal and regulatory underpinnings of the instruments, the broad methods of valuation and calculation of fees or assessments, and a primer on debates or controversies surrounding the instruments.

This chapter is intended to provide an initial set of LVC instruments in the LVC practitioner's toolkit and allows mixing and matching of more than one LVC instrument. A few LVC instruments are new to developing countries, and the current legal systems may not allow their implementation, but the chapter provides details from legal, financial, and application perspectives, so that local jurisdictions can consider how to apply the concept in their local context.

9.4 DEVELOPMENT EXACTIONS AND IMPACT FEES (DE&IF)

9.4.1 What Are They?

Development exactions and impact fees (DE&IF) are "front-end," "up-front," one-time charges levied on a new development in the local land market. They are also called "linkage fees," and they are intended to make developers pay for the costs of service expansion directly. It is a one-time charge that a city levies on a developer, in exchange for the right to build on a site, the revenue of which is used to finance either on- or off-site public infrastructure (Peterson, 2009). The fee is popular in rapidly growing communities in the United States where local citizens do not want to bear the cost burden of new entrants. Impact fees often face lower public opposition than other forms of revenue generation, such as property taxes, because they place the burden on developers, rather than the public (FHWA, 2019).

The principle behind the DE&IF charges is that new residential developments, especially, do not pay for their fair share of the burden on the community and developers should pay an upfront cost to cover the added costs to the community that would come with the project and its new users (NAHB, 2016). These could include expanding facilities like water and sewage treatment or financing new facilities like fire stations or supporting schools (Evans-Cowley, 2006). Additionally, implementing more costly impact fees toward the peri-urban area of a city and keeping them low or nonexistent in areas where the urban infill is desirable may incentivize developers to focus construction activity on denser areas that are easier to service.

The DE&IF can be paid as a land contribution (typically for roads, parks, schools), or a one-time payment. The on-site public investments could include improvements such as streetlights, traffic lights, turning lanes, access ramps, electrical or sewer connections. Cities can also charge cash payments that go into a dedicated city fund for infrastructure improvements. Revenue from the fees, however, could also be used for off-site public projects that would be affected in some way by the new development. The off-site improvements may include developing nearby roads that would absorb traffic from the site, schools, public safety buildings, trails, or other projects (Center for Land Use Education, 2008; Germán and Bernstein, 2018).

Impact fees have a long legal history and are central to two landmark planning cases from the U.S. Supreme Court: Nollan vs. California Coastal Commission, 483 U.S. 825 (1987), and Dolan vs. City of Tigard, 512 U.S. 374 (1994). These cases established standards for the development exactions that required local governments to ensure that fees collected are "roughly proportional" to the costs and "dual rational nexus" meaning that a) impact fees must be spent to minimize the impact of the new development on the city's infrastructure, and b) the fee must be used to benefit the new development itself (NAHB, 2016). For example, a developer of a car wash is not required to donate funds to improve a school building ten kilometers away. However, a developer of single-family homes (SFHs) may be required to do so, because the children living in the new homes will bring an extra burden on the school system.

9.4.2 How Are They Implemented?

In the United States, many states have enabling legislation that allows impact fees and prescribes procedures, and classifies the type of infrastructure that may be improved through the use of impact fees, how charges should be levied, etc. (NAHB, 2016). The DE&IF can be part of the local zoning code or could be used as a condition for rezoning or variance. Once impact fees are authorized at the state level, cities can implement the fees by writing an impact fee ordinance and passing it through the local government. The ordinance should include a schedule of improvements to be made in each area where fees are to be charged and include a plan for collection and distribution of the funds to ensure they are used specifically for improvements justified by the new developments. The creation of the ordinance should be conducted with the inclusion of key stakeholders in the community, including developers, utilities, and local, regional, and state agency representatives, if applicable. See Figure 9.1 for the implementation process.

9.4.3 How Are They Calculated?

Impact fees may be calculated in two different ways: inductive and deductive (Ross and Thorpe, 1992). The inductive calculation estimates the cost of a new improvement in the community due to additional users, such as a new water tower per x residents, and then it divides the cost proportionally for each new development or user until the improvement is made. The process starts again with a new project. This method has advantages of simplicity but may not reflect the unique circumstances and needs of individual neighborhoods on the ground. The deductive calculation is more complicated and administratively costly but allows those unique needs to be taken into consideration in a planned way. For fee calculation, the city reviews its current land-use plan and zoning codes, determines areas where there may be pressure on services, and prices its fees accordingly. These calculations are updated periodically as specified in the ordinance.

9.4.4 Debates and Controversy

There is a significant conflict in the literature about the efficacy of impact fees. Impact fees and exactions have been used for decades to finance infrastructure improvements, but there is continuing debate over when to use them and what effects they have on the larger development market, particularly when applied to new housing stock.

Critiques of the DE&IF argue that exactions and impact fees depress new entrants and developments and inflate the new development's cost, resulting in higher property prices for the end

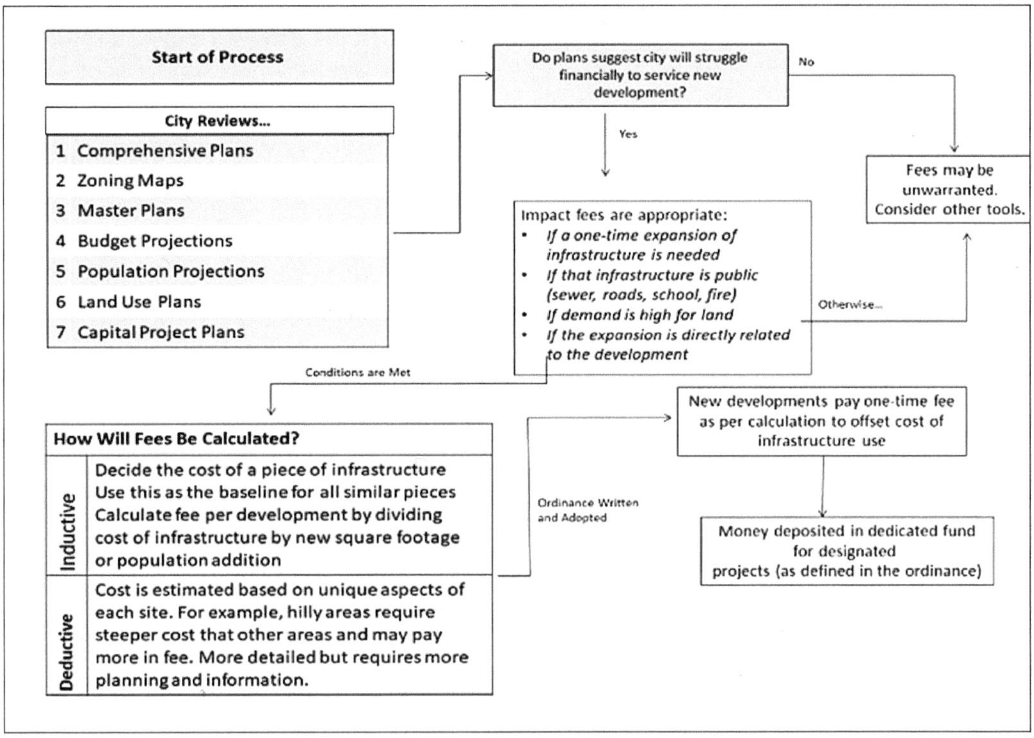

FIGURE 9.1 Development exaction and impact fees (DE&IF).

(Source: Authors.)

consumers (Evans-Cowley et al., 2009; Mathur et al., 2004; NAHB, 2016; Singell and Lillydahl, 1990). The price rise due to impact fees affects developers' profitability, and they may either steer their project to a different jurisdiction or pass the increased costs to the consumers. Supporters of impact fees have argued that DE&IF promotes growth and can be an effective instrument for mitigating the externalities associated with private new developments, facilitates jobs and new developments, increases the supply of buildable land, and ensures benefits to those who pay them (Burge et al., 2007; Nelson and Moody, 2003).

Impact fees have different effects on commercial and residential uses. Burge (2014) found that impact fees for schools decreased the price of residential land but increased it for commercial land, while impact fees for water and sewers had no effect on the commercial land but decreased the price of undeveloped residential land. Burge also found that impact fees for roads, parks, and other general infrastructure depressed the value of commercial land more than residential. The reasoning appears to be based on whom the results of the fees benefit: educational impact fees are paid by residents, but commercial businesses reap benefits from a higher-educated workforce. On the other hand, fees for water or sewers depress the value of undeveloped land, as they simply add an extra cost and provide a more limited benefit.

Efforts to use impact fees to lessen sprawl can also backfire, according to research from Burge et al. (2013), which found little evidence that impact fees increased core development and discovered negative spillover effects wherein sprawl development simply moved to other communities. A more regional program, or a coordinated effort among neighboring cities, may reduce this effect. It may be important to plan for equity goals by making these fees flexible depending on usage. For example, municipalities may exempt affordable housing units from the impact fees, allow payment plans, or vary the cost by the intensity of land use.

9.5 TAX INCREMENT FINANCING

9.5.1 What Is It?

Tax increment financing (TIF) is a popular economic development instrument of financing infrastructure, initiating area redevelopment, and stimulating growth and private investments in underperforming areas. TIFs are used to restore run-down, blighted areas, in historic preservation, for conservation, or jump-starting development in economically challenged neighborhoods. TIFs use two funding mechanisms: bond funding or "debt financing" by accessing the capital market; and "pay as you go" financing, when a developer undertakes some risk by investing equity in the development project, while the city creates the TIF and reimburses the developer as funds are generated by increased property values.

The financial success of a TIF depends on the local market conditions, and its viability is highly reliant on the anticipated increments in the property values in the future to pay off the costs or incentivize developers today. Cities' efficient fiscal management, creditworthiness, and high ratings play important roles in the success of a bond. Some challenges to TIF come from statutes that limit the amount of debt a municipality can take on at any one time, which will be important for cities to consider in the preplanning stage (Selby and Hunter, 2004).

TIF started in the United States in the early 1950s in California, and now over 10,000 TIF districts operate in all states of the US except Arizona (Merriman, 2018; Schneider, 2019). The city of Chicago has used the TIF instrument extensively with 136 active TIF districts as of 2020, and 184 TIF districts created since 1983. Chicago generated over US$840 million in increment, annually (City of Chicago, 2020). TIF in its variant forms is also used in Canada, Europe, and the United Kingdom, and is being explored in developing-world cities.

In the TIF process, cities take on debt to finance infrastructure improvements on the assumption that future values will rise. The increased property values and taxes (increments) resulting from the new investments (and new development attracted by that infrastructure) are then used to pay down the initial debt over the life of a TIF – 20 to 30 years (Farris and Horbas, 2009). The TIF expires after 20 to 30 years, and the city starts collecting full property taxes (base + increment) of the parcels to its general fund. Cities can also offer incentives to lure private developments and investments to the TIF area; as they do so, theoretically, property values in the TIF increase, and the increment is earmarked to pay off the bonds, incentivize businesses, or in continuous improvements. The taxes on the base value continue to be directed to the city's general fund, while just the incremental taxes are collected in a separate fund (Greuling, 1987).

9.5.2 How Is It Implemented?

As Figure 9.2 shows, TIFs must be first authorized through enabling legislation at the state level or its equivalent, after which it may be implemented by local governments. In the United States, the enabling State Acts authorize and set qualifying conditions for the use of TIF. The city determines if the targeted area qualifies for the preset conditions as per the state's Act for TIF. The mechanism works by first identifying an area for redevelopment and designating it as a TIF district. The city evaluates areas deemed blighted, stressed, or in a state of disrepair, and if the designation matches the requirements set by the state laws, the city establishes a TIF district, and the district boundaries are demarcated. This is done following a public engagement process and a vote by the city leadership (Center for Land Use Education, 2009). At this time, the base property value, known as the equalized assessed value (EAV), is noted for all properties in the TIF. The EAV is the initial taxable property value before any TIF-related initiative is undertaken, and is frozen on the city's general tax roll. The city continues to collect EAV-based property taxes for its general funds, but all future incremental property taxes (IPTs) over and above the EAV are reinvested in the TIF area. The IPT is spent funding public infrastructure such as capital construction, demolition, clearing, environmental contamination, grading, renovation, repair, placemaking, etc., to compensate (subsidize) private

TIF

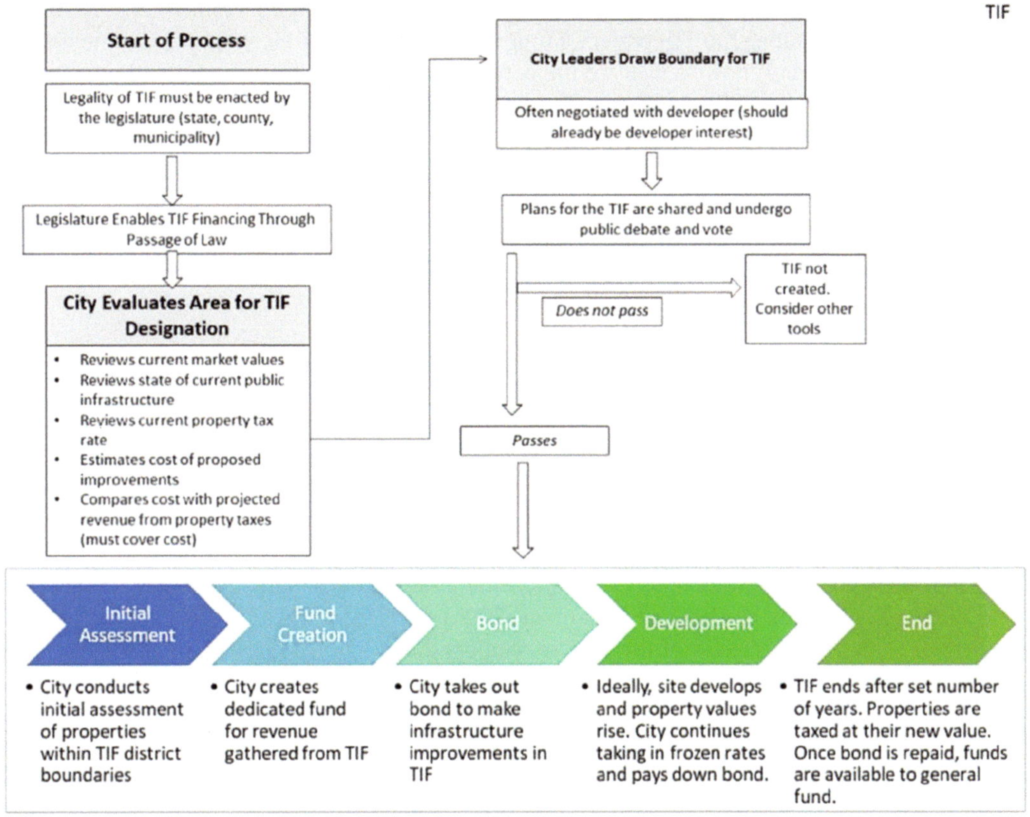

FIGURE 9.2 Tax increment financing (TIF).

(Source: Authors.)

developers for investments (on a pay-as-you-go basis), and to provide collateral and repayment for bond-based debt financing (Selby and Hunter, 2004). The city issues TIF bonds to pay for the infrastructure investments in the TIF area and begins to make improvements. Once the financial obligations are met, TIFs expire in accordance with the State Acts. If a TIF district has not undertaken any redevelopment project for a specified time, the city has to repeal the TIF. Several major private projects in the United States such as Hudson Yard in New York City and Amazon's Hq2 in Arlington, Virginia, have received TIF funding support.

TIF financing follows a "but for" argument – meaning, the assumption is that new development would not have occurred "but for" the investments the city made in the TIF area. The exact mechanism for determining this assumption is spelled out in the statute (Greuling, 1987). Thus, any future increases in the taxable property value are assumed to be due to the new infrastructure improvements and are diverted into a TIF fund to either pay for the continued improvements or to pay down the debt (Dye and Merriman, 2006; Farris and Horbas, 2009). The city periodically assesses property values and adjusts its tax rates accordingly during the lifetime of the TIF.

9.5.3 How Is It Calculated?

Before implementing a TIF district involving bond funding, cities undertake a bond feasibility study estimating the expected future property values across the lifetime of the TIF. The revenue from the tax increment is expected to pay for debt servicing (principal and interest) of the bond that is used to fund the improvements in the TIF district.

The value increment is a key part of the TIF calculations, as only the increment is used for servicing the bond and funding the TIF project. Future projections of this increment over time through the use of discounted cash flow to analyze the performance of the TIF are performed by qualified assessment staff in the feasibility analysis. These estimations, assumptions, and analysis of realistic market conditions are key to determining whether the TIF project can become self-financing. Typically, the city's professional assessment staff undertake the process of future projections of expected property tax receipts. They consider the current and projected tax revenues and compare them to the expected debt service cost of a bond. This analysis may also include the "but for" analysis that also determines how much of a "gap" there is for a development project to only be possible with assistance from the TIF (Sullivan et al., 2002).

9.5.4 Debates and Controversy

The debates related to TIF are around lack of fiscal transparency, undue incentives to private investments, lack of fiscal oversight, and potential for gentrification (Merriman, 2018). Additionally, it remains unclear how much growth in TIF-designated areas has occurred just because of the TIF designation – and if there is a difference, whether that difference is statistically significant enough to justify the use of the TIF designation. Some studies have found no significant difference in economic indicators for TIF developments (Agboola et al., 2018; El-Khattabi and Lester, 2019; Hicks et al., 2019). Critics such as Dye and Merriman (2006) have found that TIF districts can depress commercial values in other areas of the city, as retailers direct their business to the TIF area at the expense of other locations.

Other studies point to the importance of the actual amount and type of capital investment provided by the municipality as a key determinant (Drucker et al., 2019; Overton and Bland, 2014). Still others (Greenbaum and Landers, 2014) conclude that the "but for" analysis must be considered by asking whether the diversion of funds to, say, local schools, is worth the theoretical growth in the overall tax base over the length of the TIF period. These questions are highly dependent on local context and should be considered based on the priorities and needs of each city's unique situation.

9.6 SPECIAL ASSESSMENT DISTRICTS AND BETTERMENT LEVIES

9.6.1 What Is It?

The special assessment district (SAD) is also known as benefit assessment district (BAD) or local improvement district (LID). These quasi-government sub-districts are created to finance urban infrastructures such as mass transit lines (transit districts), water and sewerage lines (sewer districts), roads, parks, trails, and business areas (business improvement districts) in the United States (Mathur et al., 2004). SADs explicitly charge an extra fee called a betterment levy over and above the prevailing property tax to the direct beneficiaries of public action or investment (Peterson, 2009). Private developments near a transit line, for instance, may be assessed a betterment fee to help defray the costs of maintaining the transit line. It is the charge on the gain in unearned property value that the private development accrues due to added convenience simply by being near a transit line. The value gains are significant and have been quantified to varying degrees by several studies (Li, 2018; Yang et al., 2020; Zhong and Li, 2016; Zolnik, 2020).

One thing to keep in mind when considering a SAD is that the affected properties must be benefiting from the tax they are paying. The extra levy funds cannot be used to, say, fund a redevelopment project across town. In this way, SADs are subject to a similar "rational nexus" standard, much like DE&IF. They must show a clear and logical connection between the fee charged and the development project that the fee is used to fund (Burger, 1993).

Property owners in a designated SAD area pay an extra fee on top of the base property taxes based on location, size, and assessment of benefits. The rationale of the fee is that the properties in a SAD accrue direct benefits and value gains through access to transit, better services, or recreation facilities as provided by the quasi-government special district, and nearby private properties thus can be charged above and beyond regular property tax levels (Zhao and Larson, 2011).

9.6.2 How Is It Implemented?

As described in Figure 9.3, the ability to form a SAD is enabled at the state level, after which a local government may impose the additional tax. It may also be enacted through the petition of most property owners in an area who decide to fund an improvement (Zhao and Larson, 2011). Like other instruments, implementing an SAD requires public support and engagement with public notices, periodic meetings, and hearings, which will likely have different requirements by geography (California State Legislature, 2011). Many statutes require property owners to vote on the creation of the SAD (Zhao and Larson, 2011). Some states allow special assessment funds to be used for general improvement funds, such as street cleaning, graffiti removal, beautification, and more (Kogan and McCubbins, 2008). These are sometimes also called Business Improvement Districts (BIDs).

9.6.3 How Are They Calculated?

The dollar amount of the betterment levy can be calculated in several different ways, as Hayes (1977) and Mathur (2015) have described. Different betterment levies may be charged to different properties within the same SAD depending on its use and location. For example, a commercial establishment may be charged additional taxes because it experiences increased foot traffic due to passengers traveling on an adjacent transit station (Bladikas and Pignataro, 1990). The cost may be

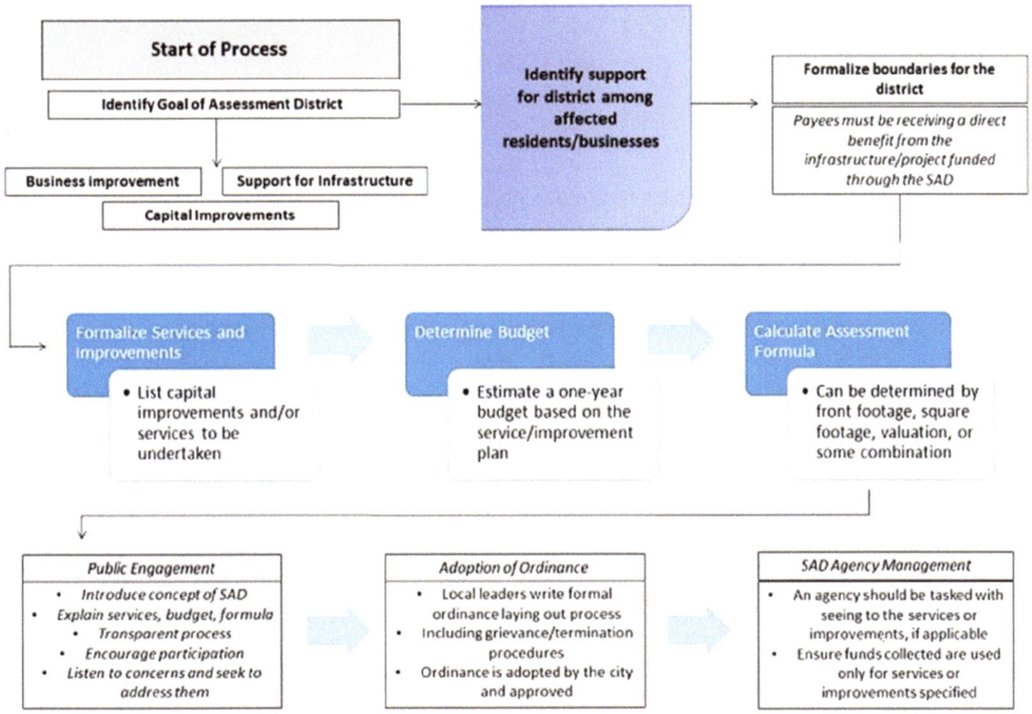

FIGURE 9.3 Special assessment districts (SADs).

(Source: Authors.)

shared evenly across an SAD area, or it may be specific to property in a particular SAD, based on individual lot features, parcel size, street frontage, property values, and use (Mathur, 2015). It may also be based on a simple proportion of the assessed increased value of the property based on the amenity provided, or it could be based on distance (for example, lots closer to a transit line pay more than those further away).

9.6.4 Debates and Controversy

SADs can provide a steady stream of revenue for local improvements, but there are risks. They can be unpopular and politically perilous, as citizens may be averse to paying a targeted tax on top of other taxes. Additionally, the process of determining incremental value created due to government action in the SAD can be subject to legal challenge. Many statutes require property owners to vote on the creation of the district (Zhao and Larson, 2011) to mitigate this risk.

The success of an SAD also varies with the trust and confidence that property owners have in their cities. A Lincoln Institute study authored by Ochoa (2011) examined decades of betterment levy use in Columbia, South America, and found that political acceptance of the levy is dependent on whether the public believes the project is a legitimate need, and whether they feel the administrative agency is trustworthy and competent.

Zhao and Larson (2011) also note that enforcement of SAD, once agreed, is relatively easy, as only property owners, who directly benefit through the value boost (and are easily tracked through assessment records), are subject to the tax.

There may be questions of fairness and equity in the creation of SADs, and Mathur (2015) identifies several ways of advancing fairness and reducing inequity. One is to charge less than the fully calculated increment to those experiencing negative effects such as noise and pollution, especially for residences – reducing the amount or negating it for those who live directly next to a busy new highway, for example. Discount adjustments may also be made to low-income households or seniors.

9.7 DENSITY BONUSES

9.7.1 What Are They?

Density bonuses allow private owners to create denser developments if they do something in return that benefits the public – usually create or reserve space for affordable housing or open space, or incorporate a public plaza. The term is often used synonymously with "inclusionary zoning." Density bonuses may be considered a type of LVC because they require a contribution from a private developer in exchange for an increase in value for the developer's property through upzoning (Rosen et al., 2017). It is like "inverse incentive FSI (floor space index)," which rewards developers offering smaller-sized affordable apartments who are allowed more FSI to use on larger apartments (WEF, 2019, 21).

Density bonuses are often associated with transit-oriented development (TOD) and encourage developers to build within the catchment area of stations. Developers may be willing to pay as much as a 10% premium for space within the catchment area of a well-designed transit station, and density bonuses may further encourage development (Yang et al., 2016). The developer may be incentivized by further bonuses, such as a waiver of impact fees, smaller parking requirements, expedited project approvals, and other rewards (NMHC, 2019).

9.7.2 How Are They Implemented?

Density bonuses are usually created as an ordinance and implemented into a larger zoning code at the local level, as shown in Figure 9.4. That ordinance will include definitions, standards, and procedures for claiming the bonus. It will also include definitions of "affordable" and "low income."

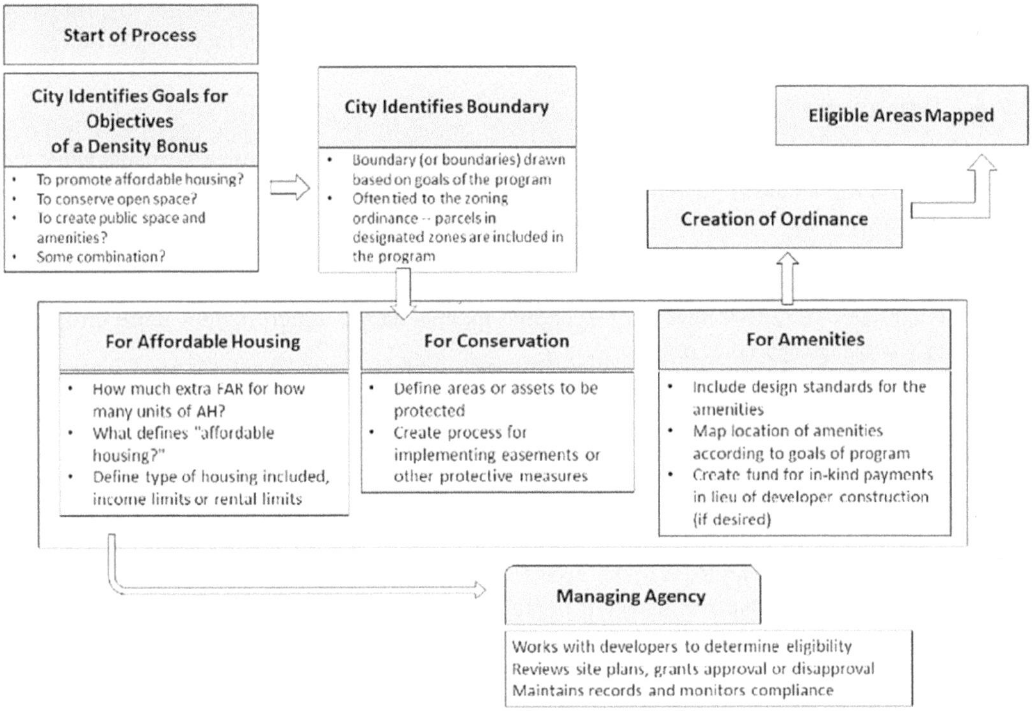

FIGURE 9.4 Density bonus.

(Source: Authors.)

It should include information about how large the development must be to qualify, how many units, or what percentage of the building is to be reserved for lower-income occupants (renters or owner occupied). It also has standards for defining low income by income levels that must be targeted for the reserved housing, clear rules for minimum locked-in time to avoid house flipping and arbitrage, mandating the units remain affordable to low-income persons (even if property ownership changes). A blend of incentives may be in place to further encourage the participation of a developer. The ordinance may also include relevant standards for building and construction of the units, though this may be a reference to a broader zoning plan. For a more detailed breakdown of ordinance creation, see (RISPP, 2006).

9.7.3 How Are They Calculated?

In areas of significant and intense development pressure, and with a major mandate for providing affordable housing for a low-income population, the bonus may be significant. The general formula is that if a developer reserves a certain number of units for low-income renters, they will be allowed more overall units than they otherwise would have been able to build. The exact ratio is chosen at the local level depending on government goals and local conditions.

For example, say a developer wishes to build in an area where the maximum density is 100 units. The hypothetical ordinance in this instance states that if they reserve 10% of their units for low-income renters, they are given a "boost" of 15% to their density. So, they can build a total of 115 units.

Whom the government decides is the target renter for housing assistance often determines is the number of units required to be affordable to receive the bonus. To maintain a project's commercial viability, developers may be required to reserve only 5% of units for very low-income renters to receive the bonus, as they will receive very little revenue from those units. Conversely, they may need to reserve as much as 15% of units for moderate-income renters, or some combination.

9.7.4 Debates and Controversy

Inclusionary zoning has seen a mixed reception in the literature. Several studies have concluded that inclusionary zoning modestly increases multi-family housing construction as developers chase the bonus, but depresses the single-family housing market and may lead to higher costs overall in cities with mandatory inclusionary zoning requirements, as developers pass the cost of compliance on to the price of their market-rate housing (Bento et al., 2013; Mukhija et al., 2015; Schuetz et al., 2009). Others (Mukhija et al., 2010) have concluded that inclusionary zoning provides more affordable/ multi-family housing than other programs.

Inclusionary housing policies vary considerably. Program participation may be mandatory for developers, voluntary as an incentive, or mixed (meaning requirements are waived for developments of smaller size, for example). This variation makes it difficult to conduct broad research on their effectiveness as a general tool (Mukhija et al., 2015; Schuetz et al., 2011).

While they can be a good tool for cities to use, they are still bound to the functioning of the market – meaning there must be a market for development in the area. Comparing the return on the base density of the development to the expected return after a density bonus can help determine the feasibility of density bonuses (Seyfried, 1991). Cities may also look to other incentives, such as fast-tracking permits or loosening design regulations (Garde, 2016).

9.8 LAND READJUSTMENT (LR) OR LAND POOLING (LP)

9.8.1 What Is It?

Land readjustment (LR) or land pooling (LP) is a multi-step planning-based LVC method in which local governments and private landowners work together to redevelop an area by first demarcating contiguous areas for planning, pooling all irregular-shaped inaccessible private lands, creating a layout plan, reconstituting the original land parcels to fit the layout, and installing infrastructure and providing direct access to all newly created parcels. The process ends with returning smaller, more developable, reconstituted parcels to the owners that have greater value than the original larger lots (Mittal, 2013, 2014). It is particularly useful because it allows governments to undertake infrastructure projects without expropriating land from private owners (U.N. Habitat, 2019). LR requires the least amount of displacement, dispossession, and compensation and is partially self-financing (Mittal and Kashyap, 2015).

Variants of LR are practiced in several countries such as Japan, Germany, the Netherlands, Spain, S. Korea, Western Australia, Israel, Turkey, India, Bangkok, Taiwan, Nepal, and Sri Lanka, to name a few (Mittal and Byahut, 2019)

Figure 9.5 shows an example of the land readjustment process that has been common in Gujarat, India. Landowners will agree to pool their properties together and allow infrastructure improvements to be made, such as utility installation, road or transit construction, and other such developments. Once the improvements are made, the landowners receive their land back but in a different (usually smaller) shape more suitable for development, and at a higher value (World Bank, 2015a). During the process, the government may retain some land for public use, such as for parks, schools, or low-income housing. It may also hold some land in a land bank for future sale.

9.8.2 How Is It Implemented?

The process is intensive and requires significant administrative oversight. Mittal (2014) identifies six steps in the process: the intent declaration stage, collection and organization of land records, preparation of a draft layout plan, estimates of costs and finances (which may include additional

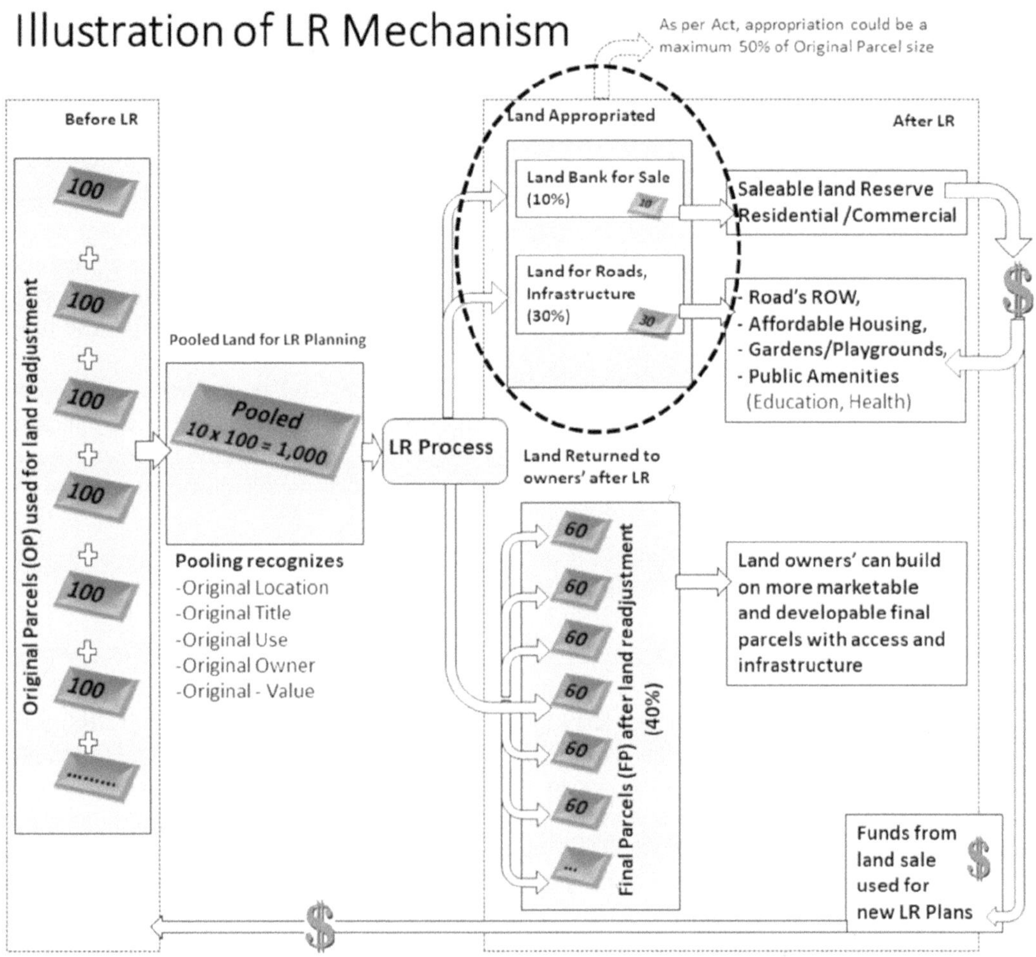

FIGURE 9.5 Land readjustment and land pooling (LR & LP).

(Source: Mittal and Kashyap, 2015.)

betterment levies on the reconstituted parcels), the public participation and legal process, and the final modification of land records and return to original owners once the readjustment is complete.

Land readjustment has been implemented to great effect in India, Turkey, Hong Kong, China, Portugal, Spain, and many other countries (Almeida et al., 2018). The process can vary in some cases, such as in Japan, where private landowners initiate the process and petition the government to make improvements – usually the construction of transit (Suzuki et al., 2015). In other countries, a government body will initiate the project, such as in the case of a regional ring road project in Ahmadabad (Mittal, 2013).

9.8.3 How Is It Calculated?

The land readjustment process is predicated on considerable public engagement and a systematic estimation of the current value of land and the expected value of land post adjustment. This ensures property owners have a clear picture of their expected value gains, even as they consider the potential loss of part of their property during the adjustment process. The current and potential value is carried out by appraisers, and there should be ample opportunity for owners to appeal those appraisal decisions (Alterman, 2007).

9.8.4 Debates and Controversy

Land readjustment has been popular in countries experiencing rapid urbanization because it bypasses onerous and lengthy forcible taking processes (Hong and Brain, 2012). Yet complications may arise from perceived or actual inequities which are realized from the process. During the redrawing of parcel lines, some parcels may receive a greater proportional value boost than otherwise "deserved." For example, properties that are adjusted to be directly adjacent to a new road (but which were originally some distance away) may receive a greater proportional "boost" in value than other nearby parcels, through no real value creation of their own (Alterman, 2007; Mahto, 2018). These inequities may be remedied through the charging of redistributive "balance fees" that compel those property owners to pay for the excess value. Alterman (2007) argues, however, that extracting these funds can be difficult.

Muñoz and Mualam (2019) studied the differences between land readjustment schemes in Germany, Spain, and Israel to develop a framework for analyzing their effectiveness by measuring the timeliness of delivering the new serviced plot, the competitive benefit the landowner would receive, and the effectiveness of the administrative body that handles the readjustment process.

9.9 TRANSFERABLE DEVELOPMENT RIGHTS (TDR)

9.9.1 What Are They?

Transferable development rights (TDR) is a market-based tool that is used for spatially transferring unused air-right credits of one property to another one in a different location. TDRs in the United States are used to preserve historical buildings, open spaces, natural areas, and farmlands, and to encourage infill developments (Dyca et al., 2020). Its uses in Mumbai, India even include slum redevelopment, affordable housing, road widening, and compensating land acquisition-affected landowners with TDR credits. TDR requires the presence of active real-estate markets, identification of areas to be preserved and promoted, and quantitative estimation of TDR credits. As a process, the city demarcates TDR "sending areas," areas that need to be preserved, and TDR "receiving areas," areas that need growth, for geographically transferring the TDR credits. Developers in the receiving area can buy TDR credits to build over and above the permissible floor space index (FSI). TDR certificates are available for open trading in the market for cash without the government determining the price. Its price is based on demand, supply, and availability of credits (HDFC, 2021). Cities can use TDR to guide urban growth – to promote in some areas and discourage in others.

TDR allows developers to sell unused development rights to other developers, who can buy the rights and apply to build taller or larger buildings in a receiving area. This helps to protect sensitive areas such as historically significant areas or conserved farmlands, while encouraging infill. Any associated fees from the TDR transactions can provide additional LVC by supporting the administration of the program (Germán and Bernstein, 2018).

The TDR is a market-based LVC instrument. Consider a scenario in which a city may want to increase the density of one area while improving services. If the city were to simply "upzone" the area to allow for higher densities, property developers in the area would receive a windfall profit, based mostly on the zoning decision, rather than through any value creation of their own.

The TDR program relegates value back to a market-based system by allowing these developers to purchase extra development rights from others and allowing property owners in areas where development is discouraged to receive compensation, while allowing developers to create larger buildings in areas where denser development is encouraged.

TDR programs have an additional advantage over simple upzoning, in that it is a more durable system that is protected over a long period. Zoning is relatively transient and could be changed on a political whim (Hanly-Forde et al., 2011). The TDR program is often administered by a dedicated agency or TDR "bank" which facilitates transactions and keeps records. Because of this, it is one of the more administratively intense LVC programs (Center for Land Use Education, 2005; Pruetz and Standridge, 2008).

9.9.2 How Are They Implemented?

The process of implementing a TDR program varies by jurisdiction but is generally authorized by enabling legislation at state level as part of an overall zoning code (Chiodelli and Moroni, 2016), as described in Figure 9.6. As part of the TDR process, municipalities create "sending – TDR export zones" and "receiving – TDR import zones" for spatial TDR transactions. The sending zone is the area from which development rights will be sold, and the receiving zone is where they will be applied to increase FAR (floor-area ratio) and densities. These areas are either zoning designations or an overlay.

The city determines the exchange rate for TDRs – meaning, how many units (floors, square feet, residences, etc.) one "unit" of TDR rights entitles a buyer to build. Additional needs are to determine how many TDR rights landowners are permitted to sell, and how large is the maximum density bonus allowed for the developer on the receiving end (Walls and McConnel, 2007). Sometimes it may be a simple proportion, but the city could add bonuses to encourage people to buy more (Morley, 2018).

Governments must also determine whether TDR rights are allowed automatically within the designated zones or if there should be extra requirements, such as minimum lot sizes (Walls and McConnel, 2007).

9.9.3 How Are They Calculated?

The exact rate of development rights exchange is determined locally in each area depending on the goals of the program and the local context. If the receiving area is primarily residential, a unit may be as simple as one potential residential unit to one residential unit. For commercial areas,

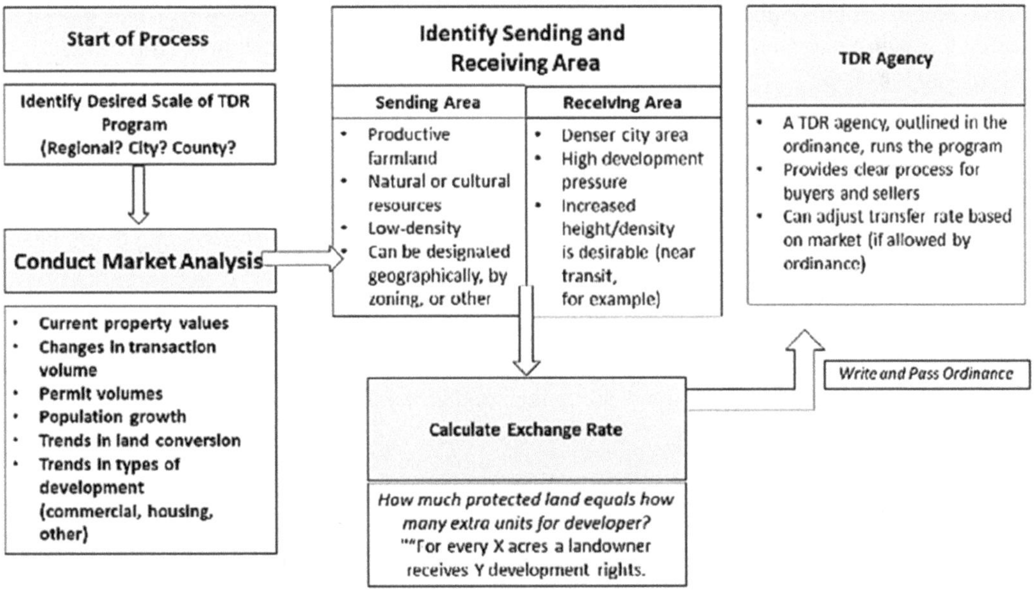

FIGURE 9.6 Transferrable development rights (TDR).

(Source: Authors.)

the exchange rate may be one "development unit" that is equal to a certain ratio of additional floor space (CRCOG, 2016). Cities may place additional restrictions on the exchange if they anticipate potential issues: for example, a city may place a limit on the number of development rights a single developer may purchase, or they may impose more stringent restrictions in areas of special historic value (Been et al., 2014).

9.9.4 Debates and Controversy

As of 2018, 36 states in the U.S.A. had enacted TDR programs (Shahab et al., 2018). TDRs have a substantial international presence and are particularly common in countries like Taiwan, China, Peru, and Italy. However, there can be significant costs for administering a TDR program which can make the system prohibitively expensive, particularly for smaller areas, and transaction costs can be as much as 21% of the total TDR cost (Colavitti and Serra, 2018; Shahab et al., 2018).

The market conditions of the area must also be taken into account. If receiving areas, for example, are already places of high density, the market for increased density may be limited and the volume of transactions may be small, which McConnell et al. (2006) identify as one of the greatest problems for many TDR programs.

Because of all these factors, administrative complexity can be prohibitive. Colavitti and Serra (2018), when comparing TDR programs in Italy and the United States, identified common measures of success: a lively real-estate market, flexible receiving areas, bonus density incentives, clear rules and procedures, promotion and awareness campaigns to generate public support, and the creation of an intermediate agency such as a TDR bank. Pruetz and Pruetz (2007) also provide an excellent breakdown of factors common in successful TDR programs.

9.10 AIR RIGHTS

9.10.1 What Are They?

The selling of air rights is a form of value capture where space above a public investment, like a railway track, station, or highway, is sold or leased to a developer. The selling of air rights has a long history in English common law, which imparts development rights to a landowner from the soil "to the sky and the depths" (Goldschmidt, 1964).

The revenue from leasing or selling air rights can be invested in building a new piece of infrastructure, and the area above it can be sold to a private developer to help finance that infrastructure (or to build more). This generates a dual benefit: the property owner gets desirable development space, the government receives payment, and the government receives even more payment over time through higher property taxes on the new development and its impact on surrounding development (Suzuki et al., 2015).

Air rights can be sold over many types of development, including railway lines, parking lots, highways, streets, and even bodies of water such as rivers or canals (Goldschmidt, 1964). Allam and Jones (2019) point out that speculation over air rights is of increasing interest for developers, particularly in major cities, where rents are rising amid increasing scarcity of developable space in trendy urban centers.

9.10.2 How Are They Created?

Chen (2020) identified several mechanisms by which air rights are commercialized. Specific legislation authorized by local governments can be written which allows for air rights to be converted into a volumetric property and then sold (they may or may not be transferable). See Figure 9.7 for a representation of this concept. The rights may be sold for that site or converted into "bonus credits" for a developer to build elsewhere, like a transfer of development rights program.

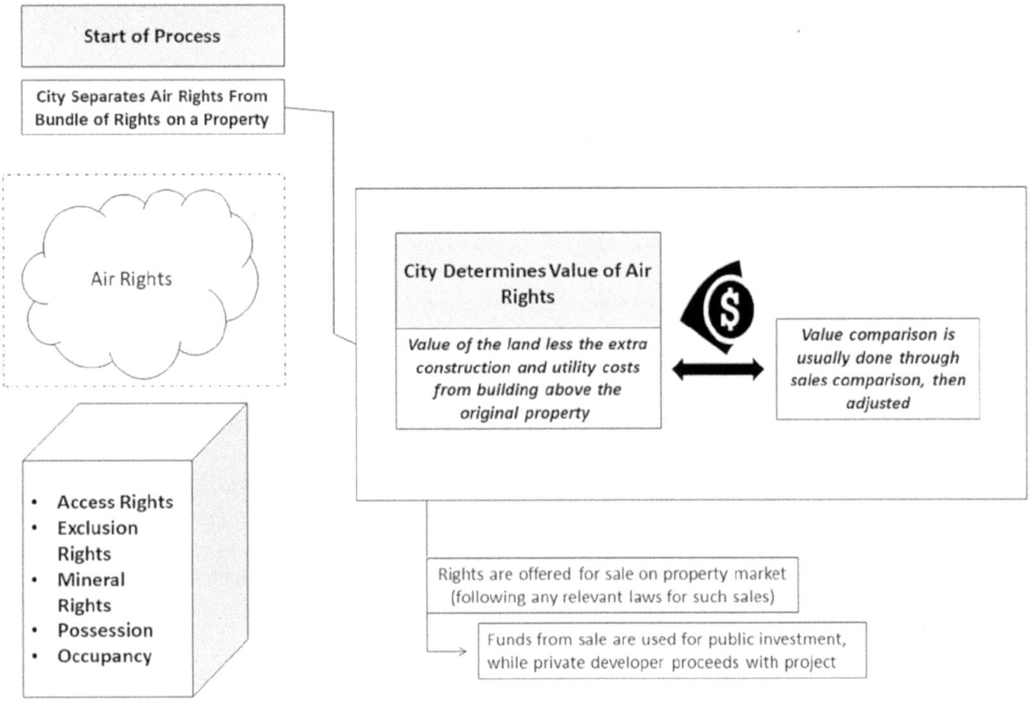

FIGURE 9.7 Air rights.

(Source: Authors.)

Goldschmidt (1964) noted that the rights are often leased instead of sold, for periods of a few decades to nearly 100 years. They may be called right-of-way use agreements, and can take several forms: one-time lease payments, long- or short-term leases with renewals, a direct sale to a private buyer who then allows an easement for the public infrastructure, or the opposite, where the government sells only the air rights and grants an easement for the private company to construct its development.

9.10.3 How Are They Valued?

Air rights valuation can become complicated by the specifics of the site and the intricacies of negotiations between the government and developers. However, air rights can generally be evaluated through a fee-simple comparative method, and then the value is adjusted for the cost increases inherent in building taller structures (U.S. Department of Housing and Urban Development, 1979). A property is valued as if it is on land, then the costs of additional construction time and material, additional legal and organizational cost and additional design costs are subtracted, leaving a final valuation. A far more detailed explanation of air rights valuation is described in Machen (1966).

9.10.4 Debates and Controversy

Air rights sales are often feasible only in areas of high development demand where space for development is limited and expensive but still highly desirable. Additional costs associated with building over existing infrastructure may make air rights unattractive to developers in all but the most intense land markets. Machen (1966) identified three determinants for whether air rights were salable in each development. One is whether the site is of sufficient size to interest a developer because scale

TABLE 9.1
Comparison Chart for Land Value Capture Tools

Tool	Definition	Benefits	Drawbacks
Development Exactions and Impact Fees	Up-front fees charged to developers to defray the cost of new infrastructure	Simple in concept, pays for needed improvements	Can raise prices for end consumers, fee usually does not cover future **maintenance**
Tax Increment Financing	A temporary, spatial designation that dedicates a portion of tax revenue increases in an area to infrastructure improvement within that area	Opens up capital to fund revitalization projects	"But for" analysis can be hard to prove, making efficacy bard to measure
Special Assessment Districts	An additional levy in a defined area to support additional benefits or services	Predictable revenue stream for ongoing service and maintenance	Difficulty assigning levy value, possible political opposition
Density Bonus	Allowing additional density in an area in exchange for reservation of affordable living **units**	Market-based incentives for providing housing	May raise prices in other housing
Land Readjustment	The collective pooling and reconstitution of private land to organize and finance infrastructure projects	Largely self-financing, can fund major projects	Complexity
Transfer of Development Rights	A market-based system for trading development rights from areas of conservation concern to denser areas	Market-based, allows targeted conservation and development	Cost and complexity
Air Rights	T1le selling or leasing of space above existing infrastructure, like mil lines	Relative ease of sale and implementation	Limited applicability

returns must be higher due to the complexity of the construction. The second is whether there are competitive sites developers are more likely to choose, and third, there must be a lively land market with high prices. Additionally, the administrative legal costs associated with an air rights program can be complex and costly to manage, potentially slowing development progress (Zhao et al., 2012) (Table 9.1).

9.11 CONCLUSION

LVC is likely to become an increasingly vital instrument as the world continues to urbanize, especially in developing countries. It is important to note that these instruments are alternative financing mechanisms and can offer substantial and reliable revenues, but they are not a panacea. The successful use of LVC requires several enabling conditions, such as a supportive real-estate market and an entrepreneurial public agency with visionary and strong leadership. It also means that there is a need for a cultural shift in cities in the developing world, as it is more likely to be successful when applied in areas with supportive land-use regulations, statutory authority, and legal backing.

Recently, in India, the Ministry of Finance (MOF) OM No. 66(20)/ PF-II (2016), dated 7 March 2017, has issued instructions to include LVC-based financing as an integral part of all detailed project reports (DPRs) of all central government projects (MOUD, 2017. The challenge with countries like India, however, is that with land being a state subject, land and urban development-related legislation varies by state (MOUD, 2017; Peterson, 2009). Accuracy and transparency in land and property records and precision in value assessments and property appraisals are important prerequisites for the effective implementation of LVC instruments. Any future LVC policies must be formulated keeping both the local and state governments in mind, and not all instruments are applicable in every situation.

The overreliance on LVC as a source of financing could also be somewhat risky in some cases as real-estate markets are cyclical (FHWA, 2019). If instruments-related laws and policies are not transparent and equitable, LVC is susceptible to litigation and public mistrust. Additionally, not every local jurisdiction can utilize these tools. If cities are using public land assets, they should prepare a detailed municipal land-based real-estate asset management plan and carry out a detailed inventory and valuation of public land assets. To monitor and appraise the effectiveness of LVC instruments meeting their stated goals, implementing agencies should also take a balance-sheet approach in using the cost of asset utilization and how it meets the goals of infrastructure–finance gaps.

The seven instruments discussed above are a starting point for public agencies, in that cities facing dire challenges may consider experimenting with self-financing ways of using the land market to fund infrastructure to make cities economically competitive, more equitable, pleasant, and livable. While these instruments provide a necessarily incomplete picture of LVC finance, this chapter is an attempt to provide an entry point for cities seeking new ways to generate funds they can use to benefit their citizens.

REFERENCES

Agboola, Alirat Olayinka, T. O. Ayodele, and A. Olofa. "Tax Increment Financing in the UK and USA: Its Prospects for Urban Regeneration in Nigeria." *Smart and Sustainable Built Environment* 7, no. 3/4 (2018): 277–292.

Ahluwalia, Isher J. "Urban Governance in India." *Journal of Urban Affairs* 41, no. 1 (2019): 83–102.

Allam, Zaheer, and David Sydney Jones. "The Potential of Blockchain within Air Rights Development as a Prevention Measure against Urban Sprawl." *Urban Science* 3, no. 1 (March 1, 2019): 38. doi: 10.3390/urbansci3010038.

Almeida, J., J. Ferreira, B. Condessa, and R. Tome "Improving Land Readjustment Practice. Application of Management Models to Portugal." *European Planning Studies* 26, no. 7 (2018): 1431–1449.

Alterman, R. "Much more than land assembly: Land readjustment for the supply of urban public services." Chapter 3 in Yu-Hung Hong and B. Needham Eds., *Analyzing Land Readjustment: Economics, Law and Collective Action*. 57–86, Cambridge, MA: Lincoln Institute for Land Policy (2007).

Ameen, R., and M. Mourshed "Urban Environmental Challenges in Developing Countries—A Stakeholder Perspective." *Habitat International* 64 (2017): 1–10.

Arfanuzzaman, M., and B. Dahiya "Sustainable Urbanization in Southeast Asia and Beyond: Challenges of Population Growth, Land-Use Change, and Environmental Health." *Growth and Change* 50, no. 2 (2019): 725–744.

Been, Vicki, J. Infranca, and J. Yager. "Unlocking the Right to Build: Designing a More Flexible System for Transferring Development Rights," *Furman Center for Real Estate and Urban Policy*, (March 2014). Accessed on January 29, 2021, from https://furmancenter.org/research/publication/unlocking-the-right-to-build-designing-a-more-flexible-system-for-transferr

Bento, A., S. Lowe, G. J. Knaap, and A. Chakraborty. "Housing market effects of inclusionary zoning." From cityscape (2009) in R. Tighe Eds., *The Affordable Housing Reader*. 438–451, Abingdon: Taylor and Francis (2013). doi: 10.4324/9780203722671.

Bladikas, Athanassios K., and L. Pignataro "Efficiency and Equity Issues in Special Fees for Financing Transportation Infrastructure." *Transportation Research Part A: General, Special Issue Financing Transportation Infrastructure*, 24, no. 4 (1990): 283–292.

Burge, Gregory. "The Capitalization Effects of School, Residential, and Commercial Impact Fees on Undeveloped Land Values." *Regional Science and Urban Economics*, 44 (2014): 1–13.

Burge, Gregory S., A. C. Nelson, and J. Matthews. "Effects of Proportionate-Share Impact Fees." *Housing Policy Debate*, 18, no. 4 (2007): 679–710.

Burge, Gregory S., Trey L. Trosper, Arthur C. Nelson, Julian C. Juergensmeyer, and James C. Nicholas. "Can Development Impact Fees Help Mitigate Urban Sprawl?" *Journal of the American Planning Association* 79, no. 3 (July 3, 2013): 235–248.

Burger, Alan M. "Innovative Uses for the Special Assessment: Building Infrastructure Without Raising Taxes," *U. Miami Business Law Review* 129 (1993): 15.

California State Legislature. "*Assessing the Benefits of Benefit Assessments*." (2011). Accessed on December 27, 2020, from https://sgf.senate.ca.gov/sites/sgf.senate.ca.gov/files/Assessing%20the%20Benefits%20of%20Benefit%20Assessments%20801-S.pdf

Center for Land Use Education. "*Transfer of Development Rights*." (2005). https://www.uwsp.edu/cnr-ap/clue/Documents/PlanImplementation/Transfer_of_Development_Rights.pdf

Center for Land Use Education. "*Impact Fees*." (2008). https://www.uwsp.edu/cnr-ap/clue/Documents/PlanImplementation/Impact_Fees.pdf

Center for Land Use Education. "*Tax Increment Financing*." (2009). https://www.uwsp.edu/cnr-ap/clue/Documents/implementation/TIF.pdf

Chen, Hung-Ying. "Cashing in on the Sky: Financialization and Urban Air Rights in the Taipei Metropolitan Area." *Regional Studies* 54, no. 2 (2020): 198–208.

Chiodelli, Francesco, and S. Moroni. "Zoning-Integrative and Zoning-Alternative Transferable Development Rights: Compensation, Equity, Efficiency." *Land Use Policy* 52 (March 2016): 422–429.

City of Chicago. "Tax Increment Financing—Program Guide," *Department of Planning and Development, City of Chicago*, 2020. Accessed online on June 25, 2021, from https://www.chicago.gov/content/dam/city/depts/dcd/general/2020_tif_program_guide.pdf

Colavitti, Anna Maria, and S. Serra. "The Transfer of Development Rights as a Tool for the Urban Growth Containment: A Comparison between the United States and Italy: Transfer of Development Rights." *Papers in Regional Science* 97, no. 4 (2018): 1247–1265. doi: 10.1111/pirs.12295.

CRCOG. "Livable Communities Toolkit: A Best Practices Manual for Metropolitan Regions: Transit-Oriented Development: Fact Sheet." (2016): 1–9. https://crcog.org/wp-content/uploads/2016/07/Ch03_FactSheet_TDR.pdf

Dolan vs. City of Tigard, 512 U.S. 374 (1994).

Drucker, Joshua, G. Kim, and R. Weber. "Did Incentives Help Municipalities Recover from the Great Recession? Evidence from Midwestern Cities." *Growth and Change* 50, no. 3 (2019): 894–925.

Dyca, B., K. Muldoon-Smith, P. Greenhalgh "Common Value: Transferring Development Rights to Make Room for Water." *Environmental Science & Policy*, 114 (2020): 312–320. doi: 10.1016/j.envsci.2020.08.017.

Dye, Richard F., and David F. Merriman. "Tax Increment Financing: A Tool for Local Economic Development. (Cover Story)." *Land Lines*, 18, no. 3 (January 2006): 2–7.

El-Khattabi, Ahmed Rachid, and T. William Lester. "Does Tax Increment Financing Pass the 'But-for' Test in Missouri?" *Economic Development Quarterly* 33, no. 3 (August 2019): 187–202.

Evans-Cowley, J. *Development Exactions: Process and Planning Issues*. Cambridge, MA: Lincoln Institute of Land Policy (2006).

Evans-Cowley, Jennifer, Larry Lockwood, Ronald Rutherford, and Thomas Springer. "The Effect of Development Impact Fees on Housing Values." *Journal of Housing Research* 18, no. 2 (2009): 173–193.

Farris, Sherri, and J. Horbas. "Creation vs. Capture: Evaluating the True Costs of Tax Increment Financing." *Journal of Property Tax Assessment & Administration* 6, no. 4 (2009): 5–28.

FHWA. "Value Capture: Capitalizing on the Value Created by Transportation." *Federal Highway Administration (FHWA)*, Report no. FHWA-HIN-19-004 (2019). Accessed on December 18, 2020, from https://ntlrepository.blob.core.windows.net/lib/74000/74800/74824/FHWA-HIN-19-004.pdf

Garde, Ajay. "Affordable by Design? Inclusionary Housing Insights from Southern California." *Journal of Planning Education and Research* 36, no. 1 (2016): 16–31.

Germán, Lourdes, and A. E. Bernstein. "Land Value Capture Tools to Finance Our Urban Future." *Lincoln Institute of Land Policy* (2018): 4. Accessed from https://www.lincolninst.edu/sites/default/files/pubfiles/land-value-capture-policy-brief.pdf

Goldschmidt, Leopold. "Air Rights." American Planning Association. (1964). Accessed on December 18, 2020, from https://www.planning.org/pas/reports/report186.htm

Greenbaum, Robert T., and J. Landers. "The TIFF over TIF: A Review of the Literature Examining the Effectiveness of the Tax Increment Financing." *National Tax Journal* 67, no. 3 (September 2014): 655–674.

Greuling, J.E. "Tax Increment Financing: A downtown development tool." *Economic Development Review* (1987): 23–27.

Hanly-Forde, Jason, G. Homsy, K. Lieberknecht, and R. Stone. "*Transfer of Development Rights Programs: Using the Market for Compensation and Preservation*." (2011). Accessed from http://www.mildredwarner.org/gov-restructuring/privatization/tdr

Hayes, David J. "Rapid Transit Financing: Use of the Special Assessment."*Stanford Law Review* 29, no. 4 (1977): 795–818. doi: 10.2307/1228261.

HDFC. "*Transferable Development Rights (TDR) – The Concept*," (2021). Accessed on December 26, 2020, from https://www.hdfc.com/blog/transferable-development-rights-tdr-concept

Hicks, Michael J., D. Faulk, and S. Devaraj. "Tax Increment Financing: Capturing or Creating Growth?" *Growth and Change* 50, no. 2 (2019): 672–688.

Hong, Yu-Hung, and Isabel Brain. "Land Readjustment for Urban Development and Post-Disaster Reconstruction." *Lincoln Institute of Land Policy* (January 2012). Accessed on December 26, 2020, from https://www.lincolninst.edu/publications/articles/land-readjustment-urban-development-post-disaster-reconstruction

Kogan, Vladimir, and Mathew McCubbins. "The Problem with Being Special Democratic Values and Special Assessments." *Public Works Management & Policy* 14 (October 2008): 4–36.

Li, Zheng. "The Impact of Metro Accessibility on Residential Property Values: An Empirical Analysis." *Research in Transportation Economics* 70 (2018): 52–56.

Machen, J. C. "Air Rights Development." *Appraisal Journal* 34, no. 2 (1966): 288.

Mahto, Prakash. "A Critical Review on Land Readjustment Technique" *Journal of Emerging Technologies and Innovative Research* 5, no. 12 (2018): 8.

Mathur, Shishir. "Funding Public Transportation Through Special Assessment Districts: Addressing the Equity Concerns." *Public Works Management & Policy* 20, no. 2 (April 2015): 127–145.

Mathur, Shishir, Paul Waddell, and Hilda Blanco. (2004). "The Effect of Impact Fees on the Price of New Single-family Housing." *Urban Studies* 41. doi: 10.1080/0042098042000214806.

McConnell, Virginia, Elizabeth Kopits, and Margaret Walls. "Using Markets for Land Preservation: Results of a TDR Program." *Journal of Environmental Planning & Management* 49, no. 5 (September 2006): 631–651.

McKinsey Institute. "India's Urban Awakening: Building Inclusive Cities, Sustaining Economic Growth." *McKinsey* (2010). Accessed on December 21, 2020, from https://www.mckinsey.com/featured-insights/urbanization/urban-awakening-in-india#

Merriman, David. "Improving Tax Increment Financing (TIF) for Economic Development." *Lincoln Institute of Land Policy*, (2018), Policy Focus Report/Code: PF050.

Ministry of Finance (MOF). "Value Capture Financing as part of PIB format per appraisal of project of the Central Government." OM No. 66(20)/ PF-II (2016).

Mittal, Jay. "Extending Land Readjustment Schemes to Regional Scale: A Case Study of Regional Ring Road via Mosaicking Neighborhood Level Plans." *Real Estate Finance* 30, no. 2 (Fall 2013): 62–73.

Mittal, Jay. "Self-Financing Land and Urban Development via Land Readjustment and Value Capture." *Habitat International* 44 (2014): 314–323.

Mittal, Jay, and S. Byahut. "Scenic landscapes, visual accessibility and premium values in a single-family housing market: a spatial hedonic approach." *Environment and Planning B: Urban Analytics and City Science*, 46, no. 1 (2019): 66–83. Sage. doi: 10.1177/2399808317702147.

Mittal, Jay, and Anil Kashyap. "Real Estate Market Led Land Development Strategies for Regional Economic Corridors – A Tale of Two Mega Projects." *Habitat International* 47 (2015): 205–217.

Morley, David. "Transfer of Development Rights." *PAS Quicknotes* 74 (2018).

MOUD. "Value Capture Finance Policy Framework." (2017). Accessed on December 23, 2020, from http://mohua.gov.in/upload/whatsnew/59c0bb2d8f11bVCF_Policy_Cook_FINAL.pdf

Mukhija, Vinit, Ashok Das, Lara Regus, and Sara Slovin Tsay. 2015. "The Tradeoffs of Inclusionary Zoning: What Do We Know and What Do We Need to Know?" *Planning Practice & Research* 30, no. 2 (2015): 222–235.

Mukhija, Vinit, Lara Regus, Sara Slovin, and Ashok Das. "Can Inclusionary Zoning Be an Effective and Efficient Housing Policy? Evidence from Los Angeles and Orange Counties." *Journal of Urban Affairs* 32, no. 2 (2010): 229–252.

Muñoz, Gielen Demetrio and Nir Mualam. "A Framework for Analyzing the Effectiveness and Efficiency of Land Readjustment Regulations." *Land Use Policy* 87 (2019): 104077.

NAHB. *Impact Fee Handbook*. National Association of Home Builders, Washington, DC (2016). Accessed on July 12, 2021, from https://www.nahb.org/-/media/NAHB/advocacy/docs/industry-issues/land-use-101/infrastructure/impact-fee-handbook.pdf

Nelson, Arthur C., and Mitch Moody. "Paying for Prosperity: Impact Fees and Job Growth." Brookings Institution (June 1, 2003). Accessed on July 12, 2021, from https://www.brookings.edu/wp-content/uploads/2016/06/nelsonimpactfees.pdf

NMHC. "Tool: Inclusionary Zoning." *Housing Affordability Toolkit* (2019): 74–85.

Nollan, V. California Coastal Commission, 483 U.S. 825 (1987).
Ochoa, Oscar. "Betterment Levy in Colombia: Relevance, Procedures, and Social Acceptability." *Lincoln Institute of Land Policy* (2011). Accessed on December 26, 2020, from https://www.lincolninst.edu/es/publications/articles/betterment-levy-colombia
Overton, Michael, and Robert L. Bland. "The Great Recession's Impact on Credible Commitment: An Analysis of Private Investment in Tax Increment Financing Districts." *State and Local Government Review* 46, no. 4 (2014): 282–297.
Peterson, George E. "*Unlocking Land Values to Finance Urban Infrastructure*." The World Bank, (2009) Trends and Policy Options; no. 7. doi: 10.1596/978-0-8213-7709-3.
Peterson, George E. "Unlocking Land Values for Urban Infrastructure Finance: International Experience — Considerations for Indian Policy." *Policy Research Working Papers. The World Bank*, Washington, DC (2013). doi: 10.1596/1813-9450-6683.
Pruetz, Rick, and Erica Pruetz. "Transfer of Development Rights Turns 40." *Planning & Environmental Law* 59, no. 6 (2007): 3–11.
Pruetz, Rick, and Noah Standridge. "What Makes Transfer of Development Rights Work? Success Factors From Research and Practice." *Journal of the American Planning Association* 75, no. 1 (2008): 78–87.
RISPP. *Handbook On: Developing Inclusionary Zoning*. State of Rhode Island, USA (2006).
Rizvi, Zaigham M. "Population Growth, Urbanization and Slums: Challenges for Developing Low-Income Affordable Housing." *Housing Finance International*, (2016): 37–41.
Rosen, David, Nora Lake-Brown, and Bryan Glascock. "How Value Capture can Create Affordable Housing." *Lincoln Institute of Land Policy* (February 2017). Accessed on December 26, 2020, from https://www.lincolninst.edu/publications/multimedia/how-value-capture-can-create-affordable-housing
Ross, Dennis H., and S. I. Thorpe. "Impact Fees: Practical Guide for Calculation and Implementation." *Journal of Urban Planning and Development* 118, no. 3 (1992): 106–118.
Schneider, Benjamin. City Lab University: Tax Increment Financing, *Bloomberg City Lab* (October 24, 2019). Accessed on April 3, 2021, from https://www.bloomberg.com/news/articles/2019-10-24/the-lowdown-on-tif-the-developer-s-friend
Schuetz, Jenny, Rachel Meltzer, and Vicki Been. "31 Flavors of Inclusionary Zoning: Comparing Policies From San Francisco, Washington, DC, and Suburban Boston." *Journal of the American Planning Association* 75, no. 4 (2009): 441–456.
Schuetz, Jenny, Rachel Meltzer, and Vicki Been. "Silver Bullet or Trojan Horse? The Effects of Inclusionary Zoning on Local Housing Markets in the United States." *Urban Studies* 48, no. 2 (2011): 297–329.
Selby, Douglass P., and Claire Hunter. "Tax Increment Financing: How Public-Private Partnerships Are Financing Urban Redevelopment." *Real Estate Finance (Aspen Publishers Inc.)* 21, no. 2 (2004): 3–7.
Seyfried, Warren R. "Measuring the Feasibility of a Zoning Bonus." *Journal of the American Planning Association* 57, no. 3 (Summer 1991): 348. doi: 10.1080/01944369108975506.
Shahab, Sina, J. Peter Clinch, and Eoin O'Neill. "Estimates of Transaction Costs in Transfer of Development Rights Programs." *Journal of the American Planning Association* 84, no. 1 (2018): 61–75.
Sharma, Rohit, and Peter Newman. "Does Urban Rail Increase Land Value in Emerging Cities? Value Uplift from Bangalore Metro." *Transportation Research Part A: Policy and Practice* 117 (2018): 70–86.
Singell, Larry D., and Jane H. Lillydahl. "An Empirical Examination of the Effect of Impact Fees on the Housing Market." *Land Economics* 66, no. 1 (1990): 82–92.
Sullivan, Gary L., Steve A. Johnson, and Dennis L. Soden. "Tax Increment Financing (TIF) Best Practices Study." IPED Technical Reports. Paper 20 (2002). http://digitalcommons.utep.edu/iped_techrep/20
Suzuki, Hiroaki Jin Murakami, Yu-Hung Hong, and Beth Tamayose. *Financing Transit-Oriented Development with Land Values: Adapting Land Value Capture in Developing Countries*. The World Bank, (2015). doi: 10.1596/978-1-4648-0149-5.
U.N. *World Urbanization Prospects: 2018: Highlights*, United Nations, Department of Economic and Social Affairs, and Population Division. (2019).
U.N. Habitat. "Global Experiences in Land Readjustment." *Urban Legal Case Studies* 7 (2019). Accessed on December 26, 2020, from https://unhabitat.org/sites/default/files/documents/2019-05/global_experiences_in_land_readjustment_urban_legal_case_studies_volume_7_1.pdf
United States Department of Housing and Urban Development. "Valuation Analysis for Project Mortgage Insurance." (1979).
Walls, Margaret A., and Virginia McConnel. "Transfer of Development Rights in U.S. Communities." (2007). Accessed on December 27, 2020.

WEF. "Making Affordable Housing a Reality in Cities." *World Economic Forum*, (June 2019). Accessed on March 2021, from http://www3.weforum.org/docs/WEF_Making_Affordable_Housing_A_Reality_In_Cities_report.pdf

World Bank. "*Land Readjustment*." (2015a). https://urban-regeneration.worldbank.org/node/31

World Bank. "*Tax Increment Financing*." (2015b). https://urban-regeneration.worldbank.org/node/17

Yang, Jiawen, Jige Quan, Bin Yan, and Canfei He. "Urban Rail Investment and Transit-Oriented Development in Beijing: Can It Reach a Higher Potential?" *World Transit Research*, (January 1, 2016). https://www.worldtransitresearch.info/research/6080

Yang, Linchuan, Chen Yang, Nenglai Xu, Rui Zhao, K. W. Chau, and Shijian Hong. "Place-Varying Impacts of Urban Rail Transit on Property Prices in Shenzhen, China: Insights for Value Capture." *Sustainable Cities and Society* 58 (2020): 102140. doi: 10.1016/j.scs.2020.102140.

Zhao, Z., Iacono, M., Lari, A., & Levinson, D. (2012). "Value Capture for Transportation Finance." *Procedia-social and behavioral sciences*, *48*, 435–448.

Zhao, Z. J., & Larson, K. (2011). "Special Assessments as a Value Capture Strategy for Public Transit Finance." *Public Works Management and Policy*, *16*(4), 320–340. https://doi.org/10.1177/1087724X11408923

Zhong, Haotian, and Wei Li. "Rail Transit Investment and Property Values: An Old Tale Retold." *Transport Policy* 51 (2016): 33–48.

Zolnik, Edmund. "A Longitudinal Analysis of the Effect of Public Rail Infrastructure on Proximate Residential Property Transactions." *Urban Studies* 57, no. 8 (2020): 1620–1641.

10 Urban Simulation Model and Urban Growth, Agglomeration, Gentrification

The Case of Lucknow, India

Sukanya Koner, T. Athulya Satheesh, and Kalyan Nath Somavarapu
Indian Institute of Technology, Roorkee, India

Anugya Shandilya
School of Planning and Architecture, Bhopal, India

CONTENTS

DOI: 10.1201/9781003131922-12

10.1 INTRODUCTION

Cities are being profoundly reshaped, often triggered by the links between population growth, socio-economic transformations and urbanization (Allen and Lu 2003, Hegazy and Kaloop 2015, Dinda, Chatterjee and Ghosh 2021). They play their role in determining the hierarchical structure of the urban system, as well as attracting agglomerations and associated spatial problems (Chen and Zhou 2008, Fan, Yu and He 2017, Ma et al. 2019, Shi et al. 2021). Many researchers have discussed concepts such as *pseudo-urbanization* and *over-urbanization*, claiming that the "fulcrum of the urban growth" is shifting to Asian countries like India (Davis and Golden 1954, McGee 1967, Kundu 2011). Thus, cities in developing countries like India act as a major contributor to the *incremental urban populace*, mostly striving to achieve a balance between tremendous demographic pressure and the dynamics of urbanization (Cobbinah, Kwasie and Amoateng 2015, Haque and Patel 2017). Lucknow, as one of India's 'million plus' cities, is no exception to this phenomenon and its rapid change continues to be a critical issue with respect to the city's future developmental path (Singh, Kikon and Verma 2017, Shukla and Jain 2019b).

Urban growth is among the prime concerns for decision makers, planners, economists, ecologists and policy makers (Aburas et al. 2016). Simulation models for predicting urban growth patterns have become an essential tool to understand spatial and temporal changes, especially those using remote sensing (RS) and geographic information system (GIS) techniques (Punia and Singh 2012, Pijanowski et al., 2014, Mubea, Goetzke and Menz 2014). The course of urban growth can be determined by the driving factors of the urban settlement, including socio-economic developments, physical characteristics, neighborhood factors and governmental policies (Zhang et al. 2013, Li, Weiqi and Ouyang 2013, Tan et al. 2014, Jiang and O'Neill 2018). The present study aims to test the hybrid urban growth model with an assessment of different determinants for detecting spatial and temporal changes in the city of Lucknow, India.

It is essential to understand the nexus of urbanization and economic growth as it expedites the process of urban agglomeration, especially in fast-growing cities like Lucknow. Theories of agglomeration economies that deliver proximity of large labor supply, supply chains, customers and competitors establish a relationship between urban growth and the flow of the economy through goods, products and services (de Bok and van Oort 2014, Giuliano, Kang and Yuan 2019). Such externalities of urban agglomerations accelerate socio-economic transformations in the character of the neighborhood, especially in view of the concentration of commercial activities (Snir and van Ham 2019, Ibrahim et al. 2020). The phenomenon of the transition of residential areas to cafés, clothing stores, large chain stores and restaurants is recognized as 'commercial gentrification', and causes skyrocketing rents and changes to the composition of the residents of the neighborhood (Jeong, Heo and Jung 2015, Ryu, Kim and Park 2020). Taking the changing socio-cultural pattern in Lucknow's neighborhoods into account, the research further entails studying commercial gentrification in the Patrakarpuram area of the city.

The research is divided in two sections, a macro-level study and a micro-level study. At macro level, there are four stages. First, inter- and intra-spatial interactions using gravity-based models are illustrated. Second, various urban simulation models are assessed, and a hybrid-based model, cellular automata and fuzzy urban growth modeling (CAFUGM), is adopted. Third, the required data are assembled and analyzed using ArcMap 10.8 (GIS software) (for example, land-use classification for years 2010 and 2020, socio-economic data, major driving urban growth factors, etc.) Finally, the simulation process derives the size, shape and pattern of urban growth till 2020, using a hybrid simulation model (CAFUGM). Pixel matching is applied to validate the model for both the actual and simulated maps for the year 2020. Finally, growth predictions for the year 2030 are derived using the driving factors in the year 2020. Next is the micro-level study which encapsulates the economic growth in Lucknow by focusing on two major indicators, division of labor and economies of scale. The commercial gentrification phenomenon in Lucknow city's neighborhoods is studied through a qualitative research method with an interpretive approach to consolidate social reality through experience. The social survey is conducted using the participant observation research technique to represent decadal temporal changes in neighborhood land use in the study area. The neighborhood character analysis helps to determine both positive and negative effects of the phenomenon. Assessment of property value changes in the area and their impact on small businesses are also discussed. Lastly, innovative approaches implemented in metropolitan cities like San Francisco and Toronto are recommended to build a resilient neighborhood.

10.1.1 Lucknow: City Profile

The Garden City, Golden City of the East, Constantinople of India, City of the Nawabs, Shiraz-e-Hind or a city known for its sophistication, courtesy and etiquette (unique Tehzeeb) are among the names given to the glorious monumental heritage of Lucknow (Khan 2003, Taqui 2005, Markel 2010, Hasnain 2016, Varshneya 2017). It is the capital city of the Indian state of Uttar Pradesh, geographically situated on 26°51′N and 80°57′E coordinates, with an elevation of 123m MSL. The city is located in the central part of India, 550 km from the national capital, Delhi. The municipal area of Lucknow was 91.7 km^2 in 1987, which increased to 162.70 km^2 in 2004–05 and further to 244.82 km^2 in 2010–11 (LDA 2014). Currently the area under Lucknow Municipal Corporation is 350 km^2, with 712.20 km^2 under the 2031 development boundary (Senes Consultants 2015). It is the 11th most populous city in India, with a total population for Lucknow Urban Agglomeration of 2,902,920 in 2011, a decadal increase of 28.26% from 2001. The average population density is 50 persons per hectare (pph) with the highest density in the central old city area. The current population of the city is more than ten times that of 1901 and is estimated to reach 17 times by 2031 (LDA 2014). The city is following a radial pattern of growth equally on both sides of the river Gomati, with sufficient road networks and railways. It lies at the intersection of five national highways – NH-24, NH-25, NH-28, NH-24B and NH-56 – enhancing its connectivity and regional distribution network (Kumari 2015). Figure 10.1 is a representation of the study area for the research. The boundary of the study has been adopted from the Master Plan 2031. The study area includes the local planning area of Lucknow.

10.1.2 Lucknow as a Growth Pole

Lucknow acts as a growth pole, as suggested by Perroux[1] (1949), "an economic space" for five districts – Sitapur, Rae Bareli, Barabanki, Hardoi, and Unnao district – and six rural towns – Malihabad-Dushheri Mango orchard town, historic Kakori, Sarswan, Gosaiganj, Clay Market Chinhat, and Itaunja (Chopra 2012, Shukla and Jain 2019a. An area of approximately 1000 ha is under industrial use constituting major clusters – Talkatora, Chinhat, Sarojini nagar, Amousi, Kursi Road and Sitapur Road (District Industrial Centre 2019). The informal sector, almost 5% of the population, is engaged in street vending, weekly markets, roadside stalls, etc. (Senes Consultants 2015). Inter-spatial interactions between predominant cities in a radius of 250 km and intra-spatial interactions within Lucknow district are represented in Figures 10.2 and 10.3 respectively, using a gravity-based model.[2] Lucknow–Kanpur has the maximum inter-spatial interactions, and the Lucknow–Kanpur Corridor has also been identified as the counter-magnetic area for the NCR region (Tech Mech International (P)Ltd. 2015). Sarsawan in the south-east and Bakshi ka Talab in the north have the highest intra-spatial interactions within the district. Major developments with many shopping malls, high-rise apartment buildings, UP state police headquarters and the Medanta Hospital are situated in the Saraswan area along the Amar Shaheed Route, a 23-km four-lane expressway that connects NH-25 (Lucknow–Kanpur) with NH-28 (Lucknow–Faizabad).

10.1.3 Urban Growth and Land-Use Classification

The expansion of a village covering 48 sq km to a metropolis with a population of almost 3 million over an area of 350 sq km has grown equally on both side of the river Gomati (Kumari 2015). The once linear town has amalgamated into a radial-patterned city which can now be distinguished as old Lucknow and new Lucknow, and has been acknowledged as the 17th fastest-growing city in India and 74th in the world (Mayors 2010). The city witnessed edge-and-ribbon development from

FIGURE 10.1 Study area for the research.

(Source: www.mapsofindia.com (Map of India), www.indiainmaps.com (Map of state of Uttar Pradesh), Compiled by authors (Map of Lucknow).)

1990 to 1999, whereas in 1999–2009, a mixed pattern of infill with edge-and-ribbon expansion was experienced Shukla and Jain 2019a. The spatial growth of Lucknow from 1975 to 2015 is shown in Figure 10.4 and the land-use classification in Table 10.1.

According to the data in Table 10.1, the percentage increment from 1975 to 2015 is approximately 220% in urban areas, with a 45% reduction in water bodies, according to the assessment of RS data.[3]

With growing population demand, new residential colonies in Gomtinagar, Rajajipuram and Indiranagar, along with a real-estate boom, led to peri-urbanization with integrated townships like

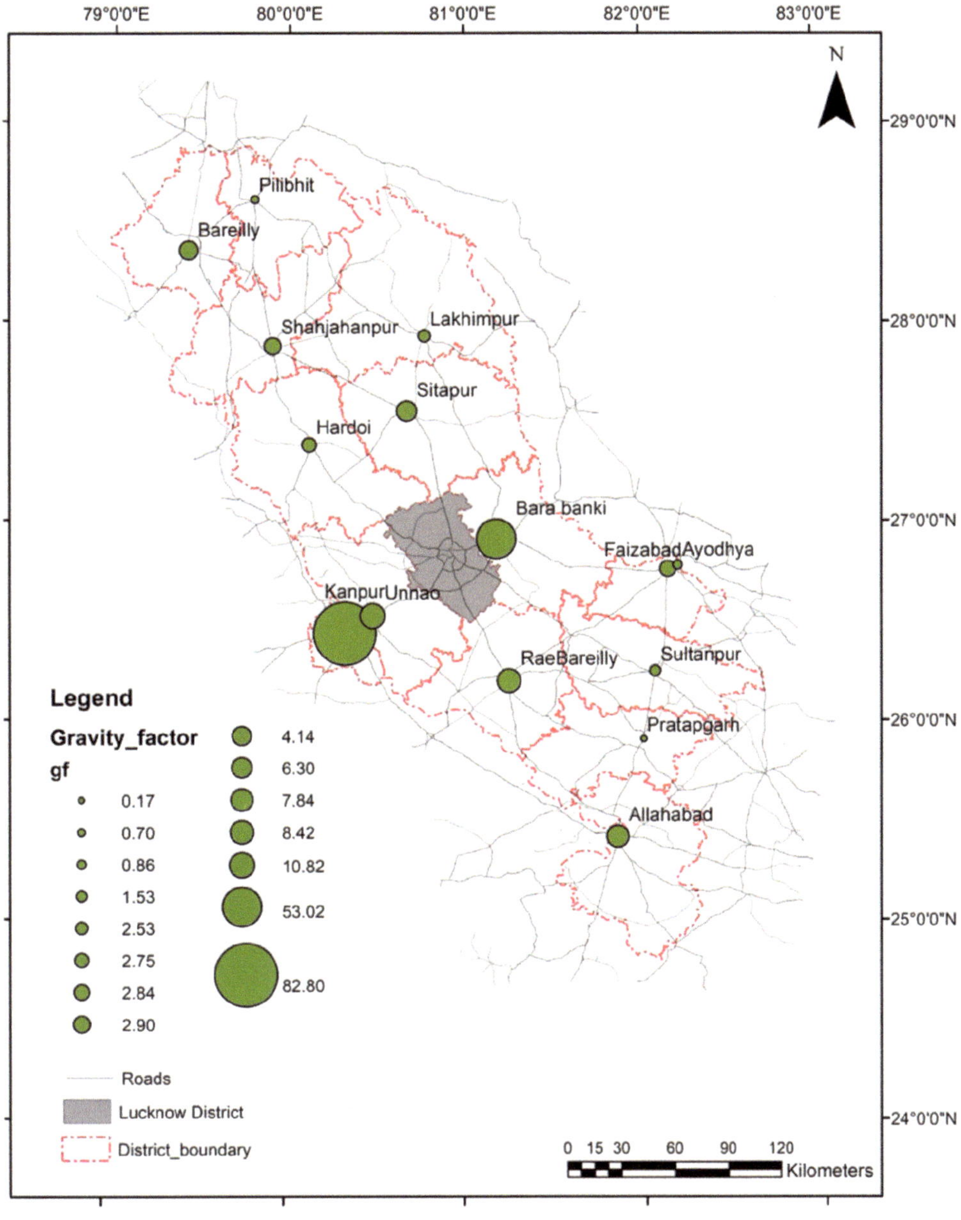

FIGURE 10.2 Inter-spatial interaction between cities within 250 km of Lucknow.

(Source: Compiled by authors.)

Sushant Golf city, Omaxe Township, and BBD green (Dutta 2012). The changes in land-use patterns are analyzed in Table 10.2, where the land-use distribution classifications for 1987, 2004–05 and 2010–11 are compared with the URDPFI guidelines.

The drastic increase in residential use between 1987 and 2005 is mainly due to the development of new colonies by public and private developers. The reduction in the residential area in 2010–11 was because 197 villages were added in the master plan dated 27 January 2009, resulting in 98 sq km of total developed and developable area (Senes Consultants 2015). The city needs to strike a balance between urban areas, water bodies, transportation services and recreation facilities,

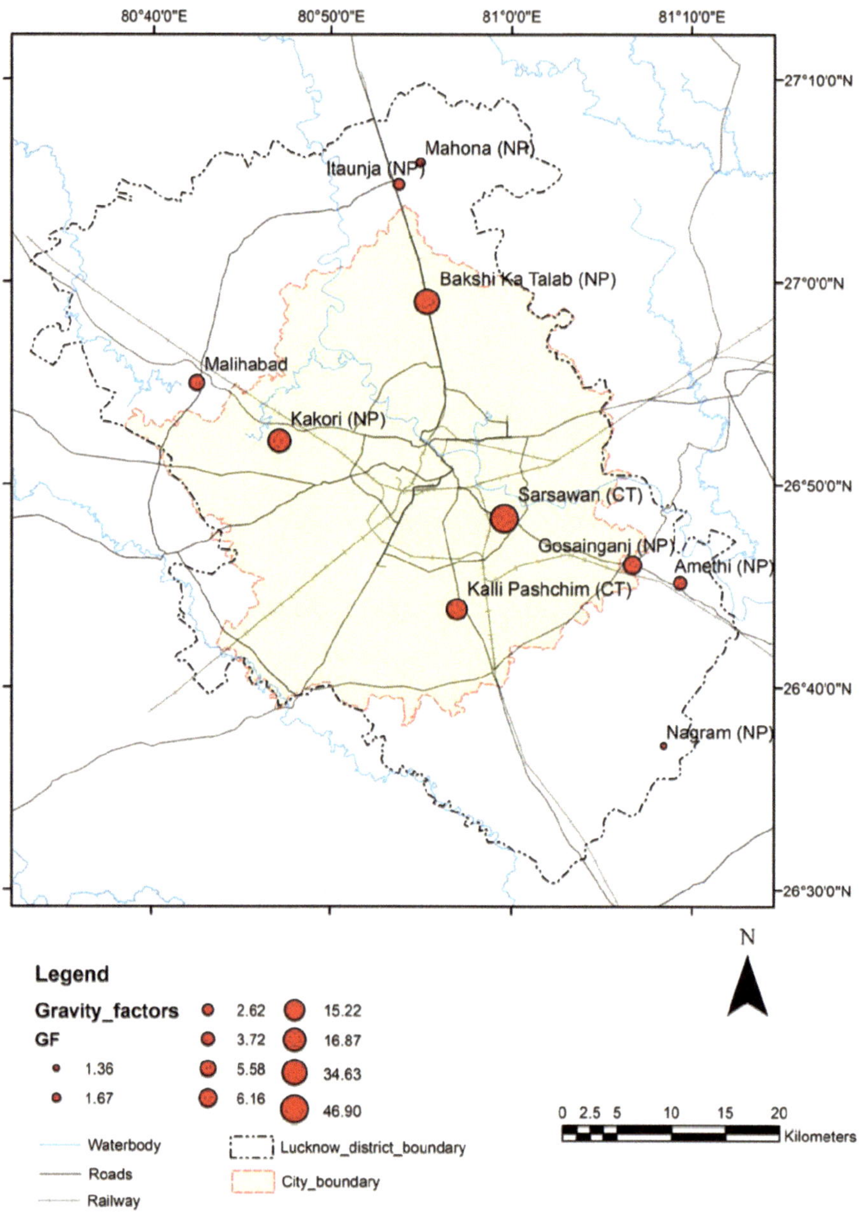

FIGURE 10.3 Intra-spatial interaction between the core city and towns with Lucknow district.

(Source: Compiled by authors.)

according to assessment of RS data and URDPFI guidelines, especially in order to reduce ecological vulnerability and major changes in the local climate (Singh, Kikon and Verma 2017).

10.1.4 Economic Growth in Lucknow

The two aspects of economies of scale are internal economies and external economies (Loesch 1954). Internal economies of scale or 'traditional economies' are subject to the internal functions of the firm reducing average unit costs to expand to larger-scale production (Hariharan and Biswas, 2020a).

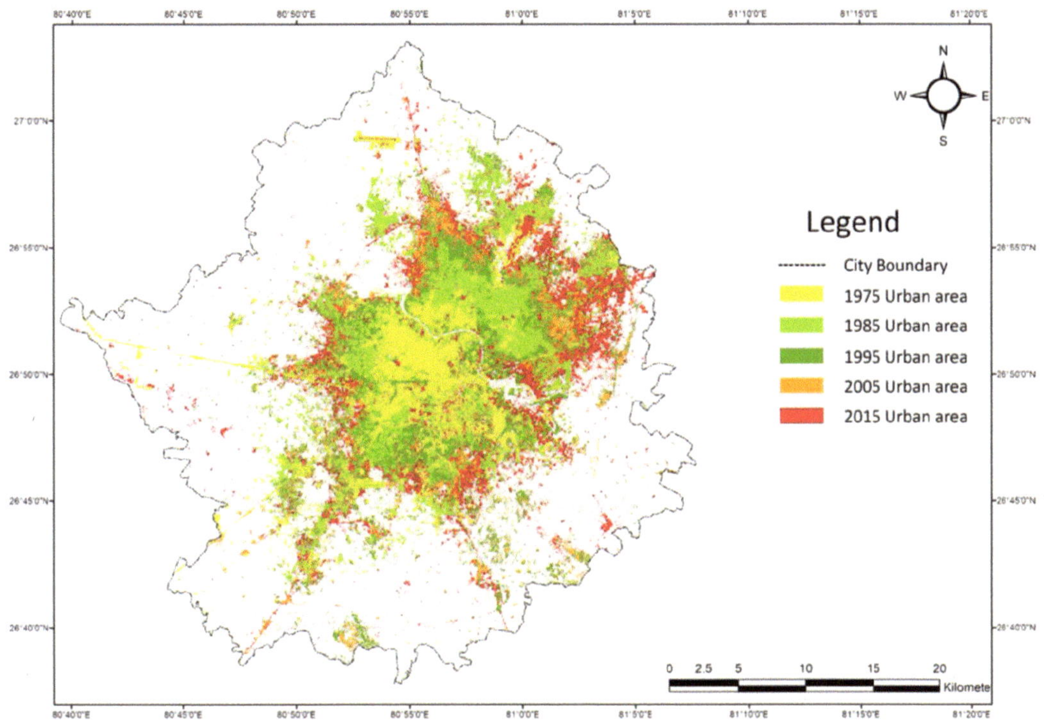

FIGURE 10.4 Spatial growth of Lucknow from 1975 to 2015.

(Source: Compiled by authors.)

TABLE 10.1
Land-Use Classification from 1975 to 2015

Year	Urban (Built-up) (Sq km)	Water (Sq km)	Vegetation (Sq km)
1975	112.94	65.47	1038.96
1985	240.47	75.96	900.95
1995	303.61	35.51	878.18
2005	315.33	30.92	871.05
2015	363.42	35.69	818.21

Source: Compiled by authors.

External economies of scale or 'agglomeration economies' are the economic network effects between firms and the locational benefit of being close to each other (Pfluger and Tabuchi 2018). These networks enhance the flow of information and increase competitiveness (Glaeser et al. 1992). Geographical proximity and the varying composition of economic activities play critical roles in classifying agglomeration economies into localization economies and urbanization economies (Dwivedi and Arora 2019). Localization, internal to an industry within a geographic region, indicates industry-specific input factors stimulating growth, whereas urbanization economies are internal to a spatial unit with diverse economic agents leading to urban agglomeration (Jacobs, 1969, Turok and McGranahan 2013). Mydral elaborated a phenomenon where clusters of economic activities within a region accumulate growth, known as 'cumulative causation' (Myrdal 1957). Accessibility of good

TABLE 10.2
Land-Use Classification for the Years 1987, 2004–05 and 2010–11

Land Use	1987 (%)	2004–05 (%)	2010–11 (%)	% Metropolitan Region (URDPFI Guidelines)
Residential	48.9	55	42	36–38
Commercial	2.4	2.2	2.55	5–6
Industrial	6.5	6.1	3.84	7–8
Public	5.2	3.4	3.63	10–12
Community services	9.8	8.7	8.72	–
Recreation & parks	3.8	2.7	4.08	14–16
Transportation	10.4	7.6	32.04	12–14
Water bodies	2.1	1.9	2.36	Balance
Others	10.9	12.4	0.79	–

Source: Master Plan 2031 (LDA 2014), (Town and Country Planning 2015).

markets, forward and backward linkages amongst firms, a concrete labor market and technological support are among the determinants of regional economic growth (Marshall 1890, Ricci 1999).

The Chikankari Embroidery, a 400-year-old Persian work of art accorded a geographical indicator (GI) tag in December 2008, is a symbolic cultural asset of Lucknow and its localization economy (Agrahari and Brar 2017, Textile Value Chain 2020). The city has its roots in the international handicraft market with the largest Chikankari artisan cluster in India, producing unique styles of the art (Ahmad and Anees 2016, Chantia 2017). The state government of Uttar Pradesh initiated One District One Product (ODOP), emphasizing boosting economic development and encouraging young people to secure new employment opportunities. Lucknow has been identified as an economic cluster for Chikankari. 250,000 artisans are associated with this cluster, disseminated in and around Lucknow within a radius of 125 km (Chantia 2017). A turnover of Rs 600 crores is the economic output from the cluster, with 4000 manufacturers in small, medium and large enterprises (Arya and Sadhana 2007, Pandey and Pawar 2020). The structure of enterprises in the cluster, based on their turnover, is given in Table 10.3. The contribution of Lucknow's Chikankari industry to Uttar Pradesh state's GDP is 8.12%, with an annual revenue of $1.2 billion (Arya and Sadhana 2007).

Lucknow is a prime economic base in northern India and a pre-eminent trading center. The economic diversity representing 'urbanization economies' comprises various sectors – tourism, health services, pharmaceuticals, aerospace, design and culture (Feedback Ventures 2006, Cyrill 2016). Major produce within the city region includes Chikankari garments, Zardozi works, aircraft peripherals, steel, wrought-iron furniture and asbestos in the production category; wheat, rice and barley

TABLE 10.3
Composition of Chikankari Enterprises in the Cluster

Turnover (in Rupees per Annum)	Number of Enterprises
40–50 million	10
10–40 million	250
5–10 million	500
0.5–5 million	2000
0.3–0.5 million	1000

Source: Chantia (2017).

TABLE 10.4
Major Commodities within Lucknow Region

	Most Important Commodity	
Area	**Manufactured**	**Exported**
Lucknow UA	Chikankari garments, airplane parts and tools	Chikankari garments, airplane parts and tools
Gosaiganj NP	Rice, peppermint oil, Chikankari work and garments	Rice, oil seed
Kakori NP	Chikankari garments, earthenware pots	Mango, vegetables, Chikankari garments
Malihabad NP	Mango, nursery, Chikankari garments	Vegetables, mango, Chikankari garments
Bakshi ka Talab	Chikankari garments, dairy, ice-cream	Chikankari garments, dairy, processed foods

Source: Lucknow District Industrial Profile (MSME, GOI 2020).

in agriculture; and pulse milling, vegetable oil processing and dairy products in processing (MSME, GOI 2020). The traditional trade in Lucknow also includes cultivation of mangoes known as *dusheri* in Malihabad (Rajan, Lal and Lamers 2016). Table 10.4 illustrates important commodities produced in Lucknow district. The city was also declared a Biotechnology City in 2002 to boost the incubation facilities aiming to bring prosperity to farmers through biotechnology (KPMG 2003). A biotech park of 56,000 sq ft. in the northern part close to the ring road with an active incubator facilitates entrepreneurs and innovators with business support, technological transfer, operational links with research institutions and infrastructural facilities (Dept of Biotechnology, Govt of India 2002). The IT sector in Lucknow has witnessed tremendous growth since Software Technology Parks of India was set up in 2001. An IT SEZ (special economic zone) project in 100 acres of the Chak Gajaria IT hub and educational institutes like the Indian Institute of Information Technology, located near NH-56 and along Amar Shaheed Route, are symbolic of the emerging knowledge-based industries of the region (Sinha 2016, Hariharan and Biswas 2020b).

10.2 RESEARCH LITERATURE

10.2.1 Urban Growth Simulation Models

The emergence of information technology and datasets on urban areas led to the generation of new models (Batty 2008). The integration of GIS into spatial models has changed the way we look at urban growth by providing a realistic pattern of growth. These include spatio-temporal along with quantitative methods of mapping and monitoring urban growth patterns. Urban areas are complex and dynamic, making urban growth modeling important yet difficult. Several urban simulation techniques have been developed and used in urban studies. The efficiency of these models varies with the study area and the driving factors considered. Many modified techniques have been developed in recent years to help with urban growth modeling; however, every model has its limitations. Some of the widely used simulation models for urban growth studies are the cellular automata (CA) model (Clarke, Hoppen and Gaydos 1997, Clarke and Gaydos 1998, Liu and He, 2009), the slope, land-use, exclusion, urban extent, transportation, and hill shade (SLEUTH) model (Herold, Goldstein and Clarke 2003, Jantz et al. 2010), the land transformation model (LTM) (Pijanowski et al., 2014), cellular automata–Markov (CA–Markov) model (Sang et al. 2011, Jafari et al. 2017), cellular automata–artificial neural network (CA–ANN) model (Li and Li 2015, Gantumur et al. 2020), and the fuzzy cellular automata model (Wu 1998, Al-Ahmadi, See et al. 2009, Al-Darwish et al. 2018, Liang, Liu and Chen et al. 2020). Past research papers show that CA is the most widely used model for simulation (Aburas et al. 2016).

Hybrid models can help overcome the limitations of these models (Al-Darwish et al. 2018).

Cellular automata (CA) models have been used in recent times to simulate complex urban growth systems. They work on algorithms where transitional rules are applied to cells to simulate urban growth. It can only be done with the aid of data management software like GIS (Al-Darwish et al. 2018). Integration of CA, which has an open structure, with other models improves its capability in predicting urban growth (Aburas et al. 2016). Additional research has been done in recent years to improve the constraints of the CA model, and it has been modified in different ways to incorporate the multiple driving factors influencing urban growth to make the model more efficient and accurate. These can be land use, topological and geographical features, socio-economic factors, etc. Intelligent algorithms help establish transition rules for the CA model (Liang, Liu and Li et al. 2018). Agent-based models with multi-criteria analysis are also used to simulate urban growth patterns (Arsanjani, Helbich and Vaz 2013).

The traditional CA model uses pixel testing and does not include partial truth. To include the partial development character of urban growth, fuzzy logic is often incorporated in the transition rules of the CA model. Fuzzy logic is highly applicable to urban areas as the growth is continuous and not pixel based, with insufficient and vague data (Foroutan and Delavar 2012, Mandelas, Hatzichristos and Prastacos 2007). It is a better option for managing uncertainty (Mantelas et al. 2012).

Table 10.5 is a literature matrix for all the contextual literature studied for the present research on urban simulation models. It helps to understand the vitality of the cellular automata and fuzzy urban growth modeling (CAFUGM) as the simulation model adopted for the research.

The fuzzy-constrained CA approach is used for urban simulations in many research works (Mantelas, Hatzichristos and Prastacos 2010, Liu, 2012). Cellular automata and fuzzy urban growth modeling (CAFUGM), stochastic perturbation and fuzzy distance transformation (FDT), fuzzy cellular automata urban growth model (FCAUGM) etc. are hybrid models used to simulate urban growth using fuzzy logic in CA (Zhang et al. 2020, Al-Darwish et al. 2018, Al-Ahmadi, Heppenstall et al. 2009). Applications of these models showcase their effectiveness for studying urban dynamics.

10.2.2 Concept of Gentrification and Commercial Gentrification

The concept of gentrification was first identified by Ruth Glass in 1964 as transformations in housing patterns and ownership that change physical and socio-economic urban environments (Glass 1964, Seo 2002, McKinnish, Walsh and White 2010, Meltzar and Ghorbani 2017, Atuesta and Hewings 2019, Cho et al. 2020). This is 'classic gentrification', where the middle class is rehabilitated in city neighborhoods, displacing working-class or indigenous communities (Lees, 1994, Kosta 2019). Many researchers have focused on the cause of this phenomenon in cities like New York, San Francisco, Baltimore, Chicago, Washington DC, etc. (London and Palen 1984, Cohen 1983, Smith, 1996, Richardson, Mitchell and Franco 2019). The rent-gap theory, which is based on principles of potential land value and capitalized land value, was used by Smith to describe the cause of gentrification (Smith, 1996). The term 'potential land value' refers to the value of land that can be realized when it is used to its full potential. The capitalized land value is the value of the land's present use (Smith, 1996). Smith claimed that while the capitalized land value is low in older neighborhoods, the potential land value increases with the prospect of redevelopment for profit maximization, then gentrification occurs as a result of the disparity between the two divergent land values. Researchers like Ley and Moos argue that the change in the workforce to young professionals increased the city's socio-cultural activities and fostered gentrification as there was a paradigm shift from manufacturing industries to knowledge industries (Ley, 1986, Moos et al. 2018). The influx of inhabitants to the city's neighborhoods due to employment opportunities (Kennedy and Leonard 2001), post-secondary education institutions (Moos et al. 2018), and government investments supporting the middle and upper class (Morisson and Bevilacqua 2019) accelerates gentrification processes. The movement of knowledge workers into the city transforms urban neighborhoods through cultural and community displacement, with an exponential economic heft in housing and mobility costs (Richardson, Mitchell and Franco 2019).

TABLE 10.5
Literature Matrix for Different Urban Simulation Models

Authors and Year	Models	Place	Comment
Liang, Liu and Li et al. (2018)	FLUS model	Pearl River Delta region, China	Higher simulation accuracy achieved with addressing traffic planning. Spatial heterogeneity and complexity of urban growth pattern is reflected by coupling the proposed mechanisms in the simulation.
Al-Darwish et al. (2018)	Cellular automata and fuzzy urban growth modeling (CAFUGM) Multi-resolution validation (MRV) method	Ibb city (Yemen)	Urban simulation model (CA-Fuzzy) is capable of assessing the size, pattern and trends of future urban growth with an acceptable degree of accuracy.
Zhang et al. (2020)	Stochastic perturbation and fuzzy distance transformation (FDT), AHP	Wuhan (China)	FDT and the optimized stochastic variable together can obtain simulation results that are the most similar to the actual data.
Aburas et al. (2016)	Kappa, CA (literature)	–	Realistic and idealized simulation generated using integration approaches using new spatial transition rules produces more plausible results.
Mantelas et al. (2012)	CA models (literature) Sensitive sum operator	Mesogia (Greece)	Fuzzy logic is an advisable framework to deal with datasets and provides the proper tools for the management of uncertainty.
Mandelas, Hatzichristos and Prastacos (2007)	Fuzzy cellular automata-based shell	Mesogia (Greece)	The model simulates efficiently the qualitative patterns under which urban growth occurs for short periods.
Al-Ahmadi, Heppenstall et al. (2009)	Fuzzy cellular automata urban growth model (FCAUGM)	Riyadh	To a large extent successful at replicating spatial patterns over time. Offers significant benefits for simulating urban growth and change, for urban planning and decision support for policy makers and others.
Mantelas, Hatzichristos and Prastacos (2010)	Rule-based modeling, cellular automata, fuzzy logic	Mesogia (Greece)	Fitting indicators suggest that the model simulates efficiently the qualitative patterns of the urban expansion in the study area, which is further certified by visual comparison
Liu (2012)	Fuzzy-constrained cellular automata approach	Gold Coast City (Australia)	Predicts the compact smart growth scenario using fuzzy set approach.
Foroutan and Delavar (2012)	Fuzzy logic, validation, ROC, Kappa, OMA	Isfahan Metropolitan Area (Iran)	Rule-based models (fuzzy logic) are most suitable for incorporating spatial interaction effects and handling temporal dynamics.

Source: Compiled by authors.

The definition of gentrification has evolved over time, from Ruth Glass's conventional definition to a myriad other definitions such as rural gentrification and new-build gentrification (Lees, Slater and Wyly 2008). The phenomenon of the transformation of small retail shops and residential housing into luxury restaurants, boutiques, and value chain stores in the city's neighborhood has been described as commercial gentrification. It has emerged as a modern type of gentrification in major cities around the world like New York City (Zukin et al. 2009), Amsterdam (Kloosterman and Van Der Leun 1999), Edinburgh (Doucet 2009), Prague (Verwaaijen 2014), A Coruña (Barcón, Calo and

TABLE 10.6
Types of Commercial Gentrification

Type of Commercial Gentrification	Definition	Features
Retail upscaling	The process that changes an area's retail composition by 'boutiquing' of streetscapes (Zukin et al. 2009) and supermarkets that appeal to high-income group consumers. (Sullivan 2014)	• Development of selective product-based consumerism. • Inequitable distribution to retail access due to high prices. • Unable to serve low-income residents.
Corporatized Commercial Gentrification	The phenomenon of changing an area's historical ethnicity by commercializing ethnic cultures to promote tourism-based consumerism (Gotham 2005, Hackworth and Rekers 2005, Burnett 2014)	• Use of business improvement techniques to package ethnicity of neighborhoods • Escalated housing prices due to changes in residential and commercial real estate. • Transition in focus from ethnic population to tourist population, creating a sense of exclusion and alienation in ethnic residents.
Transit-Oriented Commercial Gentrification	Commercial gentrification aided by the characteristics of transit proximate neighborhoods (Ong, Pech and Ray 2014, Chapple et al. 2017)	• Changes in retail employment due to accessibility. • Increase in commercial and residential property values. • High risks to pedestrians and cyclists as these areas attract major vehicular traffic.
Residential-Commercial Gentrification	The phenomenon of conversion of residential housing to cafés, restaurants, big retail outlets, and apparel stores (Yoon and Park 2018, Cho et al. 2020)	• Mounting rent and property values. • Transformation in the composition of residents and character of place • Large chain stores may disrupt social bonds and cohesiveness. • High competitiveness in the area

Source: Compiled by authors.

Enríquez 2018), Toronto (Rankin and McLean 2014) and Seoul (Cho et al. 2020). Different forms of commercial gentrification based on their mechanisms are synthesized from the literature and listed in Table 10.6. These typologies have been identified based on previous academic research, and the individual features mentioned in the table help to understand the socio-economic impact of these gentrifications.

10.3 RESEARCH METHODOLOGY

The chapter is primarily based on both macro-level and micro-level study. The macro-level study of the urban growth of Lucknow is simulated through the adopted model, CAFUGM. The micro-level study discusses the socio-economic impact of rapid urbanization in Lucknow's neighborhood, where rapid urban growth has stimulated the phenomenon of commercial gentrification. Both these studies are presented in this chapter and a detailed research methodology is illustrated in Figure 10.5.

10.3.1 Macro-Level Study

The research in the macro-level study has been broadly divided into three levels. The first level deals with preparation of data and processing it using an Esri GIS software program called ArcMap 10.8, along with maps that showcase elements that affect the growth of urban areas (Dewan et al. 2014). The next level is the cellular automata and fuzzy urban growth hybrid model (CAFUGM), in which

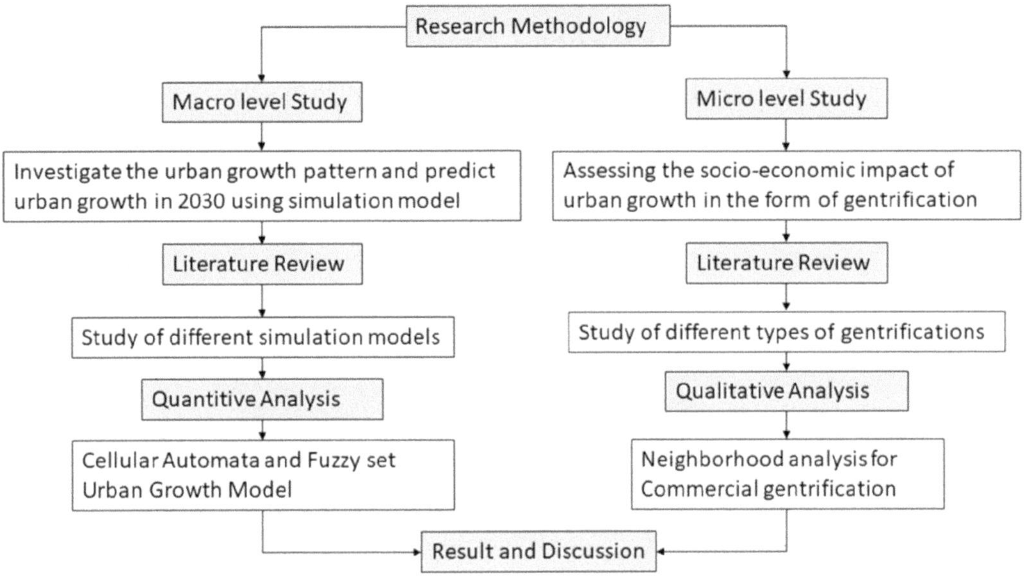

FIGURE 10.5 Research Methodology.

(Source: Compiled by authors.)

data has been used to carry out data simulation in the LanduseSim (Pratomoatmojo 2018) software program for forecasting subsequent urban growth (Gharbia et al. 2016). The software can be assessed from https://www.landusesim.com/. The third level consists of data visualization and presentation of results.

10.3.2 Micro-Level Study

Commercial gentrification is one of the most crucial forces transforming the contemporary urban environment of Lucknow. The study, which follows a qualitative research method, seeks to understand how commercial gentrification is emerging in the Patrakarpuram area of Lucknow. Qualitative research is generally descriptive, exploratory or inductive, and aims to reveal new phenomena by examining the available data (Booth, Colomb and Williams 2008, Gabriel 2013, Hamilton and Finley 2020). The social reality is constructed by seeing, interpreting and experiencing the elusive character of the urban neighborhood (Willis 2007, Haralambos et al. 2017). The research techniques used to obtain qualitative data are participant observation and semi-structured interviews (Booth, Colomb and Williams 2008). The interviews were conducted with 11 retail business owners and seven landlords in Sahara Plaza, Patrakarpuram market.

10.4 ANALYSIS AND FINDINGS

10.4.1 Macro-Level Study

10.4.1.1 Data Preparation

In city planning it is very important to use spatial data from different times to understand the growth patterns of the city (Alsharif et al. 2015, J. Li et al. 2021). The false-color composite remote sensing data from Lucknow for the years 2010 and 2020 are displayed in Figures 10.6 and 10.7. The general land use supervised classification without details of urban types (residential, industrial, commercial,

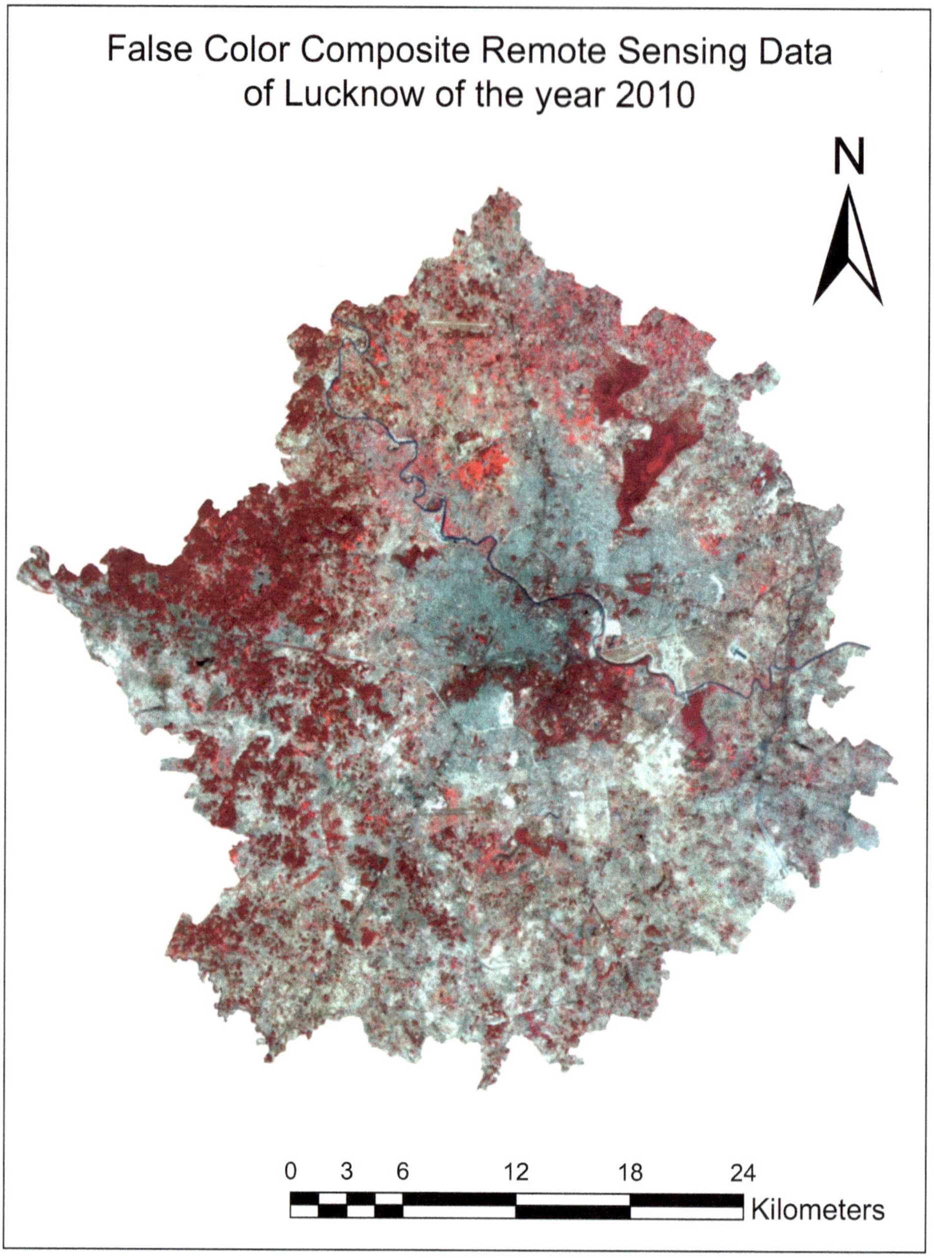

FIGURE 10.6 False-color composite remote sensing data from Lucknow for 2010.

(Source: USGS Earth Explorer-LANDSAT 5 FCC Remote Sensing data from Thematic Mapper retrieved on 17-12-2010.)

etc.) for the years 2010 and 2020 are illustrated in Figures 10.8 and 10.9. The data, gathered from multiple sources, comprise three land-use components: urban areas, non-urban areas (agricultural areas, vegetation and barren land) and reserved areas (water bodies). As Table 10.7 shows, the urban areas witnessed a decadal growth of 58.41%.

The study creates Euclidean distance (ED) maps for analysis of the socio-economic drivers behind urban growth, that is, transportation networks, hospitals, higher education, recreation spaces,

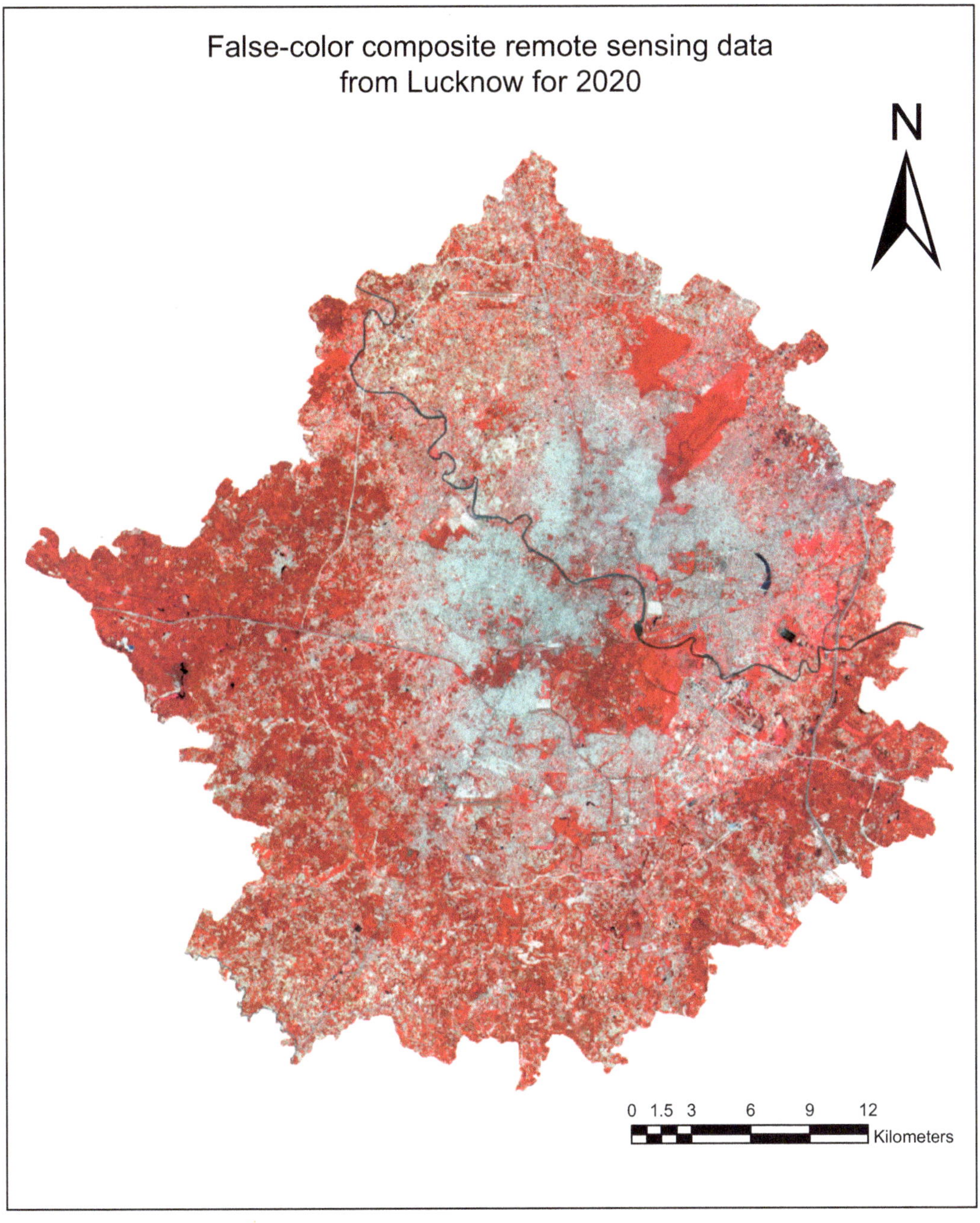

FIGURE 10.7 False-color composite remote sensing data from Lucknow for 2020.

(Source: USGS Earth Explorer- LANDSAT 8 FCC Remote Sensing data from Operational Land Imager (OLI) and Thermal Infrared Sensor (TIRS) retrieved on 09-10-2020.)

shopping complexes/commercial spaces (Yoon and Park 2018, Kim, Newman and Güneralp 2020). The data mentioned is processed in a GIS environment with satellite images of Lucknow for the year 2010 (with resolution of 30m) and for 2020 (with resolution of 30m). The socio-economic drivers in the year 2010, demonstrated in Figure 10.10(i)–(v), are used to simulate urban growth in 2020, and the drivers in the year 2020, as shown in Figure 10.11(i)–(v), are taken to predict urban

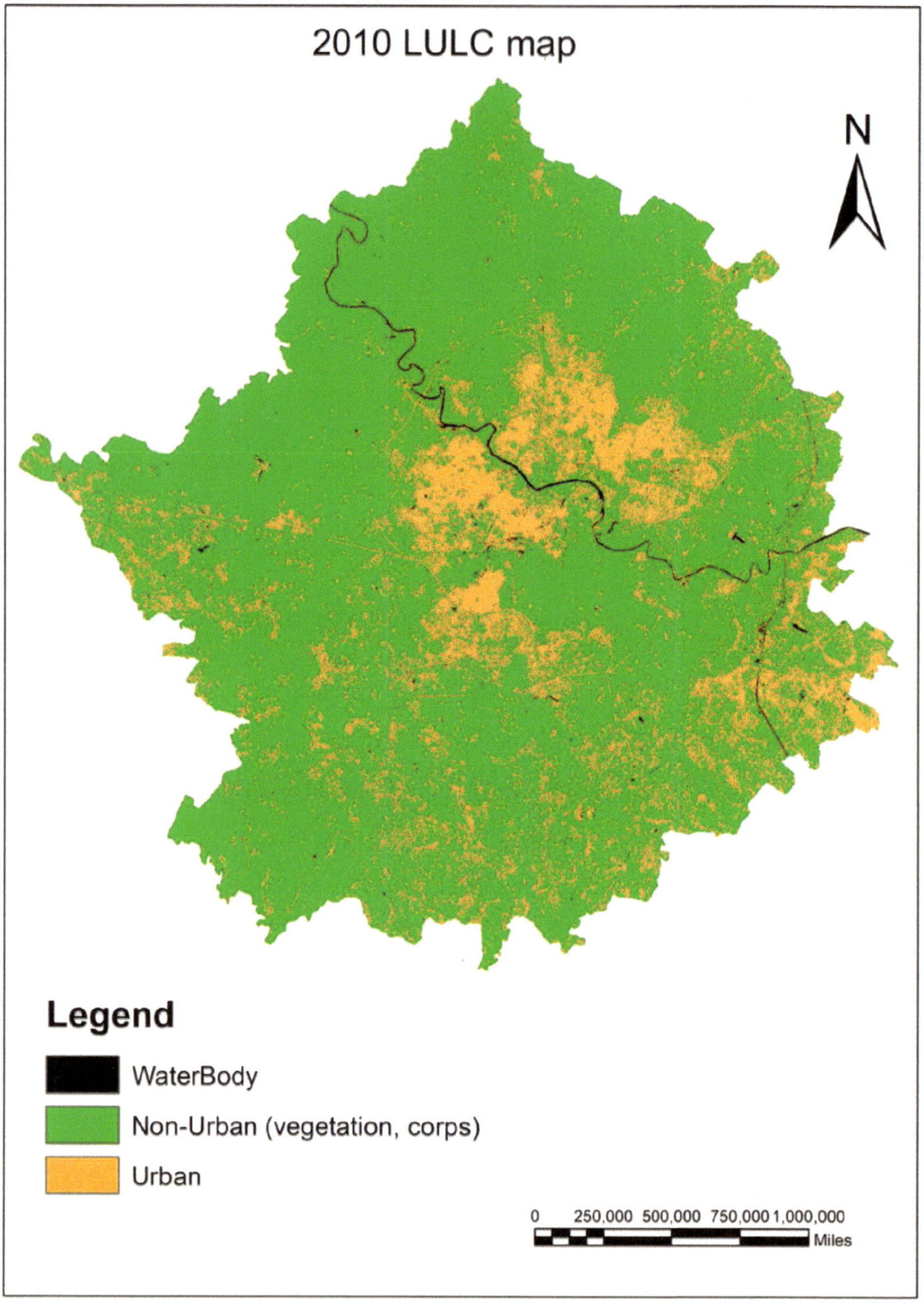

FIGURE 10.8 Supervised land-use classification for the year 2010.

(Source: Compiled by the authors.)

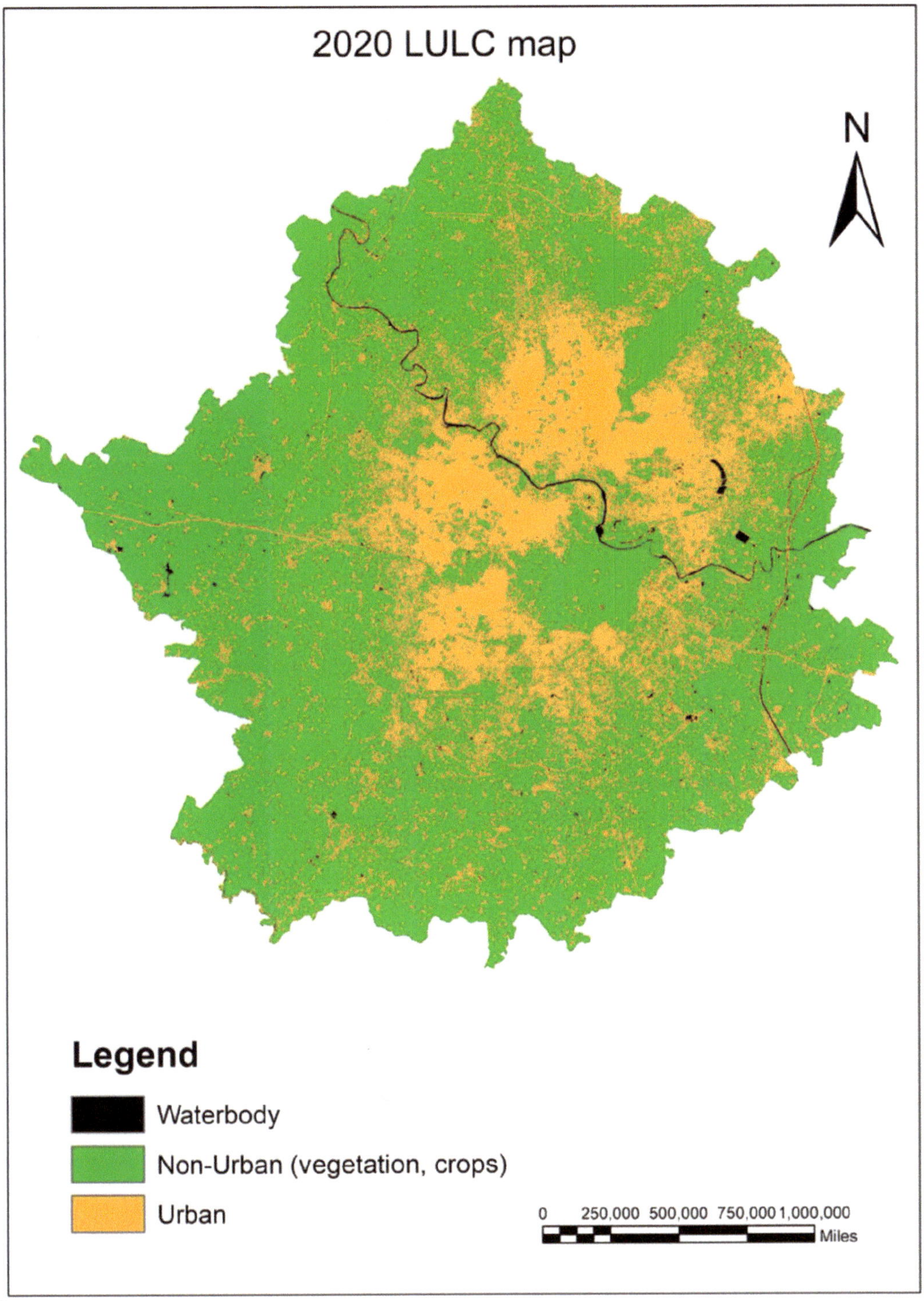

FIGURE 10.9 Supervised land-use classification for the year 2020.

(Source: Compiled by the authors.)

TABLE 10.7
Land-Use Components for the Years 2010 and 2020

Major Land Use (LU)	LU 2010 Area (ha)	LU 2020 Area (ha)
Urban area	23315.76	36934.65
Non-urban area (Agricultural, vegetation, barren land)	97708.05	83941.11
Reserved areas (Water bodies)	704.88	852.93

Source: Compiled by Authors using ArcGIS software program.

growth in 2030. The final data is obtained in raster format using ArcGIS Land Use Polygon to Raster software.

10.4.1.2 Data Simulation

All the steps at this stage are processed in the LanduseSim software environment. The processed dataset is imported and standardized using a fuzzy approach for the simulation model (Al-Ahmadi, 2018). Further, the weights for each factor are determined by preparing potential transition maps using transition rules. Thus, a CA model is used to simulate the future growth of urban Lucknow. Figure 10.12 represents the general framework for the simulation model. As illustrated in the framework, Euclidean distance maps have been created using a linear method of fuzzy class membership to obtain the distance of the driving force dealing with different factors (Dong et al. 2008, Ergen 2014, Gopal et al. 2016). These EDMs (Euclidean distance maps) are then converted to real numbers with a set of values between 0 and 1, where 0 is the farthest and 1 is the nearest distance. This evaluation is done to provide the best value for urban growth potentials and when the cells are closest to the driving forces, then we can see the value gradually diminishing as the cells move away from the driving forces to represent the effect of diminution and decay. The chances of the cell becoming an urban land increase as it gets closer to any of the driving variables (Sridhar 2016, Hwang and Woo 2020).

Since urban growth is unpredictable and does not follow a set pattern, an analytical hierarchy process (AHP) weighted method has been used to give weight to each deciding factor, which are then compared with expert opinions (Feng and Qi 2018, Saha and Roy 2021). However, before determining the weightings, the magnitude of the impact created by each category (for instance, roads) is assessed via simulations independently of other driving factors. Similarly, each factor – hospital, recreational, educational institutions, commercial centers/shopping complexes – is simulated separately and these are then checked as a part of ground truthing (Aithal, Vinay and Ramachandra 2018). Depending on the similarity of the factor and its actuality, a ratio is obtained and the most similar one becomes the most important factor, and so on. Weights calculated by the AHP method are expressed in Table 10.8. The maximum weight calculated is for commercial centers/shopping complexes followed by hospitals. Thus, the role of commercial activities predominates in the urban growth of Lucknow.

An initial transition potential map (ITPM) is obtained by overlaying the driving factors which shape the urbanization of the city and help determine the key factors in cell development. Following this analysis, a cellular automata model is used and a transition rule established using three cellular cases – urban, non-urban (agriculture and vegetation), and reserved area (water) – where urban areas are the primary basis and non-urban areas negate the equation (Deep and Saklani 2014, Gharbia et al. 2016). Integration of ITPM and the land-use map results in a 3x3 neighborhood cell, then the growth model for Lucknow is generated for the simulation process time from 2010 to 2020. The process is based on transition rules, which includes the weights for driving factors, restricted areas for urban growth and the estimation of developing cells determined through land-use maps.

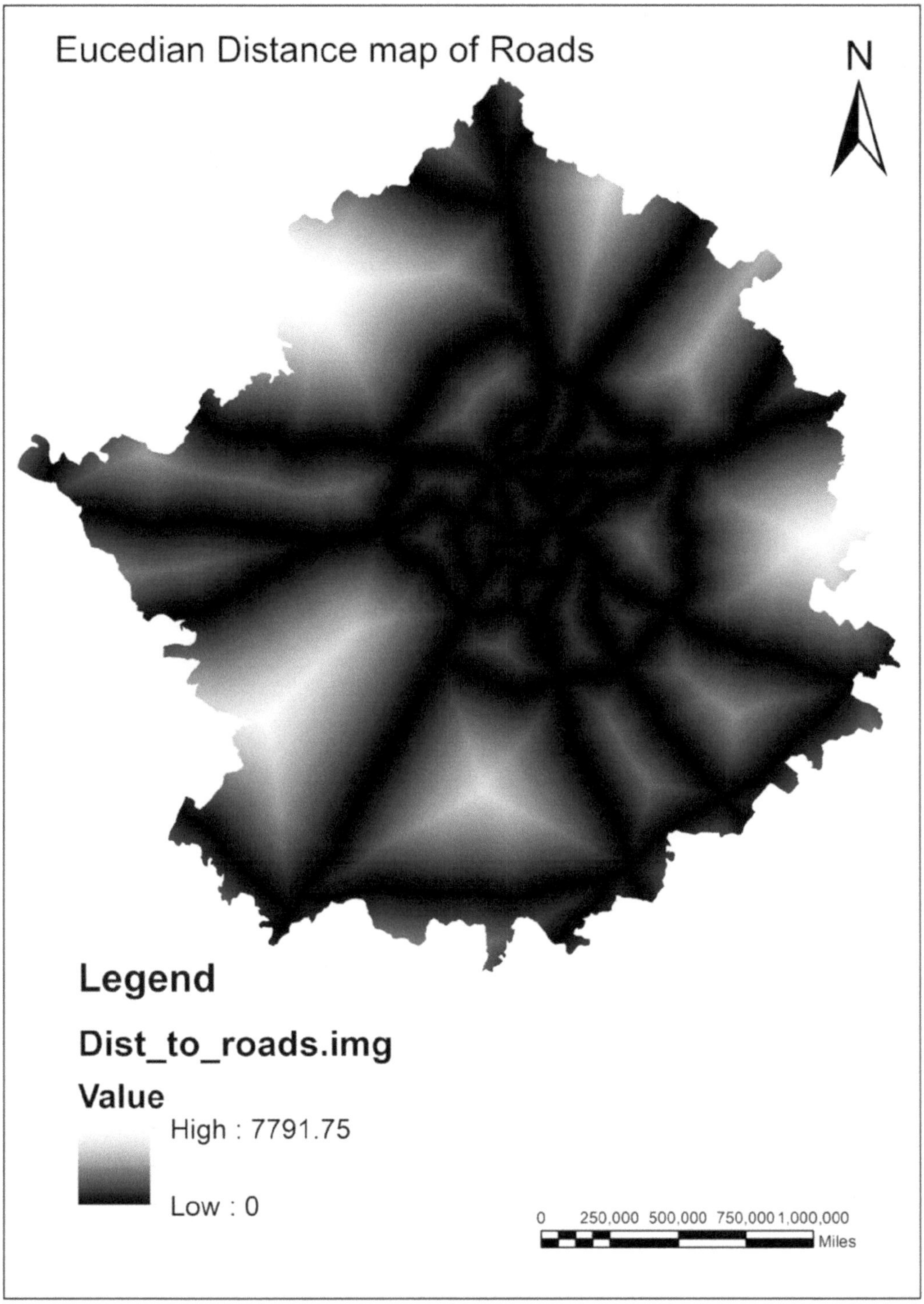

FIGURE 10.10 (i) Driving Factor in 2010 – Primary and Secondary Roads.

(Source: Compiled by the authors.)

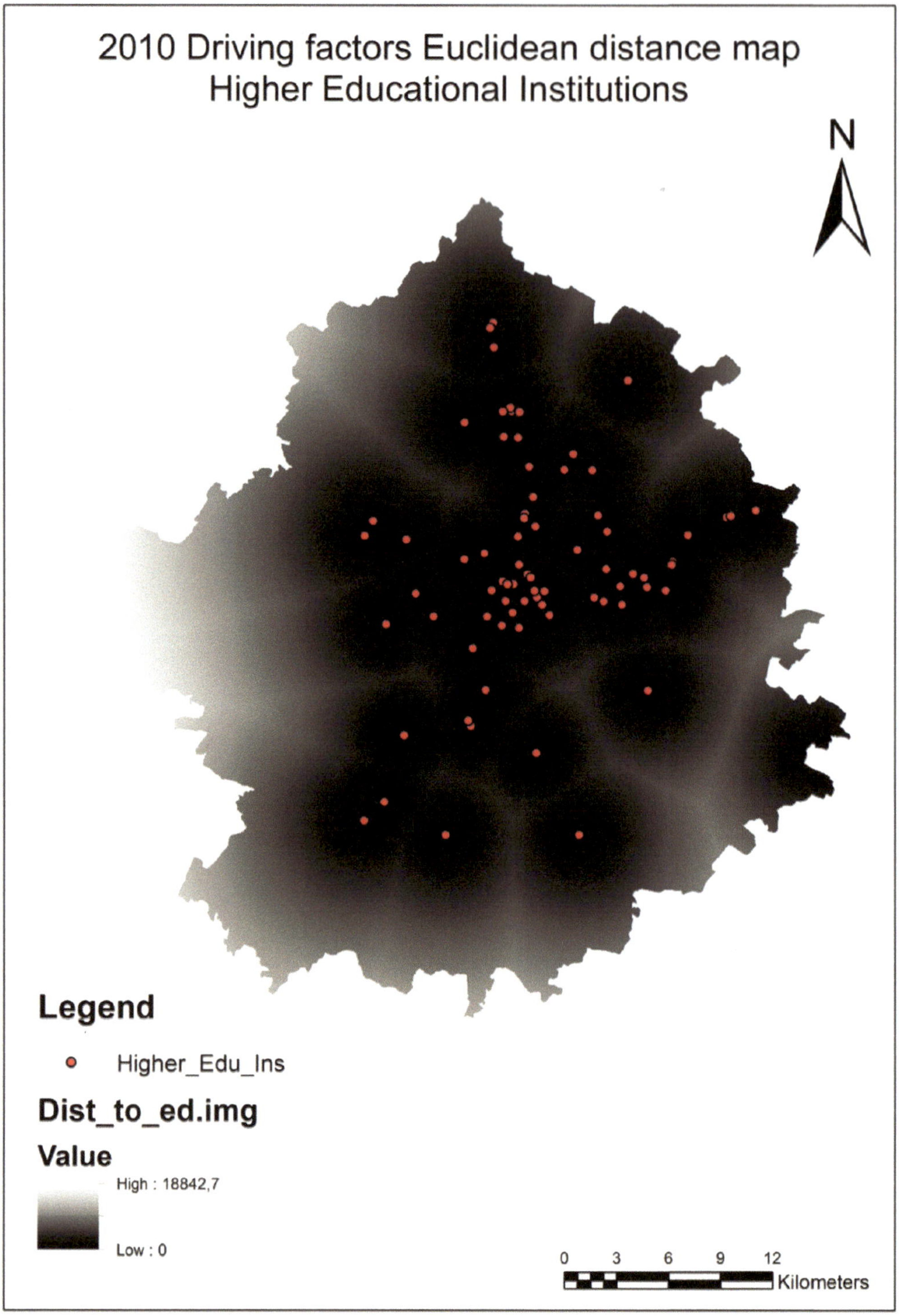

FIGURE 10.10 (Continued) (ii) Driving Factor in 2010 – Higher education institutions.

(Source: Compiled by the authors.)

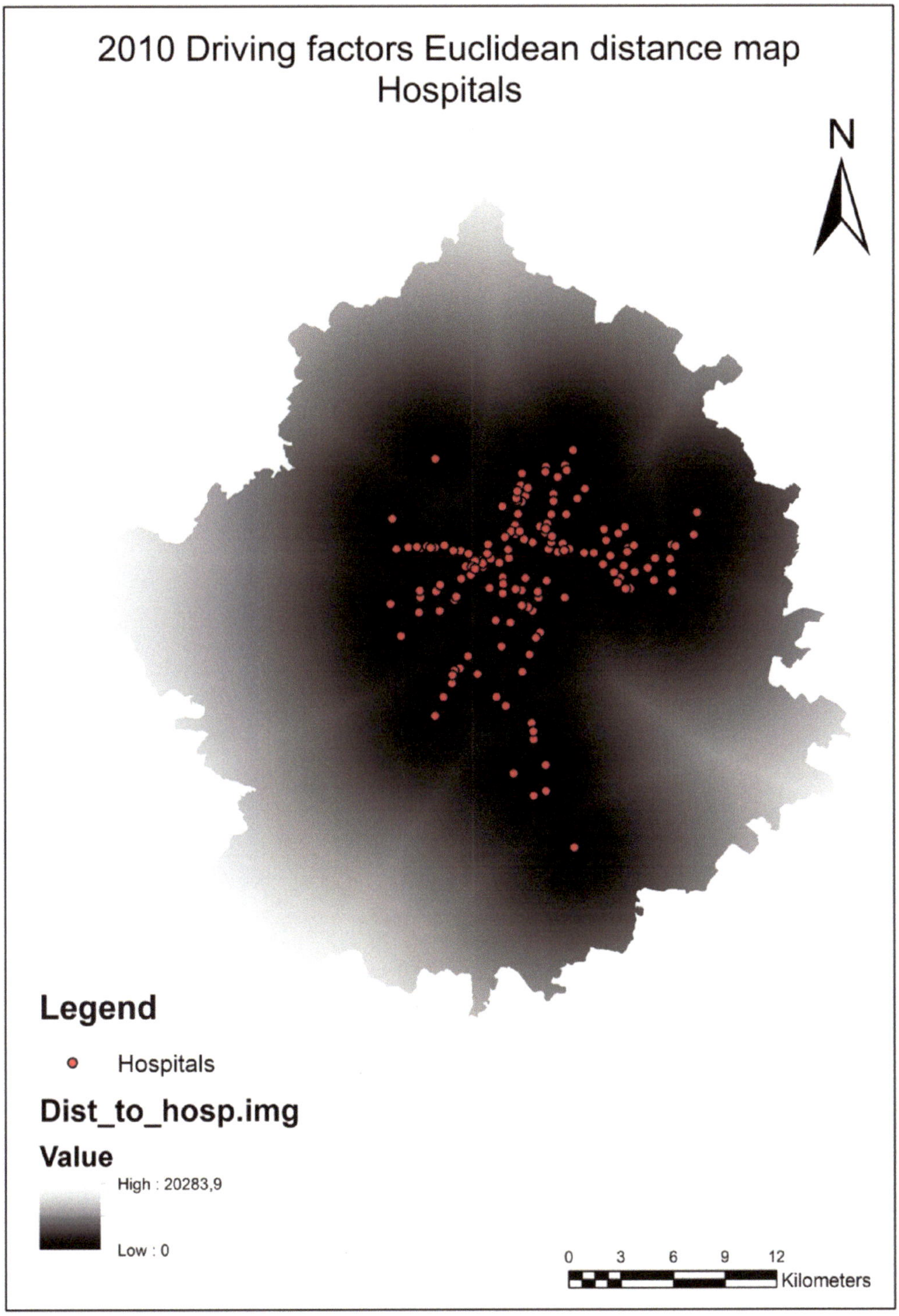

FIGURE 10.10 (Continued) (iii) Driving Factor in 2010 – Hospitals.

(Source: Compiled by the authors.)

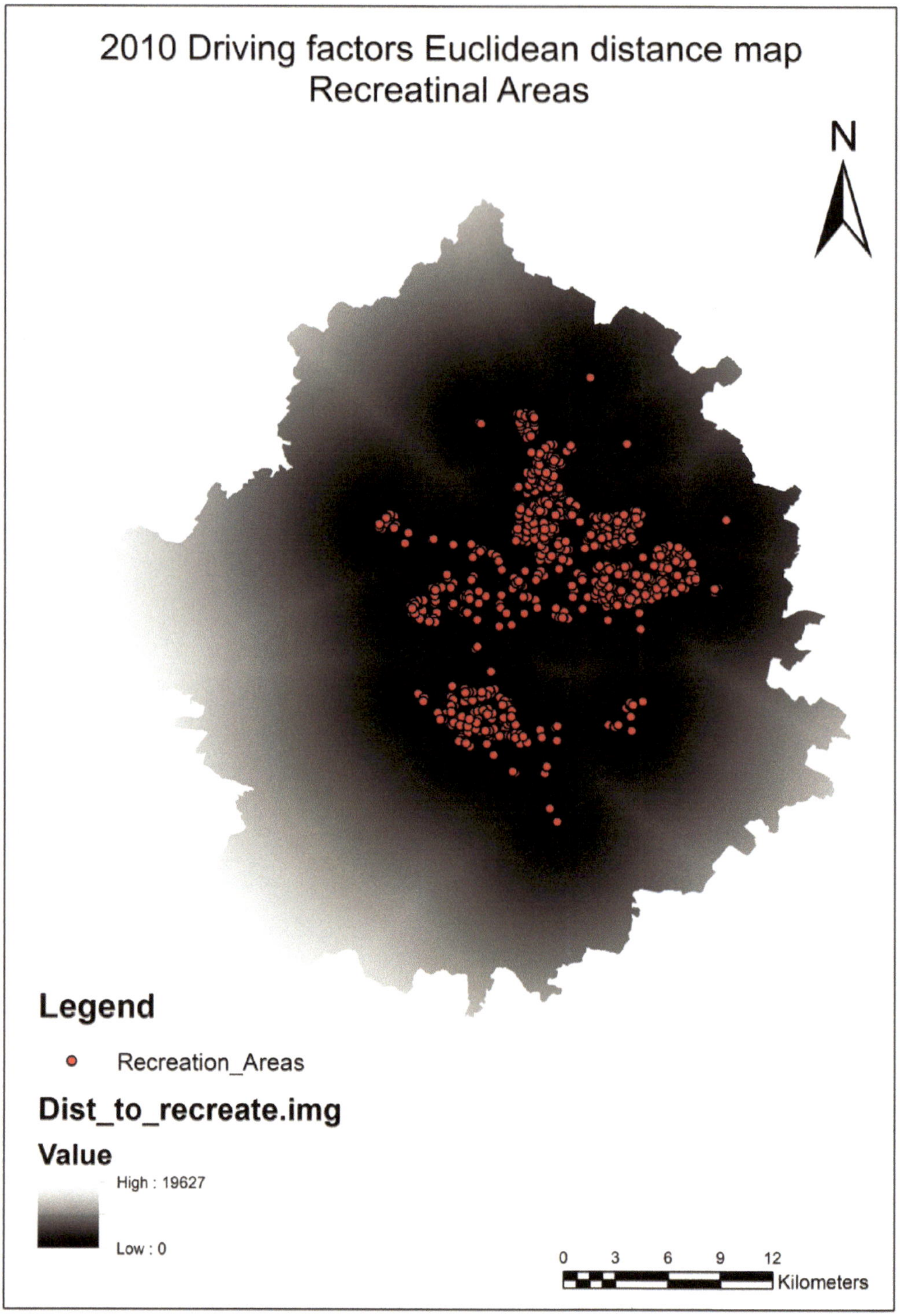

FIGURE 10.10 (Continued) (iv) Driving Factor in 2010 – Recreational areas.

(Source: Compiled by the authors.)

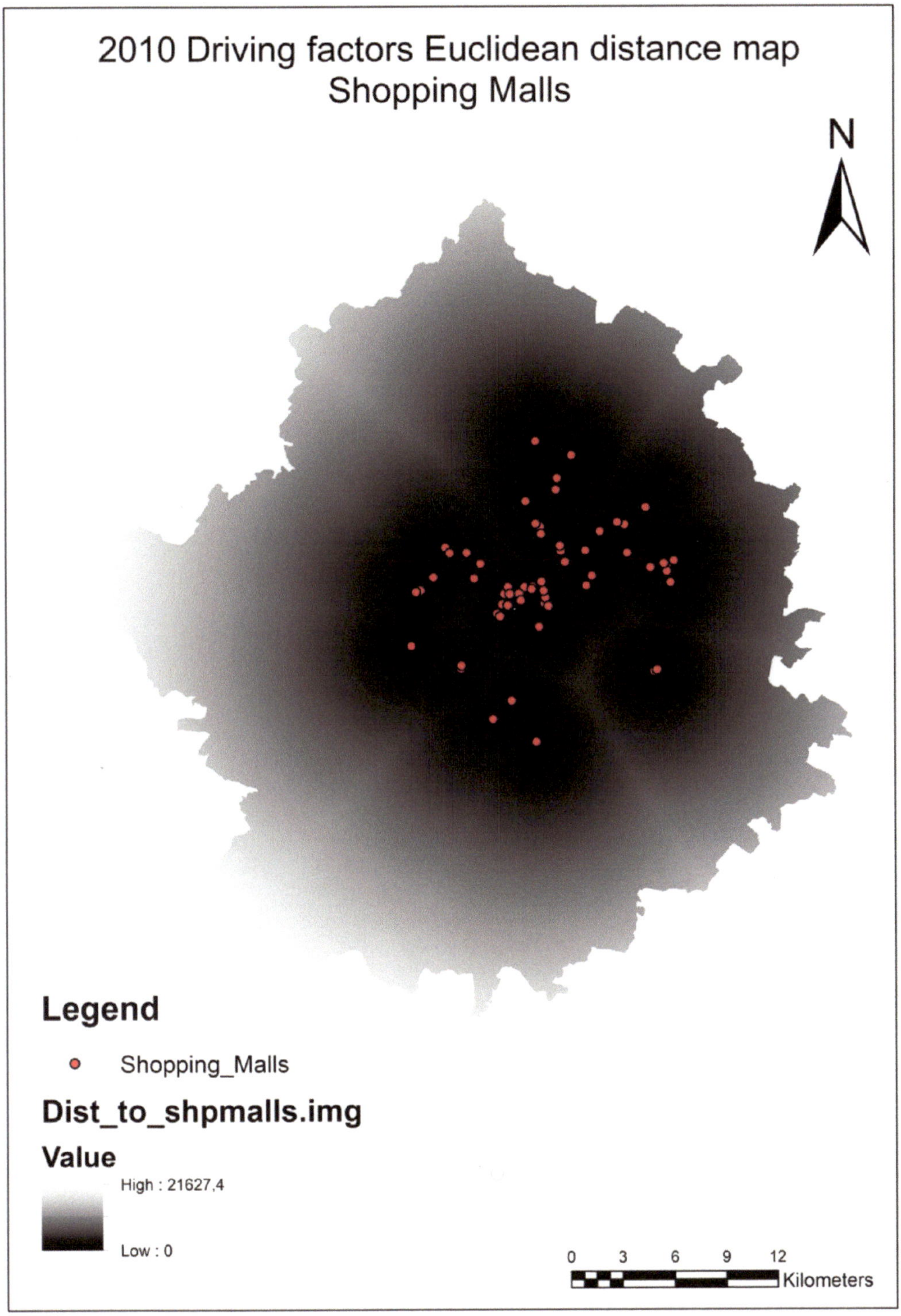

FIGURE 10.10 (Continued) (v) Driving Factor in 2010 – Commercial centers.

(Source: Compiled by the authors.)

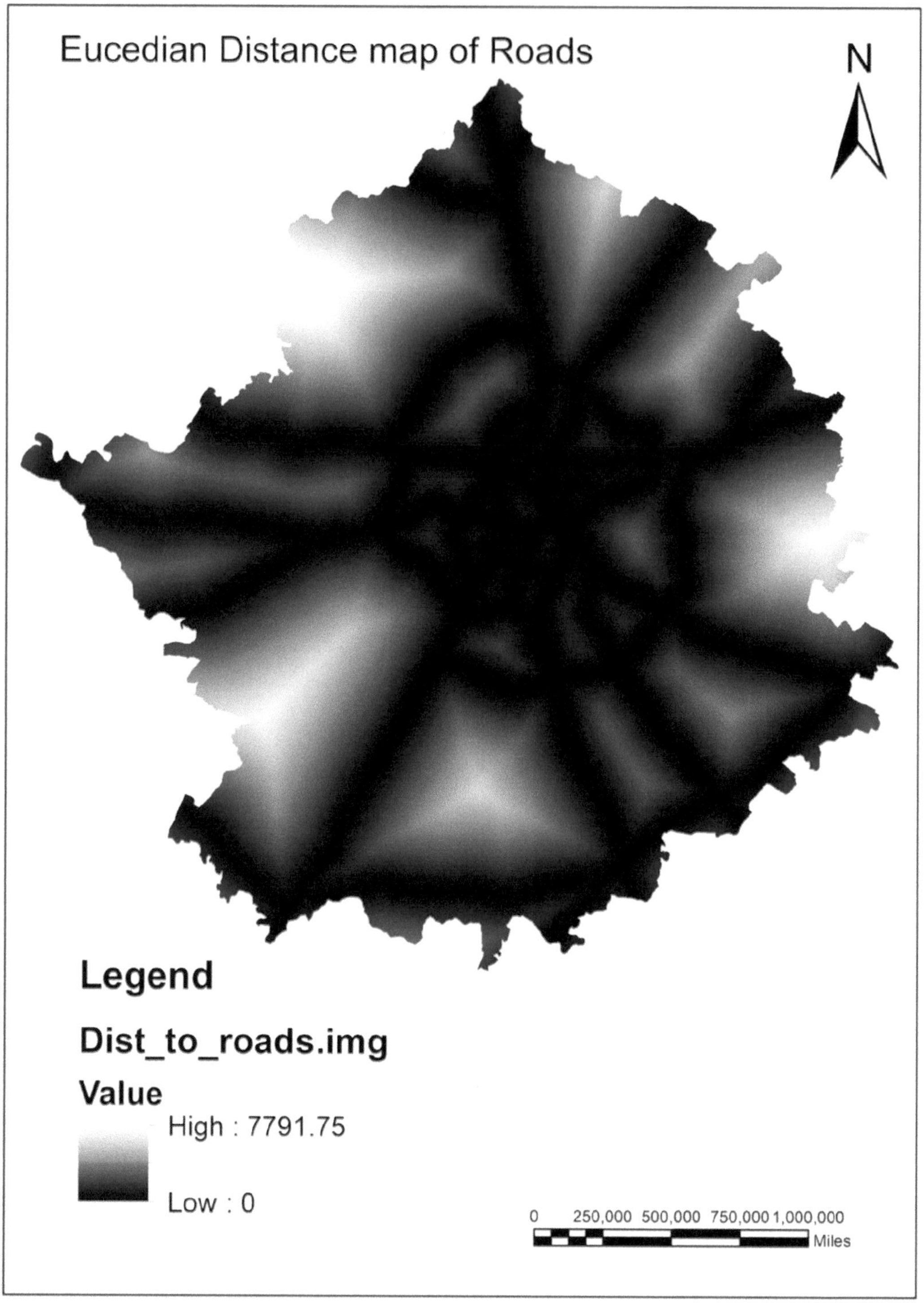

FIGURE 10.11 (i) Driving Factor in 2020 – Primary and Secondary roads.

(Source: Compiled by the authors.)

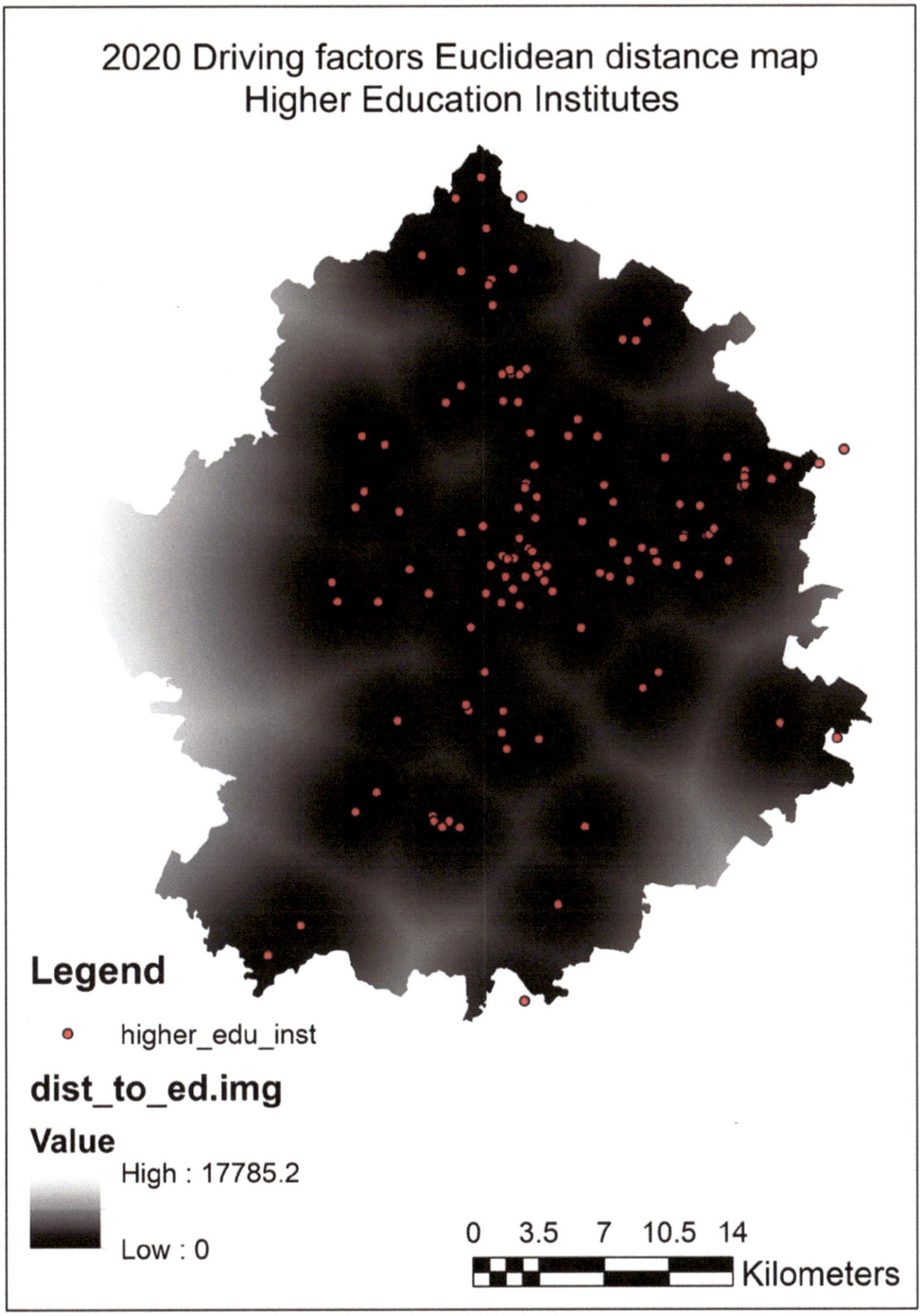

FIGURE 10.11 (Continued) (ii) Driving Factor in 2020 – Higher education institutions.

(Source: Compiled by the authors.)

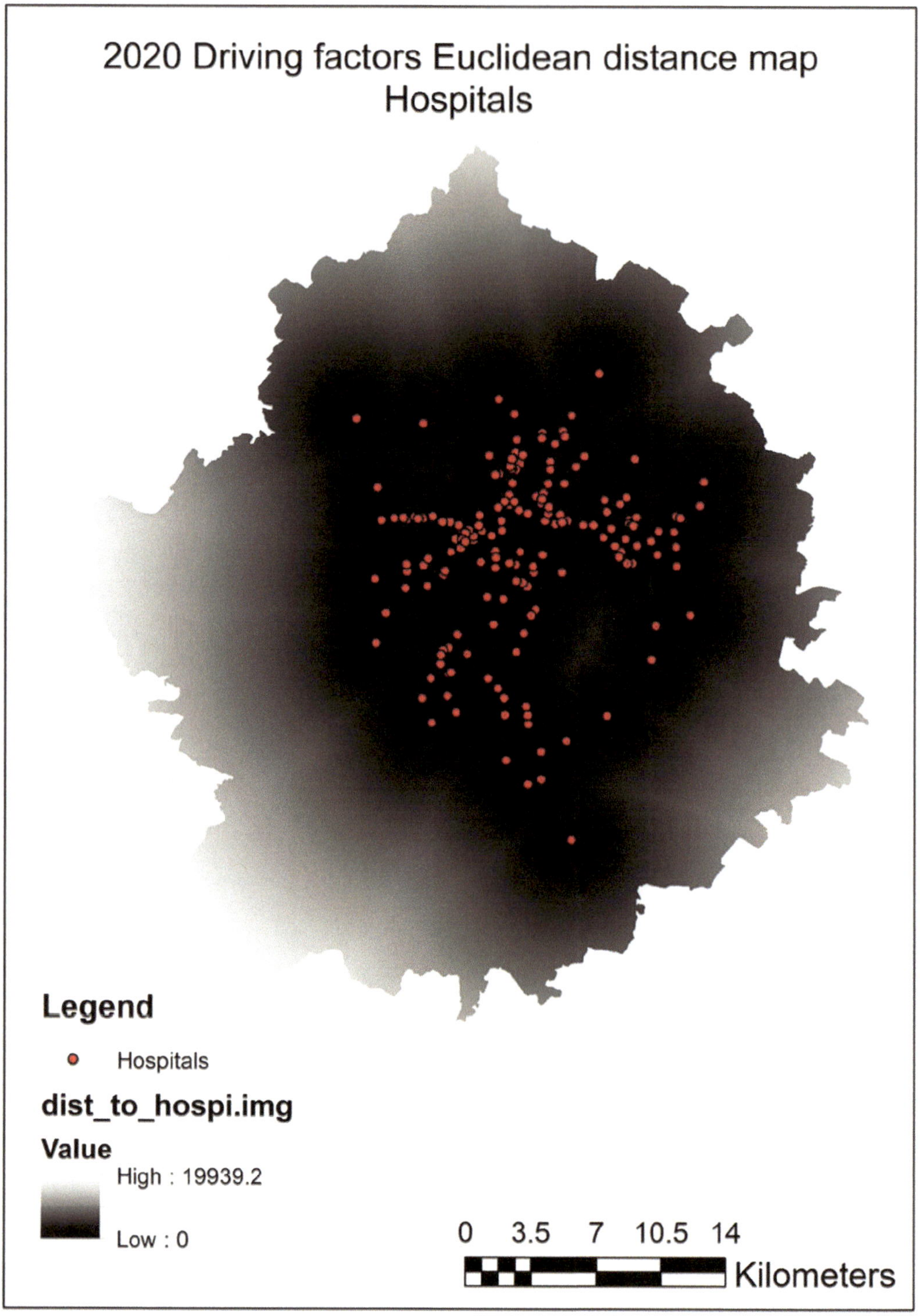

FIGURE 10.11 (Continued) (iii) Driving Factor in 2020 – Hospitals.

(Source: Compiled by the authors.)

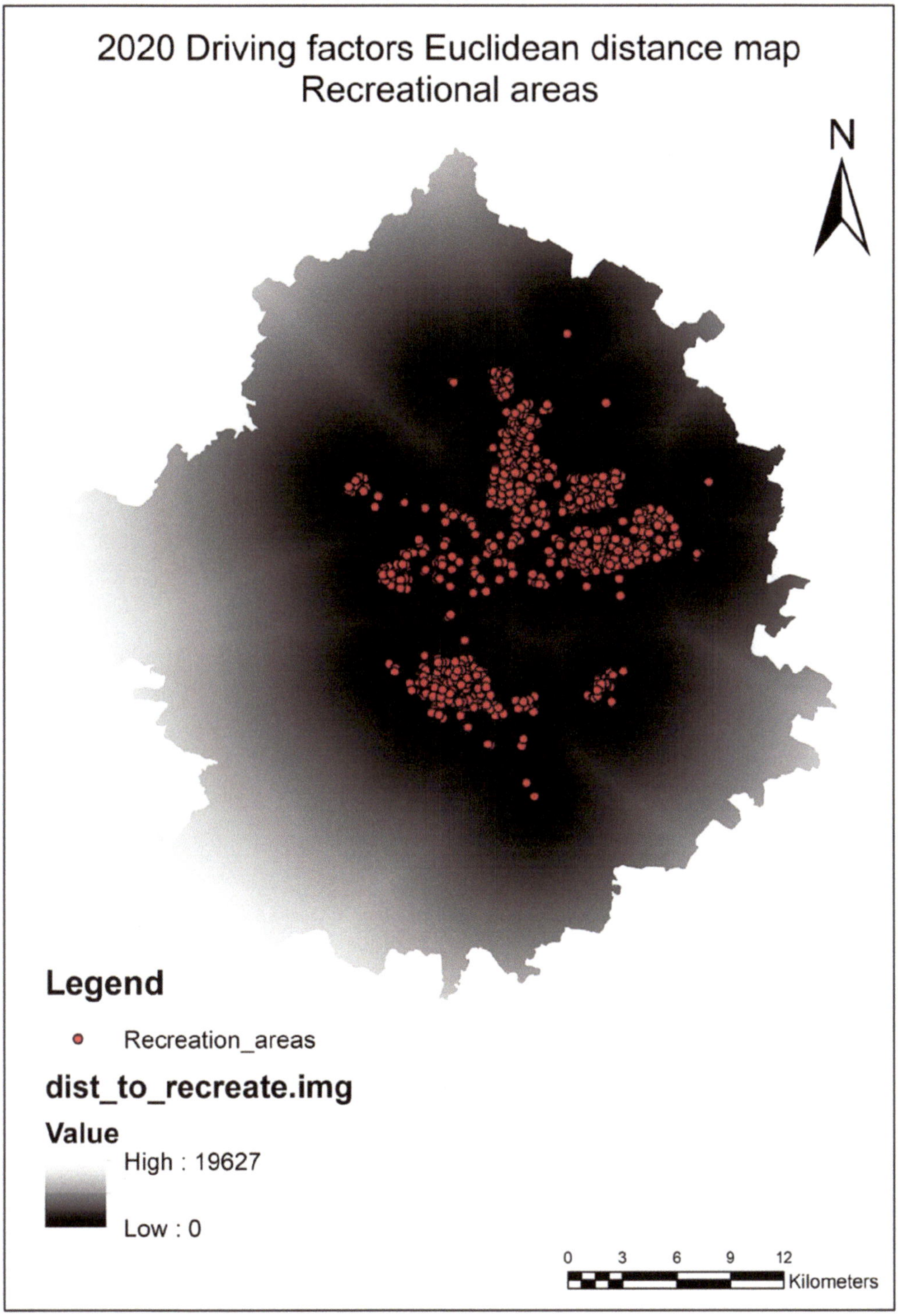

FIGURE 10.11 (Continued) (iv) Driving Factor in 2020 – Recreational areas.

(Source: Compiled by the authors.)

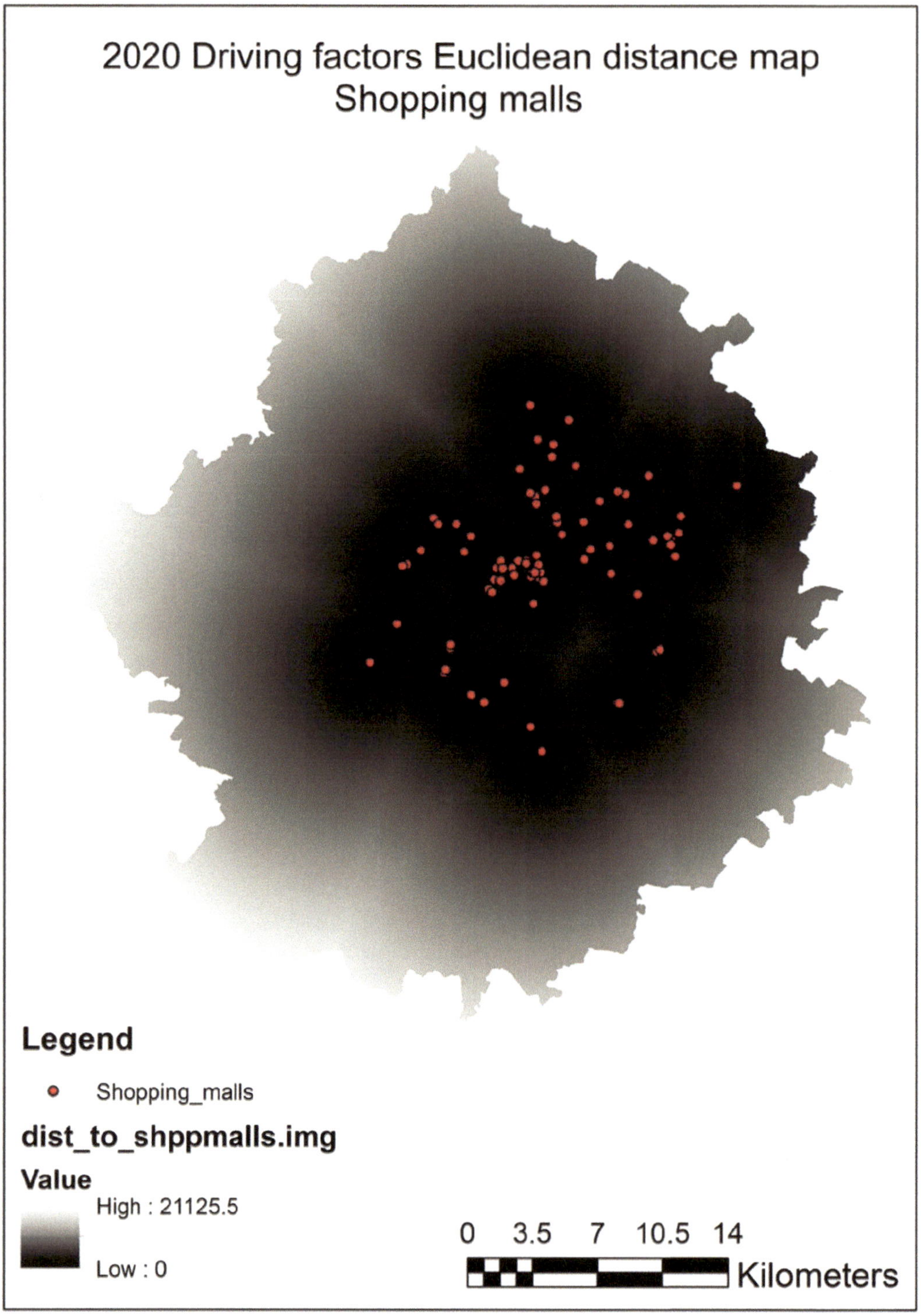

FIGURE 10.11 (Continued) (v) Driving Factor in 2020 – Commercial centers.

(Source: Compiled by the authors.)

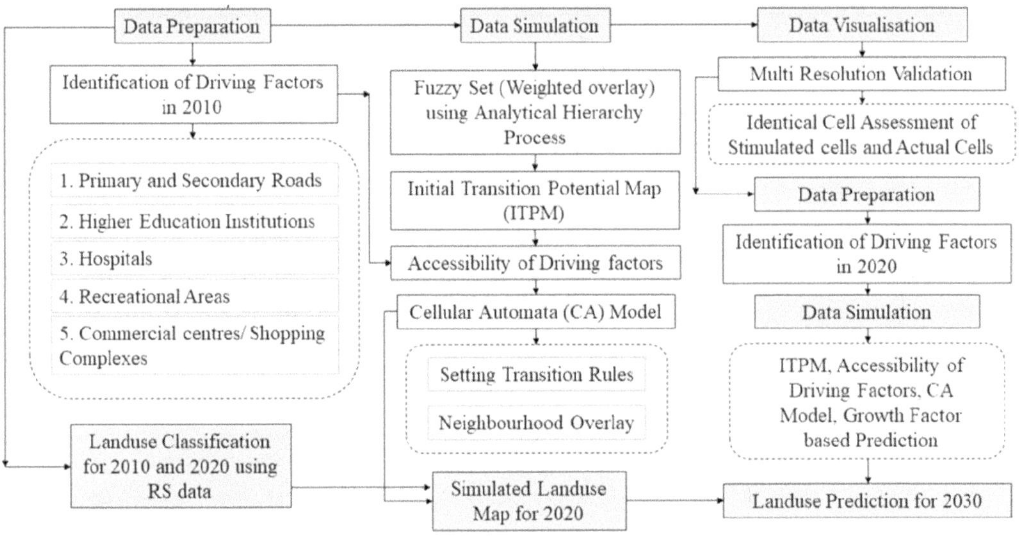

FIGURE 10.12 General Model Framework.

(Source: Compiled by the authors.)

TABLE 10.8
Weights Calculated by Analytical Hierarchy Process for Driving Factors

Driving Factors	Weights Calculated by AHP
Roads	0.035938
Higher educational institutions	0.094238
Hospitals	0.403583
Commercial centers/ shopping complexes	0.424386
Recreational spaces	0.041855

Source: Primary (Electronic Mail) Survey 2021.

10.4.1.3 Data Visualization

Initial simulations are carried out on 2010 land-use data to achieve 2020 urban growth predictions, as depicted in Figure 10.13 (CA simulated urban growth for 2020). These data help to measure the accuracy of the simulation process based on the actual scenario on the ground.

The model can also be verified by assessing its relation to the 2020 land use. A multi-resolution validation (MRV) technique is used where the pixels of the simulated land use are compared with the corresponding pixels in the actual land use for the year 2020 (see Table 10.9). The identical cell assessment of the simulated model with the actual land use is found to be 82.92% and the identical percentage for the urban land use is 88.92%. This model has been calibrated multiple times to improve simulation accuracy. Some adjustments are made at each step of the calibration by adding or eliminating various factors affecting the urban land-use change and by adjusting the weights of the parameters. The model can be further improved by including more factors that affect urban development, such as slope, soil character, land prices, etc., based on data availability and the city character. Finally, a simulation is run to predict the urban areas of 2030 by taking 2020 as the driving factor, and using the same model with exactly the same transitional rules as the 2020 simulation from 2010 data.

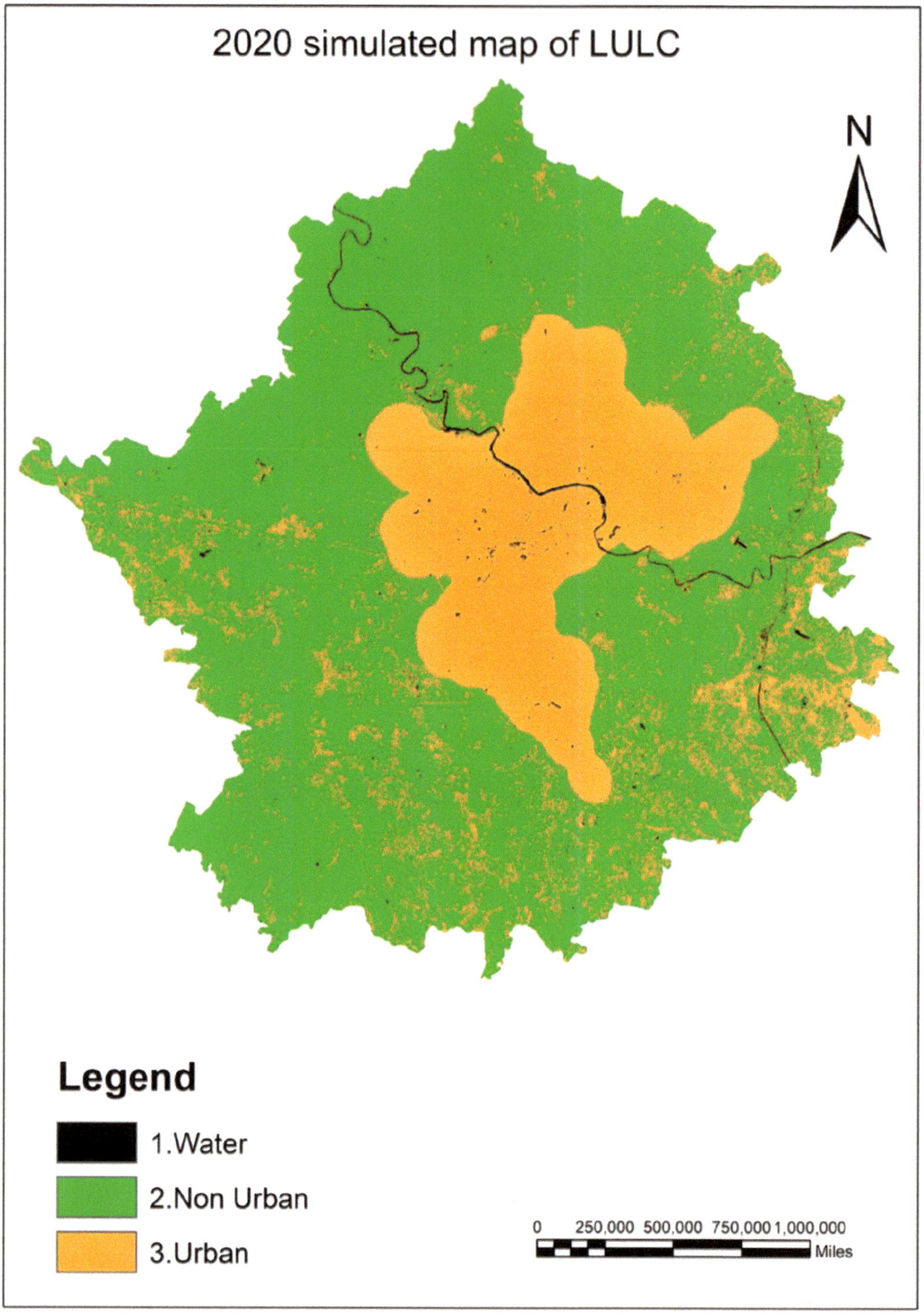

FIGURE 10.13 Cellular automata simulated urban growth for 2020.

(Source: Compiled by the authors.)

TABLE 10.9
Simulation Accuracy under All Transition Rules over Time (2010–2020)

Land Use Name	LU_ Code	LU-2010 Actual Cells	LU_2020 Actual Cells	LU_2020 Stimulated Cells	Identical cells between Actual and Stimulated Cells for the Year 2020	Identical Percentage
Urban	3	259064	410385	410385	364953	88.92%
Non-Urban	2	1085645	932679	934324	747464	80.00%
Reserved areas	1	7832	9477	7832	9025	95.42%
Total	--	1352541	1352541	1352541	1121442	82.92%

Source: Compiled by authors using ArcGIS and LanduseSim Software programs.

10.4.2 Micro-Level Study

The distribution of commercial spaces and their high densities in a city are important indicators of economic growth (Porta et al. 2009, Lin, Chen and Liang 2018). Chronological development of the major commercial streets in Lucknow reflects the geographic concentration and the pattern of socio-economic interactions with the urban built area in these commercial spaces. Commercial streets are the source of a neighborhood's character and vitality, and play a crucial sociological role in catering to the needs of local residents (Jacobs 1961, Keller 1968). Lucknow's neighborhoods have experienced rapid growth in commercial activities, primarily due to expansion of agglomeration economies of scale and focused transit-oriented developments. These neighborhoods are becoming major social hubs because of exotic restaurants, large chain stores and cafés. This transformation has entailed shifts in the characteristics of retail shops, conversion of urban land uses and huge escalation in housing as well as commercial land or rent pricing.

Lucknow has a total of 50,056 commercial establishments that generate employment for approximately 300,000 to 400,000 people (Senes Consultants 2015). In key commercial markets (see Table 10.10), wholesale as well as retail trade and commerce thrive in the city as one of the prime daily activities. Major urban growth was witnessed in 2020 in the south-east and north-east of the

TABLE 10.10
Timeline for Development of Major Commercial Centers in Lucknow

Time Period	Markets/Commercial Spaces	Features
Pre- Colonial Time (17th century–18th century)	Chowk, Nakhas (Nadan Mahal Road), Yahiyaganj, Aminabad, Alambagh, Halwasiya, Narhi	Social segregation based on caste and occupation, compact urban form, irregular and narrow street patterns, less open space, very dense urban fabric and attracts low- and medium-income group.
Colonial Time (18th century–20th century)	Lalbagh, Latouche Road, Bhoothnath, Hazratganj	Commercial street buildings in Hazratganj display Victorian-style architecture, accommodate many institutional spaces, have well-defined parking spaces, high traffic movement, multi-story buildings, wide streets with less dense urban fabric and attract middle- and high-income groups.
Post-Colonial Time (20th century to present)	Kapoorthala, Gol Market, Patrakarpuram Market	Fully developed, regular street pattern, mixed land use with commercial and residential units, wide streets with contemporary urban design forms like cycle tracks and plantations, transit-oriented development markets and attracts middle- and high-income groups.

Source: Compiled by authors (Malhotra 2017, Mishra 2019).

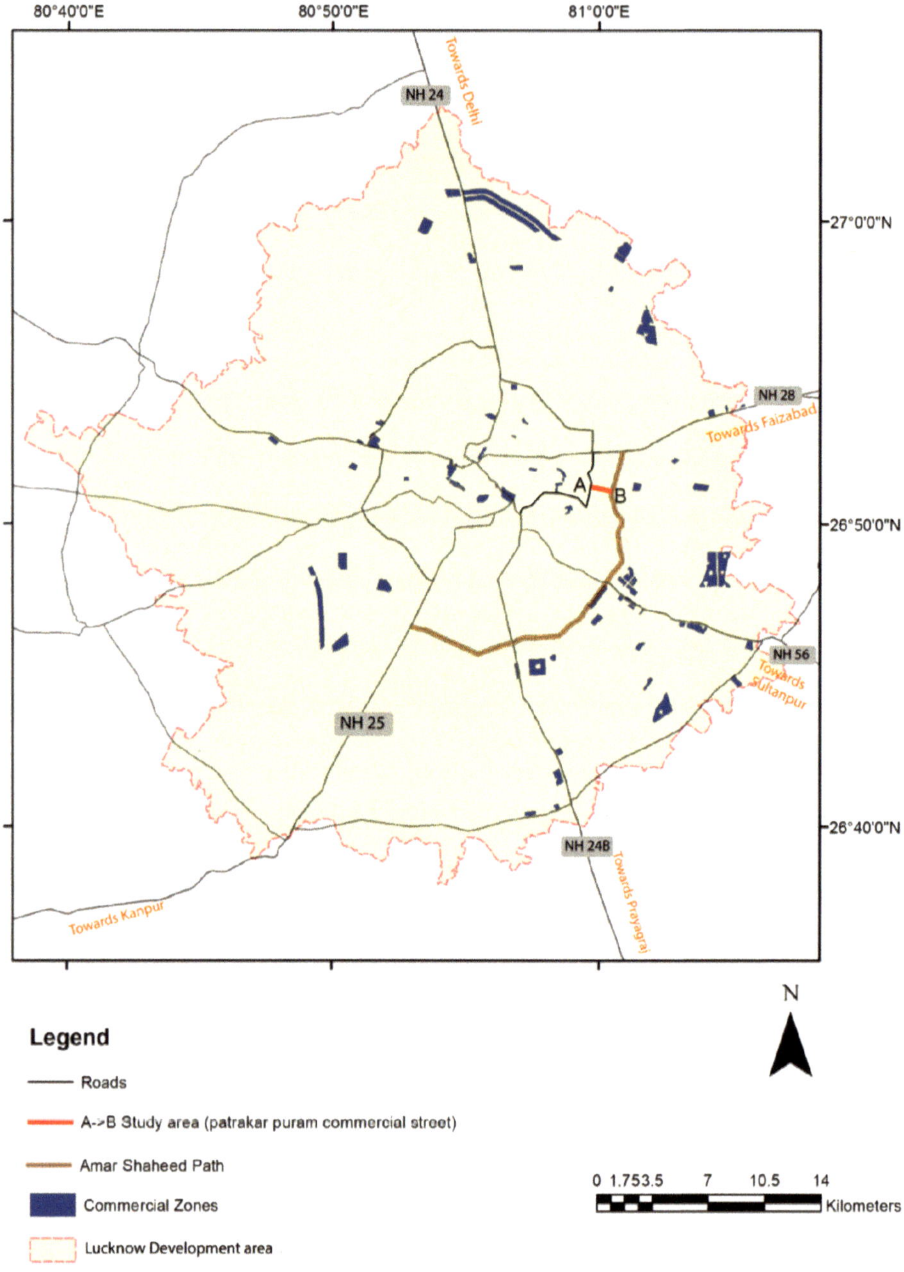

FIGURE 10.14 Study area for micro-level study.

(Source: Compiled by the authors.)

city because of the transit-oriented development and the presence of the Amar Shaheed Route as a connector between the NH-28, NH-56, NH-24-B and NH-25 roads. One of the main commercial markets in this area is Patrakarpuram market. It caters to the residential area of Gomtinagar. A commercial street 1.5 km long connecting Patrakarpuram market to Amar Shaheed Path was defined as the study area (Figure 10.14) along with other major commercial centers in the city. A research framework for micro-level study is outlined in Figure 10.15.

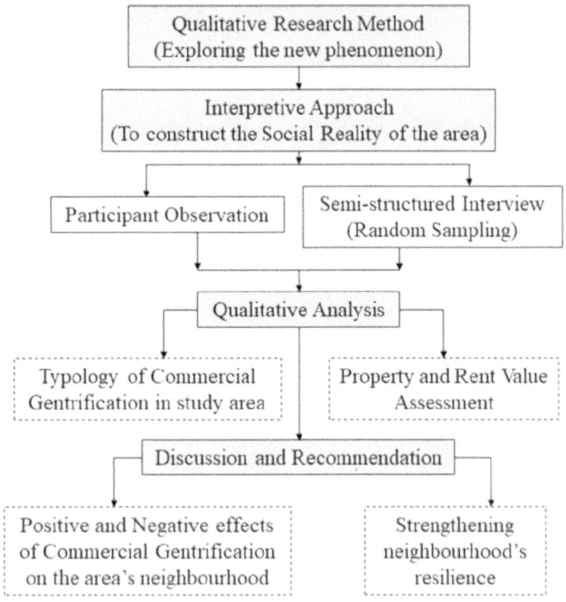

FIGURE 10.15 General framework for micro-level study.

(Source: Compiled by the authors.)

An empirical understanding of the role of business in gentrifying neighborhoods is crucial, considering the accelerated importance of consumption patterns for shaping urban neighborhoods (Clark 2004, Parker 2018). Commercial gentrification is observed in various locations of the study area. A social survey based on participant observation highlighted that 52 properties were commercially gentrified from the year 2010 to 2020. The land-use change is illustrated in Figures 10.16 and 10.17 for the years 2010 and 2020 respectively.

Two types of commercial gentrification were identified: residential-commercial gentrification and retail upscaling. Sixty-two percent of the properties were previously in residential use, elaborating the dominance of residential-commercial gentrification. Figure 10.18 shows the classification of commercial gentrification based on the category of commercial activity. The highest share of retail store conversions is into cafés/bars/lounges, whereas residential houses are mostly converted into large clothing and footwear chain stores such as Levi, Puma, Jockey, Raymond, U.S. Polo, Sketchers, etc. A unique mixed-use commercial and residential character was also observed in the study area, with housing in the upper storys and commercial activities, mainly large chain stores, in the rest. A new phenomenon of conversion of residential houses to hotels has been witnessed, mainly Oyo Rooms, an Indian hotel chain. Supermarkets are also growing in number in the area.

Semi-structured interviews conducted with retail shop owners helped to understand the exponential growth in property values in Sahara Plaza. Figure 10.19 shows an exorbitant rise of 310% in the average commercial property value of a shop in Sahara Plaza from 2002 to 2010. The excessive escalation in commercial real-estate values from Rs 3900 per square foot in 2002 to its present value (in 2021) of Rs 54,500 per square foot is mainly attributed to the urban growth experienced in this area. Semi-structured interviews were also conducted with commercial landlords to establish the present rental price of a shop in the plaza. The current average rental value is Rs 150 per square foot. The landlords interviewed owned cafés/bars/lounges or restaurants, and 100% of the previous tenants had been retail shops. This intensifies the phenomenon of the 'retail upscaling' type of commercial gentrification in the area. The average property value for a residential plot in the Gomtinagar area has also increased by a factor of three, from Rs 2000 per square foot in 2010 to Rs 6000 per square foot in 2021(Magicbricks 2021) (Figure 10.20).

FIGURE 10.16 Land use of Patrakarpuram, 2010.

(Source: Primary survey 2021.)

FIGURE 10.17 Land use in Patrakarpuram, 2020.

(Source: Primary survey 2021.)

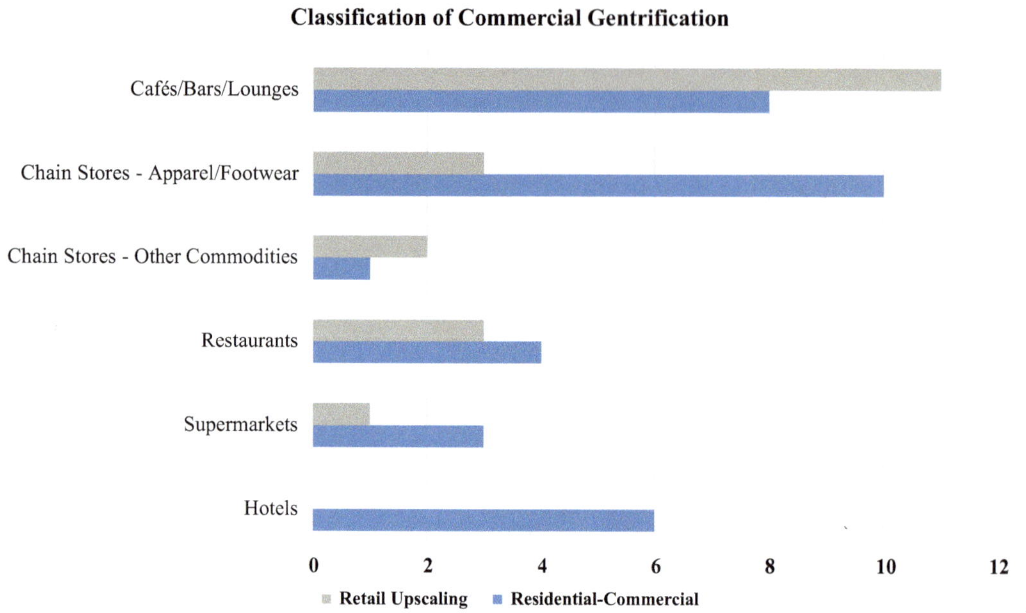

FIGURE 10.18 Classification of commercial gentrification in the study area.

(Source: Compiled by the authors.)

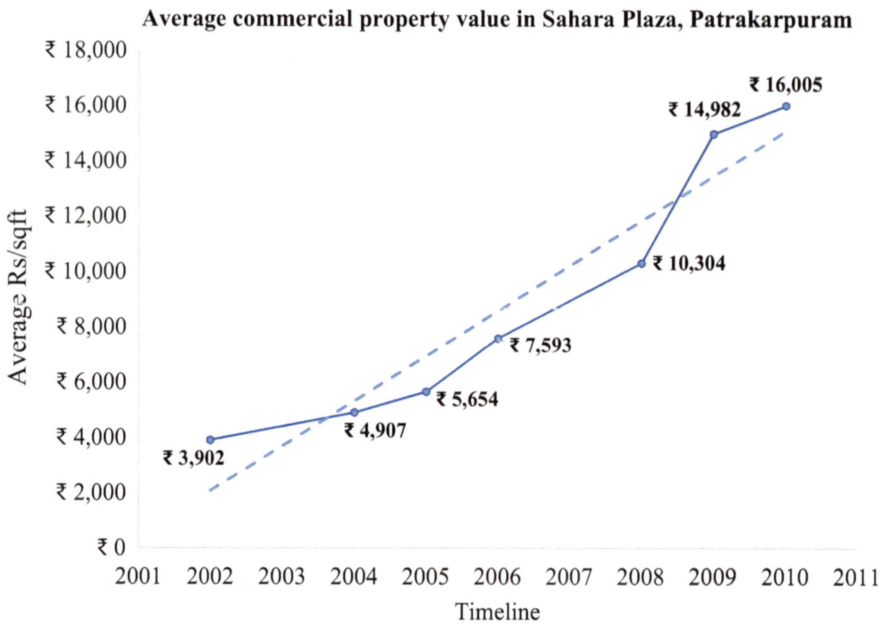

FIGURE 10.19 Average commercial property value in Sahara Plaza, Patrakarpuram, according to primary survey.

(Source: Primary survey 2021.)

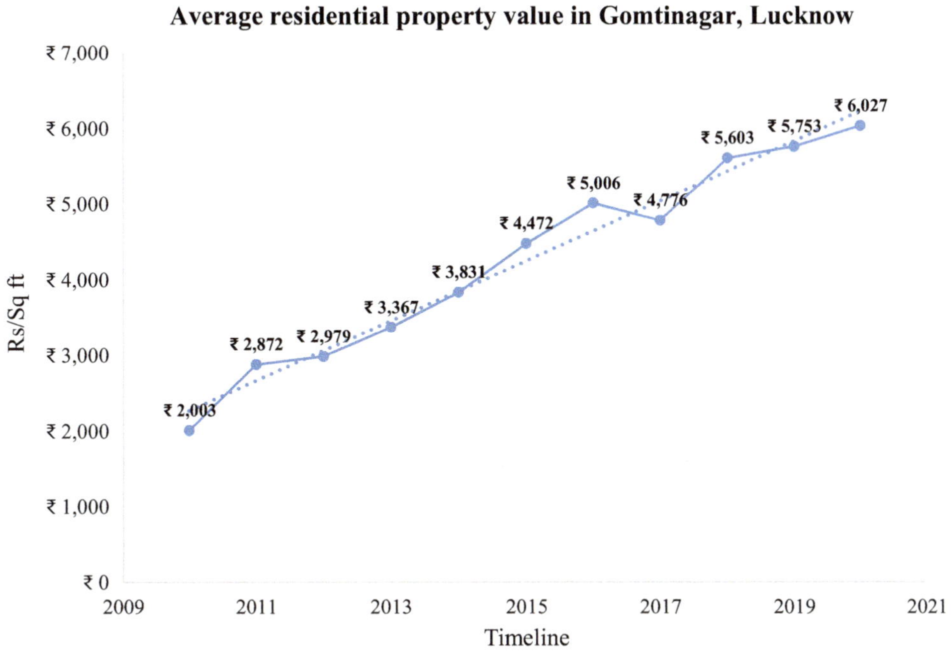

FIGURE 10.20 Average residential property value in Gomtinagar, Lucknow.

(Source: Magicbricks (2021).)

10.5 RESULT AND DISCUSSION

10.5.1 Macro-Level Study

The cellular automata and fuzzy urban growth hybrid model (CAFUGM) used in this research is capable of urban simulations, based on the accuracy achieved in the trial simulation of 2020 showcased in Table 10.9. The same simulation model is used to predict the 2030 urban scenario, using 2020 data as the driver (Figure 10.21).

The predicted growth in 2030 follows a pattern different from the 2020 model in terms of intensity and direction. This is attributed to the change in the 2020 driving factors in comparison to the 2010 driving factors. There is a significant change in these factors due to the increase in population in the past decade and the resultant increase in urban facilities, with rapid expansion of roads, hospitals, educational institutions, health facilities, shopping malls and recreational areas. These changes in the driving factors denote the chances of different and substantial growth in the urban area as predicted by the simulation model. According to the predicted model for 2030, the urban area will expand in a radial pattern mainly in the south-east and north-east directions, along the major roadways such as Amar Shaheed Route connecting the four major national highways NH-28, NH-56, NH-24B and NH-25. The land-use transformation from 2010 to 2020 to the simulated 2030 scenario is illustrated in Table 10.11, showing a significant proportion being converted to urban use.

Note that the urban area has grown from 19.15% to 30.16% between 2010 and 2020, and is predicted to increase to 53.52% in 2030. Non-urban land uses, mainly consisting of agricultural land, are reducing as a result from 80.26% in 2010 to 68.56% in 2020, and predicted to drop to 45.77% in 2030. Urban land use is predicted to increase and non-urban land use is predicted to decline by 28,224.63 ha from the 2020 scenario, accounting for a change in 23.2% of the total land area. This indicates an extensive urban agglomeration process around Lucknow, with 53.52% of the total land under urban use. The agricultural land will be consumed for urban uses, and to accommodate the

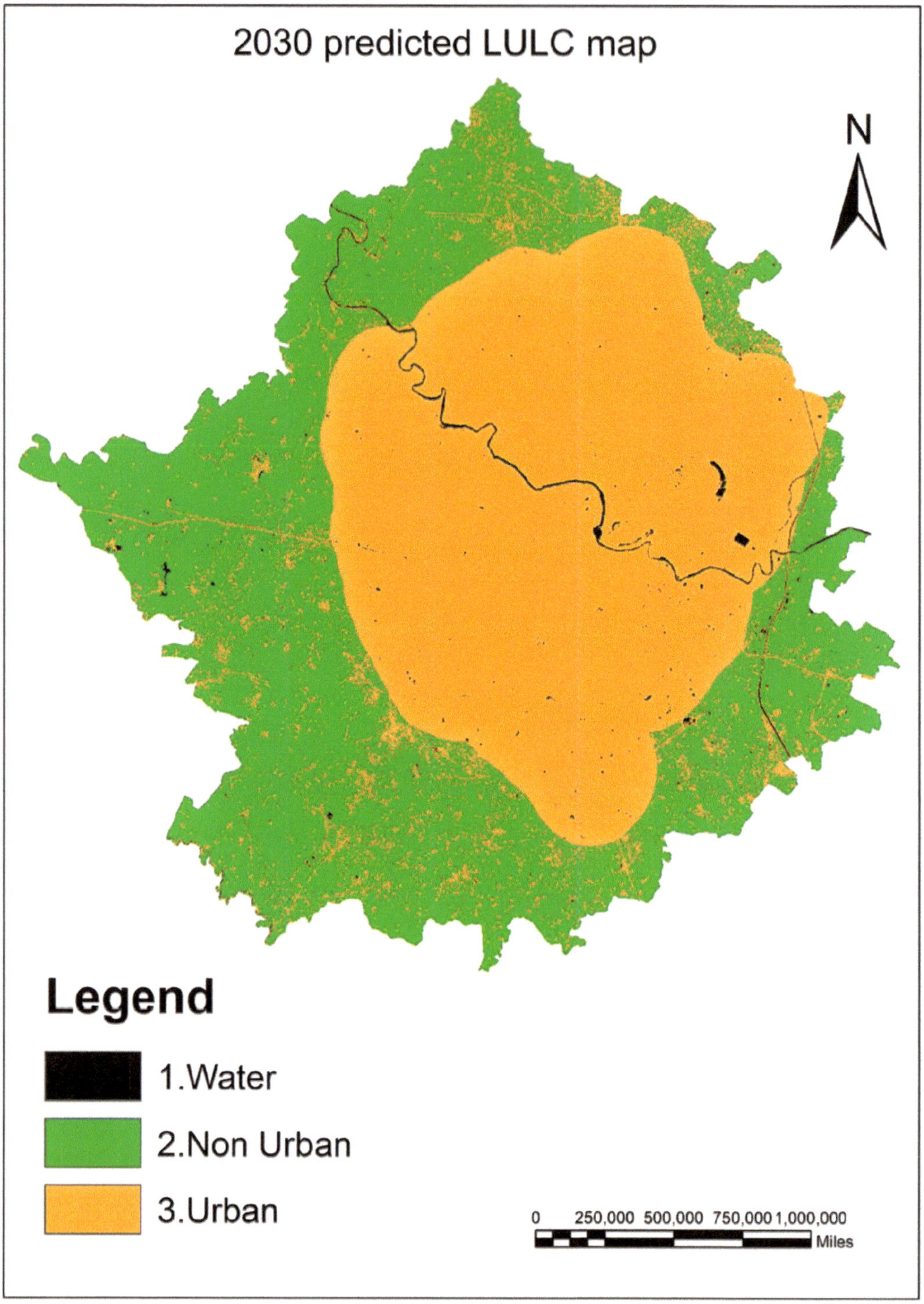

FIGURE 10.21 Predicted urban growth in Lucknow for 2030.

(Source: Compiled by authors.)

TABLE 10.11
Simulated Area Percentage for Land-Use Components in 2020 and 2030

Land Use Name	LU_ Code	LU_2010 Actual Area (ha)	LU_2010 Actual Area %	LU_2020 Actual Area (ha)	LU_2020 Actual Area %	LU_2020 Simulated Area(ha)	LU_2020 Simulated Area %	LU_2030 Simulated Area	LU_2030 Simulated Area %
Urban	3	23315.8	19.15%	36934.7	30.16%	36934.7	30.34%	65159.3	53.52%
Non-Urban	2	97708.1	80.26%	83941.1	68.56%	84089.2	69.07%	55716.5	45.77%
Reserved areas	1	704.88	0.57%	852.93	0.69%	704.88	0.57%	852.93	0.7%
Total		121728	100%	121728	100%	121728	100%	121728	100%

Source: Compiled by authors using ArcGIS and LanduseSim Software programs.

increasing population, mainly due to this urban growth. This change in the urban landscape is also dependent on regional economic growth. There will be massive pressure on land and resources in the next decade.

10.5.2 Micro-Level Study

Commercial gentrification creates demand, and hence economic growth, due to the purchasing power of new consumers. This kind of upscale commercialization can be regarded as a seed for a gentrified marketplace (Cho et al. 2020). This phenomenon in commercial gentrified places like Patrakarpuram market in Lucknow has some beneficial consequences, such as urban market diversity, enhanced neighborhood perceptions, strengthened local economies, and new employment opportunities, but there are still many drawbacks or negative consequences, such as increased product prices, a shift in commodity-style usage, escalating rent or property values, and changes in the composition of the residents. Social issues like commercial gentrification expedite regional changes, thus predicting changes in the location of residential areas will be a cumbersome urban planning exercise (Ryu, Kim and Park 2020). Neighborhood commercial facilities are being phased out or replaced by large-scale capital stores for profit maximization to meet the needs of new external consumers. Residents have experienced inconveniences such as noise, invasion of privacy, traffic congestion, and general neighborhood degradation as a result of the excessive revitalization of the area. Soaring rents for residential and commercial property force tenants and small businesses to relocate.

These developments have the greatest impact on small businesses, as explained by the rent-gap and consumption side theories. According to the former theory, as the rent gap – the difference between potential land value and capital land value – widens, property owners or developers decide either to sell, redevelop or reassess the use of their properties to benefit from the hikes in real-estate prices (Smith 1987, Hill and Pozzo 2005). This threatens the sustainability of small businesses as the real-estate market is run on the highest and best value principle (Sutton 2010). The consumption side theory describes the shift in consumption pattern in a neighborhood caused by demographic transition (Ley 1994). The influx of higher-income groups stimulates new businesses to enter the market to meet their preferences.

An innovative approach to regulating the density thresholds of bars, cafés, restaurants and fast-food establishments can check the socio-economic polarization in the neighborhood creating congestion and noise. Such regulations are implemented in the neighborhoods of San Francisco, where a special use permit is issued to any business if there is overconcentration (Cohen 1983). This strengthens neighborhood resilience by preventing commercial development that caters for a narrow spectrum of consumers. Another approach is the 'retail curation strategy' in California, Maine and Maryland, which promotes small retail businesses that commit to the goals and objectives of the city's development plan and regulates the number of 'formula businesses' or chain stores (Sutton 2010, Flood 2018). Municipalities can create co-working spaces in publicly owned buildings and underutilized buildings for small businesses at affordable rates; for example, Parkdale Neighborhood Land Trust in Toronto rents spaces to non-profit partners by acquiring land in the gentrified areas. Boston City's Small Business Plan prioritizes minorities, female or immigrant business owners as commercial tenants in under-utilized properties (LaVecchia and Mitchell 2016). The trend of soaring rents for commercial spaces is threatening many independent businesses; the problem is not limited to these businesses but also creates critical barriers to affordable housing.

10.6 CONCLUSION

The unprecedented speed of Lucknow's expansion is a major challenge for urban planners and policy makers. The urban simulation model (CAFUGM) for studying spatio-temporal urban growth trajectories integrated with socio-economic driving factors indicates a dominant urban pattern with a predicted growth of approximately 23% in the next decade. Balancing land-use components, according to URDPFI guidelines, is critical for sustainable growth. The city's economic development has been a magnet for the nearby population, fostering immigration because of inter- and intra-spatial interactions. The predicted labor force of 2.2 million by 2031 needs to be catered for with adequate physical and social resources.

Lucknow's performance in the ease of living index is below the national average, and the city only performs above average on the economic sub-index (Kapoor 2020). The city performs worst in both the social and physical sub-indices, scoring far below average, which reflects the city's inefficiencies in achieving sustainability. Strategies for identifying potential hubs based on urbanization economies and localization economies will help to achieve economic as well as social prosperity while retaining Lucknow's intrinsic cultural identity.

The transformation of residential areas or retail shops into cafés/bars/lounges, restaurants or hotels is setting the growth orientation towards cosmopolitan urbanism (Diepen and Musterd 2009, Rössel and Schroedter 2015). This cosmopolitan lifestyle encapsulates freedom, diversity, egalitarianism, independence and universalism (Sevincer, Kitayama and Varnum 2015). The emerging knowledge economy tends to drive the process of youthification and studentification in any area (Moos et al. 2018), and flourishing knowledge-based industries in Lucknow indicate the onset of cosmopolitan urbanism. However, fine-grained urbanism is also important to the character of a neighborhood as a sustainable microcosm (Jacobs 1961). Jane Jacobs argued that the value of retail and small businesses is that they act as community hubs and a reason for using sidewalks, thus improving the safety of the streets. Small businesses are integral to a city neighborhood's character, serving a large population of low-to-middle-income groups and creating a social bond because of the various forms of credit they offer, an option that is not available in supermarkets or large capital stores (Zukin et al. 2009). Lucknow's ethnicity and culture resides in small businesses, and not in large chain stores. The developmental pressure does leverage commercial gentrification, but local businesses contribute to the socio-economic fabric of the neighborhood. A balance between regional growth and the sustainability of a neighborhood's character is a solution to addressing the challenges of urban dynamism.

10.7 SCOPE AND LIMITATION

The following limitations to the study should be noted.

a) Only five drivers – transportation networks, hospitals, higher education, recreation spaces, and shopping complexes/commercial spaces – are considered in the macro-level study, due to availability of data.

b) The trend of increase in the average property value (Figure 10.19) applies only till 2010 because of the data collected from retail shop owners who had established their shops in the respective year.
c) Due to the Covid-19 pandemic, the interviews conducted were limited to Sahara Market, Patrakarpuram.

The present research seeks to understand how and where commercial gentrification has occurred in Patrakarpuram area. Future research might include a study of why this phenomenon has happened in the area, with more empirical evidence.

NOTES

1 Francois Perroux introduced the idea of growth poles in 1949, as a fusion of an economic plan, a field of gravity force or influences and an aggregate of linkages.
2 The equation for the gravity-based model used in the study is - G (Interactions between two settlements) = g (Proportional constant) X P1*P2 / (Distance between settlements)^2, P1 is Population of city 1, P2 is population of city 2 and g is taken as 10^7.
3 Satellite images used for 1975 are Landsat 2 (resolution 80m), for 1985, 1995 and 2005 Landsat 5 (30m) and for 2015 Landsat 8 (30m).

REFERENCES

Aburas, Maher Milad, Yuek Ming Ho, Mohammad Firuz Ramli, and Zulfa Hanan Ash'aari. 2016. "The simulation and prediction of spatio-temporal urban growth trends using cellular automata models: A review." *International Journal of Applied Earth Observation and Geoinformation* 52: 380–389.
Agrahari, Ritu, and Kanwaljit Kaur Brar. 2017. "Impediments countenanced by Lucknow Chikankari artisans." *Asian Journal of Home Science* 12 (2): 339–348. doi:10.15740/HAS/AJHS/12.2/339-348.
Ahmad, Yusairah, and Mohammed Anees. 2016. "Issues of The Lucknow Chikan Handicraft Industry." *International Journal of Multidisciplinary Research in Social Science* 2 (1): 1–9.
Aithal, Bharat H., S Vinay, and T.V. Ramachandra. 2018. "Simulating urban growth by two state modelling and connected network." *Modeling Earth Systems and Environment* 4 (4): 1297–1308. doi:10.1007/s40808-018-0506-1.
Al-Ahmadi, Khalid. 2018. "Modelling Driving Forces of Urban Growth with Fuzzy Sets and GIS." In *Spatial Analysis, Modelling and Planning*, edited by Jorge Rocha and José António Tenedório. Intech Open. doi:10.5772/intechopen.78646.
Al-Ahmadi, Khalid, Alison Heppenstall, Jim Hogg, and Linda See. 2009. "A Fuzzy Cellular Automata Urban Growth Model (FCAUGM) for the city of Riyadh, Saudi Arabia. Part 1: Model structure and validation." *Applied Spatial Analysis and Policy* 6(2): 65–83.
Al-Ahmadi, Khalid, Linda See, Alison Heppenstall, and James Hogg. 2009. "Calibration of a fuzzy cellular automata model of urban dynamics in Saudi Arabia." *Ecological Complexity* 2: 80–101.
Al-Darwish, Yazid, Hany Ayad, Dina Taha, and Dina Saadallah. 2018. "Predicting the future urban growth and it's impacts on the surrounding environment using urban simulation models: Case study of Ibb city – Yemen." *Alexandria Engineering Journal* 57(4): 2887–2895.
Allen, Jeffery, and Kang Lu. 2003. "Modeling and prediction of future urban growth in the Charleston region of South Carolina: A GIS-based integrated approach." *Ecology and Society* 8 (2): 2. doi:10.5751/ES-00595-080202.
Alsharif, A.A., B. Pradhan, S. Mansor, and H.Z.M. Shafri. 2015. "Urban expansion assessment by using remotely sensed data and the relative Shannon entropy model in GIS: A case study of Tripoli, Libya." *Theoretical and Empirical Researches in Urban Management* 10 (1): 55–70.
Arsanjani, Jamal Jokar, Marco Helbich, and Eric de Noronha Vaz. 2013. "Spatiotemporal simulation of urban growth patterns using agent-based modeling: The case of Tehran." *Cities* 32: 33–42.
Arya, Pankaj, and Shilp Sadhana. 2007. *Diagnostic Study Report of Chikan Embroidery, Lucknow*. Report for Cluster Development Programme, India, Lucknow: SIDBI and UNIDO.
Atuesta, Laura H., and Geoffrey J.D. Hewings. 2019. "Housing appreciation patterns in low-income neighborhoods: Exploring gentrification in Chicago." *Journal of Housing Economics* 44: 35–47. doi:10.1016/j.jhe.2018.08.005.
Barcón, Alberto Rodríguez, Estefanía Calo, and Raimundo Otero Enríquez. 2018. "Commercial gentrification in a medium-sized city: An ethnographic look at the transformation process of the historic centre of a Coruña (Spain)." *Urbanities* 8 (2): 87–96.

Batty, Michael. 2008. "Fifty Years of Urban Modeling: Macro-Statics to Micro-Dynamics." In: Albeverio S., Andrey D., Giordano P., Vancheri A. (eds) *The Dynamics of Complex Urban Systems*. Physica-Verlag HD. https://doi.org/10.1007/978-3-7908-1937-3_1

de Bok, Michiel, and Frank van Oort. 2014. "Agglomeration economies, accessibility, and the spatial choice behavior of relocating firms." *Journal of Transport and Land Use* 4(1): 5–24.

Boote, David N., and Penny Beile. 2005. "Scholars before researchers: On the centrality of the dissertation literature review in research preparation." *Educational Researcher* 3–15. doi:10.3102/0013189X034006003.

Booth, Wayne C., Gregory G. Colomb, and Joseph M. Williams. 2008. *The Craft of Research*. Chicago: The University of Chicago Press.

Burnett, Katherine E. 2014. "Commodifying poverty: Gentrification and consumption in Vancouver's Downtown Eastside." *Urban Geography* 35 (2): 157–176. doi:10.1080/02723638.2013.867669.

Chantia, Shikha. 2017. "Sustainable Chikankari Art in contemporary Indian fashion industry: An observation." *International Journal of Advance Research in Science and Engineering* 6 (9): 1765–1774.

Chapple, Karen, Anastasia Loukaitou Sideris, Silvia R. González, Dov Kadin, and Joseph Poirier. 2017. *Transit-Oriented Development & Commercial Gentrification: TOD Study*. Los Angeles: UC CONNECT.

Chen, Yanguang, and Yixing Zhou. 2008. "Scaling laws and indications of self-organized criticality in urban systems." *Chaos, Solitons & Fractals* 35 (1): 85–98. doi:10.1016/j.chaos.2006.05.018.

Cho, Wonjun, Minho Kim, Hyunjung Kim, and Younsang Kwon. 2020. "Transforming housing to commercial use: A case study on commercial gentrification in Yeon-nam District, Seoul." *Sustainability* 12: 1–17. doi:10.3390/su12104322.

Chopra, Sandeep. 2012. *Lucknow- The City of Culture*. Allahabad: Point Publication.

Clark, Terry Nicholas. 2004. *The City as an Entertainment Machine*. Amsterdam, Netherlands; Boston, MA: Elsevier.

Clarke, Keith, and Leonard J. Gaydos. 1998. "Loose-coupling a cellular automaton model and GIS: Long-term urban growth prediction for San Francisco and Washington/Baltimore." *International Journal of Geographical Information Science* 12: 699–714.

Clarke, Keith, Stacy Hoppen, and J.L. Gaydos. 1997. "A Self-Modifying Cellular Automaton Model of Historical Urbanization in the San Francisco Bay Area." *Environment and Planning B: Planning and Design* 24: 247–261.

Cobbinah, Patrick Brandful, M.O.E. Kwasie, and Paul Amoateng. 2015. "Rethinking sustainable development within the framework of poverty and urbanisation in developing countries." *Environmental Development* 13: 18–32. doi:10.1016/j.envdev.2014.11.001.

Cohen, Mark. 1983. " San Francisco's neighborhood commercial special use district ordinance: An innovative approach to commercial gentrification." *Golden Gate University Law Review* 13: 367–397.

Cyrill, Melissa. 2016. *Investing in Lucknow – A Leading Tier II City in India*. 08 25. https://www.india-briefing.com/news/lucknow-city-spotlight-12677.html/

Davis, Kingsley, and Hilda Hertz Golden. 1954. "Urbanisation and the development of pre-industrial areas." *Economic Development and Cultural Change* 3 (1): 6–26.

Deep, Shikhar, and Akansha Saklani. 2014. "Urban sprawl modeling using cellular automata." *The Egyptian Journal of Remote Sensing and Space Science* 17 (2): 179–187. doi:10.1016/j.ejrs.2014.07.001.

Dept of Biotechnology, Govt of India. 2002. *Biotech Park, Lucknow*. Accessed April 19, 2021. http://www.biotechpark.org.in/index.php

Dewan, Ashraf, Robert Corner, Sohel J. Ahmed, Glen Bramley, and Peter H. Verbung. 2014. "Key Driving Factors Influencing Urban Growth: Spatial-Statistical Modelling with CLUE-s." In *Dhaka Megacity*, pp. 123–145. Dordrecht: Springer Netherlands. Accessed March 29, 2021. doi:10.1007/978-94-007-6735-5_7.

Diepen, Albertine M.L., and Sako Musterd. 2009. "Lifestyles and the city: Connecting daily life to urbanity." *Journal of Housing and the Built Environment* 24: 331–345. doi:10.1007/s10901-009-9150-4.

Dinda, Shantanu, Nilanjana Das Chatterjee, and Subrata Ghosh. 2021. "An integrated simulation approach to the assessment of urban growth pattern and loss in urban green space in Kolkata, India: A GIS-based analysis." *Ecological Indicators*. doi:10.1016/j.ecolind.2020.107178.

District Industrial Centre, Lucknow, interview by Sukanya Koner. 2019. *Key Industrial Estates in Lucknow* (09 22).

Dong, Jiang, Dafang Zhuang, Xinliang Xu, and Lei Ying. 2008. "Integrated evaluation of urban development suitability based on remote sensing and GIS techniques – A case study in Jingjinji area, China." *Sensors (Basel, Switzerland)* 8 (9): 5975–5986. doi:10.3390/s8095975.

Doucet, Brian. 2009. "Living through gentrification: Subjective experiences of local, non-gentrifying residents in Leith, Edinburgh." *Journal of Housing and the Built Environment* 24: 299–315. doi:10.1007/s10901-009-9151-3.

Dutta, Dr Vankatesh. 2012. "War on the Dream – How Land use Dynamics and Peri-urban Growth Characteristics of a Sprawling City Devour the Master Plan and Urban Suitability? A Fuzzy Multi-criteria Decision Making Approach." *Urbanization and Development: Delving Deeper into the Nexus*. Budapest: Global Development Network (GDN) in partnership with Central European.

Dwivedi, Aasheerwad, and Aakanksha Arora. 2019. "Economic geography of innovation in India: An empirical investigation." *Innovation and Development* 395–412. doi:10.1080/2157930X.2019.1679952.

Ergen, Barış. 2014. "Euclidean distance mapping and the proposed greenway method in Malta." *Journal of Urban Planning and Development* 140 (1): 1–11. doi:10.1061/(ASCE)UP.1943-5444.0000163.

Fan, Yong, Guangming Yu, and Zongyi He. 2017. "Origin, spatial pattern, and evolution of urban system: Testing a hypothesis of "urban tree"." *Habitat International* 59: 60–70. doi:10.1016/j.habitatint.2016.11.012.

Feedback Ventures. 2006. *Lucknow City Development Plan*. Lucknow: Government of Uttar Pradesh.

Feng, Yanlei, and Yi Qi. 2018. "Modeling patterns of land use in Chinese cities using an integrated cellular automata model." *ISPRS International Journal of Geo-Information* (IJGI) 7 (10): 403. doi:10.3390/ijgi7100403.

Flood, Anna. 2018. *A Study of Commercial Gentrification and the Displacement of Small businesses in the downtown Young street neighbourhood*. Master's Research, Toronto, Canada: Ryerson University.

Foroutan, Ehsan, and Mahmoud Reza Delavar. 2012. "Urban Growth Modeling using Fuzzy Logic." *ASPRS 2012 Annual Conference*. Sacramento, California: ASPRS.

Gabriel, Deborah. 2013. *Inductive and Deductive Approaches to Research*. March 17. Accessed May 3, 2021. http://deborahgabriel.com/2013/03/17/inductive-and-deductive-approaches-to-research/#:~:text=Inductive%20approaches%20are%20generally%20associated%20with%20qualitative%20research%2C,some%20qualitative%20studies%20may%20have%20a%20deductive%20orientation

Gantumur, Byambakhuu, Falin Wu, Battsengel Vandansambuu, Bazarkhand Tsegmid, Enkhjargal Dalaibaatar, and Yan Zhao. 2020. "Spatiotemporal dynamics of urban expansion and its simulation using CA-ANN model in Ulaanbaatar, Mongolia." *Geocarto International*. doi: 10.1080/10106049.2020.1723714

Gharbia, Salem S., Sara Abd Alfatah, Laurence Gill, Paul Johnson, and Francesco Pilla. 2016. "Land use scenarios and projections simulation using an integrated GIS cellular automata algorithms." *Modeling Earth Systems and Environment* 2 (3): 1–20. doi:10.1007/s40808-016-0210-y.

Giuliano, Genevieve, Sanggyun Kang, and Quan Yuan. 2019. "Agglomeration economies and evolving urban form." *The Annals of Regional Science* 63: 377–398. doi:10.1007/s00168-019-00957-4.

Glaeser, Edward, H.D. Kallal, J.A. Scheinkman, and A. Shleifer. 1992. "Growth in cities." *Journal of Political Economy* 100 (6): 1126–1152.

Glass, Ruth. 1964. *London: Aspects of Change*. London: MacGibbon & Kee.

Gopal, Sucharita, Xiaojing Tang, Nathan Phillips, Mallory Nomack, Valerie Pasquarella, and Joshua Pitts. 2016. "Characterizing urban landscapes using fuzzy sets." *Computers, Environment and Urban Systems* 57: 212–223. doi:10.1016/j.compenvurbsys.2016.02.002.

Gotham, Kevin Fox. 2005. "Tourism gentrification: The Case of New Orleans' Vieux Carre (French Quarter)." *Urban Studies* 42 (7): 1099–1121. doi:10.1080/00420980500120881.

Hackworth, Jason, and Josephine Rekers. 2005. "Ethnic packaging and gentrification: The case of four neighborhoods in Toronto." *Urban Affairs Review* 41: 211–236. doi:10.1177/1078087405280859.

Hamilton, Allison B., and Erin P. Finley. 2020. "Reprint of: Qualitative methods in implementation research: An introduction." *Psychiatry Research* 283. doi:10.1016/j.psychres.2019.112629.

Haque, Ismail, and Priyank Pravin Patel. 2017. "Growth of metro cities in India: Trends, patterns and determinants." *Urban Research and Practice* 11 (4): 338–377.

Haralambos, Michael, Martin Holborn, Steve Chapman, and Stephen Moore. 2017. *Sociology-Themes and Perspectives*. London: Collins.

Hariharan, Arun Natarajan, and Arindam Biswas. 2020a. "A Critical review of the Indian knowledge-based industry location policy against its theoretical arguments." *Regional Science Policy and Practice* 431–454. doi:10.1111/rsp3.12257.

Hariharan, Arun Natarajan, and Arindam Biswas. 2020b. "Global recognition of India's knowledge-based industry evolution through empirical analysis." *Journal of Knowledge Economy*. doi:10.1007/s13132-020-00673-x.

Hasnain, Nadeem. 2016. *The Other Lucknow-Am Ethnographic Portrait of a City of Undying Memories and Nostalgia*. New Delhi: Vani Prakashan.

Hegazy, Ibrahim Risk, and Mosbeh Rashed Kaloop. 2015. "Monitoring urban growth and land use change detection with GIS and remote sensing techniques in Daqahlia governorate Egypt." *International Journal of Sustainable Built Environment* 4 (1): 117–124. doi:10.1016/j.ijsbe.2015.02.005.

Herold, Martin, Noah C. Goldstein, and Keith C. Clarke. 2003. "The spatiotemporal form of urban growth: Measurement, analysis and modeling." *Remote Sensing of Environment* 86: 286–302.
Hill, Kelly, and Colette Pozzo. 2005. *Managing Neighborhood Change: Best Practices for Communities Undergoing Gentrification*. Research Report, Washington: Neighbor Works America.
Hwang, Uijeong, and Myungje Woo. 2020. "Analysis of inter-relationships between urban decline and urban sprawl in city-regions of South Korea." *Sustainability* 1656. doi:10.3390/su12041656.
Ibrahim, Hatem, Ashraf Salama, Florian Wiedmann, Bassma Aboukalloub, and Reem Awwaas. 2020. "Investigating land use dynamics in emerging cities: The case of downtown neighbourhood in Doha." *Journal of Urban Design* 25 (3): 387–411. doi:10.1080/13574809.2019.1632179.
Jacobs, Jane. 1961. *The Death and Life of Great American Cities*. New York: Random House.
Jacobs, Jane. 1969. *The Economy of Cities*. New York: Random House.
Jafari, Meisam, Hamid Majedi, Seyed Masoud Monavari, Ali Asghar Alesheikh, and Mirmasoud Kheirkhah Zarkesh. 2017. "Dynamic simulation of urban expansion through a CA-Markov model Case study: Hyrcanian region, Gilan, Iran." *European Journal of Remote Sensing* 49: 513–529.
Jantz, Claire A., Scott J. Goetz, David Donato, and Peter Claggett. 2010. "Designing and implementing a regional urban modeling system using the SLEUTH cellular urban model." *Computers, Environment and Urban Systems* 34: 1–16.
Jeong, Yeonju, Jayun Heo, and Changmu Jung. 2015. "Behind the Bustling Street: Commercial gentrification of Gyeongridan, Seoul." *Procedia-Social and Behavioral Sciences* 170: 146–154. doi:10.1016/j.sbspro.2015.01.024.
Jiang, Leiwen, and Brian C. O'Neill. 2018. "Determinants of urban growth during demographic and mobility transitions: Evidence from India, Mexico, and the US." *Population and Development Review* 44 (2): 363–389. doi:10.1111/padr.12150.
Kapoor, Amit. 2020. *Ease of Living Index*. New Delhi: Institute of Competitiveness; Ministry of Housing and Urban Affairs, GOI.
Keller, Suzzane. 1968. *The Urban Neighbourhood: A Sociological Perspective*. New York: Random House.
Kennedy, Maureen, and Paul Leonard. 2001. *Dealing with Neighbourhood change: A primer on Gentrification and Policy Choices*. Berkeley, CA: The Brookings Institution Centre for Urban and Metropolitan Policy.
Khan, Iqbal Ghani. 2003. "The Awadh renaissance and the role of the French-1750-1820." *Indian Journal of History of Science* 38: 273–301.
Kim, Youjung, Galen Newman, and Burak Güneralp. 2020. "A review of driving factors, scenarios, and topics." *Land* 9 (8): 246. doi:10.3390/land9080246.
Kloosterman, Robert C., and Joanne P. Van Der Leun. 1999. "Just for starters: Commercial gentrification by immigrant entrepreneurs in Amsterdam and Rotterdam neighbourhoods." *Housing Studies* 14 (5): 659–677. doi:10.1080/02673039982669.
Knopf, Jeffrey W. 2006. "Doing a literature review." *Political Science & Politics* 39 (1): 127–132. doi:10.1017/S1049096506060264.
Kosta, Ervin B. 2019. "Commercial gentrification indexes: Using business directories to map urban change at the street level." *City and Community* 18 (4): 1101–1122. doi:10.1111/cico.12468.
KPMG. 2003. *Indian States-Economy and Business*. New Delhi: IBEF.
Kumari, Dr Kiran. 2015. "Urban sprawl: A case study of Lucknow City." *International Journal of Humanities and Social Science Invention* 4 (5): 11–20.
Kundu, Amitabh. 2011. "Trends and Process of Urbanisation in India." In *Urbanisation and Emerging Population Issues -6*. New Delhi: Human Settlements Group; IIED; Population and Development Branch, UNFPA, 09.
LaVecchia, Olivia, and Stacy Mitchell. 2016. *Affordable Spaces: How Rising Commercial Rents Are Threatening Independent Businesses, and What Cities Are Doing About It*. Washington DC: Institute for Local Self reliance.
LDA. 2014. *Master Plan 2031*. Lucknow: Lucknow Development Authority (LDA).
Lees, Loretta. 1994. "Gentrification in London and New York: An Atlantic gap?" *Housing Studies* 9 (2): 199–217.
Lees, Loretta, Tom Slater, and Elvin Wyly. 2008. *Gentrification*. London, UK: Routledge.
Ley, David. 1986. "Alternative explanations for inner-city gentrification: A Canadian assessment." *Annals of the Association of American Geographers* 76 (4): 521–535.
Ley, David. 1994. "Gentrification and the politics of the new middle class." *Environment and Planning D: Society and Space* 12: 53–74.
Li, Jiaqian, Zhuolun Li, Shipei Dong, Minghua Wei, and Jinyan Zhou. 2021. "Spatial and temporal changes in vegetation and desertification (1982–2018) and their responses to climate change in the Ulan Buh

Desert, Northwest China." *Theoretical and Applied Climatology* 143 (3): 1643–1654. doi:10.1007/s00704-021-03522-2.

Li, Tianhong, and Wenkai Li. 2015. "Multiple land use change simulation with Monte Carlo approach and CA-ANN model, a case study in Shenzhen, China." *Environmental Systems Research* 4: 1–10.

Li, Xiaoma, Zhou Weiqi, and Zhiyun Ouyang. 2013. "Forty years of urban expansion in Beijing: What is the relative importance of physical, socioeconomic, and neighborhood factors?" *Applied geography* 13: 1–10. doi:10.1016/j.apgeog.2012.11.004.

Liang, Xun, Xiaoping Liu, Dan Li, Hui Zhao, and Guangzhao Chen. 2018. " Urban growth simulation by incorporating planning policies into a CA-based future land-use simulation model." *International Journal of Geographical Information Science* 2: 2294–2316.

Liang, Xun, Xiaoping Liu, Guangliang Chen, Jiye Leng, Youyue Wen, and Guangzhao Chen. 2020. "Coupling fuzzy clustering and cellular automata based on local maxima of development potential to model urban emergence and expansion in economic development zones." *International Journal of Geographical Information Science* 34: 1930–1952.

Lin, Geng, Xiaoxuan Chen, and Yutian Liang. 2018. "The location of retail stores and street centrality in Guangzhou, China." *Applied Geography* 100: 12–20. doi:10.1016/j.apgeog.2018.08.007.

Liu, Yan. 2012. "Modelling sustainable urban growth in a rapidly urbanising region using a fuzzy-constrained cellular automata approach." *International Journal of Geographical Information Science* 26: 151–167.

Liu, Yan, and Jin He. 2009. "Developing a web-based cellular automata model for urban growth simulation." In *International Symposium on Spatial Analysis, Spatial-Temporal Data Modeling, and Data Mining*. Wuhan: International Society for Optical Engineering.

Loesch, August. 1954. *Economics of Location*. New Haven: Yale University Press.

London, Bruce, and John Palen. 1984. *Gentrification and Neighbourhood Revitalization*. Albany, NY: State University of New York Press.

Ma, Qun, Jianguo Wu, Chunyang He, and Guohua Hu. 2019. "Reprint of "Spatial scaling of urban impervious surfaces across evolving landscapes: From cities to urban regions"." *Landscape and Urban Planning* 187: 132–144. doi:10.1016/j.landurbplan.2019.03.009.

Magicbricks. 2021. *Residential Plot Rates & Trends in Gomti Nagar, Lucknow*. Accessed May 4, 2021. https://www.magicbricks.com/Property-Rates-Trends/Residential-Plot-rates-Gomti-Nagar-in-Lucknow

Malhotra, Chanda. 2017. "Pedestrianisation of commercial area: A case of Amindabad, Lucknow." *International Research Journal of Engineering and Technology* 4 (8): 930–936.

Mandelas, Eleftherios A., Thomas Hatzichristos, and Poulicos Prastacos. 2007. "A Fuzzy Cellular Automata Based Shell for Modeling Urban Growth – A Pilot Application in Mesogia Area." *10th AGILE International Conference on Geographic Information Science*. Denmark: Aalborg University.

Mantelas, Lefteris A., Thomas Hatzichristos, and Poulicos Prastacos. 2010. "A Fuzzy Cellular Automata Modeling Approach–Accessing Urban Growth Dynamics in Linguistic Terms." *International Conference on Computational Science and Its Applications, ICCSA 2010*. Berlin: Springer, pp. 140–151.

Mantelas, Lefteris A., Poulicos Prastacos, Thomas Hatzichristos, and Kostis Koutsopoulos. 2012. "Using fuzzy cellular automata to access and simulate urban growth." *Geo Journal* 77:13–28.

Markel, Stephen. 2010. "The Dynastic History of Lucknow." In *India's Fabled City: The Art of Courtly Lucknow*, edited by Stephen Markel and Tushara Bindu Gude, pp. 14–23. Los Angeles: LACMA and DelMonico Books.

Marshall, Alfred. 1890. *Principles of Economics*. London: MacMillan.

Mayors, City. 2010. *World Fastest Growing Urban Areas*. http://citymayors.com/statistics/urban_growth1.html

McGee, T.G. 1967. *The Southeast Asian City*. London: Bell and Son.

McKinnish, Terra, Randall Walsh, and T. Kirk White. 2010. "Who gentrifies low-income neighborhoods?" *Journal of Urban Economics* 67 (2): 180–193. doi:10.1016/j.jue.2009.08.003.

Meltzar, Rachel, and Pooya Ghorbani. 2017. "Does gentrification increase employment opportunities in low-income neighborhoods?" *Regional Science and Urban Economics* 66: 52–73. doi:10.1016/j.regsciurbeco.2017.06.002.

Mishra, Neha. 2019. "Social life on commercial street: A case of commercial markets of Lucknow, Uttar Pradesh." *International Journal of Engineering Research & Technology* 8 (7): 209–221.

Moos, Markus, Nick Revington, Tristan Wilkin, and Jean Andrey. 2018. "To Knowledge economy city: Gentrification, studentification and youthification, and their connections to universities." *Urban Studies* 56: 1–18.

Morisson, Arnault, and Carmelina Bevilacqua. 2019. "Balancing gentrification in the knowledge economy: The case of Chattanooga's innovation district." *Urban Research & Practise* 12 (4): 472–492. doi: 10.1080/17535069.2018.1472799.

MSME, GOI. 2020. *District Industrial Profile*. Lucknow: MSME, GOI.

Mubea, Kenneth, Roland Goetzke, and Gunter Menz. 2014. "Applying Cellular Automata for Simulating and Assessing Urban Growth Scenario Based in Nairobi, Kenya." *International Journal of Advanced Computer Science and Applications* 5 (2): 1–13. doi:10.14569/IJACSA.2014.050201.

Myrdal, Gunnar. 1957. *Economic Theory and Underdeveloped Regions.* London: Duckworth.

Ong, Paul M., Chhandara Pech, and Rosalie Ray. 2014. *TOD Impacts on businesses in Four Asian American neighbourhoods.* Los Angeles: UCLA Centre for the study of Inequality.

Pandey, Reena S., and Subhash Pawar. 2020. "Contemporary trends in Chikankari". *International Journal od Research and Analytical Reviews* 7 (1): 436–440.

Parker, Jeffrey Nathaniel. 2018. "Negotiating the spacce between Avant-Garde and " Hip Enough": Businesses and Commercial Gentrification in Wicker Park." *City and Community* 1–23. doi:10.1111/cico.12294.

Pfluger, Michael, and Takatoshi Tabuchi. 2018. "Comparative advantage, agglomeration economies and trade costs." *Journal of Urban Economics* 1–13. doi:10.1016/j.jue.2018.11.003.

Pijanowski, Bryan C., Amin Tayyebi, Jarrod Doucette, Burak K. Pekin, David Braun, and James Plourde. 2014. "A big data urban growth simulation at a national scale: Configuring the GIS and neural network based Land Transformation Model to run in a High Performance Computing (HPC) environment." *Environmental Modelling & Software* 51: 250–268. doi:10.1016/j.envsoft.2013.09.015.

Porta, Sergio, Emanuele Strano, Valentino Lacoviello, Roberto Messora, Vito Latora, Alessio Cardillo, Fahui Wang, and Salvatore Scellato. 2009. "Street centrality and densities of retail and services in Bologna, Italy." *Environment and Planning B: Planning and Design* 36: 450–465. doi:10.1068/b34098.

Pratomoatmojo, NA. 2018. "LanduseSim algorithm: Land use change modelling by means of Cellular Automata and Geographic Information System." *IOP Conference Series: Earth and Environmental Science* 202. doi:10.1088/1755-1315/202/1/012020.

Punia, Milap, and Laxman Singh. 2012. "Entropy approach for assessment of urban growth: A case study of Jaipur, India." *Journal of the Indian Society of Remote Sensing* 40: 231–244. doi:10.1007/s12524-011-0141-z.

Rajan, Shailendra, Barsati Lal, and Hugo Lamers. 2016. "A set of interconnected practices which enhance conserve mango diversity in Malihabad, India." In *Tropical Fruit Tree Diversity- Good Practices for in situ and On-Farm Conservation*, edited by Bhuwon Sthapit, Hugo Lamers, Ramanatha Rao and Arwen Bailey, pp. 172–183. New York: Earth Scan, Routledge.

Rankin, Katharine N., and Heather McLean. 2014. "Governing the commercial streets of the city: New terrains of disinvestment and gentrification in Toronto's Inner Suburbs." *Antipode* 1–24. doi:10.1111/anti.12096.

Ricci, Luca Antonio. 1999. "Economic geography and comparative advantage: Agglomeration versus specialization." *European Economic Review* 43 (2): 357–377. doi:10.1016/S0014-2921(98)00065-8.

Richardson, Jason, Bruce Mitchell, and Juan Franco. 2019. *Shifting Neighbourhoods: Gentrification and Cultural displacement in American Cities.* Washington, DC: National Community Reinvestment Coalition.

Rössel, Jörg, and Julia H Schroedter. 2015. "Cosmopolitan cultural consumption: Preferences and practices in a heterogenous, urban population in Switzerland." *Poetics* 50: 80–95. doi:10.1016/j.poetic.2015.02.009.

Ryu, Hwayeon, Donghyun Kim, and Jina Park. 2020. "Characteristics analysis of commercial gentrification in seoul focusing on the vitalization of streets in residential areas." *Sustainability* 12 (21). doi:10.3390/su12218877.

Saha, Arjun, and Ranjan Roy. 2021. "An integrated approach to identify suitable areas for built-up development using GIS-based multi-criteria analysis and AHP in Siliguri planning area, India." *SN Applied Sciences* 3 (4): 395. doi:10.1007/s42452-021-04354-5.

Sang, Lingling, Chao Zhang, Jianyu Yang, Dehai Zhu, and Wenju Yun. 2011. "Simulation of land use spatial pattern of towns and villages based on CA–Markov model." *Mathematical and Computer Modelling* 54: 938–943.

Senes Consultants. 2015. *City Developement Plan-Lucknow City-2040.* Lucknow: Lucknow Municipal Corporation.

Seo, J-K. 2002. "Re-urbanisation in regenerated areas of Manchester and Glasgow: New residents and the problems of sustainability." *Cities* 19 (2): 113–121. doi:10.1016/S0264-2751(02)00006-9.

Sevincer, A. Timur, Shinobu Kitayama, and Michael E.W. Varnum. 2015. "Cosmopolitan cities: The frontier in the twenty-first century?" *Frontiers in Psychology* 6: 1–13. doi:10.3389/fpsyg.2015.01459.

Shi, Yijun, Guofang Zhai, Lihua Xu, Shutian Zhou, Yuwen Lu, Hongbo Liu, and Wei Huang. 2021. "Assessment methods of urban system resilience: From the perspective of complex adaptive system theory." *Cities* 112. doi:10.1016/j.cities.2021.103141.

Shukla, Anugya, and Kamal Jain. 2019a. "Critical analysis of rural-urban transitions and transformations in Lucknow city, India." *Remote Sensing Applications: Society and Environment* 13: 445–456. doi:10.1016/j.rsase.2019.01.001.

Shukla, Aungya, and Kamal Jain. 2019b. "Modelling urban growth trajectories and spatiotemporal pattern: A case study of Lucknow City, India." *Journal of the Indian Society of Remote Sensing* 47: 139–152.

Singh, Prafull, Noyingbeni Kikon, and Pradipika Verma. 2017. "Impact of land use change and urbanization on urban heat island in Lucknow city, Central India. A remote sensing based estimate." *Sustainable Cities and Society* 32: 100–114. doi:10.1016/j.scs.2017.02.018.

Sinha, Arunav. 2016. *Lucknow's IT City to become operational by Sep 1*. February 6. Accessed April 19, 2021. https://timesofindia.indiatimes.com/city/lucknow/Lucknows-IT-City-to-become-operational-by-Sep-1/articleshow/50880962.cms

Smith, Neil. 1987. "Gentrification and the rent gap." *Annals of the Association of American Geographers* 77 (3): 462–465.

Smith, Neil. 1996. *The New Urban Frontier: Gentrification and the Revanchist City*. London and New York: Routledge.

Snir, Tal Modai, and Maaten van Ham. 2019. "Structural and exchange components in processes of neighbourhood change: A social mobility approach." *Applied Spatial Analysis and Policy* 12: 423–443. doi:10.1007/s12061-017-9249-z.

Sridhar, Kala Seetharam. 2016. "Costs and benefits of urbanization: The Indian case." *SSRN Electronic Journal* 1–16. doi:10.2139/ssrn.2892925.

Sullivan, Daniel Monroe. 2014. "From food desert to food mirage: Race, social class, and food shopping in a gentrifying neighborhood." *Advances in Applied Sociology* 4 (1): 30–35. doi:10.4236/aasoci.2014.41006.

Sutton, S.A. 2010. "Rethinking commercial revitalization: A neighborhood small business perspective." *Economic Development Quarterly* 24: 352–370.

Tan, Ronghui, Yaolin Liu, Yanfang Liu, Qingsong He, Licai Ming, and Shuohua Tang. 2014. "Urban growth and its determinants across the Wuhan urban agglomeration, central China." *Habitat International* 44: 268–281. doi:10.1016/j.habitatint.2014.07.005.

Taqui, Roshan. 2005. *Images of Lucknow*. Lucknow: New Royal Book Co.

Tech Mech International (P)Ltd. 2015. *Master Plan for Notified area of LIDA 2031*. Lucknow: Lucknow Industrial Development Authority.

Textile Value Chain. 2020. *Chikankari-The Inheritance of Lucknow City*. October 13. Accessed April 19, 2021. https://textilevaluechain.in/2020/10/13/chikankari-the-inheritance-of-lucknow-city/

Town and Country Planning. 2015. *Urban and Regional Development Plans Formulation and Implementation Guidelines, Volume 1*. New Delhi: Ministry of Urban Development.

Turok, Ivan, and Gordon McGranahan. 2013. "Urbanization and economic growth: The arguments and evidence for Africa and Asia." *Environment and Urbanisation* 25 (2): 465–482. doi:10.1177/0956247813490908.

Varshneya, Vipul. 2017. *Lucknow-The city of Heritage and Culture- A walk through History*. New Delhi: Niyogi Offset Pvt Ltd.

Verwaaijen, Marieke. 2014. *Commercial Gentrification in Prague: Blessing or Burden for local residents?* Master's Thesis, Utrecht: Utrecht University.

Willis, Jerry W. 2007. *Foundations of Qualitative Research: Interpretive and Critical Approaches*. Thousand Oaks, CA: Sage Publications Inc.

Wu, F. 1998. "Simulating urban encroachment on rural land with fuzzy-logic-controlled cellular automata in a geographical information system." *Journal of Environmental Management* 293–308.

Yoon, Yoonchae, and Jina Park. 2018. "Stage classification and characteristics analysis of commercial gentrification in Seoul." *Sustainability* 10 (7). doi:10.3390/su10072440.

Zhang, Bin, Haijun Wang, Sanwei He, and Chang Xia. 2020. "Analyzing the effects of stochastic perturbation and fuzzy distance transformation on Wuhan urban growth simulation." *Transations in GIS* 24: 1779–1798.

Zhang, Zhonghao, Shiliang Su, Rui Xiao, Dwei Jiang, and Jiaping Wu. 2013. "Identifying determinants of urban growth from a multi-scale perspective: A case study of the urban agglomeration around Hangzhou Bay, China." *Applied Geography* 45: 193–202. doi:10.1016/j.apgeog.2013.09.013.

Zukin, Sharon, Valerie Trujillo, Peter Frase, Danielle Jackson, Tim Recuber, and Abraham Walker. 2009. "New retail capital and neighborhood change: Boutiques and gentrification in New York City." *City and Community* 8 (1): 47–64. doi:10.1111/j.1540-6040.2009.01269.x.

11 Carrying Capacity of Air Environment and Development of a Megacity

A Case of NCT Delhi, India

Himanshhu Panwar, Jwngma Basumatary, and Meenakshi Dhote
School of Planning and Architecture, New Delhi, India

CONTENTS

11.1 INTRODUCTION

India's vulnerability to air pollution is attributed to its increasing population and vehicle ownership, coupled with rising fuel consumption, inefficient transportation systems, poor land-use patterns, industrialization, and non-implementation of environmental regulations (Zhao, 2010). As per the World Air Quality Report, nine of the world's top ten most polluted cities are in India (IQAir, 2020). The global cities ranking report is formulated using $PM_{2.5}$ data from 106 nations, estimated by ground-based monitoring stations, most of which are operated by government agencies. According to 2019 CEEW studies, air pollution was responsible for about 17,000 premature deaths and an economic loss of US$1,207 million in Delhi (Pandey et al., 2021). The average life expectancy in India would have been up to 1.7 years higher if the air pollution levels had been lower than the minimal level causing health loss (IIPH, 2020). As per D'Amato et al. (2015), respiratory diseases such as asthma and allergies show an increase with rising levels of air pollution coupled with rising temperatures due to climate change. Health deterioration due to airborne allergens and emission of pollutants can lead to decreased labor productivity, lung cancer and premature deaths.

DOI: 10.1201/9781003131922-13

Many initiatives have been taken by the government for the prevention, control and abatement of air pollution in the country. However, there remains less focus on identifying the carrying capacity with respect to the air environment and understanding the dynamics of urban air pollution, which is crucial for evaluating the extent of pollution load a particular place can withstand. This chapter focuses on understanding the growth, drivers, and development of air pollution in the city and the associated governance mechanism tasked with tackling the issue, focusing on urban planning activities. Also, it proposes a methodology for conducting carrying capacity analysis for the air environment, using Delhi as the case study. Lee and Jeong (2011) state that carrying capacity estimation can suggest the level of human activities that can keep pollution levels to the required environmental standards. A carrying capacity-based pathway can lead to sustainable development.

11.2 METHODOLOGY

The chapter uses an urban planning approach for understanding the air pollution dynamics in Delhi. To understand the causes of the rapidly increasing concentration of pollutants, the growth and drivers of air pollution in the city are discussed along with the increasing extent of the urban sprawl over the past three decades. Air pollution mitigation provisions directly or indirectly adopted or listed in the last three master plans of Delhi NCT are then discussed. A brief overview of the inclusion of mitigation measures around different land uses under the Master Plan follows. Further, to understand the effectiveness of previous legal mandates, all provisions relating to air pollution mitigation have been studied. To assess carrying capacity, the chapter further analyzes $PM_{2.5}$ and PM_{10} data from 2019 carried out for 37 monitoring stations (CPCB, DPCC and IMD) spread all over the city. Subsequently, a framework is derived to calculate the assimilative carrying capacity of the air environment and spatially represent it using GIS. Mapping has been done through raster (factor) analysis in Arc GIS using the carrying capacity formula for all the monitoring stations across NCT Delhi.

11.3 LITERATURE STUDY

Understanding the drivers and growth of air pollution in an urban atmosphere is crucial to an analysis of the causes of pollution and identifying suitable mitigation options. With the increase in the extent of the city, the associated activities also start increasing. This section covers the growth and development of Delhi city, its air pollution, provisions in the city's Master Plan, and major legislative and policy mandates related to it.

11.3.1 Growth and Development of Delhi and Air Pollution

Even during 1991–1994, the average total suspended particulate (TSP) level in Delhi was roughly five times the World Health Organization's annual average standard (Cropper, 1997). Furthermore, the total suspended particulate levels in Delhi exceeded the WHO 24-h standard on 97% of all days on which readings were taken. MOEFCC (1997) reviewed the environmental situation in Delhi over worries of degrading conditions. When compared to the 1989 pollution levels, it was observed that in 1996 every day about 3000 metric tons of pollutants were emitted. 67% of this load was contributed by vehicles and 12% by thermal power plants. During this time carbon monoxide emissions from vehicles increased by 92%. Delhi also has the highest cluster of small-scale industries in India which contribute to 12% of air pollutants along with other industrial units (Suliankatchi, 2013).

The National Capital Territory of Delhi and its planning region, the National Capital Region, is characterized by continuous growth and sprawl. With an area of 34,000 km^2 and a population of about 46 million it is the country's largest planning region. Figure 11.1 depicts this expansion of Delhi.

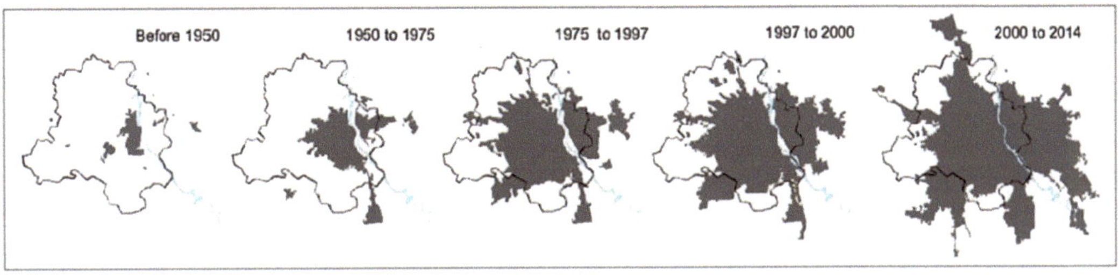

FIGURE 11.1 Urban sprawl around NCT Delhi.

(Source: Compiled from NIUA 2015.)

The city is vulnerable to elevated pollution levels because it lies in the landlocked Indo-Gangetic Plain (IGP), which encounters frequent dust storms each year during the pre-monsoon season (March–June) (Prasad & Singh, 2007). Dust storms observed in India come either from the Thar Desert in western India or from the distant Arabian Peninsula. Contingent on the wind speed, dust from the west is seen in the IGP and transported over the eastern parts of India up to West Bengal. Some of the time, depending upon the wind pattern, dust is redirected to the central parts of India (Prasad & Singh, 2007). Delhi also faces major traffic challenges. The number of private vehicles has risen steadily from 0.56 million in 1981 to 8.83 million in 2015 and is predicted to rise further (Malik, 2015). This is manifested in traffic jams. With the absence of bus corridors in many places, travel by bus becomes slow, again increasing trips by private vehicles. This seems to have become a vicious circle.

11.3.2 Master Plan of Delhi and Provisions for Air Pollution Control

The First Master Plan of Delhi (1962), with the horizon year of 1981, adopted a government-driven land acquisition policy with strict zoning of land uses. Although this Master Plan proposed closure of industries, no action was taken to implement the proposal for three decades (Narain & Bell, 2005).

The 1990 Master Plan for Delhi again proposed a need to shut down hazardous and noxious industries and heavy and large industries. The list of industries was further classified and expanded and no new industries of this type were permitted. It was further proposed to relocate industries within three years. This was also not implemented until 1997, and then only after initiation and pressure on the government from the Supreme Court (Narain & Bell, 2005). Apart from the industries, pollution arising from the growth of the city was not discussed, indicating that it did not exist as an issue at that point (Puri, 2012).

The Third Master Plan, for 2021, demarcated a number of locations where no industrial activity, including household industry, was permitted as 'no industrial activity' zones. It widened the scope of household industrial units subject to adherence to pollution control norms and proposed the redevelopment of industrial concentration in non-conforming areas. A change of fuel to gas based for power plants operating in NCT Delhi was suggested, and the importance of public transport was emphasized. But MPD 2021 did not provide any study as such of the implications of land uses for air quality. Projections of vehicle growth and modal share over the years were based on 2008 data. The severity of air quality as per the reports published by CPCB has been acknowledged in the Master Plan but no further details have been provided as to how the proposed land-use plans and existing transportation network would impact air quality. The existing zoning of industries in Delhi (as per Master Plan of Delhi, 2021) also does not support the improvement of air quality.

11.3.3 Court Orders and Directions to Control Air Pollution in NCT Delhi

In 1985, noted environmental lawyer Advocate M. C. Mehta filed public-interest litigation for controlling air pollution in Delhi in the Supreme Court of India. This was the beginning of the Court's involvement in policies and controlling air pollution. Mehta intended to tackle the issue of rising air pollution, highlighting the absence of any regulation or management efforts on the part of the government. He also requested the court to direct various departments to implement the Air Act of 1981. As a result of the appeal many laws were enacted, including the 1986 Environment (Protection) Act, the Motor Vehicles Act of 1988, amendments were made to the Air Act, and the Central Motor Vehicle Rules were formulated in 1989. The rules established principles for vehicular emissions for manufacturers and users. The 1990 vehicle exhaust emission standards led to control of smoke, visible vapor, ash, etc. (Agarwal, 1996). Air quality standards were also formulated at this time (Narain & Bell, 2005).

In December 1996, the Delhi government submitted its first action plan to control air pollution after receiving notice from the Supreme Court of India. The Supreme Court had directed the government through the Ministry of Environment, Forests and Climate Change to establish an authority to advise pollution control boards and the courts on matters relating to monitoring pollution and implementation of orders. In 1998 the MOEF established the Environment Pollution (Prevention and Control) Authority (EPCA) for NCT Delhi and the National Capital Region. The EPCA was dissolved in 2020 and a network of monitoring stations across the country was launched under the National Air Quality Monitoring program (NAMP).

In 2014, MOEFCC, with contributions from experts and civil society as well as the Union the Ministry of Health and Family Welfare, created the National Air Quality Index (NAQI) (Raman, 2019). In 2015 many orders were made by the Supreme Court in relation to control of air pollution from various sources such as vehicles, industries, construction, power plants, and agriculture (Chatterji, 2020). One such order was that all taxis were required to switch to compressed natural gas, and vehicles over 10 years old could no longer ply for trade in Delhi. The registration of diesel cars was stopped along with sport utility vehicles with an engine capacity above 2,000 cc.

Reference was also made by the National Green Tribunal to a Comprehensive Action Plan to reduce pollution by the EPCA, co-ordinated with the GRADED Response Action Plan notified by MoEFCC in 2017 (CPCB, 2018). In 2018, the Tribunal ordered all States and Union Territories with Non-Attainment Cities to prepare appropriate action plans aimed at bringing the air quality standards in line with the prescribed norms. Control of air pollution caused by waste was addressed by different orders of the court for the management of biomedical waste, plastic waste, remediation of legacy waste sites and control of polluting activity in industrial areas. A five-year action plan to address air pollution – the National Clean Air Action Plan (NCAP) – was launched by the government in 2019, focused on expanding the air quality monitoring network and raising public awareness, alongside the implementation of air pollution control measures (MOEFCC, 2019).

With these policy interventions, air quality has shown minor improvement in terms of emissions rates, but hitherto the efforts have failed to adequately address the issue. Time-bound issues across scales need to be undertaken – at local, subnational and national levels. Pollutants are transported in the city by meteorological conditions and topography. The pollutant flow patterns and resultant ground-level concentration need to be estimated to save the environment. It is for this reason that the assessment of carrying capacity is required, so that the capacity of the atmosphere to withstand further pollution concentration can be estimated.

Source Apportionment Studies for Delhi were carried out by (i) IIT Kanpur in May, 2016, (ii) The Energy and Research Institute (TERI) in August, 2018 and (iii) Centre for Science and Environment (CSE) in (2019). Subsequently, in a case filed by Shri Viswa Mohan, the National Green Tribunal directed the Delhi Pollution Control Committee to undertake a carrying capacity assessment of the air environment in NCT Delhi. This chapter describes a framework that can be used for the assessment of carrying capacity with respect to air environment, which can also be used by other cities.

11.4 THE CONCEPT OF CARRYING CAPACITY OF THE AIR ENVIRONMENT

The concept of carrying capacity needs to be understood to protect resources when planning any development. It has been observed that the air environment is considerably impacted by increasing development activities.

Carrying capacity-based development planning should identify the resources of the region; the scope of utilization can be decided based on the degree of impact (CISMHE & IITD, 2005). In the case of the air environment in urban areas, carrying capacity can be estimated based on emissions load (from various sources, man-made and natural) and on the specific conditions of resource allocation such as topography, atmospheric conditions, etc.

11.4.1 Supportive Capacity Dimensions of Air Environment

Supportive capacity determines the developmental potential of a region and its economic progress and quality of life. Supportive capacity includes urban land and shelter resources, urban social, physical and communication infrastructure, urban utilities, and human and institutional resources. Comparisons regarding these aspects are made with existing norms and standards to arrive at an assessment of adequacy/deficiency, above or below standards. The assessment is followed by recommendations.

11.4.2 Assimilative Capacity Dimensions of Air Environment

The assimilative capacity of the environment refers to the ability of the natural environment – land, air, water and biota – to carry waste without itself or any of its users being impacted adversely. Assimilative capacity includes air environment, water environment, land/soil environment, biological environment, and acoustic environment.

Carrying capacity or assimilative capacity of the atmosphere in urban areas depends upon many parameters – atmospheric stability, topography of the region etc. Solar insulation, mean wind speed, and temperature profile of an area contribute to its atmospheric stability. However, the carrying capacity gives an estimate of how much more pollution load from various sources can be sustained in the atmosphere. Thus, how the atmospheric conditions and topography contribute to containing or dispersing pollution will determine the capacity. The dispersal of pollutants is dependent on the mixing height and windspeed. The dispersal or ventilation coefficient has been found to be a product of the mixing height and the transport wind speed within the mixing layer. Thus, the pollution load and ventilation coefficient of the atmosphere need to be understood to arrive at the carrying capacity (CISMHE & IITD, 2005).

11.5 FRAMEWORK FOR CARRYING CAPACITY OF THE AIR ENVIRONMENT

Carrying capacity is the maximum capacity of environment that can support the existing emission load by a land use and its associated activities over a particular time period. The concentrations of substances in the atmosphere are determined by: 1) transport, 2) diffusion, 3) chemical transformation and 4) ground deposition.

Based on anthropogenic activity:

$$\text{Emission Concentration} \propto a * K$$

where a = activity rate (units/hour), K = emission factor (g/unit)

Based on meteorological factors:

$$\textbf{Emission Concentration} \propto \text{Activity Rate} \propto \frac{1}{\text{Wind Speed}} \propto \frac{1}{\text{Mixing Height}} \propto \frac{1}{\text{Temperature}} \propto \frac{1}{\text{R.H}} \propto \text{Pressure}$$

Therefore, we can say that

$$\textbf{Emission Concentration} \propto \frac{\text{Activity Rate} * \text{Emission Factor}}{\text{Wind Speed} * \text{Mixing Height}} \propto \frac{1}{\text{Temperature}} \propto \frac{1}{\text{R.H}} \propto \text{Pressure}$$

The carrying capacity changes over time as it depends on atmospheric stability and emission load over a certain period of time.

$$\textbf{Carrying Capacity} = \left(\frac{\textbf{Emission Standard as per CPCB}}{\textbf{Emission load from Land uses mix}} \times \textbf{Ventilation Coefficient} \right) - \textbf{Ventilation Coefficient}$$

Table 11.1 shows the monthly variation in estimated carrying capacity for 2019 for both $PM_{2.5}$ and PM_{10}. Meteorological factors such as wind speed, temperature, mixing height, pressure, relative humidity and emission loading have been considered in the preparation of the carrying capacity profile of Delhi. This, however, does not represent the true carrying capacity, as factors such as diffusion, chemical transformation and ground deposition have not been factored in. This is possible only through air quality modeling. Carrying capacity is said to be good when the estimated CC is

TABLE 11.1
Estimated Carrying Capacity (Monthly) for NCT Delhi

Month	Mean Wind Speed (m/s)	Min Mixing Height (m)	Max Mixing Height (m)	V.C. (m^2/s) Min	V.C. (m^2/s) Max	Avg V.C	Avg. $PM_{2.5}$ Conc. (μg/m^3)	Avg. PM_{10} Conc. (μg/m^3)	Estimated CC (m^2/s) PM 2.5	Estimated CC (m^2/s) PM 10	PM 2.5 C.C = E.C.C-V.C	PM 10 C.C = E.C.C-V.C
January	1.0	70	2003	76.3	2183.2	1129.7	203.8	324.2	332.46	348.41	−797.31602	−781.3655
February	1.3	90	2087.8	117.	2735.0	1426.4	140.9	240.4	607.18	593.33	−819.30072	−833.15749
March	1.3	65	2215	89.7	3056.7	1573.2	88.48	186.4	1066.8	843.62	−506.38264	−729.57065
April	1.4	70	2066.6	100.	2955.2	1527.6	85.22	236.6	1075.5	645.48	−452.09853	−882.18472
May	1.4	70	4140	102.	6044.	3073.3	85.96	244.9	2145.1	1254.8	−928.13946	−1818.4821
June	1.9	60	1685	114	3201.	657.75	62.96	208.1	1579.8	796.30	−77.937420	−861.44391
July	1.3	120	6874	164.	9418.	4791.4	46.63	141.2	6165.2	3391.9	−1373.8183	−1399.5014
August	1.3	100	1876	133	2495.	1314.0	34.62	83.35	2277.3	1576.5	−963.32568	−262.49269
September	1.1	60	1727	69.6	2003.	1036.4	40.11	98.46	1550.4	1052.6	−513.96632	−16.211135
October	0.98		2045	78.4	2004.	1041.2	123.4	241.7	506.116	430.73	−535.13366	−610.51863
November	1.0	60	1749	60.6	1766.4	913.54	204	315.7	268.68	289.29	−644.85529	−624.24707
December	0.91	50	1664	45.5	1514.2	779.87	203.6	310.	229.80	251.1	−550.06	−528.68

Source: Author.

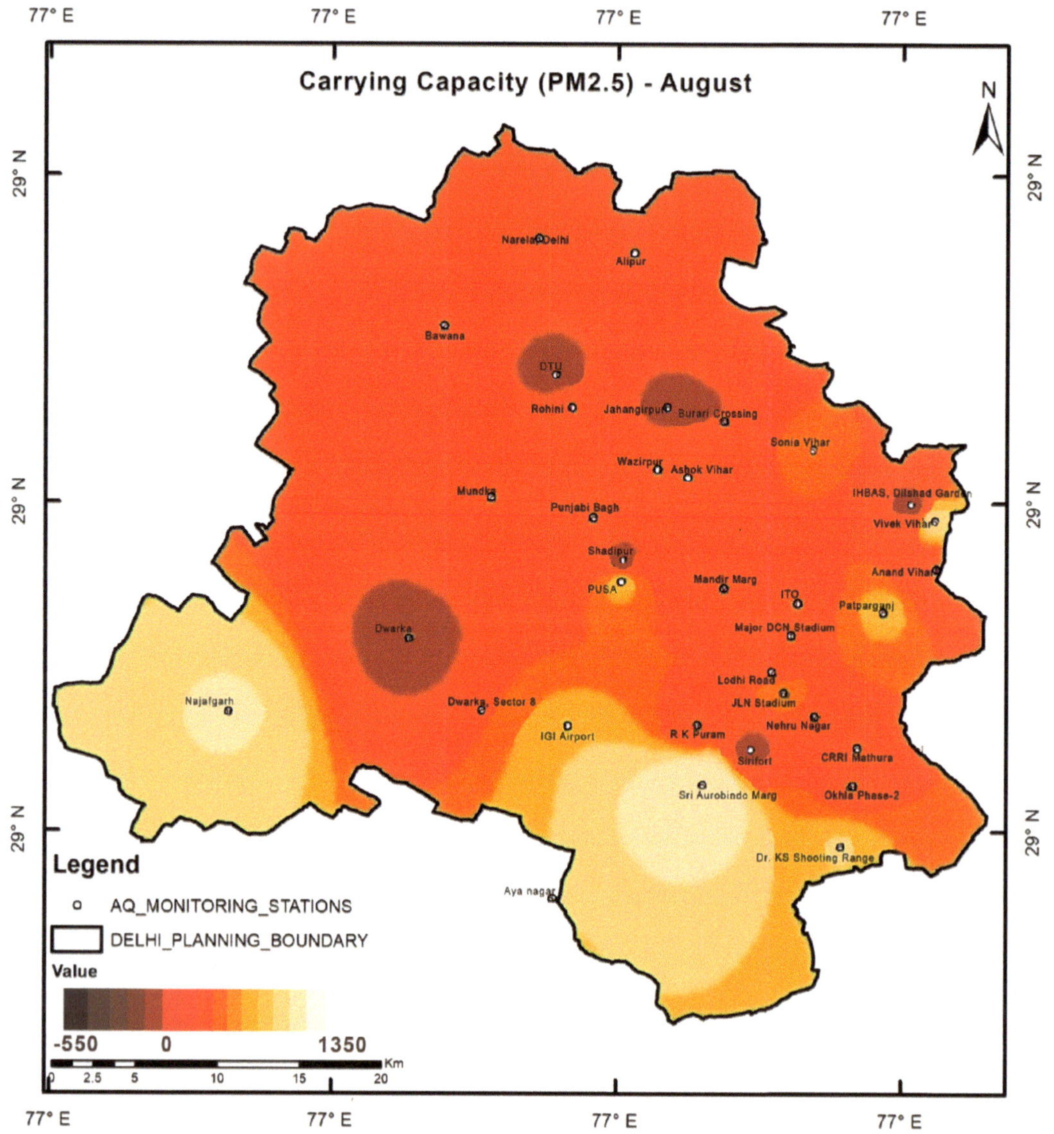

FIGURE 11.2 Maps of carrying capacity with respect to $PM_{2.5}$ from January to June.

greater than the ventilation coefficient. Thus, the carrying capacity is principally dependent on the activities, that is, the emission load per day generated by land use-based activities. This suggests that no matter what the ventilation coefficient is, the goal is to keep the emissions below the standard. Carrying capacity for August and November has been calculated for both $PM_{2.5}$ (see Figures 11.2 and 11.3) and PM_{10} (see Figures 11.4 and 11.5) and is represented by Table 11.2.

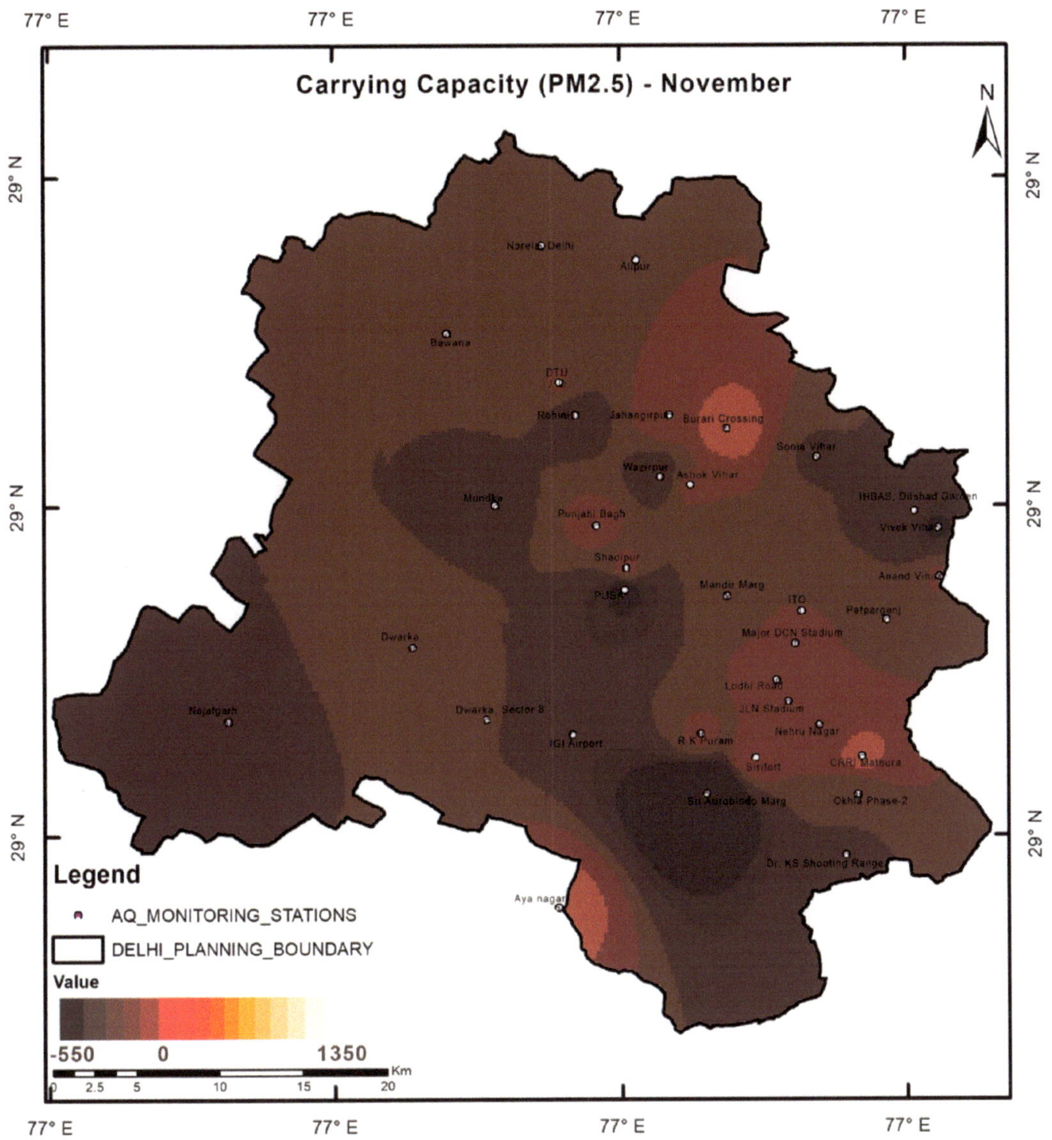

FIGURE 11.3 Maps of carrying capacity with respect to $PM_{2.5}$ from July to December.

Figures 11.2–11.5 were developed through raster (factor) analysis using the carrying capacity formula for all the monitoring stations across NCT Delhi in GIS. The carrying capacity is negative in November owing to the low temperature (winter) giving rise to low mixing height and lower wind speeds. The emission loads are high, as usual, which worsens the situation. The meteorological data supplements the situation for low carrying capacity during this month owing to high-pressure

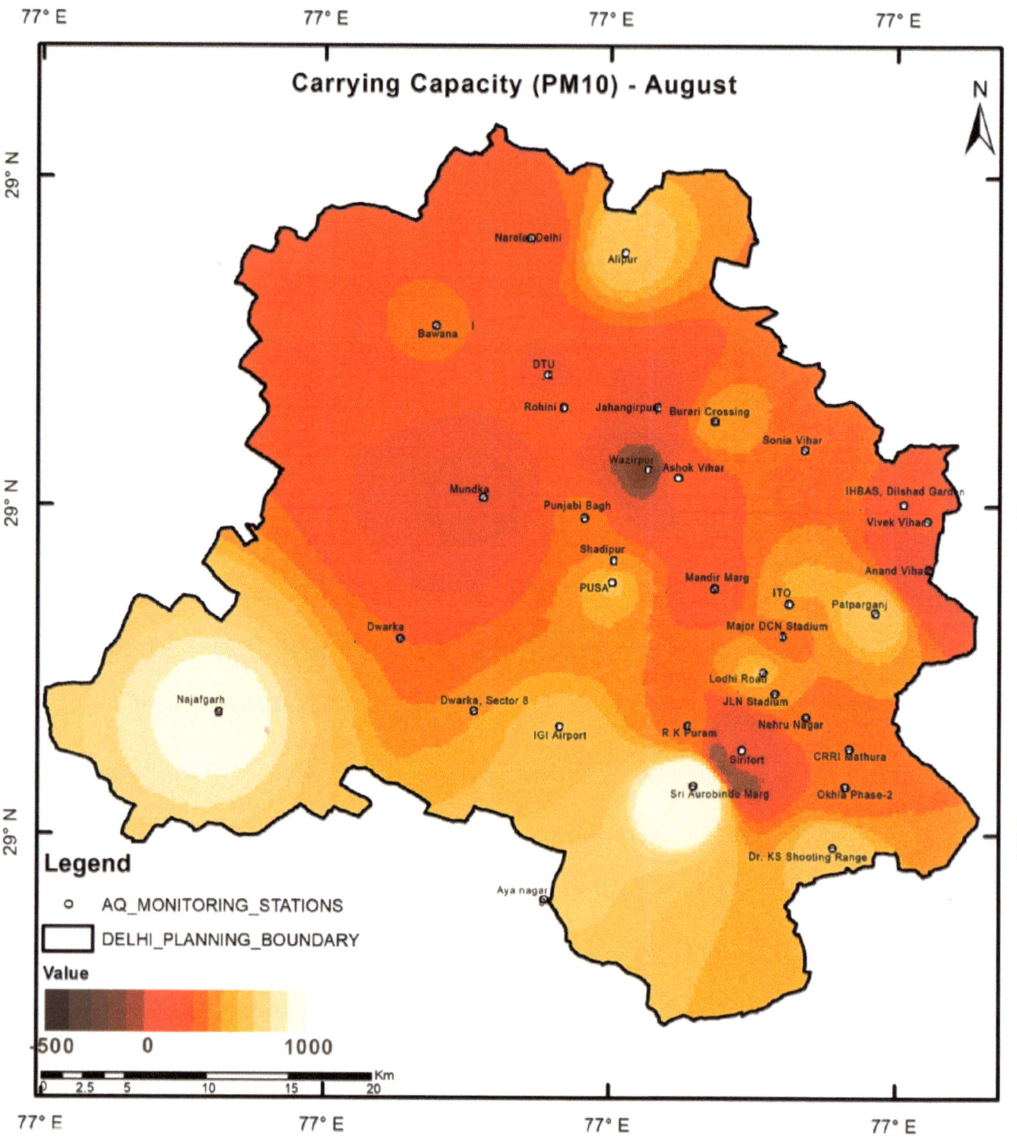

FIGURE 11.4 Maps of carrying capacity with respect to PM_{10} from January to June.

profile, low temperature and low wind speeds. The carrying capacity begins to improve from July with the onset of the monsoon season from July to September. The month of August experiences cleaner air with a high amount of precipitation to wash away the pollutants. From October onwards the carrying capacity declines, remaining negative throughout the winter season until February. The low mixing height, lower temperature and lower wind speeds worsen the air quality during the winter season. Hence it can be concluded, as per the assimilative capacity estimates, that the carrying capacity of NCT Delhi has been surpassed.

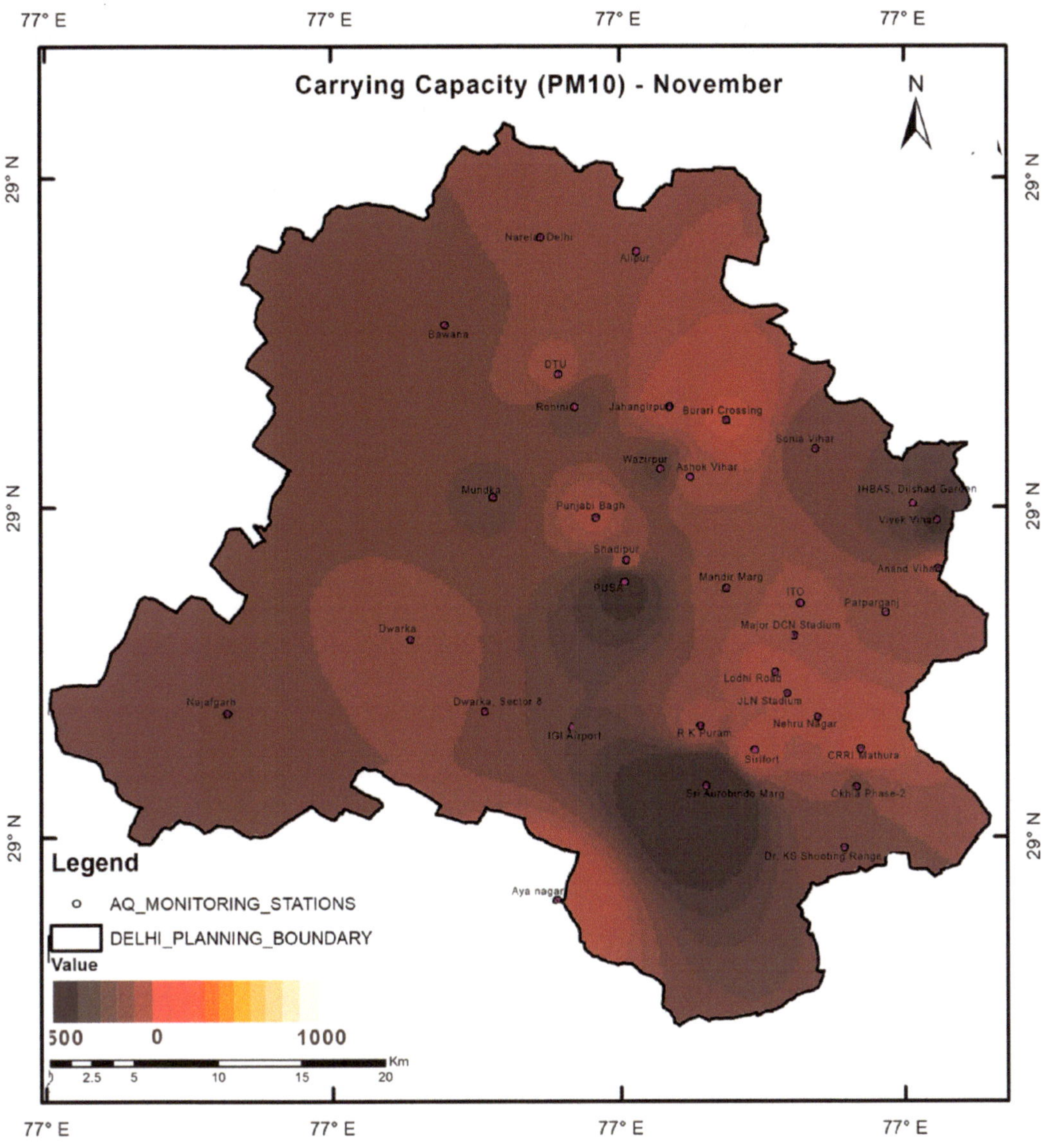

FIGURE 11.5 Maps of carrying capacity with respect to PM_{10} from July to December.

TABLE 11.2
Categorization of Carrying Capacity

Category	Condition	Code
High	Carrying Capacity > Ventilation Coefficient	
Moderate	Carrying Capacity = Ventilation Coefficient	
Low	Carrying Capacity < Ventilation Coefficient	

Source: Author.

11.6 LOCKDOWN AND IMPACT ON AIR POLLUTION

In order to contain the spread of COVID-19 the government of India announced a lockdown from 25 March 2020, which resulted in a complete stoppage of all outdoor activities (Gettleman & Schultz, 2020). This section of the chapter attempts to assess the impacts of meteorological conditions, land use, built form and activities on pollution concentrations during the lockdown phases at eight different stations. A buffer area of 1km (declared by CPCB as the influence zone of a monitoring station) around the monitoring stations has been analyzed in relation to the factors mentioned. All the stations – Dr. Karni Singh Shooting Range, Aya Nagar, IGI airport, DTU, Alipur, Bawana, ITO and Ashok Vihar – were analyzed with their different predominant land uses during the phases of lockdown, in 2019 and in 2018, and some common observations were made.

The stations on the southern side of Delhi, Dr. Karni Singh Shooting Range, Aya Nagar and IGI airport, which are mainly located on the fringes of the city, presented a common pattern. The emissions loads attributed to these zones by the monitoring station were negligible during the lockdown period, with their concentrations remaining good to satisfactory in the first three phases of lockdown. It was observed that they mainly projected the concentrations arising through the influence of regional activities. $PM_{2.5}$ concentrations in all three zones were good throughout the lockdown, whereas the PM_{10} concentration was good only in the first three phases, increasing in the fourth phase when the regional concentrations of other stations in the city increased.

The stations located in North Delhi were DTU and Alipur, which had agricultural as the predominant land use and activity. They both showed a gradual decline in pollution concentration during the first lockdown phase, but the concentrations increased in the second phase after the harvesting activity began, usually after Baisakhi on 14 April every year. The concentration of pollutants in the second and third phases were similar, as only a few restrictions were lifted in phase 2, with no additional activities permitted in these zones that would have made a significant impact on the emission loads in these zones. The concentration at Alipur was observed to be similar only in the fourth phase, just like the second and third at the station, but was found to be increased at DTU, because of the incidence of fires bringing emissions loads from the north-east to the station. Although the wind direction at Alipur was also north-easterly only, no fires that could have influenced the concentration were observed in the north-east of the station.

Concentrations at the monitoring stations at Bawana, ITO and Ashok Vihar, which had different predominant land uses – industrial, PSP and residential, respectively – showed diverse patterns throughout the lockdown phases. Ashok Vihar Station, which had residential land use, had a decrease in concentration of pollutants when the lockdown was established, remaining the same thereafter (i.e., good quality) throughout the four phases. Emission loads from residential activities were thus much reduced. In Bawana industrial area, on the other hand, the concentrations during the first three phases were similar, increasing significantly only in the fourth phase when there was a major relaxation in permitted movement of people, goods and services inside and outside the city. So the pollution levels at this station depend highly on the emission load from the movement of vehicles supporting the industrial activities. However, the concentrations at ITO station increased greatly in phase 2, because of high vehicular traffic in the area. As the phase 3 and 4 data from the station is insufficient, no pattern can be deduced.

The complete lockdown of vehicular movement and industrial operations resulted in a fall in emissions from vehicles and release of industrial exhaust. There was a significant reduction in the emission concentration for both pollutants but even though activities were halted completely, the concentration was still not close to zero. So, not only do local activities have a large impact on the emission concentrations in the city, but regional activities also influence the pollutant concentration in the city. A multiscale strategy therefore needs to be adopted to control the pollution.

11.7 RECOMMENDATIONS

NCT Delhi is located in the semi-arid biogeographic region and is characterized by low rainfall, thorny and open forests, and wetlands. It is located in the bowl-like Ganga Plains, where dust-laden winds predominantly from the west get trapped around and within NCT Delhi, contributing to almost 65% of the pollution there. So, it is important to incorporate dust mitigation measures at both the regional, i.e., National Capital Region, level (NCR) and at NCT Delhi level by planting the appropriate trees, maintaining water bodies, and maintaining moisture in urban and rural areas.

At regional scale, the demarcation of air sheds in the sub-regional plans of Uttar Pradesh, Rajasthan and Haryana should be done, on the basis of which the sources of pollution outside NCT Delhi that impact air pollution levels in Delhi should be identified. Air polluting activities identified should be regulated, including (a) burning of garbage, agricultural waste, etc. (b) relocation of brick kilns/change to clean fuel, (c) control of industrial pollution, especially in identified CEPI areas (critically environmentally polluted industrial areas), and (d) preparation of Clean Air Action Plans (CAPs) for the identified Non-Attainment Cities (NACs). A green belt along the western and eastern peripheral highways should be maintained in order to ensure trapping of dust and control of pollution by planting the right type of tree species.

At city level, the growth trajectories as specified in the Delhi Master Plan 2021 and NCR Plan should include environmental management considerations with respect to air quality. While the Master Plan Delhi recognizes the rising air pollution levels, the recommendations are primarily regulatory in nature. However, long-term land-use and built-form control measures need to be formulated to control the rising impact on health. Development controls governed by FAR should incorporate a study of wind movement towards the city and should contribute to improved pollutant dispersal.

At the planning zone level, the Master Plan NCT Delhi 2041 should discourage development in Zones E and H, and high-density parts of Zones C, M, B and F, as the AQI levels are very poor. The public/semi-public and commercial areas proposed should be so planned that they allow for dispersal of air pollution. The built form and air flow should be studied before finalizing the building heights.

At local level, the location and pattern of green areas impact pollution levels: the lower the percentage of vegetation and the more other activities, the greater the number of polluted days in a year in those areas. Adequate green infrastructure should therefore be planned to address the need for air pollution sinks. The location of land use which attracts high levels of activities should be discouraged in north-west, north-east and south-east Delhi, which show high levels of air pollution.

11.8 CONCLUSION

The chapter suggests further studies that need to be carried out related to land use, built form and vegetation for control of air pollution in NCT Delhi. The assessment of carrying capacity is essential for estimating the resulting ground-level concentration patterns in order to save the environment and, indirectly, the human race. The urban sprawl and built-form density need to be studied to control local-level increases in pollution levels. The source apportionment studies and carrying capacity analysis have identified zones reporting high pollution levels or hotspots. Detailed scientific studies need to be undertaken both in the hotspot areas and the other areas to arrive at patterns of land use, built form and vegetation types that would help to contain the rise in air pollution levels.

According to assimilative capacity estimates, the carrying capacity of NCT Delhi with respect to air pollution has been surpassed, so the city cannot sustain any additional activity or development. But that doesn't mean that efforts to reduce air pollution would be of no value. If further deterioration of pollution concentration needs to be prevented, certain ongoing steps should be continued, with the additional strategies suggested in this chapter.

REFERENCES

Agarwal, A. 1996. *Slow Murder: The Deadly Story of Vehicular Pollution*, New Delhi: Centre for Science and the Environment.

Anisha Raman, P. M., 2019. *Fighting Air Pollution in Delhi for 2 Decades: A Short but Lethal History*, Delhi: Down to Earth.

Chatterji, A., 2020. Air Pollution in Delhi: Filling the Policy Gaps. *ORF Occasional Paper*, 18 December.

CISMHE & IITD, 2005. *Carrying Capacity Study of Teesta Basin in Sikkim*, Delhi: National Hydroelectric Power Corporation Ltd.

CPCB, 2018. *Green Tribunal*. [Online] Available at: https://greentribunal.gov.in/sites/default/files/news_updates/REPORT%20BY%20CPCB%20IN%20OA%20NO.%20681%20of%202018%20(News%20Item%20Published%20in%20the%20times%20of%20india%20Vs%20NCAP%20with%20multiple%20timelines%20to%20clean%20air%20in%20102%20cities% [Accessed 20 December 2020].

Cropper Maureen, L, Simon, N. B., Alberini, A., Sharma, P.K. 1997. The World Bank Development Research Group. December.

CSE, 2019. *Fighting Air Pollution in Delhi for 2 Decades: A Short but Lethal History*. [Online] Available at: https://www.downtoearth.org.in/blog/air/fighting-air-pollution-in-delhi-for-2-decades-a-short-but-lethal-history-67585 [Accessed 11 April 2021].

Gettleman, J. & Schultz, K., 2020. Modi Orders 3-Week Total Lock-down for All 1.3 Billion Indians. *The New York Times*, 24 March, pp. ISSN 0362-4331.

IIPH, 2020. *India State-Level Disease Burden Initiative*, s.l.: The Lancet Planetary Health.

IQAir, 2020. *World Air Quality Report*, Switzerland: IQAir.

Lee, K.-H. & Jeong, Y.-W., 2011. An Assessment of Environmental Carrying Capacity by Analyzing the Emission and Concentration of Urban Atmospheric Pollutants. *LHI Journal of Land Housing and Urban Affairs*, 2(4), pp. 517–528.

Malik, A., 2015. Urban Transportation Infrastructure Development in Delhi/NCR: Trends, Challenges and Opportunities. *International Journal of Advance Research In Science And Engineering*, 4(3), pp. 332–341.

MOEFCC, 1997. *White Paper on Pollution in Delhi With an Action Plan*. [Online] Available at: http://envfor.nic.in/divisions/cpoll/delpolln.html

MOEFCC, 2019. *National Clean Air Programme*, Delhi: MOEFCC.

Narain, U. & Bell, N. G., 2005. *Who Changed Delhi's Air? The Roles of the Court and the Executive in Environmental Policymaking*, Washington DC: Resources for Future.

Pandey, A. et al., 2021. Health and Economic Impact of Air Pollution in the States of India: The Global Burden of Disease Study 2019. *The Lancet Planetary Health*, 5(1), pp. 25–38.

Prasad, A. K. & Singh, R. P., 2007. Changes in aerosol parameters during major dust storm events (2001-2005) over the Indo-Gangetic Plains using AERONET and MODIS data. *Journal of Geophysical Research: Atmospheres*, 112(D1-189).

Puri, P., 2012. *Delhi s Master Plans: 1962, 2001 and 2021: A Critical Analysis*, s.l.: Department of Geography.

Suliankatchi, R. A., Nongkynrih, B. & Gupta, S. K., 2013. Air Pollution in Delhi: Its Magnitude and Effects on Health. *Indian Journal of Community Medicine*, 38(1), pp. 4–8.

Zhao, P., 2010. Sustainable Urban Expansion and Transportation in a Growing Megacity: Consequences of Urban Sprawl for Mobility on the Urban Fringe of Beijing. *Habitat International*, 34(2), pp. 236–243.

12 Urbanization and Sustainability in the South Pacific Region from the 1990s to 2020

Mohammad Afsar Alam, Ravinesh Rohit Prasad, Sakul Kundra, and Mumtaz Alam
Fiji National University, Fiji

Md. Asif Iqubal
Aligarh Muslim University, Aligarh, India

CONTENTS

12.1 INTRODUCTION

The Pacific Ocean encompasses the largest body of water on the planet, punctuated by a slew of small and larger islands. These islands are home to over 20 countries, including 14 ADB (Asian Development Bank, 2012). Pacific developing member countries. The land area, population size, history, ethnic structure and economy of these countries are vastly different. Despite their differences, many of them face similar challenges, including geographic remoteness, increasing susceptibility to climate change and natural calamities, and a small natural resource base, every one of which limit economic growth. Pacific DMCs' (Developing Member Country) capacity to handle their affairs efficiently and the effectiveness of their government is also limited by constraints of a physical, technological, structural and sociocultural nature. Urbanization problems are among these factors.

DOI: 10.1201/9781003131922-14

National and city indicators in Pacific DMCs point to urbanization intensifying in the coming decades. While the urbanization as a percentage of the national population varies greatly across Pacific DMCs, the all-encompassing trend is that a growing population, including migrants from rural areas, are settling in the decaying urban environments of Pacific towns and cities. In 2011, the level of urbanization in certain Melanesian countries was assessed and has been defined, unsurprisingly, by leading Pacific planners as chaos, mayhem and anarchy (Kep 2011).

In this context, it has been documented that development partners, scholars, and governments alike do not properly understand Pacific urban issues. It is generally accepted that "Policy inertia on urbanization has existed for far too long" (Pacific Institute of Public Policy 2011a, 2011b). The influx of capital from major cities and business centers that have positioned themselves in the Pacific area have fueled growth in smaller Pacific cities and villages (Storey 2006). Some practitioners have used words like "everybody's but nobody's business" (Jones and Kohlhase 2002) or "flying blind" to describe anti-urban sentiments and apathy towards urban reform, urban governance, and the urban sector in general (Haberkorn 2008). As a result, urban management and broader urbanization challenges have been overlooked in national and regional growth programs (United Nations Economic and Social Commission for Asia and the Pacific and UN-Habitat 2009; Storey 2006; Jones and Lea 2007; Jones and Lea 2007). For instance, the term 'urban' occurs just once in the main communiqué (for urban youth), while the phrase 'urbanization' appears only once in the communiqué's subheadings (Pacific Islands Forum Secretariat [PIFS] 2011). As a result, the urban sector and urbanization have a hard time gaining traction on national and regional growth agendas. Urbanization – the growth of the urban population as a percentage of the total population – is unavoidable, but it can also be beneficial. The present density of poverty and slums development, and societal disorganization in urban areas is alarming; despite this, no country has ever experienced substantial economic growth without urbanization during the industrial era. "Cities are hotbeds of hardship, but they also offer the best chance of escaping it" (UN Population Fund, 2014; State of the World Population 2007). The potential for urban growth should be allowed to run wild. Consequently, urban management and larger urbanization issues have not been given the attention they need in national and regional growth strategies.

Knowing the various sub-regional settings and the diversity within them is critical to placing Pacific urbanization patterns in perspective. More than 7,500 islands of various origins make up the Pacific zone. Therefore, the area contains coralline islands, either elevated or low atolls; volcanic islands; and 'continental' islands. Melanesia, Micronesia, and Polynesia are the three geographical areas that make up the region. Despite the limitations of the borders, these sub -regional descriptors are still used today to describe the three major social, cultural, racial, linguistic, and geographical categories of the western, north-western, and central-eastern Pacific.

Land region, demographic, ethnic, geographic, and linguistic diversity, and natural resource capability are all factors to consider. Melanesia dominates the other two sub-regions. Pacific DMCs include Fiji, the Solomon Islands, Papua New Guinea (PNG), and Vanuatu, which are located north and east of Australia's eastern shore and are part of the larger Pacific DMCs. Continental islands, known as Melanesian islands, are largely fertile, resource rich, and volcanic. The grouping accounts for nearly 98% of the Pacific region's total ground surface area, which contrasts sharply with the other two Pacific sub-regions.

Micronesia refers to the western Pacific atolls, islands, and reefs to the northeast of Melanesia. Kiribati, the Marshall Islands, the Federated States of Micronesia (FSM), Nauru, and Palau are among the countries that make up the Pacific Rim and the grouping includes low islands and atolls with severe climates. Polynesia, which comprises the Cook Islands, Samoa, Tonga, and Tuvalu, is a group of islands in the central and southern Pacific (Jones 1997). Timor-Leste is situated in Southeast Asia, near the eastern end of the Indonesian archipelago, to the west of Melanesia.

Because of the colonial history of dominance in the Pacific, there is only one major city where paying jobs, government facilities, and business prospects are focused, which is rare. Besides, in contrast to its Asian neighbors, the Pacific's urbanization progression was principally distinct from its industrialization progression (Chandra 2003), making the integration of urban drifters into the paying economy deeply dependent on government activities, and the countries heavily reliant on foreign assistance and spending. The few industrial activities that have emerged are largely focused on resource extraction (e.g., mining, forestry, and fisheries), which have become ecologically unsustainable as a result of increasing economic pressures (Cocklin 1999). Industrial production is uncollateralized and clustered in a few metropolitan areas, which are dominated by oligopolistic and monopolistic enterprises (Chandra 2003). As a result, Pacific cities' economies are comparatively undiversified and sustained by underlined ecosystems, making them not so adaptable to shifts in local requirements or global demand.

12.2 OBJECTIVES OF THE STUDY

This study has been initiated to fulfill the following objectives:

- To identify the factors of urban sustainability challenges.
- To trace the effects of metropolitan change on the climate and human social orders.
- To understand the nature and potential of Pacific urban centers.
- To study the spatio-temporal dimensions of changing urban sustainability in the South Pacific region.

12.3 MATERIALS AND METHODS

This study is descriptive in nature and is based on urbanization data from secondary sources. These data are collected from government offices, published books, articles, and unpublished softcopies sourced from the internet and software. Various census reports, and website publications at different times have been also consulted. But the main source of urbanization data is the Pacific SPC and Regional Infrastructure Facility. Information gathered from various sources has been presented in the form of tables and graph.

12.4 CONCEPTUAL FRAMEWORK OF PACIFIC TOWNS AND CITIES

It was during the colonial period, which started in the sixteenth century and did not conclude till World War II, that Pacific Island peoples began to urbanize. The colonial governments, according to Burt and Clerk (1997), founded settlements to fulfil their own objectives rather than indigenous interests. Foreigners dominated the urban centers, which served as hubs for exporting natural resources and importing Western ideas about religion and development. As a result, rather than being an integral part of culture, they became an outlier on the social and cultural landscape. In certain ways, this has resulted in a cultural divide within nations between rural and urban.

Pacific urbanization is a transitional phase comprising the migration of families from rural to urban areas, followed by significant social, economic, and ecological change (Jones 2012; Connell and Lea 2002). This urbanization is a quite new trend that arose mainly during the post-colonial era and picked up steam in the early 1960s, by which time several Pacific DMCs gained their independence. The Pacific's urbanization process has been inconsistent, as populations have risen and Pacific DMCs have been pinched into a variety of globalization and regionalization trends. Human migration, both internal and external, has followed urbanization in the Pacific for the most part, with the resultant primary to higher urban population intensities, although it has caused depopulation on some islands. With the growing interconnectedness of islanders, particularly urban dwellers, with regional and international connections, their lives have become more regional and circulatory. This involves the effects of remittance flows as well as regional educational opportunities. Thus, Pacific cities have evolved into more and more complex structures of economic, social and environmental forces that are changing and improving all the time.

Demographic change, population migration, and the evolving urban-based economies have all characterized the growth of urban areas in the Pacific, all of which have occurred in the context of relatively mediocre economic results. In both rural and urban areas, technological change has been a catalyst of economic growth. The urbanization process, as well as other national development drivers, heavily influences the context in which planning must return to the increasing number of belongings in Pacific urban centers. The speed of urbanization in the Pacific has increased since the 1980s, resulting in significant shifts in demographic and population dynamics, as well as the physical configuration of towns and cities, with implications for traditional settlements and customary land ownership; demands to mobilize the three factors of development (land, labor, and capital); human nature, including beliefs, standards, behaviors, and expectations; and commuting patterns (Jones 2012).

Over the last two decades, people's perceptions of what defines Pacific urbanization have shifted. In the 1980s and early 1990s, towns and communities were mostly seen to be increasingly overburdened, with urban administration issues addressed by halting the migration of rural–urban refugees and development partners requesting that rural development projects be scaled up. Many Pacific DMCs wanted to restrict migration because they saw rural–urban drift as a pattern that was causing negative urban change. Others blamed urbanization for separating natives from their traditions and norms, while some attributed it to land, housing, and expanding settlements (Pacific Institute of Public Policy, 2011a, 2011b). Providing services and infrastructure to towns and communities across the Pacific was seen to draw more residents, creating problems in metropolitan areas. Powerful anti-urban attitudes influenced Pacific national strategies and policies, affecting how urban environments were perceived and conceptualized. In contrast to anti-urban measures such as evictions and settlement destruction, as well as requests that settlers and squatters revert to their traditional ways, the growth of Pacific cities has been unaffected.

The twenty-first century has seen a gradual change in attitudes in Pacific DMCs on what institutes urbanization. Due to a number of factors, including the fact that pro-rural ideology tends to dominate planning policies, as evidenced by large monetary allocations for healthcare and education-related infrastructure upgrades in rural areas and outer islands, Pacific urbanization is increasingly viewed as an effective means of managing the transition from traditional to modern societies. "Civilization is urban, and modern is urban, and modern is effective service delivery" (Kep 2011). According to current perceptions in the Pacific, urbanization is now seen as a process of upgrading rural communities, neighborhoods, districts, and cities in order for the benefits of urbanization to be shared rather than enjoyed by a small proportion of the population (Dekena 2010). Through their population concentrations and variety of formal and informal economic activities, urban areas are increasingly recognized as significant contributors to gross domestic product (GDP), productivity, and economies of scalc.

In this context, urbanization in Pacific towns and cities has recently been recorded in a number of publications (for example, Chand and Yala 2010; Storey 2006, 2010; Connell and Lea 2002; Jones 2005, 2012, 2016; Maebuta and Maebuta 2019; Mohanty 2006; Mawuli and Guy 2007; Mecartney and Nautong 2001; Goddard and Bruce 2017; Jones and Lea 2007), since it is characterized by rapidly expanding towns and cities. Urban growth rates are outpacing rural, and national development rates are slowing. There is a glut of demand for infrastructure and services; there is little or no structured serviced land available to meet the requirements of urban and peri-urban population growth; squatter and under-serviced settlements are on the increase; and there is a shortage of skilled workers. Land disputes and conflicts are being escalated, with individuals or select groups of customary landowners 'selling' their lands; the effects of climate change are being concentrated on towns and cities on low-lying atolls and narrow coastal hinterlands; and governance, human capital, and community development are being hampered. The distribution of limited resources in government economies, as well as the inclusion of the urban sector in national planning, are two examples.

Pacific urbanization has a distinctive appearance and identity formed by a variety of distinguishing characteristics. A few examples include the impact of large-scale migration from rural to urban areas; the rise of the rural village and, more recently, of village towns in the region; the urbanization of poverty and the resulting widening of the urban–rural divide; and the power of Pacific sociocultural orders and their cultural permeation within Pacific urban areas. This has developed as a result of the Melanesia, Micronesia, and Polynesia sub-regions' deep social, cultural, religious, linguistic, and geographical diversity.

In the Pacific, urbanization is a phenomenon that has both positive and negative consequences. If handled effectively and efficiently, concentrating citizens in one area has numerous economic, social as well as ecological advantages. National efforts to provide infrastructure and services to urban and rural areas should prove an equitable approach to managing urbanization.

Both traditional and modern planning systems exist side by side. Islanders may accept modern systems, but they are mostly ignored in practice. Tensions around an adequate balance between public interest and private rights are at the heart of successful urban management, for example resolving complex concerns when shareholders participate in urban land markets. The reality is that Pacific DMCs' hereditary colonial land-use planning structures contribute to some of the tension. Most Pacific DMCs have outdated urban and town planning arrangements based on these schemes (as opposed to urban management arrangements).

A new kind of village is emerging to complement the trend of typical (traditional) villages that has shaped the structure and form of urban buildings in the Pacific region throughout history. It is expected that the integration of villages into surrounding communities and the development of village-towns in Pacific DMCs will continue to progress and become the norm during the next 15–20 years. The next section considers the features of large towns made up of informal and squatter settlements in the Pacific region, which are experiencing high levels of legitimate unemployment, poor living standards, and widespread poverty. This is where the future Pacific town and metropolis will be located.

For a better understanding of the complexities of Pacific town and city systems, as well as for overseeing urbanization in the next era, it will be necessary to discuss the various aspects of urban cultural permeability, deprivation urban growth, and deprivation urban growth in particular rural towns. Continued anti-urbanization strategies, biases, and sentiments, such as all stakeholders averting our gaze to the problems of suburbanization, would only worsen the situation in the Pacific. It is necessary to consider the predicament of city dwellers, especially those in poverty, living in informal and squatter settlements, from diverse perspectives.

12.5 RESULT AND DISCUSSION

12.5.1 Nature and Potential of Pacific Urban Centers

Across the Pacific, urban areas play a vital role in economic development, both at the national and regional levels. They are not present as they should be specifically placed as key sponsors to gross domestic product (GDP). They continue to have an impact as gates and conduits; promoting and encouraging mining, fisheries, and forestry are examples of non-urban economic practices. They also serve as provincial and national centers for services such as health services, housing, and tertiary education. Nonetheless, insufficient foundation and facilities in terms of the quality of service, coverage, lack of access to quality property, increased housing expenses, governance problems, and increasing urban security concerns limit urban areas' potential contribution to economic and social development. In the twenty-first century, many individuals are curious about what it means to be a Pacific islander in the modern world. When it comes to characteristics associated with livability and sustainability, Pacific towns and cities are becoming more difficult to maintain. Benefits and prospects arise from the density of population in Pacific urban centers. The size, nature and pace of urbanization, on the other hand, have negative implications. The influx of poor people from rural areas and other vulnerable people in informal and squatter settlements, where they have no guaranteed land protection, accommodation, clean water, hygiene or proper sanitation, for example, calls into question the applicability of governance and urban management structures. Who benefits from planning in the first place? In the Pacific urban context, how is planning described, and how inclusive is Pacific urban development? Why is city management in the Pacific not a hot topic on national and regional growth agendas? One of many urban challenges to be tackled is the task of promoting economic development in both formal and informal sectors while at the same time dealing with worsening quality of life concerns for Pacific urban dwellers. The Asian Development Bank (ADB), in collaboration with other advancement partners, is helping to improve the quality of life for Pacific urban dwellers. The state of Pacific towns and cities involves a wide range of urban growth and organization issues; however, three major drivers of urbanization in the Pacific – the urbanization of poverty, the rise of squatter settlements, and insufficient urban governance and management – indicate the need for greater attention to be paid to Pacific urbanization and urban towns and cities in general.

Urbanization, especially in the developing world, is in its infancy. The effect of urbanization on the environment and the human social order has yet to be fully understood. Even though urbanization is linked to a slew of issues, it is only partially known. Many types of vulnerabilities, such as social and biophysical vulnerabilities, are related to urbanization in the South Pacific region. Biophysical weaknesses include precarious ecosystems, restricted land resources, a lack of essential infrastructure, and global warming issues, all of which have a straightforward impact on society. It is difficult to react appropriately to these biophysical vulnerabilities due to social vulnerabilities such as weak economies, land ownership issues, and institutional constraints. Concrete planning, structural reform, improved regional coordination, and efficient resource management are all critical at this point if South Pacific regions are to deal with urbanization issues effectively.

In developing countries, the rapid pace of urbanization and industrialization has put additional strain on available urban infrastructure such as highways, schools, hospitals, transportation systems, water and electricity supplies, and so on. Not only that, but it also places additional strain on the immediate urban environment. Urban areas exploit vast amounts of wealth and are completely reliant on their hinterlands. According to Wackernagel and Rees (1998), the ecological footprints of cities' surrounding areas are much larger than the cities themselves.

Megacities such as Tokyo, Lagos, Delhi, Shanghai, Dhaka, and others are the center of attention for the world's population, and as a result, they are rapidly growing. Contrary to popular belief, metropolitan areas, especially in developed countries, have been overlooked. However, because of the faster rate of urbanization, they face similar socio-environmental issues. The cities of the South Pacific countries are minuscule in comparison to global megacities. In reality, the population of the Pacific Island countries as a whole is much smaller than that of a single Asian megacity. Some Pacific cities (Table 12.1), on the other hand, have urbanization rates and population densities equal to Asia's, Africa's, and South America's. For example, growth rates of the urban population in some Pacific countries surpass 4%. (See Table 12.1) Just four of the 19 countries mentioned in Table 12.1 have an urban growth rate below 2%. The rates of urban population growth in the Pacific Islands are much higher than the global average. Furthermore, the proportion of people living in cities is relatively high for many Pacific countries (Table 12.1). As a result, cities in the South Pacific are struggling with problems like those faced by megacities, such as unemployment, poverty, homelessness and rural-to-urban migration. Nevertheless, these are issues embedded in very different physical and social situations, resulting in geographically unique weaknesses that warrant further study.

12.5.2 Dense and Dilapidated Pacific Island Cities

Increasing urban populations across Melanesia are resulting in sprawling settlements, many of which are un-serviced, exacerbating social, health, and economic problems. Remember that many urban communities, together with those in Kiribati, Vanuatu, and the Marshall Islands, have higher population densities than megacities like Singapore and Hong Kong, despite having a tiny proportion of the facilities and no equivalent high-rise apartment buildings (Pacific Institute of Public

TABLE 12.1
Population Statistics for Pacific Island Nations, 2020

Countries	Total Population	Population Growth Rate (%/yr.)	Urban Population (% of Total)	Urban Population Growth Rate (%/yr.)
American Samoa	55,519	−0.3	50	−0.3
Cook Islands	14,802	0.0	75	1.2
Fiji	884,887	−0.2	56	1.6
French Polynesia	275,898	−0.6	51	0.7
Guam	159,358	0.9	94	1.0
Kiribati	110,138	2.1	57	2.9
Marshall Islands	53,158	0.4	74	1.1
Micronesia (Fed. States)	102,843	0.0	22	1.5
Nauru	11,550	1.8	100	1.0
New Caledonia	271,407	1.3	67	1.9
Niue	1,591	0.0	N/A	N/A
Northern Mariana Islands	53,883	0.6	90	0.7
Palau	17,661	0.5	78	1.2
Papua New Guinea	7,275,324	2.1	13	2.6
Samoa	195,979	0.6	19	−0.5
Solomon Islands	515,870	2.3	20	4.5
Tonga	100,651	0.3	23	1.1
Tuvalu	10,566	1.1	63	2.5
Vanuatu	272,459	2.4	25	2.9

Source: SPC (2012).

Policy 2009). It is no surprise that urban destitution is on the rise, along with the health and social problems that come with it. For instance, the city of Port Moresby has the country's highest incidence of tuberculosis (25% of cases) (John 2015).

Despite growing burdens on urban property, health, and shelter, political commitment over urbanization concerns has been insufficient. Notwithstanding the growing number of urban electorates, there are limited public consultation opportunities and not many members of parliament that represent urban pursuits. When the media does discuss urban problems at all, it is almost always derogatory. Even though only Samoa, Papua New Guinea, and Fiji have government ministries committed to improving living quarters and urban growth, the pace of development has been sluggish. In most cases, donor aid services are in short supply, and when they do exist, they are seldom long term or strategic in nature. Kinship schemes are important for city dwellers unable to meet their basic needs due to patchy services and a lack of social safety nets. City villages have been referred to in Port Moresby and elsewhere in Melanesia, their order retained mostly through tradition or informal governance structures. Several urban settlements lack basic infrastructure and facilities. When new waves of city dwellers emerge, links to villages are disappearing. As the cost of living rises, informal safety lines in cities are being stretched, and many people are struggling to find regular wages (Pacific Institute of Public Policy, 2010).

The pace of urbanization in the Pacific Island countries is very high. In the South Pacific countries, the urban growth rate outstrips the national population growth rate in per cent (Figure 12.1). If the urban population continues to grow at the same rate it will take only 25 years to double. Among

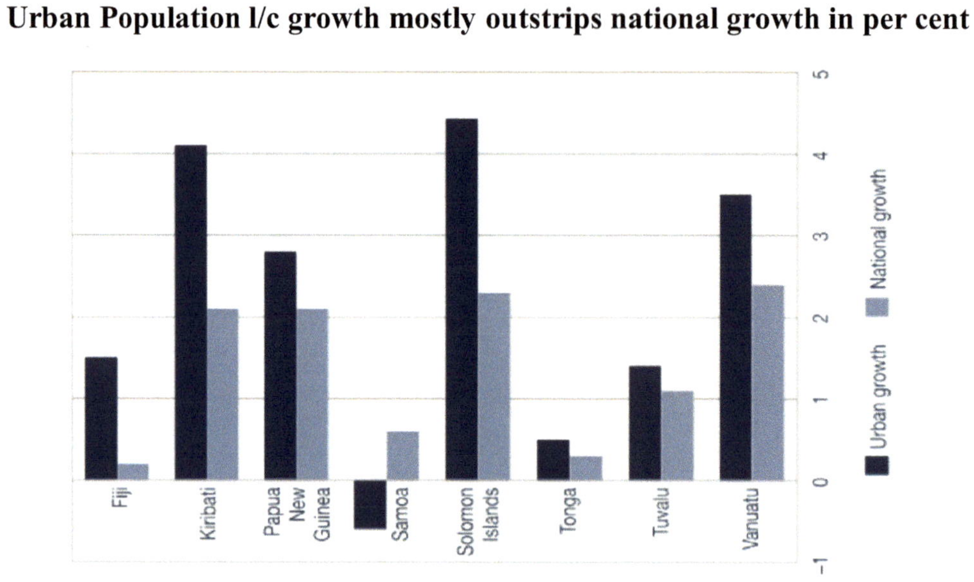

FIGURE 12.1 Urban population growth mostly outstrips national growth in per cent.

(Source: Jones (2012), SPC (2012), UNFPA (2014).)

the Melanesian countries, urban growth is highest in the Solomon Islands. Here, the population growth rate is 4.4% which is almost twice the rate of national population growth.

12.5.3 Degree of Smartness and Sustainability of Pacific Cities

Pacific cities are on the verge of being unsustainable, as their sprawl consumes precious property, depletes aquifers, and emits noxious gases from transportation. The results include lower biodiversity and bio capacity, as well as reduced ability to provide energy and consume waste. They aren't even close to being smart cities, despite the promise of smartphone technology and the widespread use of cell phones.

New communities are springing up all over the world, places of innovation and connectivity where modern technology connects people, city councils, and service providers, allowing for more efficient service delivery, quicker response to environmental problems, and even simple issues like traffic congestion. Crowdsourcing is one way for city residents to support one another, participate in planning, and press for specific improvements, with participation allowing them to make their own commitments to sustainable growth.

Quite a few Southern nations have started to adopt this vision, seeking to incorporate ICT technologies into urban management, encompassing everything from schools to the sewerage systems. In principle, ICT enables city leaders to communicate precisely with urban residents, gather input from them, and keep track of what is going on in the city, how it is changing, and how to improve quality of life. Despite this, no global city has come close to becoming a smart city; decision-making is bureaucratic and influenced by corporate interests, and technical costs are high (Yigitcanlar 2015). By incorporating emerging technology, social networks, and environmental issues, smart cities indicate a further successful effort toward sustainability and consequently equity, but the Pacific still has a long way to go. Currently, the most basic goals are to create successful municipal government and management, as well as basic interaction with all local citizens.

Urbanization in the Pacific has reached unparalleled levels of speed and size, and most of it is unplanned. Visions of tiny islands and beautiful towns with palm trees and streams do not represent reality for the vast majority of people who live in the country. Because city inhabitants are unlikely to abandon their homes, effective management and planning are required. People with insufficient incomes are living in overcrowded, unsafe, and unsanitary conditions, restricting their ability to contribute to national growth. Poverty is rapidly being moved from rural to urban zones. Even though towns and cities offer possibilities for social mobility to which many people aspire, they must be accessible to everyone, in the spirit of fairness, optimism, and citizenship, and more public space is required. Task 11 of the 2015 global Sustainable Development Goals is to "make cities and human settlements inclusive, clean, resilient, and sustainable". The focus on urbanization is current, and increasing urban resilience was a component of the Pacific Leaders' Framework for Resilient Growth, which was endorsed by the United Nations in 2016.

There is concern about disparity, but measures that integrate urban design with economic growth and poverty alleviation are notably missing from the agenda. Congestion, sprawl, and segregation all occur at the same time in the same area of the city. However, liberal policies have yet to be shown to be effective in terms of promoting social cohesion in the face of long-term urban development that requires centralized planning and administration across administrative, sectoral, and physical borders. According to the Prime Minister of Papua New Guinea, "When accountability and justice are sought, government processes are slow". On the other hand, this does not indicate a total stop to the process of doing anything.

Despite their proximity to government, cities and their residents have little effective voice. Urban elites also concentrate on their own neighborhoods, corporate interests, and patronage politics. Many local societies, on the other hand, have banded together to further their own interests. While this does not always imply higher levels of planning and management, it is possible that this is where the most positive gains can be achieved. Land will require difficult decisions, as no landowners would like to abandon customary tenancy or settlers suffer the loss of their status, and philosophies will be deeply intertwined. Among the Pacific Island nations, Fiji was the first country to recognize the rights of informal settlers, as well as the obligation to support them with services, establishing a 'policy bridge' and functional links between the informal and formal sectors (Phillips and Keen 2016). It has significant implications for other nations, as well as the implementation of a 'Pacific way' of efficiently handling urbanization with sustainability.

A limited number of Pacific islanders are also architects and city planners. Constructing towns and cities in nations with low economic development, especially with intermittent assistance and different ideas about what the cities ought to look like, is a tough endeavor. There is not enough political motivation to address the issues, and rapid political turnover, particularly in Melanesia, is detrimental to policy formulation. Better urban data is needed in Pacific states to prepare more efficiently and track progress, but a shortage of data should not prevent management improvements. The area requires urban title holders at a time when government planning is losing favor, and urban policy necessitates a responsive and progressive rural counterpart to ensure long-term development, sufficient services, and regional development centers in other parts of the nation. Environmental conservation is a priority even though climate change is not a factor. Beyond these goals, urban planning should represent a vision that reflects governments' and citizens' desires, leads to economic and social growth, and, most importantly, delivers nice places to stay and operate.

12.5.4 The Lives of Islanders in the 21st Century

When most people think of the Pacific, they imagine a peaceful rural environment surrounded by coconut trees, a calm ocean, and a way of life rich in root crops and fruit, with a sea crowded with fish. The picture-perfect scenery is changing. While not as densely populated as the world's megacities, the Pacific is experiencing rapid urbanization and strong population growth at the same time. Much of the time, we are simply unable to deal with the problems and opportunities that these pose.

Urbanization must be regulated, and urban management must be prioritized at the national level. Housing, health, education, investment, and employment policies must be carefully considered by governments; individuals must evaluate how they wish to live – to define what it means to be a Pacific islander in the twenty-first century.

12.5.5 Urban Resilience of the Pacific

The word 'resilience' has become common among donors, aid agencies, and local, national and international governments in their responses to shocks and pressures that populations, citizens, and systems face. Policy and practice approaches heavily depend on this concept, which has evolved from being coupled with climate change and disaster management to now being linked with a wide variety of issues, together with urban growth. Despite this, durability has its detractors.

Although many politicians and planners are uncertain how to solve the various challenges and problems that come with urbanization, there is no refuting the shocks and effects of the Pacific region's continued urbanization. Towards the end of the twentieth century, Pacific Island countries faced the expansion of 'villages' into 'towns', and 'towns' into 'cities', resulting in rapid, unmanaged urbanization. The desire of migrants to return to the prosperity of their rural village's livelihoods pervaded state and national urban policy, as did issues of human capital and technological ability, as well as an ingrained 'anti-urban' stigma focused on towns and cities as centers of 'immorality'. This contradiction around the definition of 'urban', and the positive and negative challenges of the broader urbanization process as complexities in the Pacific region, is expressed in a legacy of improperly sited construction, the nonexistence of knowledge of dangers, and a lack of respect for related planning doctrines like the 'public interest'. When it comes to simple urban services like sanitation, water supply, drainage, and safe and affordable land for construction, for example, it has just not been possible to keep up with the need for, or remain current with, demand (Guild and Matsumoto 2013). Furthermore, people and property in urban settings become more visible and vulnerable as a result of poverty in urban areas (ADB 2012).

In 2015, the population of the PICs was projected to be 10.9 million, with about 2.5 million living in formally identified metropolitan areas. In terms of absolute population, this equates to 23% of the entire Pacific population. To put it another way, nearly one out of every four Pacific islanders lives in a city (Jones 2016). In 2015, ten of the 21 PICs studied were classified as being more than 50% urban, with 12 having urbanization rates of more than 40%. In nearly all Pacific Island countries, cities are growing at a faster pace than their rural counterparts, which is consistent with trends seen in Pacific Island nations and the wider Pacific during the past 25 years. Exceptions to this trend include nations with decreasing populations, mostly as a result of out-migration, as well as countries with peri-urban or affluent rural suburbs that are nonetheless officially classified as rural (Jones 2016).

Contemporary urban development is the constant increase of spontaneous urban growth, with most of the transformation related to the informal economy, livelihoods and the growth of squatter and informal communities. The Melanesian capitals – Honiara, Mores, Vila, and Port Tarawa – and other Micronesian towns like the capital Majuro and South Tarawa, have the most informal and squatter settlements. There is a repeating pattern throughout Pacific towns and cities, in that they all contain different degrees of informal and squatter communities, with the highest concentration being in Port Moresby, the region's largest city, which is the most populous. Approximately 45–50% of the city's population lives in squatter or informal settlements, as well as original native village enclaves like Hanuabada, which were established during the colonial period (Table 12.2).

The increasing number of squatter and informal settlements on free hold, customary and state lands, in addition to residents seeking affordable land and housing, reflects the connection between natural hazards and environmental risk. Because they are flood-prone, subject to land slips, have steep slopes, or are considered appropriate for environmental protection, the formal planning framework has frequently classified lands where settlers take up residence as unsuitable for planned formal urban growth. As a result, they are gradually identified as vulnerable lands that must be protected

TABLE 12.2
Urban Populations and Growth of Informal Settlements in Selected Pacific Island Countries

Country and Capital City	Population of Main City 2013	Estimated Settlement Population	Annual City Population Growth Rate	Annual Settlement Population Growth Rate	Settlements as % of City Population 2013	Settlements as % of City Population 2023
Honiara, Solomon Islands	64,600	22,600	3%	6–26%	35%	64%
Suva, Fiji (Greater Suva Area)	244,000	50,000	2%	8%	20%	31%
Port Vila, Vanuatu	44,000	15,400	4%	3–12%	35%	43%
Port Moresby, PNG	500,000 to 700,000	225,000 to 315,000	2%	5–8%	45%	56%

Source: Pacific Regional Infrastructure Facility (2015).

from further development. Places such as tidal rivers and creeks, as well as waste dumping sites, are typically of low quality and dangerous. One of the most pressing challenges posed by the increasing number of squatter and informal settlements is not only a greater knowledge of how they form and change, but also a better management of their permanence and integration into Pacific towns, cities, and national development plans and policies, among other things.

The resilience approach has recently been utilized in a variety of contexts in the Pacific, indicating a rising interest in using the concept in other parts of the world to solve urban issues. Urban resilience in the Pacific can be situated in many ways:

a. *Moving from Risk to Resilience - Sustainable Urban Development in the Pacific* was published by the Asian Development Bank (ADB) in 2013. Although the assessment defines urban resilience as "the capacity to withstand and recover from unexpected shocks associated with natural disasters and climate change" (Guild and Matsumoto 2013), it claims that resilience can be improved by focusing more on its relationship and linkages to urban planning and management. The assessment suggests several actions in this context, including better natural hazard and climate change data, expanded urban service coverage tied to strategic planning, enhanced infrastructure design standards, and successful emergency disaster response and post-disaster recovery systems.

   Most notably, the report suggests that urban governance, as well as human and resource support, be improved in order to increase risk resilience. The recommendation to step away from 'top-down' land-use planning and adapt 'bottom-up' solutions is particularly noteworthy. Instead of operating in silos, the latter will use systems-based thinking rather than sectoral methods, enhancing operational coordination. As a result, moving toward sustainable urban growth implies the possibility of increased cross-sector urban resilience capability (Guild and Matsumoto 2013).

b. The Pacific Urban Agenda (PUA) was conceived at a UNESCAP sub-regional workshop in 2003 and adopted by the UN Economic and Social Commission for Asia and the Pacific (UNESCAP) at its 65th session in 2004. In 2005, the Pacific Islands Forum Leaders supported the PUA, which was then included in the Pacific Plan. The 4th Pacific Urban Forum (PUF 4) was held in April 2015, and the most recent PUA was endorsed as a result of these initiatives. Urban policymakers, local and national government officials, as well as representatives from development partners, attended the gathering.

In the 2015 PUA resolutions, four main recommendations were endorsed, with Recommendation 2, like the ADB assessment, focusing on the effect of natural hazards and climate change as a factor in rising social, economic, and environmental costs within urban settings. Those most affected were the most disadvantaged groups, such as those living in informal and squatter settlements, who lack the resources and capability to cope with disaster and climate change threats. As a result, the PUA proposed that urban planning, disaster management, and climate change be more closely aligned, and that these issues be included in federal, national, and city-level urban management strategies and infrastructure development plans. The PUA's resilience is framed in a broad policy context since it is a regional policy, and there is no concept of what resilience or being resilient implies.

c. The Honiara Urban Resilience and Climate Action Plan (HURCAP) 2016 is a collaborative strategy for the Honiara City Council and the Solomon Islands Government with the goal of improving Honiara's resilience. Honiara is historically vulnerable to a variety of natural hazards, including tropical cyclones, earthquakes, and tsunamis, due to its location. Lack of capacity and a lack of awareness of the position of urban planning have hampered the city's ability to cope with (urban) population growth in the modern period. This includes natural population growth as well as the influx of migrants seeking better living conditions in Honiara, such as jobs, services, and education.

The swift rise in squatter and informal settlements within the current urban periphery as well as the peri-urban areas of Honiara is one of the consequences of this development. Residents in such areas are particularly vulnerable to natural hazards due to inadequate living conditions and their position in environmentally sensitive areas. It is believed that addressing the resilience needs of vulnerable communities in informal settlements will not only reduce the effects of natural disasters but will also provide a useful forum for implementing climate change adaptation initiatives and improving community resilience in the future, according to HURCAP. Several topics are addressed in the plan, including the challenges that underlie the "urban resilience agenda" and the reality that all stakeholders have a role to play in a range of situations.

Therefore, housing, water, sanitation, and waste, as well as health, well-being, ecosystem services, and disaster risk mitigation and management, are all addressed in urban resilience action plans (Trundle and McEvoy 2016). The word 'resilience' appears 126 times in the HURCAP, with uses ranging from "inform resilience strengthening behaviour" to "resilience needs to be recognized as a complex feature of a community, one that will continue to evolve as a result of urbanization processes and local and global environmental change" (Trundle and McEvoy 2016).

12.5.6 Urban Planning and Management in the Pacific

The Pacific's future is clearly one of urbanization. Rapid urbanization can be better shaped and managed if countries recognize it as a national development problem and respond to the multi-sector challenges using a multi-tiered strategy including the government, the private sector, and the community. To achieve a safe physical, social, and economic climate for all Pacific peoples, it is critical to address the main issues of land, housing, and governance. The Pacific Urban Agenda (PUA), a regional urban policy project, focused on highlighting urban problems and committing planners to solving challenges. In the last five years, some Asian Development Bank Pacific developing-country members have made significant efforts to put urban plans into action, including implementing development restrictions. The successful execution of actions has been constrained in a context marked by a lack of solid political engagement, minimal policy direction, insufficient funding, and poor institutional and government frameworks. Consequently, the intended beneficiaries of urban management and planning, namely urban and peri-urban residential neighborhoods, the economy, and the environment, have achieved little progress in recent years. We need to better understand the significance of cities in maintaining a stable and successful people as well as an economically thriving economy, and include it in national and regional development strategies. Every stakeholder has a role to play in urban management, and a greater emphasis on effective urban government that involves a broad coalition of interested stakeholders, including participation from traditional authority, is long overdue.

Proper urban planning should consist of operational, developmental, and restorative planning, and should be implemented properly so that damage to both environment and population is minimized. The upgrading of urban infrastructure, such as roads, traffic, and transportation, should be considered as part of operational planning. The development of newly acquired urban areas should be the primary focus of development planning. Residents' capacity to adapt and adjust to their new living situation as independently as feasible is encouraged via restorative planning (Mandal et al. 2020).

12.6 CONCLUSION

Urbanization is one of the most critical of the demographic, fiscal, and environmental issues concerning communities and habitats. While many Western countries have already gone through the urban transformation, nations and the vast majority of the world's people have yet to do so, but the wait will be lengthy. Poverty between rural and urban populations, sociocultural homogenization, resource shortages, and agrarian reform are all putting additional strain on the issue of the rural population shifting to urban settings. Whilst rural–urban migration may recognize better prospects in cities, many of the realities of the developing world are based on the fact that cities' capacity to keep up with population expansion is insufficient; as a result, societal, monetary, and organic disruptions are unavoidable.

There is a justification for it – the rapidly increasing urban populations of Africa, Asia, and South America have piqued foreign concentration in the potential live ability of cities. The population pressures in these areas are immense, the urban infrastructures are unable to cope, and the future challenges are overwhelming. The Pacific islands are less evident in discussions of urban futures, but as seen here, they are also subject to rapid rates of urbanization, with equally serious social, economic, and environmental consequences.

This chapter has investigated the Pacific islands' 'urban crisis' in the context of human protection. The definition, as described by the IHDP project on global environmental change and human security (GECHS), emphasizes the concept of options, specifically the ability to eliminate, reduce, or adapt to threats and vulnerabilities, as well as the capacity to exercise these options. Through this analysis of urbanization and the related environmental hazards, we have attempted to demonstrate that the citizens of the Pacific islands are still vulnerable to the environmental and social challenges brought about by urbanization. The challenges are exacerbated by reduced coping capacities or social vulnerabilities, as well as natural habitats that are in many areas vulnerable and overwhelmed with energy and waste dumping imposed by rapidly expanding cities; these are vulnerabilities stemming from biophysics.

Was it possible to make sense of the conditions in the Pacific in terms of the notion of human security at the time? As we have shown, when it comes to reducing the harmful impacts of urbanization, the people of the Pacific have a number of options available. Better urban design, job creation in rural regions to decrease migration, learning how to reduce waste, increasing sanitation, and better integration of social planning with environmental protection and management are all measures that may help lessen social and biophysical vulnerabilities. The islands are also generally endowed with warm temperatures, abundant marine resources, and strong social structures; as a result, the inhabitants of the Pacific are comparatively well nourished when compared to those in many areas of Africa, Asia, and South America. In contrast, there is more uncertainty about the capacity to utilize the choices available to mitigate risks. Many nations' efforts to achieve effective urban development are hindered by difficult-to-resolve land ownership and tenure problems, weak economies, and institutions that are not sufficiently resourced or established to cope with urban changes. In order to enhance human security in the face of urbanization, it is necessary to prioritize resource development, which includes, among other things, the elimination of social vulnerabilities. This should be given the motivation and capability to participate in alternatives for minimizing the negative effects of urbanization's growth.

12.7 RECOMMENDATIONS

This study finally concludes with the following recommendations that may help in mitigating the Pacific Urban challenges.

1. To address the Pacific's specific urban challenges, an urban resilience approach is needed – one that encompasses growth, climate change adaptation, and disaster risk reduction. This strategy should draw on existing local and community capacities while also considering levels of vulnerability and accessibility.
2. Additional funding and resources for urban resilience should be secured in order to put the plan into action. This would help to support new ideas as well as a variety of Pacific regional policies. Furthermore, there is a compelling business case for a variety of urban resilience investments, which foreign, regional, and national stakeholders should be aware of.
3. Urban management hybrid methods should be created. These are methods for combining traditional and community-level governance with structured governance. These methods would promote local resilience efforts while still respecting the conventional structures that are so important in urban and peri-urban settings.
4. Building urban resilience necessitates a multi-disciplinary approach involving a variety of actors and collaborations. As a result, governments (both local and national), non-profits, private-sector players, multilaterals, and academics should seek out and create opportunities for deep and practical collaborations in order to advance urban resilience activities.
5. Through learning exchange, Australian NGOs and local partners must build stronger urban resilience strategies for the Pacific.
6. Since resilience is difficult to quantify, stakeholders working in this field can create resilience tools such as indices and evaluations that can be tailored to the Pacific, and look at the characteristics and costs of building resilience.
7. In order to create successful urban resilience grounded in the realities of Pacific Island towns and cities, research gaps in identifying social capital and community resilience in the Pacific urban context, especially within informal settlements, must be addressed.

REFERENCES

Asian Development Bank. *The State of Pacific Towns and Cities: Urbanization in ADB's Pacific Developing Member Countries*. Philippines: Asian Development Bank, 2012.

Burt, Ben, and Christian Clerk. "Environment and development in the Pacific Islands: introduction." *Environment and Development in the Pacific Islands* 25 (1997): 1.

Chand, Satish, and Charles Yala. "Improving access to land within the settlements of Port Moresby." (2010).

Chandra, Rajesh. *Industrialization and development in the Third World*. London: Routledge, 2003.

Cocklin, Chris. "Islands in the midst: Environmental change, vulnerability, and security in the Pacific." In *Environmental Change, Adaptation, and Security*. Dordrecht: Springer, 1999: 141–159.

Connell, John, and John Lea. *Urbanisation in the Island Pacific: Towards Sustainable Development*. London: Routledge, 2002.

Dekena, L. "Keynote address by the Hon. Minister for Lands and Physical Planning to the PNG National Urbanisation Policy Seminar." (2010).

Goddard, Michael, and Michael Bruce. *The Unseen City: Anthropological Perspectives on Port Moresby, Papua New Guinea*. Canberra, ACT: Pandanus Books, 2017.

Guild, Robert, and Ikuko Matsumoto. "Moving from risk to resilience: sustainable urban development in the Pacific." (2013).

Haberkorn, Gerald. "Pacific Islands' population and development: Facts, fictions and follies." *New Zealand Population Review* 33, no. 34 (2008): 95–127.

John, L. "Challenges of TB Control in PNG." In *13th National Rural Health Conference, Darwin*, (2015): 24–27.

Jones, P., and J. Kohlhase. "Planning for Apia—everybody's but nobody's business." *Samoa Environmental Forum* 3 (2002): 27–31.

Jones, Paul. "Managing urban development in the Pacific: key themes and issues." *Australian planner* 42, no. 1 (2005): 39–46.

Jones, Paul. "Searching for a little bit of utopia–understanding the growth of squatter and informal settlements in Pacific towns and cities." *Australian Planner* 49, no. 4 (2012): 327–338.

Jones, Paul. "The emergence of Pacific urban villages: urbanization trends in the Pacific islands." (2016).

Jones, Paul, and John P. Lea. "What has happened to urban reform in the island Pacific? Some lessons from Kiribati and Samoa." *Pacific Affairs* 80, no. 3 (2007): 473–491. https://www.jstor.org/stable/40023394

Kep, M. "Urbanisation in the PNG Context." In *Public Seminar Presentation in the Faculty of Architecture, Design and Planning, University of Sydney*, vol. 21 (2011).

Maebuta, Helen Esther, and Jack Maebuta. "Generating livelihoods: A study of urban squatter settlements in Solomon Islands." (2019).

Mandal, R. K., M. A. Alam, U. N. Baba, T. Raiula, and S. Kumar. Status and Velocity of Urbanization in Arunachal Pradesh, India: A New Direction, *Elementary Education Online* 19, no. 4 (2020): 3040–3054.

Mawuli, A., and R. Guy. "Informal social safety nets: Support systems of social and economic hardships in Papua New Guinea (NRI Special Publication No. 46)." *Port Moresby, Papua New Guinea: National Research Institute* (2007).

Mecartney, Sarah, and Anne Nautong. "Blacksands settlement: a case for urban permanence in Vanuatu." Master's thesis, University of Sydney, 2001.

Mohanty, Manoranjan. "Urban squatters, the informal sector and livelihood strategies of the poor in the Fiji Islands." *Development Bulletin* 70 (2006): 65–68.

Pacific Institute of Public Policy. Fast growth in the Pacific is possible—look at Vanuatu. Briefing Notes. 10 July 2009.

Pacific Institute of Public Policy. Pacific Regional Millennium Development Goals Tracking Report. Suva. July 2010.

Pacific Institute of Public Policy. Communiqué of the Forty-Second Pacific Islands Forum. Auckland, September 2011a.

Pacific Institute of Public Policy. Urban Hymns—managing urban growth. Discussion Paper No. 18. July. Port Vila. Pacific Islands Forum Secretariat. 2007. Revised Pacific Plan. Suva. October 2011b.

Pacific Islands Forum Secretariat. Communiqué of the Forty-Second Pacific Islands Forum. Auckland. September 2011.

Pacific Region Infrastructure Facility (PRIF) Unsettled: Water and Sanitation in Urban Settlement Communities in the Pacific, Sydney, 2015.

Phillips, Tarryn, and Meg Keen. "Sharing the city: Urban growth and governance in Suva, Fiji." (2016).

SPC (Secretariat of the Pacific Community), Pacific Island Population Estimates and Projections, April, 2012.

Storey, Donovan. "Urbanisation in the Pacific." (2006).

Storey, Donovan. *Urban Poverty in Papua New Guinea*. Port Moresby, Papua New Guinea: National Research Institute, 2010.

Trundle, A., and D. McEvoy. "Honiara urban resilience and climate action plan." *Prepared for UN-Habitat. RMIT University Climate Change Adaptation Program, Melbourne* (2016).

United Nations Human Settlements Programme. *Planning Sustainable Cities: Policy Directions: Global Report on Human Settlements 2009*. UN-HABITAT, 2009.

United Nations Population Fund. "Population and development profiles: Pacific Island countries." (2014).

Wackernagel, Mathis, and William Rees. *Our Ecological Footprint: Reducing Human Impact on the Earth*. Vol. 9. New society publishers, 1998.

Yigitcanlar, Tan. "Smart cities: An effective urban development and management model?" *Australian Planner* 52, no. 1 (2015): 27–34.

Part III

Environmental Management and Ecosystem Services

13 Walkable Urban Neighborhoods
The Adverse Effects of Industrialization and Climate Change in Developing Countries

Nestor Asiamah
University of Portsmouth, Portsmouth, United Kingdom
Africa Centre for Epidemiology, Accra, Ghana

CONTENTS

13.1 INTRODUCTION

The United Nations (2019) reveals that there are currently 703 million older persons aged 65 years or more in the world and that 34 countries including the United Kingdom (UK) and United States (US) will be super-aged by 2030. It adds that the current global population of older people will increase by more than 100%, reaching over 1.5 billion people in 2050. Every country will experience an increase in its older adult population over the next three decades (Asiamah 2017; United Nations 2019), with developing countries likely to serve as home to 1.1 billion older persons (United Nations 2019). This trend points to grim ripple effects such as an increased old-age dependency ratio, disease burden, and healthcare expenditure (Asiamah 2017; Kpessa-Whyte 2018; Asiamah et al. 2020). On the other hand, aging is accompanied by benefits such as elderly volunteering and participation in the labor market (Windsor et al. 2016), but the realization of these benefits depends on how well interventions are put in place to ensure that older adults maintain functional capacity and health into later life.

The World Health Organization (WHO) and other stakeholders have developed relevant interventions to reduce the rate of aging and ensure that people maintain health as they age (WHO 2013; Asiamah et al. 2020). Noteworthy among such interventions are three closely related programs: (1) physical activity (PA) counseling for older adults, (2) supporting older adults to live successfully in their neighborhoods, and (3) the provision of neighborhoods that support lifelong participation in

DOI: 10.1201/9781003131922-16

social and physical activities (Koohsari et al. 2018; Stoutenberg et al. 2017; Brandt and Janse van Vuuren 2019; Lindeman et al. 2020). These interventions are public health flagship programs aimed at encouraging community-dwelling older adults to maintain health as they age. They are rolled out in harmony with the 'aging in place' initiative (Federmeier and Kutas 2005; Wiles et al. 2011), which emphasizes the need for older adults to age successfully in their neighborhoods rather than in residential care homes where opportunities for social and physical activities are limited (Wiles et al. 2011). As such, the aging-in-place initiative is a timely response to global population aging and a paradigmatic strategy to realize 'healthy aging', a term used in gerontology to describe the process of developing and maintaining the functional capacity that enables well-being in old age (WHO 2013, p. 5).

Gerontologists have established that aging is associated with physiological limitations and a gradual decline in the individual's functional capacity (Asiamah 2017). More so, older adults are generally frail and are less able to perform physical tasks independently (Asiamah 2017; Asiamah et al. 2020). As a consequence, older adults face the highest risk of social isolation and physical inactivity (PI) in neighborhoods lacking essential services, sidewalks, sanitation, and opportunities for active transportation. If so, community-dwelling older adults need neighborhoods that support mobility to live normally and age in optimal health. Arguably, this idea has encouraged the provision of smart cities and highlights the importance of improving neighborhood design to enhance opportunities for social participation by older adults. For the above reasons, any situation that deprives older adults of a PA-supported neighborhood needs to be understood and curbed. As relevant as it is to wealth creation and economic development, industrialization and its interaction with climate change can constrain interventions aimed at improving the capacity of neighborhoods to support mobility in older adults. Since functional capacity reduces as people age, the worse impacts of industrialization and climate change on neighborhoods will be felt by older adults. This thinking informs the focus and scope of this chapter and is expounded in the next section.

13.2 BACKGROUND

The provision of walkable neighborhoods is a central outcome of aging-in-place campaigns that stakeholders consider one of the best ways forward in environmental gerontology (Koohsari et al. 2018; WHO 2013; Brandt and Janse van Vuuren 2019). A walkable neighborhood is a residential community that allows residents to engage in community-level social and physical activities (Koohsari et al. 2018, p. 2). A walkable neighborhood is characterized by essential services (e.g., healthcare, security, sanitation), sidewalks, traffic lights in the right places, overpasses, gardens and parks, and resources for active transportation (Koohsari et al. 2018). Neighborhood walkability is the extent to which a community allows residents to participate in social and physical activities (Koohsari et al. 2018; Lee and Tan 2019). Research has shown that social activity and PA in older adults increase as neighborhood walkability increases (Merom et al. 2015; Asiamah et al. 2020; Brandt and Janse van Vuuren 2019). There is no doubt that the provision of walkable neighborhoods by stakeholders is an effective strategy for tackling population aging.

Many studies have reported factors that counter the efforts of stakeholders to design neighborhoods to encourage social inclusion (Leyva et al. 2017). Some of the most frequently reported factors concern urbanization and industrialization, which have become the priority of federal governments, and an apex approach to poverty reduction and livelihood enhancement (Michel 2020; Parvaneh et al. 2016). For example, the construction of commercial buildings, manufacturing hubs, depots, warehouses, and service delivery channels in the neighborhood is reportedly a strong force of industrialization that limits or curtails the inclusiveness of neighborhoods (Mandeli 2019; Parvaneh et al. 2016). On the other hand, climate change and its associated extreme weather conditions such as heatwaves, rainstorms, hurricanes, landslides, and earthquakes have been reported to make everyday life more difficult for community-dwelling older adults and other vulnerable groups (Frumkin et al. 2012; Leyva et al. 2017). Nevertheless, the discourse on the after-effects of industrialization does not include ways in which neighborhood walkability is being compromised by agents of urbanization. This discourse is also isolated from debates on the impacts of climate change on the health of seniors living in the neighborhood. Thus, threats to neighborhood walkability that are the result of the interaction between climate change and industrialization are not well understood. Though several studies (Mandeli 2019; Parvaneh et al. 2016; Wijaya et al. 2018) have been conducted on the adverse impacts of industrialization on urban neighborhoods, no identifiable study has analyzed the emerging threats of industrialization and its interaction with climate change to neighborhood walkability in line with aging-in-place initiatives. Therefore, the author's first objective is to discuss how urban neighborhood walkability in developing African countries can be hindered by emerging industrialization concerns as well as the interaction between climate change and industrialization.

Efforts to improve neighborhood walkability in the interest of healthy aging have been informed by bio-ecological models, particularly Lawton's (1982) person–environment (P–E) fit model (Lak et al. 2020). This theory has shaped the way researchers frame research in environmental gerontology. The emerging threats to neighborhood walkability have unreported implications for the use of the P–E fit model in research and interventions. That being so, industrialization and its interaction with climate change can affect the relevance of paradigms developed from the P–E fit model to improve neighborhood walkability. Similarly, the relevance of Lawton's P–E fit model to the design of urban neighborhoods may be altered by emerging industrialization and climate-change factors. The second objective of this chapter is, therefore, to discuss these possible changes and highlight their research and policy implications for healthy aging programs. The problem being addressed is that industrialization and its interplay with climate change may thwart the effort of stakeholders to provide walkable urban neighborhoods for older adults, which may counteract healthy aging interventions. The above objectives are addressed by relating insights from key literature to real-life situations (depicted in photographs) in an African context.

13.3 THEORETICAL FOUNDATIONS OF NEIGHBORHOOD WALKABILITY PROGRAMS

The community development approach to healthy aging draws on the imports of two dominant models, namely Lawton's (1982) P–E fit model and Lewin's (1951) bio-ecological framework. The P–E fit model posits that there are *environmental demands* that every individual should meet in order to live a normal life in the neighborhood. Key demands include the provision of walkable resources (e.g., lorry parks, gardens, sidewalks, gyms, and services) in the community by politicians to facilitate social and economic activities. The inability of people to meet these demands in the way of community engagement has consequences, which include social isolation, sedentary behavior, and PI. The model adds that whether residents would meet demands depends on *competence*, a term used to describe physical functional abilities, financial ability, and social support needed by the individual to utilize community services and other walkable attributes. Though more prominent in environmental gerontology, Lawton's P–E fit model draws on Lewin's framework, which argues that behaviors such as social activity and PA in the community are a function of environmental and personal factors. To illustrate, whether older residents would participate in PA and social activities in their neighborhood depends on their ability to perform physical tasks unaided and the availability of walkable factors.

The import of the above models is that individuals who live in sufficiently walkable neighborhoods are more likely to maintain PA and social activity into later life. Similarly, community-dwelling older adults who have access to essential community services, recreational centers, and well-designed road networks are more likely to participate in social and physical activities in their neighborhoods. What could, therefore, be inferred from Lawton's (1982) theory is that there is a positive association between neighborhood walkability and social activity and PA in community-dwelling older adults. This implied relationship has been confirmed by several studies. In a study conducted by Kim et al. (2020) in Korea, for instance, it was found that high neighborhood walkability was associated with active transportation, which includes walking for a purpose, cycling, and skating. In Australia, Merom et al. (2015) also found in a 12-month randomized controlled trial that neighborhood walkability is associated with walking and other active behaviors in the community. Several other studies (Koohsari et al. 2018; Stockton et al. 2016; Zhang et al. 2019) have reported a positive association between PA and neighborhood walkability in older adults.

The above-stated empirical evidence is complemented by studies (Lohne-Seiler et al. 2014; Van Dyck et al. 2015; Lee et al. 2017; Merom et al. 2015) that have affirmed the relationship between PA and health. Some other studies have evidenced that older adults living in more walkable neighborhoods better maintain optimal health into later life than their colleagues in areas of low neighborhood walkability (Merom et al. 2015; Koohsari et al. 2018; Stockton et al. 2016; Zhang et al. 2019). These pieces of evidence back Lawton's (1982) P–E fit model and, therefore, provide a reliable basis for aging-in-place initiatives that focus on the provision of walkable neighborhoods. As such, environmental gerontologists and other stakeholders can rest assured that investments in walkable neighborhoods would produce desirable outcomes such as improved community engagement, PA, and economic agility. As the next section indicates, however, these investments and their related interventions can be counteracted by the side-effects of industrialization and their interactions with climate change.

13.4 THE NEGATIVE EFFECTS OF INDUSTRIALIZATION ON WALKABLE NEIGHBORHOODS

Accompanied by industrialization in developing African countries is the proliferation of commercial infrastructure in cities and neighborhoods, which is due to a gradual increase in investor activities, including the construction of commercial buildings (Mandeli 2019; Wijaya et al. 2018). African cities undergoing industrialization are characterized by two main events, the first of which is the expansion of industrial areas as more investors enter the local economy. The size of the public space is naturally fixed; hence, the expansion of industrial areas, also called official towns (Abubakar and Doan 2017), necessitates the use of walkable public spaces (e.g., parks, gardens) for industrial purposes. This understanding lies behind the second event, which is the replacement of residential properties (i.e., flats) with commercial buildings and skyscrapers to provide factories, warehouses, and office complexes while optimizing land use (Agyemang et al. 2018). More so, modernist city design models used in developing African countries draw on Western paradigms, whereby stakeholders tend to economize public space by replacing flats with multi-story buildings and skyscrapers. With this approach, neighborhoods are transformed into official towns where flats and other small buildings are replaced with larger and taller commercial buildings.

Industrialized neighborhoods, no doubt, offer residents several opportunities, including employment for the young and enhanced access to suppliers of raw materials, finished goods, and services. Industrialization, however, may cause more harm than good to older adults and other vulnerable populations. An entrenched anomaly that has not been documented as a threat to geriatric health is the low disability-friendliness of commercial infrastructure in developing countries (Yarfi et al., 2017). Commentaries (Asiamah et al. 2020; Agyemang et al. 2018; Abubakar and Doan 2017) reveal that commercial buildings in Ghana, Nigeria, Kenya, and other African countries are not disability friendly, suggesting that industrialization in the neighborhood is likely to replace flats (which are more suited for older adults) with multi-story buildings and skyscrapers that are not fully accessible to older adults and people with disabilities because staircase, lift, and evacuation system designs do not take into account the physiological limitations of older adults with limited functional capacity (Asiamah et al. 2020). As essential service delivery channels at the peak of neighborhood industrialization, where public space is being used to construct factories and warehouses, these buildings are expected to exclude older adults from essential service providers. Research (Asiamah et al. 2020;

Koohsari et al. 2018; Lee and Tan 2019) has shown that essential services form the core of neighborhood walkability; they encourage walking and other means of active transportation in the community. Because emerging commercial infrastructure (e.g., multi-story buildings, skyscrapers) would eventually serve as sources of key neighborhood services, older adults may be unable to maintain access to essential services. The import of the foregoing assertions is that neighborhoods where commercial infrastructure is not disability friendly are less walkable to older adults and people with physical disabilities.

Tema, an industrial city in Ghana, for instance, has seen a gradual increase in industrialization, manufacturing activities, and employment (Stoler et al. 2014), but residents have had to grapple with increasing noise, air, and land pollution (Stoler et al. 2014; Odonkor et al. 2020). This can be said of similar cities in Nigeria, Kenya, and South Africa where intensive mining, refinery, and construction activities are carried out at the expense of residents' comfort and safety. As depicted in Figure 13.1, land pollution in developing African countries mars environmental sanitation, which is a key component of neighborhood walkability (Koohsari et al. 2018; Asiamah et al. 2020). Land pollution can serve as a major threat to neighborhood walkability because residents are less likely to participate in social activities in the community and use active means of transportation (e.g., walking) if their neighborhoods are not kept neat. With African countries having low waste management and recycling capacity (Stoler et al. 2014), industrialization can reduce neighborhood walkability for older adults by heightening waste generation and land pollution. Noise pollution from industrial activities can also be expected to make neighborhoods less walkable in developing countries. Some researchers (Koohsari et al. 2018; Asiamah et al. 2020; Lee and Tan 2019) have reported that noise from industrial activities in the neighborhood is disliked by residents, reduces satisfaction with the neighborhood, and discourages social engagement. Noise pollution can render communities less walkable for residents, especially for older adults with significant auditory impairment. Last but not least, air pollution can be a stronger threat to neighborhood walkability than land and noise pollution because older adults face multiple morbidities that are aggravated by air pollution (Rhoades et al. 2018). In a study by Leyva et al. (2017), older adults reported reduced walking and social activity

FIGURE 13.1 Poor environmental sanitation as a consequence of land pollution along Ingonyama Street, Johannesburg.

(Source: Gilili (2019).)

linked to their resolve to protect themselves from environmental air contamination. Participant older adults explained that they had to stay indoors because they had respiratory disorders (e.g., asthma) that could be triggered by poor air quality.

In the face of industrialization, roads in the neighborhood are more frequently used by trailers and other heavy-duty vehicles owned by manufacturers (Wijaya et al. 2018). In this situation, road infrastructure including streets and minor roads for private and smaller commercial vehicles may be destroyed. In most African developing-country communities, roads are not tarred and regularly serviced (Odonkor et al. 2020). Even those that are, are not designed for large commercial vehicles and are therefore likely to deteriorate more quickly and become less supportive of active behaviors (walking, cycling, skating) as their use by heavy-duty vehicles increases. In neighborhoods where the roads are not tarred, residents struggle to cope with dust from vehicle movement. Figure 13.2 shows an untarred road in Sakomono, a suburb of Accra, where air pollution from manufacturing is aggravated by dust from vehicle movement. Exacerbating this situation are gullies resulting from road erosion and misuse by trailers (Odonkor et al. 2020). Because African countries have the poorest road safety profile and the largest number of cases of road accidents (Odonkor et al. 2020; Tran 2017), the foregoing industrialization factors are more likely to discourage PA in the neighborhood.

Though mainly caused by population growth and poor city planning (Parvaneh et al. 2016), urban sprawl is significantly affected by industrialization. Industrialization, with its employment opportunities, encourages rural–urban migration, which gradually increases urban population size. At the peak of neighborhood industrialization, residents are compelled to relocate to the countryside and outskirts where pollution is less or non-existent, and services and goods are more affordable (Pateman 2011; van Hoof et al. 2018). The movement of residents into the outskirts in response to industrialization has been reported in some studies (Parvaneh et al. 2016; Ajaero and Onokala 2013) as a behavioral response to several stimuli (e.g., increased cost of living, pollution) of industrialization in residential areas. These stimuli can cause more harm than good to older adults. For instance, older adults or their social network members are likely to permanently move out of the

FIGURE 13.2 An untarred road in Sakumono leading to Tema Railway Station in Accra, Ghana.

(Source: Author's fieldwork.)

neighborhood, which suggests that industrialization does not necessarily support aging in place. Individuals who cope with the said stimuli and remain in the community may lose friends to the gradual relocation process. It is, therefore, probable that older adults will lose social capital in the form of social ties and support. The social capital theory of health suggests that social ties and supports are the primary facilitators of active behaviors in the community. That is, seniors would receive no or little encouragement to utilize walkable factors if they were isolated from their peers by industrialization through what Mandeli (2019) refers to as social segregation.

Industrializing neighborhoods face not only social and cultural disengagement, as Mandeli (2019) observes, but also experience a change in what the author refers to as *neighbourhood service mix*, which is operationally defined in this chapter as the basket of services provided in a neighborhood and the extent to which these services meet the survival needs of residents. It comprises utilities and basic services needed daily by the individual, examples being catering, milling, laundry, recreation, and health and social care. In developing countries, these services are provided by ordinary people in small shops and under canopies located near the road and on public space. The space occupied by these service providers is often owned by private developers or the government (Deininger et al. 2017; Yin et al. 2020). As industrialization intensifies, they lose their spaces to radical private estate developers or government agencies (Deininger et al. 2017), which forces them to relocate or stop providing services. This event, arguably, can deprive residents of more essential services or replace these services with non-essential ones (e.g., distribution services) that do not meet the normative needs of individuals. Even if the vendors affected can afford legitimate and well-structured shops and offices in their neighborhoods, the cost of their services may increase, rendering those services less accessible to older adults with relatively low financial security. Finally, economic activities and outsiders in the neighborhood can change or at least modify community culture, leading to a major change in neighborhood service mix. Older adults are used to their neighborhood's culture and service mix, and a change in traditional essential services provided would put their adaptive capacity to the test and reduce their life satisfaction. Some studies (Rhoades et al. 2018; Mandeli 2019) suggest that older adults isolate themselves in communities facing a cultural change. Needless to say, social isolation in the neighborhood is one of the main ways that older adults abandon walkable factors and confine their lives to their homes. Regrettably, the above side-effects of industrialization on neighborhood walkability prevail, alongside the harsh climate change events discussed in the next section.

13.5 THE ADVERSE IMPACTS OF CLIMATE CHANGE ON WALKABLE NEIGHBORHOODS

Climate change affects human beings in different ways, including its obliteration of ecosystem stability. One of the most frequently reported outcomes of climate change is atmospheric temperature fluctuation (Chineke and Chiemeka 2009), the basis of recurrent heatwaves and dry winds in developing African countries (Chineke and Chiemeka 2009; Okeahialam 2016). Sub-Saharan Africa is well known for its dry season (Harmattan), which used to appear in the fourth quarter of each year. Cold, dry winds, which are stronger in the morning and evening (Okeahialam 2016), were the hallmark of this African season. As climate change intensifies, the Harmattan is being replaced with a harsher season characterized by extreme heatwaves that occur in the fourth quarter with intermittent and unexpected rain showers (Azzarri and Signorelli 2020). What makes this weather condition a potential threat to neighborhood walkability is its heatwaves and unexpected rain showers. According to some researchers (Chineke and Chiemeka 2009; File and Derbile 2020; Rhoades et al. 2018), heatwaves are consequences of climate change that cause social isolation, skin burns and wrinkly skin, and withering of ornamental plants that are aesthetic components of parks and gardens. When atmospheric temperatures are high, even highly walkable neighborhoods are unlikely to be used by older residents who are already battling with skin disorders. The destruction of ornamental plants by heatwaves in the community can also reduce the aesthetic properties of gardens, roads, and parks. As such, residents may have no reason to use parks and gardens as well as streets for PA and social activity. From this viewpoint, social segregation and disengagement are a direct repercussion of climate change.

What makes the gradual replacement of the Harmattan with the foregoing season an impediment to community engagement and walkability is that its heatwaves and isolated rain showers are erratic and unpredictable. Naab et al. (2019) support this observation with their acknowledgment of the inability of climate service providers (e.g., meteorological service departments) in many African countries to accurately forecast rainfalls and changes in atmospheric temperature. Even in settings where weather forecasts are accurate, stakeholders are not able to efficiently communicate future weather conditions. As a result, many people would be unable to prepare for extreme changes in weather conditions and schedule their events for dates free of climate anomalies. It can be contended from this perspective that unexpected weather conditions (e.g., rains and heatwaves) can make residents anxious and, in turn, compel older adults with limited functional and adaptive capacity to stay indoors. Of course, older adults will not utilize walkable attributes such as parks, gardens, and streets for PA and social activity if they cannot tell when the next rainfall or unwanted weather condition will occur. In developing African countries, unwanted or unexpected weather conditions can discourage the utilization of walkable factors because streets and public spaces in most neighborhoods are untarred (Odonkor et al. 2020). Most neighborhoods are, therefore, prone to muddy floors, stagnant water bodies, debris, and dust, which sensible residents would always want to avoid. Dust in the neighborhood that develops from prolonged drought and heatwaves can aggravate poor environmental air quality.

Floods, earthquakes, landslides and rainstorms are other climate-change outcomes that are frequently experienced in developing African countries. Somalia, for example, has experienced more floods and destructive rainstorms in the last decade than in the preceding one (BBC News 2020). On June 3, 2015, Ghana lost 150 people to an explosion that occurred simultaneously with torrential rain, flood, and storm around a gas filling station (BBC News 2013). No doubt memories of the disaster will discourage people from using services around that filling station. Flooding in some other African countries such as Zimbabwe, Congo Brazzaville, and Uganda have destroyed myriads of commercial buildings and closure of key public offices, including hospitals, tourist sites, and parks (BBC News 2020; Ficchì and Stephens 2019). What can be inferred from these events is that core walkable attributes in the neighborhood are being destroyed by floods and stormy rains. Because they are not tarred in most neighborhoods in developing African countries, walkable streets are eroded by heavy rains. At worse, road gullies and potholes filled with water make walking and active transportation less enjoyable and convenient. Figure 13.3 shows water-filled potholes in an urban neighborhood in the Tema area of Ghana; illustrating what happens after rainfall. Figure 13.4 also shows debris created by a rainstorm accompanied by a flood in Chipinge, Zimbabwe.

Stakeholders including environmental gerontologists aim to facilitate the creation of walkable neighborhoods that make aging in place possible and appealing to older adults. The fulfillment of this goal depends on how well walkable neighborhoods are utilized by older adults to meet recreational, social, and economic needs. If neighborhoods are made highly walkable but are not used, can they be said to be beneficial to healthy aging and aging-in-place programs? Of course not! One of the problems environmental gerontologists have with climate change is that the above-reported abnormal weather conditions can cause post-traumatic stress disorders, heat sensitivity, progressive skin disorders, and a loss of social ties and support. Studies (Rhoades et al. 2018; Mandeli 2019; Chineke and Chiemeka 2009) have reported cardiovascular diseases, mortality, as well as the other conditions mentioned, as consequences of climate change events such as heatwaves, floods, and storms. Deductively, climate change can reduce the quality of life, physical function, and capacity of older adults to utilize walkable factors. Thus, the idea that climate change can make walkable neighborhoods less beneficial to aging-in-place initiatives is irrefutable, more so in an environment where climate change events interact with industrialization.

FIGURE 13.3 Water-filled potholes in a neighborhood in Accra experiencing industrialization.

(Source: The Author's Fieldwork.)

FIGURE 13.4 The scene in Chipinge, Zimbabwe, after a rainstorm accompanied by a flood.

(Source: Davies (2020).)

13.6 THE JOINT EFFECTS OF INDUSTRIALIZATION AND CLIMATE CHANGE ON WALKABLE NEIGHBORHOODS

The notion that climate change is a direct outcome of industrialization is globally shared by researchers. The combustion of fossil fuels and the rampant emission of carbon are caused or facilitated by industrial activities. This being the case, some of the indirect side-effects of industrialization reported in this chapter are extreme weather conditions and related events. This understanding is further expressed in Figure 13.5, which depicts the subjection of neighborhood walkability to climate change and industrialization. In the figure, neighborhood walkability and climate change are located within the sphere of industrialization because they are both attributable to industrial activities. This idea changes slightly in Figure 13.6, which includes the shaded region where climate change and industrialization overlap. The shaded region represents different ways in which climate change and industrialization interact to disorientate community life. In the next few paragraphs, these specific interactions between climate change and industrialization in developing African countries are identified and discussed.

Drawing on earlier discussions, the researcher can identify four major ways in which climate change interacts with industrialization to affect neighborhood walkability. Of utmost importance is the interplay between floods (caused by climate change) and the radical use of public space by individuals and private real-estate developers. This situation is the consequence of the government and its agencies having little or no control over how land is owned and used in developing African countries (Abubakar and Doan 2017; Agyemang et al. 2018). Individuals can own landed properties

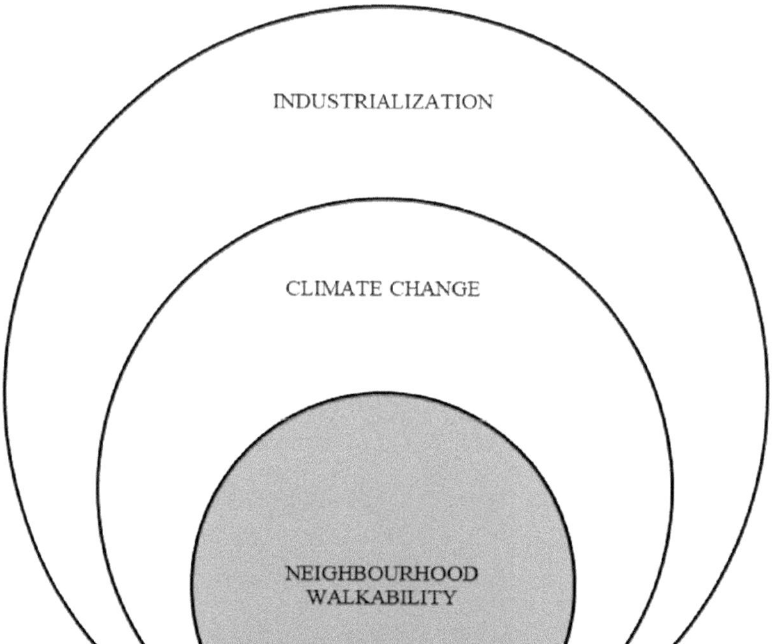

FIGURE 13.5 A conceptual framework depicting the relationship between industrialization, climate change, and neighborhood walkability.

(Source: Author's construct.)

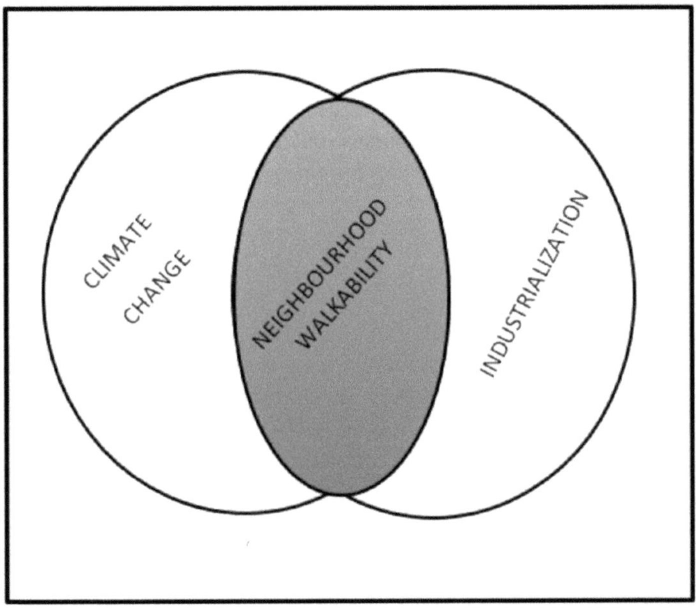

FIGURE 13.6 A conceptual model showing the exposure of neighborhood walkability to industrialization and climate change.

(Source: Author's construct.)

and do whatever they deem appropriate with these properties without being guided by the government. The consequence of this backdrop is the location of private buildings in the wrong places and public spaces, including drainage systems and water channels. It is worth noting that building on water channels and drainage systems is a recipe for entrenched disaster, including flooding. Some researchers (Ficchì and Stephens 2019; Agyemang et al. 2018) support this assertion with their idea that rain showers will always cause flooding in African cities where drainage systems are not well developed, and private and industrial buildings are wrongly located. Climate change compounds this issue with its torrential and stormy rains.

To reiterate, air pollution in the neighborhood is caused, in part, by industrial production. The daily use of untarred dusty roads by heavy-duty vehicles owned by private companies exacerbates the neighborhood air pollution problem. Worse still, climate change can prolong industry-driven air pollution with prolonged droughts and heatwaves. Air pollution would be more hazardous with prolonged drought and heatwaves because many residents would spend more time outside their homes when atmospheric temperature increases and becomes unbearable indoors. Heatwaves can also speed up the development of dust from moist sand and mud. Further to this, waste generated directly or indirectly by manufacturers in the neighborhood can end up in the community as debris from floods and strong winds. Waste generated directly by manufacturers, in this context, includes unwanted factory materials or remains of raw materials. The indirect waste (e.g., plastic bags, metal foils, and rubbers) comes from products sold by manufacturers to residents in the neighborhood. In developing countries, such waste is not properly recycled and managed (Fei-Baffoe et al. 2014) and consequently contaminates public space or is added to debris from floods and winds.

Africa is a relatively hot continent; hence, ornamental plants that constitute a major part of walkable resources (i.e., parks, gardens, and roads) in the region are always at the mercy of the weather. Ornamental plants often wither and eventually lose their aesthetic properties due to extreme temperatures and drought. Industrialization speeds up the degradation of flora and fauna by producing hazardous chemicals and waste that delay the growth of ornamental plants, and reduce or destroy the greenery of the neighborhood. It is also common to see influential individuals and real-estate developers encroaching upon parks and gardens in African countries (Wijaya et al. 2018; Yin et al. 2020). The complete use of parks and gardens for manufacturing and construction purposes is a worst-case scenario in developing countries, including Ghana, Nigeria, and South Africa (Yao et al. 2015; Yin et al. 2020). This and other negative effects of climate change and industrialization on neighborhood walkability have implications for theory, research, and policy, as reported in the next section.

13.7 RECOMMENDATIONS AND IMPLICATIONS FOR THEORY, RESEARCH, AND POLICY

Lawton's (1982) P–E fit theory has been influential in environmental gerontology for explaining the relationship between the built environment, individual factors, and healthy behaviors such as PA. By the import of this model, the adverse effects of industrialization and climate change on neighborhood walkability have dire implications for practice and research. The model's docility hypothesis emphasizes that *competence* is needed by the individual to utilize built environment factors to secure social support in the way of PA and social activities. *Competence* in this context embodies physical abilities and social supports that facilitate the individual's mobility, which is necessary for utilizing neighborhood services and other walkable factors. Physical functional and adaptive abilities are, therefore, a part of the competence needed by the individual to use a walkable neighborhood. Climate change and industrialization are expected to increase the competence needed to access and utilize walkable neighborhoods. Industrialization, on the other hand, would reduce access to walkable neighborhood resources in older populations by replacing essential services with less essential ones and flats with skyscrapers. With its floods, heatwaves, strong wings, and muddy floors, climate change would, on the other hand, demand higher functional and adaptive capability from residents, especially older adults. From this perspective, climate change and industrialization would exclude older adults from walkable communities as their physical functional and adaptive abilities decline into later life.

Drawing on the above import of Lawton's (1982) P–E fit model, the built environment loses its walkable attributes as industrialization and climate change intensify. The competence needed by older adults to use walkable attributes is also heightened by industrialization and climate change, which means that older adults would eventually be isolated and compelled to leave their neighborhoods, causing a major setback to aging-in-place programs. No doubt, interventions maintaining or improving neighborhood walkability while industrialization advances are needed urgently. Going forward, because climate change and industrialization are not modifiable in the short-term, programs empowering older adults to navigate natural barriers to neighborhood-level social engagement would suffice. An intervention that is already ongoing in developed countries is the empowerment of seniors (through training programs) to use skyscrapers to manoeuver out of trouble in the wake of natural disasters such as floods and hurricanes (Leyva et al. 2017; Rhoades et al. 2018). There are also training programs equipping older adults to evacuate flats and skyscrapers and cope with the after-effects of extreme temperatures, floods, hurricanes, and typhoons (Rhoades et al. 2018). The adoption of these and similar programs in African countries is urgent, given that Africa is the most vulnerable continent to climate change (Dessalegn and Akalu 2015; BBC News 2020) and has the least walkable neighborhoods (Asiamah 2017; Asiamah et al. 2020). With these training programs, older adults in Africa can learn to cope with changes in culture and the physical built environment caused by industrialization and climate change.

There is also a need for new policies to be developed or existing policies properly enforced to regulate the use of land and public space. New policies that give the government and its agencies absolute control over the acquisition and use of land are needed in earnest. With these policies, the government can preserve public realms, protect essential services, and prevent radical construction activities in the neighborhood. Existing policies emphasizing routine monitoring of construction activities in the community through local governments, as well as district and municipal assemblies, must be better enforced. In this vein, district and municipal assemblies must be adequately equipped to regulate construction activities within neighborhoods. A national task force in the form of a committee or commission ought to be established to crank down on buildings located on water channels and not backed by the necessary permits. This commission, with the support of the judiciary and legislature, should be obligated to criminalize the location of private, corporate, and public buildings in public places serving as parks, gardens, essential service delivery channels, and waterways. Before taking this step, the government and its agencies (e.g., town and city Planning Councils) must re-plan their cities and neighborhoods, provide long-term water channels and drainage, and develop a protocol for land use. Admittedly, these efforts may be rebuffed or end up being a wild goose chase if governments are not committed to public health and the welfare of vulnerable populations.

The effectiveness of the above measures would depend on research focused on an understanding of the effects of industrialization and climate change on health, community life, and aging. More specifically, research investigating the effects of climate change and industrialization on the ability of seniors to access neighborhood services and use other walkable resources in African neighborhoods is needed. This course of research could empirically evidence the author's hypothesis that Lawton's (1982) competence for using walkable neighborhoods increases as climate change and industrialization advance over time. Prospective studies assessing possible changes in neighborhood service mix and access to essential services in the community over time are also necessary. Evidence from these studies can substantiate the author's claims and affirm the urgency of the aforesaid recommended measures. To avoid being overwhelmed in future, African countries must mitigate the adverse effects of industrialization and climate change on public health and vulnerable groups while pursuing macro-economic fortunes.

13.8 LIMITATIONS AND FUTURE RESEARCH POINTERS

This chapter is a commentary based on anecdotal evidence including theories, commentaries from the literature, and photographs from the author's fieldwork. As a result, the information it presents is wholly argumentative and does not represent empirical evidence. Moreover, attention was focused on the effects of climate change and industrialization on neighborhood walkability in developing African countries, though many of the climate events (heat waves, drought, landslides) are globally felt and can be expected to reduce or even exterminate neighborhood walkability across the world. Given these limitations, future research focused on empirically substantiating key arguments in this chapter would be worthwhile. In addition to the above recommendations, future studies could explore the experiences of older adults regarding how neighborhood walkability is being affected by climate change and industrialization. The development of an aggressive research culture aimed at understanding the adverse effects of industrialization and climate change on aging and elderly health is a public health emergency.

13.9 CONCLUSIONS

Industrialization, irrefutably, plays a critical role in Africa's accelerated development, but grave consequences can be associated with the failure of stakeholders to recognize its side-effects and avoidable costs. This chapter focused on an analysis of after-effects obliterating neighborhood walkability in developing African countries where governments' pursuit of development tends to be overly ambitious and consequently undermines neighborhood culture, walkability, and sociability. Yet, public health is the heart of productivity and macro-economic growth because no one can work to contribute to the economy in poor health or disability. For this reason, interventions aimed at maintaining physical health in the general population, including investments in PA and neighborhood walkability, must be protected and made successful. As indicated in this chapter, several climate change and industrialization events threaten neighborhood walkability, suggesting that investment in community-level PA can be a waste of scarce resources and time. As aging intensifies alongside industrialization and climate change, the population of older adults in neighborhoods is expected to increase. A growing number of seniors would, therefore, face barriers to neighborhood services, parks, gardens, and other walkable resources. The principal ways forward include recognition of the dire effects of industrialization and its interaction with climate change on neighborhood culture and sociability that are being caused by African governments. Stakeholders must also prioritize public health and see it as a function of macro-economic development, laying the foundations for key measures, including the enforcement of existing and new policies to control the use of land and public space. The government must have full control over land acquisition and use by individuals and organizations in order to prevent illegal siting and construction of buildings in cities.

REFERENCES

Abubakar, I.R., and Doan, D.L. "Building new capital cities in Africa: Lessons for new satellite towns in developing countries". *African Studies* 76, no. 4 (2017): 546–565. https://doi.org/10.1080/00020184.2017.1376850

Agyemang, F.S.K., Silva, E., and Anokye, P.A. "Towards sustainable urban development: The social acceptability of high-rise buildings in a Ghanaian city". *GeoJournal* 83, no. 1 (2018): 1317–1329. https://doi.org/10.1007/s10708-017-9837-0

Ajaero, C.K., and Onokala, P.C. "The Effects of Rural-Urban Migration on Rural Communities of Southeastern Nigeria". *International Journal of Population Research* 0, no. 1 (2013): 1–11. https://doi.org/10.1155/2013/610193

Asiamah, N. "Social engagement and physical activity: Commentary on why the activity and disengagement theories of ageing may both be valid". *Cogent Medicine* 4, no. 1 (2017): 1–3. https://doi.org/10.1080/2331205X.2017.1289664

Asiamah, N., Kyriakos, K, Eduafo, R., and Borkey, R. "The influence of community-level built environment factors on active social network size in older adults: Social activity as a moderator". *International Quarterly of Community Health Education* 41, no. 1 (2020): 77–87. https://doi.org/10.1177/0272684X20915379

Azzarri, C., and Signorelli, S. "Climate and poverty in Africa South of the Sahara". *World Development* 125, no. 1 (2020): 1–19. https://doi.org/10.1016/j.worlddev.2019.104691

BBC News. "Accra in shock and mourning after deadly Ghana Fire", 2013. Retrieved on 18/11/2020 at 7:10 AM from https://www.bbc.com/news/world-africa-33049190

BBC News. "Flooding hits six million people in East Africa". 2020. Retrieved on 18/11/2020 at 7:19 AM from https://www.bbc.com/news/world-africa-54433904

Brandt, C., and Janse van Vuuren, E.C. "Dysfunction, activity limitations, participation restriction and contextual factors in South African women with pelvic organ prolapse". *South African Journal of Physiotherapy* 75, no. 1 (2019): 1–8. http://dx.doi.org/10.4102/sajp.v75i1.933

Chineke, T.C., and Chiemeka, I.U. "Particulate Concentration and Health Impacts in Sub-Saharan Africa". *African Physical Review* 3, no. 1 (2009): 125–135.

Davies, R. "Zimbabwe – Deadly Storms and Floods in Manicaland". 2020. Retrieved on 25/11/2020 at 8:47 AM from http://floodlist.com/africa/zimbabwe-floods-manicaland-november-2020

Deininger, K., Savastano, S., and Xia, F. "Smallholders' land access in Sub-Saharan Africa: A new landscape?" *Food Policy* 67, no. 1 (2017): 78–92. https://doi.org/10.1016/j.foodpol.2016.09.012

Dessalegn, O. G., and Akalu, D. S. "The impacts of climate change on African continent and the way forward". *Journal of Ecology and the Natural Environment* 7, no. 10 (2015): 256–262. https://doi.org/10.5897/jene2013.0533

Fei-Baffoe, B., Nyankson, E. A., and Gorkeh-Miah, J. "Municipal Solid Waste Management in Sekondi-Takoradi Metropolis, Ghana". *Journal of Waste Management* 0, no. 1 (2014): 1–9. https://doi.org/10.1155/2014/823752

Federmeier, K. D., and Kutas, M. "Aging in context: Age-related changes in context use during language comprehension". *Psychophysiology* 42, no. 2 (2005): 133–141. https://doi.org/10.1111/j.1469-8986.2005.00274.x

Ficchì, A., and Stephens, L. "Climate variability alters flood timing across Africa". *Geophysical Research Letters* 46, no. 15 (2019): 8809–8819. https://doi.org/10.1029/2019gl081988

File, D.J.M., and Derbile, E.K. "Sunshine, temperature and wind Community risk assessment of climate change, indigenous knowledge and climate change adaptation planning in Ghana". *International Journal of Climate Change Strategies and Management* 12, no. 1 (2020): 22–38. https://doi.org/10.1108/IJCCSM-04-2019-0023

Frumkin, H., Fried, L., and Moody, R. "Aging, Climate Change, and Legacy Thinking". *American Journal of Public Health* 102, no. 1 (2012): 1434–1438. https://doi.org/10.2105/AJPH.2012.300663

Gilili, C. "Rubbish piles up on Diepsloot Street". 2019. Retrieved on 25/11/2020 at 8:05 AM from https://www.groundup.org.za/article/rubbish-piles-diepsloot-extension-2/

Kim, E.J., Kim, J., and Kim, H. "Neighborhood Walkability and Active Transportation: A Correlation Study in Leisure and Shopping Purposes". *International Journal of Environmental Research and Public Health* 17, no. 1 (2020): 1–16. https://doi.org/10.3390/ijerph17072178

Kpessa-Whyte, M. "Aging and Demographic Transition in Ghana: State of the Elderly and Emerging Issues". *The Gerontologist* 58, no. 3 (2018): 403–408. https://doi.org/10.1093/geront/gnx205

Koohsari, M.J., Nakaya, T., and Oka, K. "Activity-Friendly Built Environments in a Super-Aged Society, Japan: Current Challenges and toward a Research Agenda". *International Journal of Environmental Research and Public Health* 15, no. 1 (2018): 1–9. https://doi.org/10.3390/ijerph15092054

Lak, A., Rashidghalam, P., Myint, P. K., and Bradaran, H. R. "Comprehensive 5P framework for active aging using the ecological approach: An iterative systematic review". *BMC Public Health* 20, no. 1 (2020): 1–13. https://doi.org/10.1186/s12889-019-8136-8

Lawton, M. P. "Competence, environmental press, and the adaptation of older people". In M. P. Lawton, P. G. Windley, and T. O. Byerts (Eds.), *Aging and the environment: Theoretical approaches* (pp. 33–59). 1982. New York: Springer.

Lee, B.Y., Adam, A., Zenkov, E., Hertenstein, D., Ferguson, M. C., Wang, P. I., Wong, M.S., Wedlock, P., Nyathi, S., Gittelsohn, J., Falah-Fini, S., Bartsch, S.M., Cheskin, L.J., and Brown, S. T. "Modeling The Economic and Health Impact of Increasing Children's Physical Activity in The United States". *Health Affairs* 36, no. 5 (2017): 902–908. https://doi.org/10.1377/hlthaff.2016.1315

Lee, J.H., and Tan, T.H. "Neighborhood Walkability or Third Places? Determinants of Social Support and Loneliness among Older Adults". *Journal of Planning Education and Research* (2019): 1–14. https://doi.org/10.1177/0739456X19870295

Lewin, K. "Field Theory in Social Science". 1951. New York: Harper and Brothers.

Leyva, E.W.A., Beaman, A., and Davidson, P.M. "Health Impact of Climate Change in Older People: An Integrative Review and Implications for Nursing". *Journal of Nursing Scholarship* 49, no. 6 (2017): 670–678. https://doi.org/10.1111/jnu.12346

Lindeman, D.A., Kim, K.K., Gladstone, C., and Apesoa-Varano, E.C. "Technology and Caregiving: Emerging Interventions and Directions for Research". *The Gerontologist* 60, no. 1 (2020): 41–49. https://doi.org/10.1093/geront/gnz178

Lohne-Seiler, H., Hansen, B. H., Kolle, E., and Anderssen, S. A. "Accelerometer-determined physical activity and self-reported health in a population of older adults (65–85 years): A cross-sectional study". *BMC Public Health* 14, no. 1 (2014): 1–11. https://doi.org/10.1186/1471-2458-14-284

Mandeli, K. "Public space and the challenge of urban transformation in cities of emerging economies: Jeddah case study". *Cities* 95, no. 1 (2019): 1–11. https://doi.org/10.1016/j.cities.2019.102409

Merom, D., Gebel, K., Fahey, P., Astell-Burt, T., Voukelatos, A., Rissel, C., and Sherrington, C. "Neighborhood walkability, fear and risk of falling and response to walking promotion: The Easy Steps to Health 12-month randomized controlled trial". *Preventive Medicine Reports* 2, no. 1 (2015): 704–710. https://doi.org/10.1016/j.pmedr.2013.08.011

Michel, J.P. "Urbanization and Ageing Health Outcomes". *Journal of Nutrition, Health, and Aging* 24, no. 5 (2020): 463–465. http://dx.doi.org/10.1007/s12603-020-1360-1

Naab, F. Z., Abubakari, Z., and Ahmed, A. "The role of climate services in agricultural productivity in Ghana: The perspectives of farmers and institutions". *Climate Services* 13, no. 1 (2019): 24–32. https://doi.org/10.1016/j.cliser.2019.01.007

Odonkor, S.T., Mitsotsou-Makanga, H., and Dei, E.N. "Road Safety Challenges in Sub-Saharan Africa: The Case of Ghana". *Journal of Advanced Transportation* (2020): 1–9. https://doi.org/10.1155/2020/7047189

Okeahialam, B. N. "The Cold Dusty Harmattan: A Season of Anguish for Cardiologists and Patients". *Environmental Health Insights* 10, no. 1 (2016): 143–146. https://doi.org/10.4137/ehi.s38350

Parvaneh, M., Hajipour, K., and Hosseinpour, M. 2016. "Assessing Impact of Industrialization on Urban Expansion in Surrounding Cities (Case Study: Assalouyeh, Iran)". *Journal of Applied Sciences* 16, no. 4 (2016): 167–177. https://doi.org/10.3923/jas.2016.167.177

Pateman, T. "Rural and urban areas: Comparing lives using rural/urban classifications". *Regional Trends* 43, no. 1 (2011): 11–86. https://doi.org/10.1057/rt.2011.2

Rhoades, J.L., Gruber, J.S., and Horton, B. "Developing an In-depth Understanding of Elderly Adult's Vulnerability to Climate Change". *Gerontologist* 58, no. 3 (2018): 567–577. https://doi.org/10.1093/geront/gnw167

Stockton, J.C., Duke-Williams, O., Stamatakis, E., Mindell, J.S., Brunner, E.J., and Shelton, N.J. "Development of a novel walkability index for London, United Kingdom: Cross-sectional application to the Whitehall II Study". *BMC Public Health* 16, no. 1 (2016): 1–12. https://doi.org/10.1186/s12889-016-3012-2

Stoler, J., Weeks, J.R., and Fink, G. "Sachet drinking water in Ghana's Accra-Tema metropolitan area: Past, present, and future". *Journal of Water, Sanitation and Hygiene for Development* 2, no. 4 (2014): 1–24. https://doi.org/10.2166/washdev.2012.104

Stoutenberg, M., Shaya, G.E., Feldman, D.I., and Carrol, J.K. "Practical Strategies for assessing patient physical activity levels in primary care". *Mayo Clinical Proceedings and Innovation in Quality Outcomes* 1, no. 1 (2017): 8–13. https://doi.org/10.1016/j.mayocpiqo.2017.04.006

Tran, F. "Patchwork in an interconnected world: The challenges of transport networks in Sub-Saharan Africa". *Transport Reviews* 38, no. 6 (2017): 710–736. https://doi.org/10.1080/01441647.2017.1414899

United Nations. "World Population Ageing 2019: Highlights". Department of Economic and Social Affairs. 2019. Retrieved on 29/11/2019 at 11:09 AM from https://www.un.org/en/development/desa/population/publications/pdf/ageing/WorldPopulationAgeing2019-Highlights.pdf

Van Dyck D., Teychenne M., McNaughton S.A., De Bourdeaudhuij I., and Salmon, J. "Relationship of the Perceived Social and Physical Environment with Mental Health-Related Quality of Life in Middle-Aged and Older Adults: Mediating Effects of Physical Activity". *PLoS ONE* 10, no. 3 (2015): 1–16. https://doi.org/10.1371/journal.pone.0120475

van Hoof, J., Kazak, J.K., Perek-Białas, J.M., and Peek, S.T.M. "The Challenges of Urban Ageing: Making Cities Age-Friendly in Europe". *International Journal of Environmental Research and Public Health* 15, no. 1 (2018): 1–17. https://doi.org/10.3390/ijerph15112473

Wijaya, H.B., Kurniawati, H., and Hutama, S.T.E.W. "Industrialization Impact on Worker Mobility and Land Use in Peri Urban Area (Case study of Semarang District, Indonesia)". *Earth and Environmental Science* 123 (2018): 1–10. https://doi.org/10.1088/1755-1315/123/1/012037

Wiles, J. L., Leibing, A., Guberman, N., Reeve, J., and Allen, R. E. S. "The Meaning of 'Aging in Place' to Older People". *The Gerontologist* 52, no. 3 (2011): 357–366. https://doi.org/10.1093/geront/gnr098

Windsor, T.M., Curtis, R.G., and Luszcz, M.A. "Social engagement in late life". In: Population Aging and Australia's Future, ANU Press, Sydney, Australia, 2016: 185–204.

World Health Organization. "World report on ageing and health: WHO Library Cataloguing-in-Publication Data". 2013. Retrieved on 01/12/2019 at 5:59 PM from https://apps.who.int/iris/bitstream/handle/10665/186463/9789240694811_eng.pdf;jsessionid=C3785C78DA41AB9A853BA9863D5784BF?sequence=1

Yao, X., Wang, Z., and Wang, H. "Impact of Urbanization and Land-Use Change on Surface Climate in Middle and Lower Reaches of the Yangtze River, 1988–2008". *Advances in Meteorology* 2015: 1–10. http://dx.doi.org/10.1155/2015/395094

Yarfi, C., Ashigbi, E.Y.K., and Nakua, E.K. "Wheelchair accessibility to public buildings in the Kumasi metropolis, Ghana". *Africa Journal of Disability* 6, no. 1 (2017): 1–8. https://doi.org/10.4102/ajod.v6i0.341

Yin, J., Zhao, X., Zhang, W., and Wang, P. "Rural Land Use Change Driven by Informal Industrialization: Evidence from Fengzhuang Village in China". *Land* 9, no 1 (2020): 1–17. http://dx.doi.org/10.3390/land9060190

Zhang, J., Tan, P.Y., Zeng, H., and Zhang, Y. "Walkability Assessment in a Rapidly Urbanizing City and Its Relationship with Residential Estate Value". *Sustainability* 11, no. 1 (2019): 1–20. http://dx.doi.org/10.3390/su11082205

14 Urban Resilience to the Challenges of Climate Change

The Case of Africa

Joan Nyika and Megersa Olumana Dinka
Technical University of Kenya, Nairobi, Kenya
University of Johannesburg, South Africa

CONTENTS

14.1 INTRODUCTION

The apparent growth in urbanization and climate variation and change are interrelated, as stipulated in the sustainable development goals (SDGs) of Agenda 2030. According to the Department of Economic and Social Affairs (2018) of the United Nations, 55% of the global population resided in urban areas by 2018, up from 30% in 1950. The department also noted that projections show that in 2050 more than 68% of the population will have urbanized. Regionally, urbanization levels were 82% in North America, 81% in Latin America and the Caribbean, 74%, 68%, 50% and 43% in Europe, Oceania, Asia and Africa, respectively, as of 2018 (Department of Economic and Social Affairs, 2018). The effects of climate change, such as storms, heat waves, sea level rise, water stress, drought and floods, pose some of the greatest risks to the socio-economic growth of many cities around the globe (World Bank, 2019). Seto, Guneralp and Hutyra (2012) projected that urban land cover will increase three-fold to about 1.2 million km^2, as will the increased vulnerability of the population to anthropogenic climate change, according to IPCC (2014). Similarly, Argueso et al. (2015) and Oleson et al. (2015) agree that the risks to cities and their populations due to climate change and, in particular, heat-related stress and mortality, is rising as these areas expand.

Africa is also transitioning to an urban age from a previously predominantly rural population tendency. Guneralp and others (2017) noted that the continent is one of the fastest urbanizing regions of the world. Africa's urban population is set to grow more than three times its 2010 size,

DOI: 10.1201/9781003131922-17

from 395 million to 1.339 billion by 2050, a rise of 21%. This is a significant rise considering that the continent only had 10% of its population living in urban areas in the 1950s. The continent has more than seven megacities with a population of over 5 million, including Nairobi in Kenya, Johannesburg-Pretoria in South Africa, Accra in Ghana, Lagos in Nigeria, Kinshasa in Democratic Republic of Congo and Cairo in Egypt (United Nations Department of Economics and Social Affairs, 2014). In Africa, the city has become the easiest escape route from the agricultural effects of climate change that are souring rural areas. According to Henderson et al. (2017), climate change has resulted in low crop productivity and a preference for urbanization in the search for alternative income sources in Africa. Additionally, the phenomenon has disrupted the financial sector by interfering with production and consumption patterns, as Nyika (2021) has noted. Consequently, African cities are thriving in the 'grow dirty now, clean up later' trend that is further worsening the already apparent climate change crisis (White et al., 2017). For instance, higher rainfall amounts and intensities have been perilous for slum dwellers of cities such as Maputo in Mozambique and Dhaka in Ghana, according to Zehra et al. (2019). In cities of South Africa such as Cape Town and Johannesburg, extended drought due to climate variation is associated with energy and water insecurities (Nyika, 2020a). Fombe (2019) and Siegel (2020) agree with these observations, claiming that Africa is experiencing more severe and frequent droughts and floods and these have resulted in water-borne diseases such as diarrhea, cholera and dysentery, as well as rodent-borne diseases, and consequently, worsened the region's poverty levels. Other African cities, including Addis Ababa (Ethiopia), Kampala (Uganda), Mombasa (Kenya), Alexandria (Egypt) and Niamey (Niger), among others, have recorded analogous climate change effects, which have resulted in socio-economic deterioration of their population (Taylor et al., 2014).

Adapting to the climate change crises and building socio-economically empowered and environmentally sustainable cities across Africa is therefore a priority. According to Schaeffer et al. (2013), 'the need for adaptation is expected to be high in Africa, especially in light of the existing deficit in adaptation to current climate variability' (2013, p. 14). Chersich and Wright (2019) express similar sentiments when assessing the status of climate change adaptation in South Africa, highlighting the need to act now to salvage the future of the world. Current efforts to adapt to climate change in the continent are hampered by inadequate research, limited finances and lack of reliable data to quantify the negative impacts of the phenomenon and the economic costs associated with adaptation (Nyika, 2020b). Additionally, the focus on climate change adaptation must now, unlike in the past, inculcate direct and short-term actions aimed at enhanced resilience building and better preparedness (Chersich and Wright, 2019). The initiatives on climate change adaptation must deal with the twin challenges of rising global heat and the rapid rural-to-urban migration predominant in the cities (Patel et al., 2014; Kareem et al., 2020). The interplay between global and local forces such as geopolitics, technological developments, economic predispositions and municipal regulations, and up/downstream interventions that are sometimes contradictory should be considered (World Bank, 2019).

This chapter focuses on the climate change challenge among African cities and explores their resilience levels using a systematic review to gain insights. Furthermore, the chapter highlights the science of climate change crises and the interventions that enhance resilience to bridge the existing knowledge gap in Africa. The research approach in this chapter uses a transparent, rigorous and systematic review methodology that involves assessing, synthesizing and identifying available literature to generate an accurate and empirical-based answer to a research query (Mallet et al., 2012). Systematic reviews are important tools for advising policy makers and donors on appropriate corrective measures to take with regard to humanitarian interventions (Mallet et al., 2012). Systematic reviews also serve to clarify and strengthen inferences from previously published findings through collective analysis (Edwards, 2014). It is with these justifications that the research approach in this study is applied to defining the state of climate change adaptation in African cities and suggesting measures towards improved resilience to the phenomenon, using findings from previous studies.

14.2 METHODOLOGY

14.2.1 Approach

This research used a systematic literature review approach, which has advantages over the commonly used narrative reviews. According to Nunez-Mir et al. (2016), systematic reviews cover a larger scope of studies in a repetitive approach. The evaluated studies in this case also had varied study designs, time frames, climate change and adaptation scenarios. These characteristics, according to Vetter et al. (2013), make it difficult to conduct a summative review of their findings. The search for relevant academic literature was conducted qualitatively, since this is the preferred method of assessing social and ecological dimensions of climate change adaptation and resilience in urban centers, according to Berrang-Ford et al. (2015). The Web of Science was used to source the academic material and keywords were used as the search strategy in line with the topic of this research. In this case, the question to be answered was on the status of climate change in African cities and what strategies the cities are adopting to enhance preparedness and resilience to climate change while at the same time reducing the apparent vulnerability resulting from this phenomenon. Keywords used include climate change adaptation and climate change resilience in African cities. Only literature published in English was considered; therefore material selected excluded Francophone and Anglophone cities of Africa and was not representative.

14.2.2 Criterion for Research Inclusion and Exclusion

Using the keywords mentioned above, 2745 initial Web of Science documents were found for the period 1990–2021. The results were subjected to the inclusion and exclusion criterion summarized in Figure 14.1. Based on various aspects, the number of documents was reduced to 19 articles based on the relevance of their specific methodologies, results, texts, abstracts and titles to the current research objective. Only articles published between 2012 and 2021 and available in open access were selected to provide the most relevant information. Papers from all regions of Africa including Northern, Central, East, West and Southern Africa were considered by refining the regions in which various documents originated. The search further considered articles based on Web of Science categories and research areas, the most relevant of which were environmental science, environmental studies, meteorology atmospheric science, water resources and green sustainable science technology.

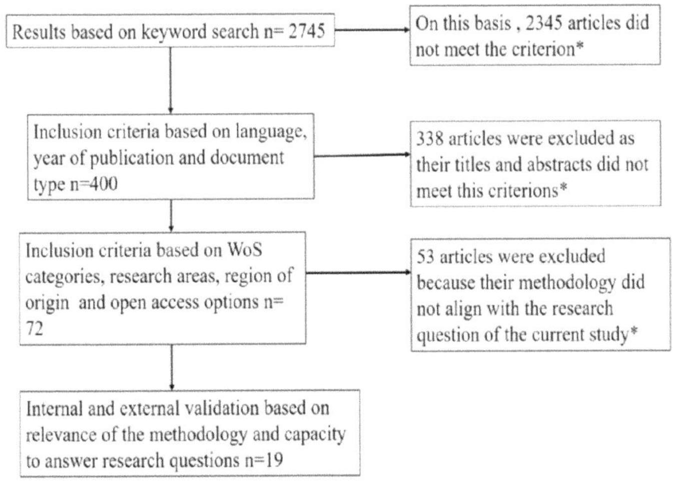

FIGURE 14.1 Procedure used for inclusion and exclusion of searched documents.

Note: The asterisk indicates exclusion for more than one reason.

This systematic review included articles describing the current state of the climate change crises, climate change adaptation and resilience enhancement approaches focusing on technological innovations at localized levels, interaction of various relevant stakeholders, climate change plans and policies, and interventions by local communities.

Independent screening of the texts, abstracts and titles of the documents was autonomously done by the authors. Further, they were reviewed using three complementary approaches: (1) identification of abstracts in titles on climate change resilience and adaptation in African cities; (2) internal validation of how the specific research questions were answered and the methodological approach used; and (3) external validation of how well the current research question was answered by the validated documents in (2) above. The author then reviewed the full texts to make a final decision on inclusion or exclusion of the refined articles. Using the ten-item critical appraisal skills program (CASP, 2019) checklist, quality appraisal was conducted on the 19 articles. The CASP tool was preferred in this context due to its common application when conducting qualitative research, according to Hannes et al. (2010). This exercise did not result in any further exclusions as the documents were deemed relevant.

14.3 RESULTS

14.3.1 Climate Change Crisis in African Cities

Evidence indicated that climate change effects have the potential to disrupt urban systems. As shown in Table 14.1, climate change alters city infrastructure and morphology (Barau et al., 2015), and increases vulnerability of cities to energy (Nduhuura et al., 2020), water (Modonsela et al., 2019; Millington and Scheba, 2021) and food insecurities (Karg et al., 2016). Additionally, the phenomenon disrupts wastewater systems (Friedrich and Kretzinger, 2012) and biodiversity (Ziervogel et al., 2014), and reduces greening efforts in urban areas (White et al., 2017; Du Toit et al., 2018). The effects of climate change are experienced in all parts of Africa: Northern (Cairo), Central (Kinshasa), Southern (Maputo, Johannesburg, Windhoek, Cape Town and Lusaka), West (Accra, Abidjan, Kano, Luanda, Ndjamena and Kumasi) and East (Nairobi, Dar es Salaam and Addis Ababa) African cities have been shown to be adversely affected. The effects of climate change span across inland and coastal towns. The effects are not only limited to big and densely populated cities such as Accra, Nairobi, Cape Town and Lusaka among others, but also to local towns, as was evident in Mkomazi (Grothmann et al., 2017), Faya and Ati (Pattnayak et al., 2019) and Port Saint Johns (Musekiwa et al., 2016), to name just a few. The evidence agrees with previous studies suggesting that Africa and its cities are the most vulnerable to climate change effects but the least prepared with regard to adaptation (Schaeffer et al., 2013; Taylor et al., 2014; Henderson et al., 2017; Spaliviero et al., 2019; Kareem et al., 2020; Nyika, 2020a, 2020b). The poor state of African cities, characterized by food, energy and water insecurities as well as environmental degradation, has been exacerbated by climate change, according to Mpandeli et al. (2020).

14.3.2 Resilience and Adaptation to Climate Change

The selected articles showed the relentless efforts to enhance climate change resilience and preparedness to deal with the challenge and at the same time reduce the associated vulnerability. As shown in Table 14.2, some of the pathways to enhance climate resilience include the expansion of green spaces in urban centers to enhance food security (Karg et al., 2016), effective planning on energy use and distribution, effective use of land and water resources, redesigning and reimagining cities to adapt to changes caused by extreme weather effects (Ziervogel et al., 2017), and scientific solid waste and wastewater management (Friedrich and Kretzinger, 2012). Suggestions on reducing water surges such as afforestation and infrastructural modifications at coastal areas due to sea level rise are included by Cavan et al. (2014) and Musekiwa et al. (2016). Freshwater resources

TABLE 14.1
Climate Change States and Effects in Various Cities of Africa

No.	Citation and Year	Geographical Area and City of Focus	Ecology of the City	Described Climate Change State
1.	Nduhuura et al. (2020)	Accra and Kumasi cities of Ghana	Coastal and inland	Regions and cities of East, Southern and Central Africa are vulnerable to climate change. Climate change is also attributable to energy insecurities in the cities.
2.	Spaliviero et al. (2019)	Addis Ababa (Ethiopia), Arusha (Tanzania) Maputo (Mozambique) among other Sub Saharan African cities	Inland and coastal cities	Interventions to develop urban regions of Africa have been worsened by climate variation and change.
3.	Millington and Scheba (2021)	Cape Town (South Africa)	Coastal city	Infrastructural and budgetary priorities of South African cities have changed due to climate change that has resulted in water insecurities. Cape Town reached its day zero in 2018 due to climate change effects.
4.	Ndebele-Murisa et al. (2020)	Harare (Zimbabwe), Durban of (South Africa), Lusaka (Zambia) and Windhoek (Namibia)	Inland and coastal cities	Densely populated Southern African cities are more vulnerable to climate variability and its resultant effects than rural areas.
5.	Meyer and Auriacombe (2019)	Cairo (Egypt), Lagos (Nigeria), Kinshasa (Democratic Republic of Congo), Johannesburg (South Africa), Dar es Salaam (Tanzania), Nairobi (Kenya), Luanda (Angola) and Abidjan (Ivory Coast).	Inland and coastal towns	Climate change, among other urban problems such as poor spatial planning, inequalities and poverty, have deterred resilience growth of urban cities of Africa.
6.	Pattnayak et al. (2019)	Cities of Faya, Ati, Abeche, Mongo, Lere, Moundou, Sarh and Ndjamena (Chad)	Coastal and inland	African cities are increasingly becoming vulnerable and sensitive to climate change.
7.	Madonsela et al. (2019)	Cape Town (South Africa)	Coastal	Severe water shortages worsened by climate change are rampant in Cape Town with water levels of its reservoirs lowering to critical states.
8.	Du Toit et al. (2018)	Sub Saharan Africa cities	N/A	Climate change effects are going to severally affect populated African cities unless they are re-imagined and made green.
9.	Grothmann et al. (2017)	Mkomazi (Tanzania)	Inland	Climate change remains one of the greatest modern-day challenges to both urban and rural areas.
10.	Ziervogel et al. (2017)	Urban areas of Africa in general	Coastal and inland	Climate change results in risks and disasters that hinder the growth of urban areas.
11.	Karg et al. (2016)	Tamale (Ghana) and Ougadougou of (Burkina Faso)	Inland	Accelerating climate change has negative effects on food systems, prices and demand, which is worsening food insecurity.

(Continued)

TABLE 14.1 (*Continued*)
Climate Change States and Effects in Various Cities of Africa

No.	Citation and Year	Geographical Area and City of Focus	Ecology of the City	Described Climate Change State
12.	Musekiwa et al. (2016)	Port Saint Johns, East London and Cape Town (South Africa)		Coastal towns of South Africa are vulnerable to wave erosion and inundation due to climate change effects of sea level rises.
13.	Barau et al. (2015)	Kano city (Nigeria)	Inland	Climate change effects resulting from environmental degradation and morphological modifications have enhanced the vulnerability of Kano city to flooding.
14.	Ziervogel et al. (2014)	Cities of South Africa	Inland and coastal	Climate change has negative effects on biophysical aspects including biodiversity, agriculture and water resources.
15.	Cavan et al. (2014)	Dar es Salaam (Tanzania) and Addis Ababa (Ethiopia)	Coastal and inland	Urban areas of Africa are the most vulnerable to climate change effects of erosion, sea level rise, flooding, droughts and heat waves.
16.	Pasquini and Shearing (2014)	A municipality in Western Cape (South Africa)	Inland	Hefty and costly developmental challenges result from climate change effects in South African towns.
17.	Lopez et al. (2014)	Praia (Cape Verde)	Coastal	Cities of Cape Verde suffer from heat stress due to climate variations and changes.
18.	Silver et al. (2013)	Saint-Louis (Senegal) and Bobo-Dioulasso (Burkina Faso)	Coastal and inland	In West African cities, climate change is a major developmental hindrance.
19.	Friedrich and Kretzinger (2012)	eThekwini Municipality (South Africa)	Coastal	Climate change effects of sea level rise have had a negative effect on wastewater systems of coastal towns.

that are on a diminishing trend due to climate change should be used effectively and efficiently using sound allocation plans (Millington and Scheba, 2021). Urban areas of Africa should adopt water-sensitive designs by increasing the value of fresh water and modifying their respective water tariffs (Pattnayak et al., 2019). Land-use plans that accommodate greening, reduced carbon emissions through environmentally friendly mobility systems, green building and minimization of waste towards sustainability were suggested. These modifications must be rooted in good governance that supports strategic planning, power decentralization, information sharing and the involvement of all stakeholders, including urban residents (Meyer and Auriacombe, 2019; Spaliviero et al., 2019). Furthermore, good governance should nurture political championing and partnerships that foster increased climate resilience through effective institutional management and policy re-evaluation and modification to enhance environmental sustainability (Du Toit et al., 2018; Madonsela et al., 2019).

Generally, the studies described the current state of climate change and some emerging aspects that support climate adaptation and resilience. Seventeen of the studies gave suggestions on improvements to be made in urban areas while the other two only assessed the state of climate change in the specified towns. Across the evaluated studies in Table 14.2, resilience is emphasized in urban aspects that concern architectural technology, energy use, water allocation, and resource management and civic engagement via information sharing in decentralized systems. The implementation of the resilience measures is dependent on administrative structures at municipal and city levels,

TABLE 14.2
Research Aims of Examined Articles, Their Methodological Approaches and Suggested Measures to Enhance Climate Resilience

No.	Citation and Year	Research Aim	Methodological Approach	Suggested Adaptation Measures
1.	Nduhuura et al. (2020)	Exploring the effects of climate change on energy inconsistencies resulting in frequent load shedding in Accra.	Quantitative	Effective planning by urban energy regulating agencies on better energy distribution amidst shortages.
2.	Spaliviero et al. (2019)	To assess the prospects of developing a city's resilience action planning tool to promote adaptation to climate change in Africa.	Qualitative	Culturing of good governance that channels finances, improves policies of climate change adaptation while including all stakeholders such as urban dwellers.
3.	Millington and Scheba (2021)	To establish a relationship between climate change and municipal governance in Cape Town.	Qualitative	Modification of water allocation plans to provide free basic water for all. Modifying water tariffs to increase the resource's value amidst climate change-based shortages.
4.	Ndebele-Murisa et al. (2020)	To explain how Southern African cities are sharing information to promote sound practices on urban ecosystem improvement and climate change resilience.	Qualitative	By engaging in knowledge and information sharing, improved water, energy and waste management can be realized, hence integrated climate change planning and monitoring towards adaptation and resilience.
5.	Meyer and Auriacombe (2019)	To establish a relationship between good governance and climate change resilience.	Qualitative	Good governance involving power decentralization is an indispensable prerequisite for climate change adaptation in urban areas and must be adopted.
6.	Pattnayak et al. (2019)	To investigate the climate patterns of cities in Chad and establish the presence/absence of climate variation.	Quantitative	Climate variations in the form of temperature rises, increased precipitation and high drought and storm frequencies were identified and mitigation strategies based on advanced research and development were advocated.
7.	Madonsela et al. (2019)	To provide a roadmap on the methods to implement a city blueprint approach based on good governance to deal with Cape Town's water crises worsened by climate change.	Qualitative	There is a need to take up water-sensitive urban designs to improve water allocation plans and manage scarcity, flooding and wastewater systems amidst climate change effects.
8.	Du Toit et al. (2018)	To understand the effects of adopting green infrastructure on ecosystem services that are vulnerable to climate change in Sub Saharan African cities.	Qualitative	Adaptation and resilience could be enhanced by overcoming challenges for governance and planning on resource use with climate variation vulnerability.
9.	Grothmann et al. (2017)	To explore the social-ecological systems of two cities and assess their vulnerability to climate change.	Qualitative	The vulnerability of the towns to climate change can be alleviated by intensive research on suitable mitigation measures and adaptation technologies.

(*Continued*)

TABLE 14.2 (*Continued*)
Research Aims of Examined Articles, Their Methodological Approaches and Suggested Measures to Enhance Climate Resilience

No.	Citation and Year	Research Aim	Methodological Approach	Suggested Adaptation Measures
10.	Ziervogel et al. (2017)	To explore the options of climate resilience, risk management, and entitlements and rights involved.	Qualitative	There is a need to expand resilience notions to create diverse adaptation measures that promote justice for affected urban residents while reimagining the morphology of African cities.
11.	Karg et al. (2016)	To assess the geographical sources of food to named urban areas and assess supply fluctuations due to climate change.	Quantitative	The need to diversify foodsheds to promote resilience amidst climate-based food insecurities is advocated.
12.	Musekiwa et al. (2016)	To assess the vulnerability of South African coastal cities to climate change.	Quantitative	N/A
13.	Barau et al. (2015)	To assess how urban land administration institutions can enhance Kano city's climate resilience.	Qualitative	Increased number of green spaces and modifications in building, adopting green mobility measures and architecture to reduce carbon emissions of urban areas can improve climate change adaptations.
14.	Ziervogel et al. (2014)	To explore current measures and recent advancements in climate resilient enhancement for South African towns.	Qualitative	Policy and institutional improvements are needed in climate-resilient planning and development.
15.	Cavan et al. (2014)	To describe the urban morphology types of two Eastern African cities.	Quantitative	Typing of urban morphologies is essential in land-use planning and ecosystem improvement towards climate adaptation.
16.	Pasquini and Shearing (2014)	To investigate the influence of politics in managing environmental problems such as climate change towards sustainable development.	Qualitative	To drive climate resilience, political champions, partnerships among stakeholders and collaborators to benefit urban residents are pre-conditions.
17.	Lopez et al. (2014)	To quantify the bio-thermal state and assess the microclimatic characteristics that promote heat stress in Praia city, Cape Verde.	Quantitative	N/A
18.	Silver et al. (2013)	To evaluate the emergent perspectives on climate change adaptation in West African cities.	Qualitative	Adopting urban planning measures that are environmentally friendly, adaptive to water-prone environs and green building coupled with better legislation is advised.
19.	Friedrich and Kretzinger (2012)	To investigate the vulnerability of solid waste and wastewater systems to sea-level rise in coastal South African cities.	Quantitative	Redesigning infrastructure for both solid waste and wastewater systems with considerations of climate change is highly recommended.

since it is they who are involved in disseminating responsibilities. Although there is evidence that modifications towards climate adaptation are aimed at reducing emissions, limited research efforts exist to enhance flexibility and interagency networking in the management of, in particular, the energy, water and transport sectors, despite the effectiveness of adopting greener measures in urban areas as suggested by Frantzeskaki et al. (2014).

Some of the articles evaluated focused on energy, linking it to health interventions such as green building and preferential use of natural ventilation. Health interventions reduce heat stress in hot areas. Rao et al. (2013) highlighted that these natural interventions minimize indoor temperatures to prevent dampness, mold and heat strokes, at the same time reducing the costs and emissions associated with domestic air conditioning. Planning therefore should take into account the energy, water distribution, transport, health and associated interlinkages, as these aspects influence greenhouse gas emissions and the severity of some climate change effects. Lwasa et al. (2015) observed that such a planning approach, when rooted in good policies, provides opportunities to co-benefit or understand trade-offs that enhance resilience in multiple sectors.

The suggested climate resilience measures are mute on how to leverage opportunities related to the agile nature and differing degrees of informality in towns and cities of Africa. The evidence advocates for institutional reforms when planning for urban redesign and transformation accompanied by sound planning to enhance resilience. Additionally, participation of all stakeholders, including urban residents and information exchange, is given precedence. According to Lindell et al. (2019), these initiatives are transformative as they have resulted in better waste recycling, the production of energy-efficient cooking alternatives, and greener building and transportation approaches in the growing cities of Uganda in Africa. Based on the results of this research, cities and urban spaces of Africa have begun transformation towards better understanding of the disruptions of climate change and ways to withstand their inevitability. Related studies have made similar observations, citing how effective planning with the involvement of urban residents has facilitated the uptake of environmentally friendly measures in African cities (Luque-Ayala and Maia, 2019; Buyana et al., 2019; Monteith and Camfield, 2019).

14.4 DISCUSSION

Climate resilience is the ability of ecological and social systems to withstand disturbances of extreme weather and retain their ability to function while at the same time adapting to the resultant changes and stresses (Chirisa et al., 2016). The resilience capacities of cities that have many ecologies and many informal settlers, despite being less built, as is the case in Africa, are different (Nagendra et al., 2018). This predisposition could be a result of the rights and justice channels supporting the risks associated with changes towards climate resilience, as Ziervogel et al. (2017) have observed. Although most of the studies evaluated in this case did not focus on understanding the term 'resilience', they are convinced that this agenda is rooted in developed-country policies and is therefore practical for cities of developing regions such as Africa, though the issues therein are yet to be addressed satisfactorily. This observation necessitates theoretical perspectives from Africa that can inform urban redesign to make resilient towns and cities, guided by the measures suggested by intensified research and development on climate resilience issues. Kareem et al. (2020) propose a different direction, applying three viewpoints to resilience: (1) social, (2) ecological and (3) institutional. The social approach to resilience focuses on involving different actors learning, sharing information and experimenting with different urban scenarios that could enhance climate resilience. Ecological approaches involve the interactions of cities and towns with others at local and regional levels to learn from one another on measures to enhance climate change adaptation. Lastly, institutional resilience deals with coordinated planning and response to climate risks and effects founded on good policies and governance to maintain the functionality of African cities.

14.4.1 Future Prospects

It is evident from this research that Africa is facing and will continue to face serious challenges with regard to climate change and its resultant effects. However, the pathways to resilience suggested at ecological and regional urban scales of Africa in this case come in handy to enhance the current call for climate action. This is because they will enhance understanding of climate patterns and devise new and climate-sensitive measures to adapt to climate variations and change. Inter-regional differences were evident in Southern, East and West African cities but limited evidence from Central and Northern Africa was found. Based on Table 14.1, four studies drew conclusions from African cities in general while seven, six and two focused on Southern, West and East African cities, respectively. Therefore, Central and Northern African cities were scarcely represented compared to other regions. Based on the city's ecology, the focus was on the coastal and inland region while mountainous and semi-arid regions were scarcely represented. This could be because the latter regions focused on the severity of climate change risks rather than on climate adaptation and resilience measures. Evidently, in the interests of holistic understanding and accurate generalization, it is important to expand the ecological and inter-regional measures being taken to enhance climate change resilience in Africa.

Based on the reported state of climate change in urban regions of Africa and the growing threat affiliated to the phenomenon, stakeholders and agencies involved must collaborate both in capital and secondary cities to enhance coordinated planning towards climate resilience. Muller (2019) made this suggestion following an observation of the effects of tropical cyclone Idai whose effects were felt not only in Maputo but also in neighboring towns of Mozambique and Zimbabwe. This review represented mainly leading tourism, administrative and financial capital cities of Africa. Examples include Cape Town, Accra, Nairobi, Johannesburg, Windhoek, Maputo, Lusaka, Praia, Addis Ababa and Ouagadougou. A number of studies focused on collaboration by smaller towns and municipalities towards climate resilience. This observation necessitates more research on the links between network and climate resilience planning at various urban administrative levels, including local, intra-metropolitan and inter-city connectivity.

14.5 CONCLUSION

The climate change phenomenon remains one of the greatest modern-day challenges, with its effects traversing the economic, environmental and equality aspects of sustainability. Africa, which comprises mainly poor countries, is most vulnerable and least prepared for the crises associated with climate change. Current research on climate change adaptation and resilience in Africa were reviewed in this study to gain insight into the vulnerability and preparedness level of the continent as well as suggest affirmative measures towards adaptation and mitigation. It was established that most studies emphasized the seriousness of the phenomenon and, in most cases, went on to cite some of its effects at global, regional and local levels. Further results showed that most studies also suggested varied measures towards resilience focusing on improving resource management, policy making, institutional remodeling and technology inclusion in reimagining the greening of urban areas. The studies focused on coastal and inland cities, most of which were economic or tourism hubs in the specified countries. The majority of the studies examined focused on towns in West and Southern Africa. Evidently, the climate change challenge is growing in Africa and so is the awareness of its effects, the mitigation and adaptive measures to alleviate them. However, the focus of future research should include secondary towns, and cities in mountainous and arid areas particularly in Northern and Central Africa. This will make the results generalizable at continental level and improve future understanding of the implications of the suggested resilience measures at the planning and implementation stages.

REFERENCES

Argueso, D., J. Evans, A. Pitman, and A. Di Luca. "Effects of city expansion on heat stress under climate change conditions." *PLoS One* 10, no. 2 (2015): e0117066.

Barau, A., R. Maconachie, A. Ludin, and A. Abdulhamid. "Urban morphology dynamics and environmental change in Kano, Nigeria." *Land Use Policy* 42 (2015): 307–317.

Berrang-Ford, L., T. Pearce, and J. Ford. "Systematic review approaches for climate change adaptation research." *Regional Environmental Change* 15 (2015): 755–769.

Buyana, K., D. Byarugaba, H. Sseviiri, G. Nsangi, and P. Kasaija. "Experimentation in an African neighbourhood: reflections for transitions to sustainable energy in cities." *Urban Forum* 30 (2019): 191–204.

Cavan, G., S. Lindley, F. Jalayer, K. Yeshitela, S. Pauleit, F. Renner, S. Gill, P. Capuano, A. Nebebe, T. Woldegerima, D. Kinassa, and R. Shemdoe. "Urban morphological determinants of temperature regulating ecosystem services in two African cities." *Ecological Indicators* 42 (2014): 43–57.

Chersich, M., and C. Wright. "Climate change adaptation in South Africa: a case study on the role of the health sector." *Globalization and Health* 15, no. 22 (2019): 1–16.

Chirisa, I., E. Bandauko, E. Mazhindu, N. Kwangwama, and G. Chikowore. "Building resilient infrastructure in the face of climate change in African cities: Scope, potentiality and challenges." *Development Southern Africa* 33 (2016): 113–127.

Critical Appraisal Skills Programme. "CASP qualitative checklist: critical appraisal skills programme." Available at: (https://casp-uk.net/casp-tools-checklists/), (2019, February).

Department of Economic and Social Affairs. "World Urbanization Prospects: The 2018 revision." New York: United Nations (2018).

Du Toit, M., S. Cilliers, M. Dallimer, M. Goddard, S. Guenat, and S. Cornelius. "Urban green infrastructure and ecosystem services in sub-Saharan Africa." *Landscape and Urban Planning* 180 (2018): 249–261.

Edwards, T. "What is the value of a systematic review?" *Journal of Shoulder and Elbow Surgery* 23 (2014): 1–2.

Fombe, L. "Diseases and health; case of wetland populations in the Bamenda urban area, North West Region of Cameroon." *Advances in Social Sciences Research Journal* 6 (2019): 220–231.

Frantzeskaki, N., J. Wittmayer, and D. Loorbach. "The role of partnerships in 'realising' urban sustainability in Rotterdam's City Ports Area, The Netherlands." *Journal of Cleaner Production* 65 (2014): 406–417.

Friedrich, E., and D. Kretzinger. "Vulnerability of wastewater infrastructure of coastal cities to sea level rise: A South African case study." *Water South Africa* 38, no. 5 (2012): 755–764.

Grothmann, T., M. Petzold, P. Ndaki, V. Kakembo, B. Siebenhuner, M. Kleyer, P. Yanda, and N. Ndou. "Vulnerability assessment in African villages under conditions of land use and climate change: case studies from Mkomazi and Keiskamma." *Sustainability* 9 (2017): 976.

Guneralp, B., S. Lwasa, H. Masundire, S. Parnesll, and K. Seto. "Urbanization in Africa: challenges and opportunities for conservation." *Environmental Research Letters* 13 (2017): 015002.

Hannes, K., C. Lockwood, and A. Pearson. "A comparative analysis of three online appraisal instruments' ability to assess validity in qualitative research." *Qualitative Health Research* 20 (2010): 1736–1743.

Henderson, V., A. Storeygard, and U. Deichmann. "Has climate change driven urbanization in Africa?" *Journal of Development Economics* 124 (2017): 60–82.

IPCC. "Climate change 2014: impacts, adaptation, and vulnerability. Part A: global and sectoral aspects. Contribution of working group II to the fifth assessment report of the intergovernmental panel on climate change" [Field, C.B., V.R. Barros, D.J. Dokken, K.J. Mach, M.D. Mastrandrea, T.E. Bilir, M. Chatterjee, K.L. Ebi, Y.O. Estrada, R.C. Genova, B. Girma, E.S. Kissel, A.N. Levy, S. MacCracken, P.R. Mastrandrea, and L.L. White (eds)]. Cambridge, United Kingdom and New York, NY: Cambridge University Press (2014).

Kareem, B., S. Lwasa, D. Tugume, P. Mukwaya, J. Walubwa, S. Owuor, P. Kasaija, H. Sseviiri, G. Nsangi, and D. Byarugaba. "Pathways for resilience to climate change in African cities." *Environmental Research Letters* 15 (2020): 073002.

Karg, H., P. Dreschsel, E. Akoto-Danso, R. Glaser, G. Nyarko, and A. Buerkert. "Foodsheds and city region food systems in two West African cities." *Sustainability* 8, no. 12 (2016): 1175.

Lindell, I., C. Ampaire, and A. Byerley. "Governing urban informality: Re-working spaces and subjects in Kampala, Uganda." *International Development Planning Review* 41, no. 1 (2019): 63–84.

Lopez, A., E. Correia, J. Nascimento, and P. Canario. "Urban bioclimate and comfort assessment in the African city of Praia (Cape Verde)." *Finisterra* 98 (2014): 33–48.

Lwasa, S., F. Mugagga, B. Wahab, D. Simon, J. Connors, and C. Griffith. "A meta-analysis of urban and peri-urban agriculture and forestry in mediating climate change." *Current Opinion in Environmental Sustainability* 13 (2015): 68–73.

Luque-Ayala, A., and F. Maia. "Digital territories: Google maps as a political technique in re-making of urban informality." *Environment and Planning D: Society and Space* 37, no. 3 (2019): 449–467.

Madonsela, B., S. Koop, K. van Leeuwen, and K. Carden. "Evaluation of water governance processes required to transition towards water sensitive urban design: An indicator assessment approach for the city of Cape Town." *Water* 11, no. 2 (2019): 292.

Mallet, R., J. Hagen-Zanker, R. Slater, and M. Duvendack. "The benefits and challenges of using systematic reviews in international development research." *Journal of Development Effectiveness* 4, no. 3 (2012): 445–455.

Meyer, N., and C. Auriacombe. "Good urban governance and city resilience: An Afrocentric approach to sustainable development." *Sustainability* 11 (2019): 5514.

Millington, N., and S. Scheba. "Day zero and the infrastructures of climate change: Water governance, inequality and infrastructural politics in Cape Town's water crises." *International Journal of Urban and Regional Research* 45, no. 1 (2021): 116–132.

Monteith, W., and L. Camfield. "Business as family, family as business: female entrepreneurship in Kampala, Uganda." *Geoforum* 101 (2019): 111–121.

Mpandeli, S., L. Nhamo, S. Hlahla, D. Naidoo, S. Liphadzi, A. Modi, and T. Mabhaudhi. "Migration under climate change in Southern Africa: A nexus planning perspective." *Sustainability* 12 (2020): 4722.

Muller, M. "Hydropower dams can help mitigate the global warming impact of wetlands." *Nature* 566 (2019): 315–317.

Musekiwa, C., H. Cawthra, M. Unterner, and W. van Zyl. "An assessment of coastal vulnerability for the South African coast." *South African Journal of Geomatics* 4, no. 2 (2016): 123–137.

Nagendra, H., X. Bai, E. Brondizio, and S. Lwasa. "The urban south and the predicament of global sustainability." *Nature Sustainability* 1 (2018): 341.

Ndebele-Murisa, M., C. Mubaya, L. Pretorius, R. Mamombe, K. Lipinge, W. Nchito, J. Mfune, G. Siame, and B. Mwalukanga. "City to city learning and knowledge exchange for climate resilience in Southern Africa." *PLoS ONE* 15, no. 1 (2020): e0227915.

Nduhuura, P., M. Garschagen, and A. Zerga. "Mapping and spatial analysis of electricity load shedding experiences: A case study of communities in Accra, Ghana." *Energies* 13 (2020): 1–26.

Nunez-Mir, G., B. Iannone, B. Pijanowski, N. Kong, and S. Fei. "Automated content analysis: addressing the big literature challenge in ecology and evolution." *Methods in Ecology and Evolution* 7, no. 11 (2016): 1262–1272.

Nyika, J. "The relationship between climate change and financial stability." In: M. Olarewaju and I. Ganiyu (Eds). *Handbook of Research on Climate Change and the Sustainable Financial Sector*. IGI Global, USA (2021).

Nyika, J. "Climate change situation in Kenya and measures towards adaptive management in the water sector." *International Journal of Environmental Sustainability and Green Technologies* 11, no. 2 (2020a): 34–47.

Nyika, J. "Understanding water-food-energy nexus in the climate change era and the roadmap to implementation in South Africa." In: A. Karmaoui, K. Barrick, M. Reed, and M. Barjees (Eds). *Impacts of Climate Change on Agriculture and Aquaculture*. IGI Global (2020b).

Oleson, K., A. Monaghan, O. Wilhelmi, M. Barlage, N. Brunsell, J. Feddema, L. Hu, and D. Steinhoff. "Interactions between urbanization, heat stress, and climate change." *Climate Change* 129, no. 3–4 (2015): 525–541.

Pasquini, L., and C. Shearing. "Municipalities, politics and climate change: An example of the process of institutionalizing an environmental agenda within local government." *Journal of Environment and Development* 23, no. 2 (2014): 271–296.

Patel, Z., S. Greyling, D. Simon, H. Arfvidsson, N. Moodley, N. Primo, and C. Wright. "Local responses to global sustainability agendas: Learning from experimenting with the urban sustainable development goal in Cape Town." *Sustainability Science* 12 (2014): 785–797.

Pattnayak, K., A. Abdel-Lathif, K. Rathakrishnan, M. Singh, R. Dash, and P. Maharana. "Changing climate over Chad: Is the rainfall over the major cities recovering?" *Earth and Space Science* 6, no. 7 (2019): 1149–1160.

Rao, S., S. Pachauri, F. Dentener, P. Kinney, Z. Klimont, K. Riahi, and W. Schoepp. "Better air for better health: forging synergies in policies for energy access, climate change and air pollution." *Global Environmental Change* 23 (2013): 1122–1130.

Schaeffer, M., F. Baarsch, S. Adams, K. Bruin, L. De Marez, S. Freitas, A. Hof, and B. Hare. "Africa's adaptation gap. Technical Report." UNEP, Nairobi (2013).

Seto, K., B. Guneralp, and L. Hutyra. "Global forecasts of urban expansion to 2030 and direct impacts on biodiversity and carbon pools." *Proceedings of the National Academy of Sciences of the United States of America* 109, no. 40 (2012): 16083–16088.

Siegel, R. "Disease Protection in Sea Coast (and Inland) Cities: Problems in Dense Populations with Shantytowns/Slums." In: *Adaptations of Coastal Cities to Global Warming, Sea Level Rise, Climate Change and Endemic Hazards* (Cham: Springer) Washington, DC: World Bank (2020).

Silver, J., C. McEwan, L. Petrella, and H. Baguian. "Climate change urban vulnerability and development in Saint-Louis and Bobo-Dioulasso: learning from across two West African cities." *Local Environment* 18, no. 6 (2013): 663–677.

Spaliviero, M., M. Pelling, L. Lopes, C. Tomaselli, K. Rochell, and M. Guambe. "Resilience planning under information scarcity in fast growing African cities and towns: The CityRAP approach." *International Journal of Disaster Risk Reduction* 44 (2019): 101419.

Taylor, A., A. Cartwright, C. Sutherland, and F. Grazi. "Institutional pathways for local climate adaptation: a comparison of Three South African Municipalities." *Focales Series* 18 (2014): 1–142.

United Nations Department of Economics and Social Affairs. *World Urbanisation Prospects: The 2014 Revision*. New York: United Nations (2014).

Vetter, D., G. Rucker, and I. Storch. "Meta-analysis: a need for well-defined usage in ecology and conservation biology." *Ecosphere* 4, no. 6 (2013): 1–24.

White, R., J. Turpie, and G. Letley. *Greening Africa's Cities. Enhancing the Relationship between Urbanization, Environmental Assets and Ecosystem Services*. Washington DC: World Bank Group (2017).

World Bank. "Global Risks Report." Washington DC. Available on: (https://weforum.org/reports/the-global-risks-report-2019) (2019).

Zehra, D., S. Mbatha, L. Campos, A. Queface, A. Beleza, C. Cavoli, K. Achuthan, and P. Parikh. "Rapid flood risk assessment of informal urban settlements in Maputo, Mozambique: the case of Maxaquene A." *International Journal of Disaster Risk Reduction* 40 (2019): 101270.

Ziervogel, G., M. New, E. van Garderen, G. Midgley, A. Taylor, R. Hamann, S. Stuart-Hill, J. Myers, and M. Warburton. "Climate change impacts and adaptation in South Africa." *WIREs Climate Change* 5 (2014): 605–620.

Ziervogel, G., M. Pelling, A. Cartwright, E. Chu, T. Deshpande, Harris, L. et al. "Inserting rights and justice into urban resilience: A focus on everyday risk." *Environment and Urbanization* 29, no. 1 (2017): 123–138.

15 Not a 'Peripheral' Connection
Dynamics of Peri-Urban Agriculture in Kolkata Metropolitan Area, West Bengal

Raktima Ghosh
Indian Institute of Technology Kharagpur, India

CONTENTS

15.1 INTRODUCTION

Kolkata city's mornings are busy everywhere – streets are thronged with people heading in different directions, buses race through the traffic to reach their destinations on time, workers run to catch the early local train to the city. This hustle and bustle slowly fades out along connecting roads to the outskirts of the city. The city bypass, especially on winter mornings, provides a glimpse of fields of lush green vegetation alternating with wetlands that are still enveloped in fog. Farmers start working at the earliest hour – some of them making muddy channels to harness water for the fields, some mixing manure, others picking up vegetables for the morning markets of the city. On the other side of this agrarian expanse repose giant sky-scrapers, corporate hotels and flyovers, firmly bearing a signature of stark discord within and beyond the urban fabric.

One of the concerns of contemporary urbanization scholarship is the changing pattern of interaction between city and hinterland, as well as the differential responses of urban people to the surrounding natural environment (Indraprahasta 2013). One of the key features of the urbanization process, particularly in developing countries, is the strong rural-urban partnerships in the socio-naturally co-constituted peri-urban "ecotone" (Mukherjee & Chakraborty 2016). A complex nexus of social, economic, ecological and governance processes, the peri-urban areas operate as 'transition transects' between rural and urban systems. By supporting agricultural and aquacultural activities, they emphasize the flow of produce, labor, finance and services which often contribute to the urban economy and food security. The Food and Agriculture Organization of the United Nations introduced the acronym UPA to describe urban and peri-urban agriculture, with 'urban agriculture'

DOI: 10.1201/9781003131922-18

denoting crop cultivation and horticulture amid the built-up areas, and 'peri-urban agriculture' taking place on the fringes of the cities (de Zeeuw et al. 2011). Mougeot (2000) defines UPA thus: urban agriculture is

> located within (intra-urban) or on the fringe (peri-urban) of a town, a city or a metropolis, which grows or raises, processes and distributes a diversity of food and non-food products, (re-)using largely human and material resources, products and services found in and around that urban area, and in turn supplying human and material resources, products and services largely to that urban area.

UPA is spreading out along the vacant land of the city margins as a pivotal source of food, ecosystems services and livelihoods, especially in times of economic and environmental uncertainties (McClintock 2012). The prime distinguishing feature of UPA is not so much related to its location but to its symbiotic interrelationship with the urban economic, social and ecological systems (Mougeot 2000) in using resources such as labor, land, organic wastes, wastewater etc. and in turn, ensuring fresh food for urban dwellers. Nonetheless, within rapid and asymmetrical urbanization coupled with a sizable increase in urban population, metropolitan cities of the developing world are encountering a host of challenges from food crisis, poverty and environmental decay. The UPA is explicitly shaped by different land-use practices, policies, consumer markets and prices, which in turn have strong implications for urban food security, poverty alleviation and ecological wellbeing. Apart from proximity to markets, factors such as inflow of urban organic wastes and wastewater often foster the cultivation of vegetables, milk, fish and fruits which are in comparatively high demand and have high market values in the city (Symoens & Micha 1995). As such, the UPA turns city wastes into resources which it returns to the city.

It is widely accepted that the foremost impetus to evaluate UPA was received from the Habitat II Conference which was convened by the United Nations at Istanbul in 1996 (Mougeot 2000). Since then, urban agriculture scholarship has emerged and matured with functions and significance in relation to potential environmental benefits encompassing creation of green space, diminution of dust and pollutants, recycling of urban organic wastes, circular economies, modification of urban micro-climate (such as heat islands) and biodiversity conservation. Tsuchiya and co-authors (2015) argued that a localized food system (direct marketing of produce) not only shortens the food supply chain but can also be recognized as a measure for reducing carbon emissions. Moreover, the paddy fields can play an essential role in containing urban flooding.

Compounded by public apathy about peri-urban resource recovery activities, the privatization goals of the state and dire absence of essential policy supports have silently debased small-scale UPA practices in the Kolkata Metropolitan Area (KMA) of West Bengal, India. This chapter aims to explore the dynamics and significance of UPA in relation to its proximate socio-ecologies, in the context of the sprawling KMA. By mapping current land-use and habitat conversion patterns, it examines the multiple challenges of UPA with changing socio-economic and ecological interactions at the peri-urban interface of Kolkata.

15.2 METHODOLOGY

This chapter draws on research undertaken during 2017 and 2018 in different sites of UPA in KMA. It deploys a blend of methods comprising qualitative analysis and remote sensing techniques. Insights on cropping types, land-use transformations and UPA challenges faced by farmers are derived from iterative field investigations comprising one-to-one exchanges and informal group discussions with farmers and key informant interviews. The geospatial tools entailed a comprehensive depiction of current land-use typologies and conversion patterns. The land-use land cover (LULC) map of KMA is developed on LISS-VIII satellite data using ERDAS Imagine remote sensing software. The supervised signature extraction technique based on maximum likelihood algorithm was adopted for classifying the digital data into different land-use and land cover categories in KMA.

Bridging case-study strategy and map-based geospatial analysis, nonetheless, offers a consolidated ground for evaluating and validating the complex urban and peri-urban situations of rapidly urbanizing cities such as Kolkata.

15.3 'URBAN GARDENS' IN INDIA AND BEYOND

UPA plays a crucial role by substantially contributing to total crop production and offering food security as well as nutrition to many developing countries of the world. About 90% of fresh vegetable consumption in Ghana's capital is assured by yields within the city, whereas in Hanoi, Vietnam, 80% of total fresh vegetables come from urban and peri-urban farms (Corbould 2013). Aubry and co-authors (2012) have identified disparate para-agricultural activities using farm resources, such as renting domestic animals, temporary agricultural labor at other farms, direct selling of vegetables, fish or bricks, which further boost farmers' income. Studies by Maxwell and co-authors (1998) in Kampala, Uganda, indicate that urban agriculture has a positive and effective association with high nutritional status among children. Urban agriculture in the form of community garden is popular in many cities of Australia, Canada, the United States, England, France and New Zealand. In developed countries, there are diverse land plots, such as windowsills, rooftops, basements, walls, recreation grounds and roadsides, where crops or vegetables are grown. Here, urban agriculture is predominantly a recreational activity rather than a source of food (Corbould 2013). Urban agriculture can also be an effective way of responding to a crisis, as exemplified by the Zambian Copper Belt which recognizes the role of agriculture as an urban survival strategy, particularly in times of extreme economic and social crises (Smart et al. 2015). In Madagascar, the fields are often perceived as heritage, whereas for the members of agricultural households, the farm is a refuge during times of social-political hardship (Aubry et al. 2012). In Beijing, a large number of migrants are involved in UPA, which provides a sense of place and stability in volatile circumstances (Peng et al. 2015). Urban gardens or farms can also function as nucleation points for the creation and expansion of locally transmitted knowledge (Lovell 2010). The political ecology framework, nonetheless, furthers right-based studies around urban food production and creation of urban spaces. Based on her research at Managua in Nicaragua, Shillington (2013) argues that the ecological imprint of urban agriculture (mostly cultivation of fruit trees) is a prominent feature that the marginalized urban inhabitants in Managua appropriate and thereby "demand their rights to urban metabolism".

Along different degrees of population growth, India has fairly transparent accounts on marginalization of traditional farming practices in the cities, varied food consumption patterns and differently manifested land grabbing and local evictions. Inevitably, sound capital accretion mechanisms under the neoliberal regime make the situation bleak with mounting impacts on ecosystem resources and outlying ecosystem-dependent communities such as farmers and fishers (Mukherjee & Chakraborty 2016). UPA as a renewed practice for urban food production challenges current land-use management and ownership by reinventing the urban landscape, and experimenting with "radical alternatives to the capitalist organization of urban life". On the other side, urban food security is gradually demanding attention as the acuteness of poverty is reflected in nutritional deficiency, specifically among women and children (Tornaghi 2014). Food security reports from urban India by the M. S. Swaminathan Research Foundation pointed out that about 65.5 million people live in urban slums and sprawls and experience considerable food and nutritional scarcity, and about 50% of women are anemic (MSSRF 2010). In the face of rapidly sprawling metropolitan cities, state administrative units as well as non-government agencies have started to identify UPA as not only an ecologically just substitute, but also a viable livelihood option for urban and peri-urban communities (Mohapatra 2012).

In India, urban agriculture is practiced in large cities such as Mumbai, Kolkata, Delhi, Bengaluru and Chennai as government and non-government enterprises and also by individuals (Sahasranaman 2016). Along with local organizations like City Farming, Earthoholics, Fresh and Local and Urban Leaves, inhabitants of Mumbai have been innovatively transforming balconies, terraces and common areas into vegetable gardens (Sahasranaman 2016). Moreover, by leasing lands along the railway tracks to local farmers in some areas, the Indian Railway has advocated keeping land clean and useful (Vazhacharickal & Buerkert 2012). A recent study in Hyderabad reveals that more than 4000 households in the outskirts of the city are self-sufficient for their daily vegetable needs (Awasthi, 2013). In Hyderabad, farmers along the Musi River areas use the river water for growing para grass, rice and vegetables, and 1–2% of the vegetables are sold to the nearby city market (Sahasranaman 2016). The Government of Tamil Nadu introduced a 'do-it-yourself' kit under the Urban Horticulture Development Scheme in 2014, enabling city dwellers to grow vegetables on open terraces or rooftops of individual houses as well as apartment buildings (Sahasranaman 2016). A UNEP report on Chennai, crafted jointly with M.S. Swaminathan Research Foundation, examines UPA in the context of amplifying urban pressures and climate risks. It aims at diagnosing the potential impacts of imminent threats to the long-term sustainability of the small-scale UPA in Chennai. In Karnataka, the department of horticulture has instituted the Horticultural Producers' Co-operative Marketing and Processing Society Ltd. (HOPCOMS) which ensures proper market channels for fruit and vegetables (Kolady and Krishnamoorthy 2012). Sahasranaman (2016) mentions that slum dwellers in some areas of Cuttack practice organic farming and sell the surplus produce to nearby markets. The government of Kerala has initiated the Kudumbashree project which embraces collective farming processes, aiming to ascertain food production and bring a substantial amount of fallow land back under cultivation, while in 2008, the city corporation of Pune initiated the Farming Project in which citizens were encouraged to produce vegetables as well as other staple crops on allocated land. However, the project was eventually unsuccessful (Sahasranaman 2016).

15.4 SOCIO-NATURAL HYBRIDITY AND URBAN FOOD PRODUCTION

A pool of social science researchers, including critical human geographers worldwide, has argued the possibility and necessity of eliminating the boundary between nature and social, and has demonstrated how nature and society have been materially co-productive through spatially and historically situated processes (Alkon 2013). Interdisciplinary scholars at different times have contended for relationships between social and ecological systems, rejecting the messy compartmentalization of the space we inhabit. Moreover, they have introduced 'network' as a notion of socio-natural imbrications involving multiple nodes of connections and heterogeneous (inter)relationships among human, animate and inanimate entities, along spatio-temporal scales. In this light, more "symmetrical greening" which eliminates the ecocentric as well as anthropocentric grounds of thought, defines socio-nature as a "hybrid" space emphasizing the co-constitutive features of the social and natural along various networking trajectories (Castree & MacMillan 2001).

Urban and peri-urban landscapes made up of agricultural tracts, water bodies, organisms, amusement parks, as well as the city itself, are often envisioned as a socio-natural singularity that is (re)shaped by social and ecological processes with "material and ideological consequences" (Alkon 2013; Swyngedouw 2006). Food, as "humanity's essential component", associates with the ultimate socio-nature in multiple ways. Nonhuman species and landscape attributes, excluding a few wild edibles and food species (Smith 2010), have been (com)modified by human technologies in the form of agriculture and plant breeding, to sustain human bodies. Moreover, the processes of food distribution, as well as food preferences and taboos, are inherently social phenomena (Douglas 2003; Forson & Counihan 2013). Nonetheless, disparate and overlapping interfaces within food safety, public health, food sovereignty and food justice, under the aegis of eco-agrarian ethics, have emerged in support of small-scale farmers who are considered essential for harnessing an array of socioecological benefits, such as pollution control, preparing healthy soil, augmenting rural and urban material exchanges and rendering local economic alternatives, etc. (Alkon 2013).

In current urbanization regimes, urban agriculture is perpetrated by new avenues of injustice relating to widespread corporate invasions of urban economies, land acquisition for commercial farms, and introduction of new seed varieties and manures with nuanced market logics. These are inevitably leading to sound (political) economic means of marginalization of local communities, including farmers and fishers. These symptoms are widespread in cities of developing countries where grounds of subjugation are ornated by differential interests of multiple actors, pluralities in legal systems, privatization agendas in city governance and changing social equations that also reflect a motivated art of un-recognizing and forgetting (Mukherjee 2020). In terms of forging ecologically attuned socio-economic alternatives, urban agriculture contexts, thus, relate to food justice, environmental justice and environmentalism movements for the urban poor in the face of neoliberal governmentality (Baviskar 2003).

15.5 UPA TRAJECTORIES IN KMA

The Kolkata Metropolitan Area (KMA, West Bengal) is a continually expanding urban agglomeration which constitutes municipalities, residential suburbs and peri-urban transition belts which are socio-economically tied to the urban locus. It is the third most populous metropolitan area in India, with a population of 14.11 million according to the 2011 census. The greater Kolkata region falls under the functioning of the Kolkata Metropolitan Development Authority, a statutory planning and development authority for the KMA in West Bengal. The KMA includes four municipal corporations – Kolkata (144 wards), Howrah (50 wards), Bidhan Nagar (41 wards) and Chandannagar (33 wards) – along with 39 municipalities, 2 cantonments, 75 non-municipal towns and 446 rural bodies/villages. In the last few decades, Kolkata district has experienced a substantial increase in the work participation rate from 37.6% (2001) to 39.9% (2011), whereas the percentage of 'main workers' comprising non-cultivators, household and industry workers, has dropped from 94.5% in 2001 to 87.8% in 2011 (Shaw 2016). However, the 'main work force' constitutes 94.6% of total workers. Over the last few years, the city's geographical area has invariably widened, the latest addition having occurred in 2012 when the outgrowth Joka was added to the south-western edge of the city (ibid).

Endowed with a diversity of water bodies including ponds, wetlands, lakes and rivers, KMA also accommodates small-scale aquaculture activities. In KMA, dominant water bodies include the Hugli River, the canals of Narayanpur, Bhadreswar and Ichhapur, and the East Kolkata Wetlands (EKW) occupying the south-eastern section of KMA. A prominent characteristic in most parts of KMA is that fishers are engaged in different farming practices alongside aquaculture. Hence, they are not solely dependent on aquaculture to earn a living. On the other hand, farmers often claim that aquaculture is more profitable than traditional small-scale agriculture. Consequently, they are diversifying subsistence practices by adopting fish cultivation together with the traditional paddy or vegetable farming (Mazumder et al. 2019). Field observations in different parts in KMA reveal that in many households with at least one pond as a pit, the homestead floor is raised above high-water mark to avoid the impacts of floods. In due course, such ponds are turned into a viable source for fishing, especially mixed-species cultivation dominated by carp polyculture and tilapia (*Oreochromis niloticus*). Additionally, cooperative fishing is carried out by groups of farmers and fishers in large and medium-sized ponds for commercial purposes in some areas. Recently, catfish (*Clarias batrachus*) monoculture has emerged as a flourishing practice as they are in heavy demand in the market as livestock feed. In a few UPA sites, this practice extends even to the industrial level. Some farmers are also engaged in shrimp monoculture in the eastern part of KMA.

In the south and south-eastern belt of KMA covering UPA sites like Kalipur, Baragachhia, Sarmatsapur, Sajneberia, Srikrishnapur and Jiadargot, farmers integrate aquaculture with banana plantations as well as vegetable cultivation along raised dykes. However, unimproved ponds and monsoon flooding often pose hindrances to fish culture. Rice and fish cultivation on seasonal farmlands characterize Keutia and Basudebpur UPA sites in KMA's eastern zone. Notably, this stretch is also a part of an enormous floodplain wetland called Bortir Bill, an ecologically rich and biodiverse landscape that remains inundated for a substantial period, especially after the monsoons.

Both within the KMA and at its boundaries, intensive small-scale agriculture is compelled to compete with other land uses including constructional and built-up areas. However, agriculture is the mainstay of thousands of farmers who live in the peri-urban fringes of KMA. Both Aman and Boro rice are cultivated by the farmers, who identify Boro as far more profitable than other varieties. The UPA croplands are spread widely across places like Kalyani, Hugli, Barasat, Naihati, Syamnagor, Dankuni, Srirampur, Sonarpur, Uluberia and Sankrail. As well as cereal crops, vegetables including cauliflower, cabbage, onion, potato, tomato, chili, fenugreek, cucumber, radish and beetroot are also widely cultivated. Places like Noapara and Bighati in the west and Bora, Chandigori, Dhapa in the east of KMA, are the dominant vegetable-producing units, whereas orchards growing mango, guava, banana and papaya are observed in areas such as Baruipur, Goyeshpur, Barasat, Chandernagore and Adisaptogram. Horticulture has been beneficial to farmers as it requires less labor and ensures good exchanges at the city markets. In the UPA areas of Srikrishnapur, Kalipur, Sarmatsapur, Baragachhia, Jayenpur-Hogalkuria, Sajneberia and Jiadargot, most of the land sustains dual aquaculture-orchard and vegetable farming along the water bodies.

Situated at the east of the Kolkata Metropolis, the East Kolkata Wetlands (EKW), a natural wastewater treatment plant, is a remarkable UPA site which meets the city's day-to-day vegetable and fish requirements. It contains 264 sewage-fed fisheries, agricultural units, garbage farms and built-up areas (Mukherjee 2015). Around 750 million liters of wastewater from the city is reused each day in fisheries and croplands which are dispersed within 12,500 hectares of wetland region. Wastewater from the Kolkata metropolis is channeled through interconnecting canals towards this extensive basin area where it debouches into shallow fish ponds or maturation ponds. The sewage is then treated through naturally driven bioremediation processes within the stabilization ponds which operate as solar reactors during the day. With augmentation of the photosynthesis process in presence of sunlight, plankton population starts growing into the *bheri*. The organic matter content in the wastewater serves as food for the plankton. Rapidly sprouting plankton is the source of food for the fish population which subsequently plays a dual role – they stabilize the plankton population in the *bheri*s and convert nutrients steadily into a consumable form, i.e., fish (Figure 15.1). The treated nitrogen and phosphorus-rich water in the *bheri*s is then used for both fish cultivation and irrigating paddy and vegetable fields in the adjacent areas (Mukherjee 2015). Beside this operation zone, the Dhapa dumpsite absorbs most of the municipal solid waste from the city, providing, in turn, a substrate for vegetable cultivation (Figure 15.2). The farmers who sometimes work in the dumpsite as waste pickers prepare the vegetable farm by removing garbage from the soil using hands. They grow vegetables in this garbage substrate almost throughout the year. Embedded within the memories of colonial interventions that took the form of fixing the city's riverine processes, this mechanism of converting the waste into consumable food products is one of the largest in the world (Nandeesha 2002). This robust resource recovery operation is nurtured by local peri-urban communities who derive their sustenance by selling fish and vegetables to the local markets of Kolkata. Additionally, the city receives about one-third of its everyday fish requirement, amounting to about 11,000 metric tons per annum. The EKW area also generates 150 metric tons of vegetables each day, including tomato, brinjal, potato and cabbage.

15.5.1 Practices and Encounters

During on-site conversations, the farmers mentioned that crop types and number of crops produced annually have changed little in the last 25 years. On average, 4 or 5 types of pesticides and fertilizers are used by farmers to avoid crop diseases and plant infection. Moreover, an increasing trend in applying pesticides has been confirmed by farmers from the Jamtagarh and Magra sites, whereas an opposite trend is observed in the Bijna, Uttar Panchla, Keotia, Keyapukuria and Uttar Chandmari areas. Farmers from various parts of KMA added that they adopt four main types of measures to regain soil fertility: mulching, crop rotation, cover crop generation and organic farming. However,

FIGURE 15.1 Fishery ponds of EKW beside Dhapa dumpsite.

FIGURE 15.2 Vegetable cultivation at Dhapa UPA site.

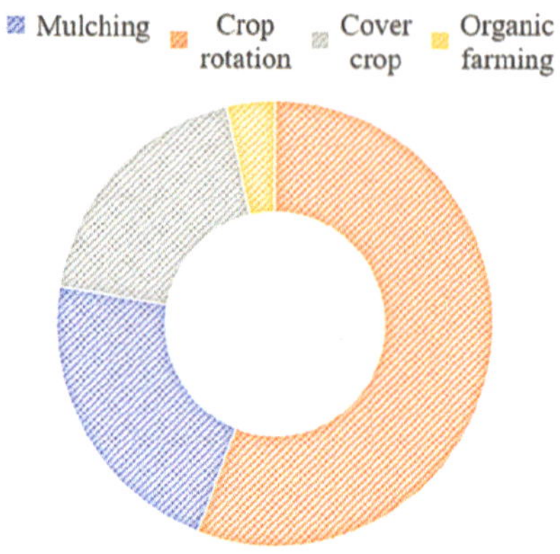

FIGURE 15.3 Methods used by farmers for maintaining soil fertility.

the zero tillage method is not yet popular in the UPA region (Figure 15.3). Organic mulching is observed commonly in the form of grass clippings and bark chips applied to the soil in Dihi, Jardah, Uttar Roypur, Koyrapur, Langolberiya, Nihata, Bijna, Jagannathbati and Noapara areas. While crop rotation is practiced in almost all the UPA sites except some areas of Dhapa, North Roypur, Baghgachhiya, Keotia and Uttar Panchla, cover cropping is exercised by growing cover plants like legumes, grasses and shelter trees to protect the crops and soil, mostly in the Jamtagarh, Madpur, Dashbhanga and Chhota Garji sites. In some UPA areas like Keyapukuria, Demrar Math, Uttar Roypur, Jardah and Dhapa, farmers often apply wastewater (above 80% of the total irrigation) to grow crops, especially vegetables. Additionally, groundwater irrigation is the main technique for irrigation during the dry months. Interviews with farmers reveal the pace and patterns of land-use conversions in the last 50 years, particularly due to the rapid urban sprawl and encroachments on peri-urban lands.

According to most of the farmers (74% of the total interviewed), conversion of farmlands to built-up areas is the prime threat to the survival of UPA. Some of the sites, such as Chandigori, Dubgariya, Chak Bansberia, Panchla and Keyapukuria, recorded substantial instances of conversions, such as farmland to water bodies (for aquaculture purposes) and built-up areas. Apart from Dubgariya, Damrar Math and Uttar Panchla, almost all UPA sites including Dhapa, Dihi, Jardah, Uttar Roypur, Kalikapur, Keotia and Jamtagarh accounted for the loss of farmland owing to the development of real-estate businesses, commercial and industrial buildings over the last few decades. Some parts of Keotia, Dubgariya, Koyrapur and Sankrail Jala are experiencing a trend of farmland being converted to roads, leading to a reduction in average landholding size (Figure 15.4).

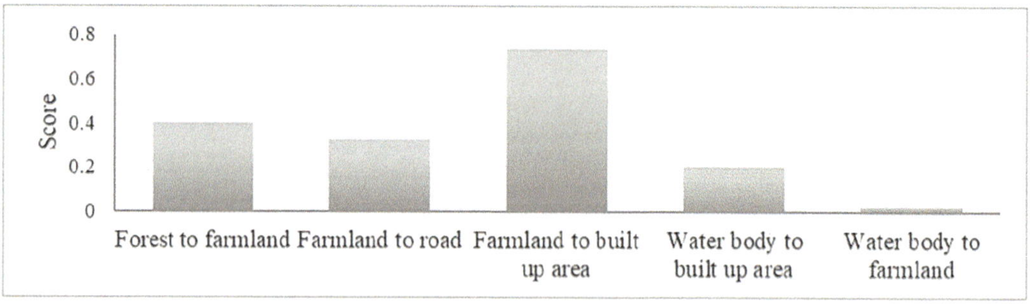

FIGURE 15.4 Land-use and habitat conversion patterns based on scores given to farmers' responses.

15.6 GEOSPATIAL ANALYSIS OF LAND-USE CHANGES

In order to identify different land-use types and trace the degree of land-use transformations and habitat modification in UPA regions, a land-use land cover (LULC) map has been produced, using the supervised signature extraction technique. The maximum likelihood algorithm was chosen for classifying Landsat-VIII digital data (April 2011). Prior to the pre-processing and classification of the satellite imagery, a rigorous field survey was conducted to accurately determine locational point data for each category included in the LULC classification scheme. This also helped create training sites for signature generation. At the processing stage, the satellite data was enhanced using ERDAS Imagine software to obtain the desired image quality and enhance classification accuracy. Based on interpretation of the remote sensing imagery and field-based experiences, seven LULC categories were established – water body, agriculture, built-up area (urban), orchard, rural area, aquaculture and bare earth.

The LULC map (Figure 15.5) informs about the dominance of the urban built-up area, whereas agriculture is primarily dotted along the fringes of northern and south-eastern KMA. Built-up

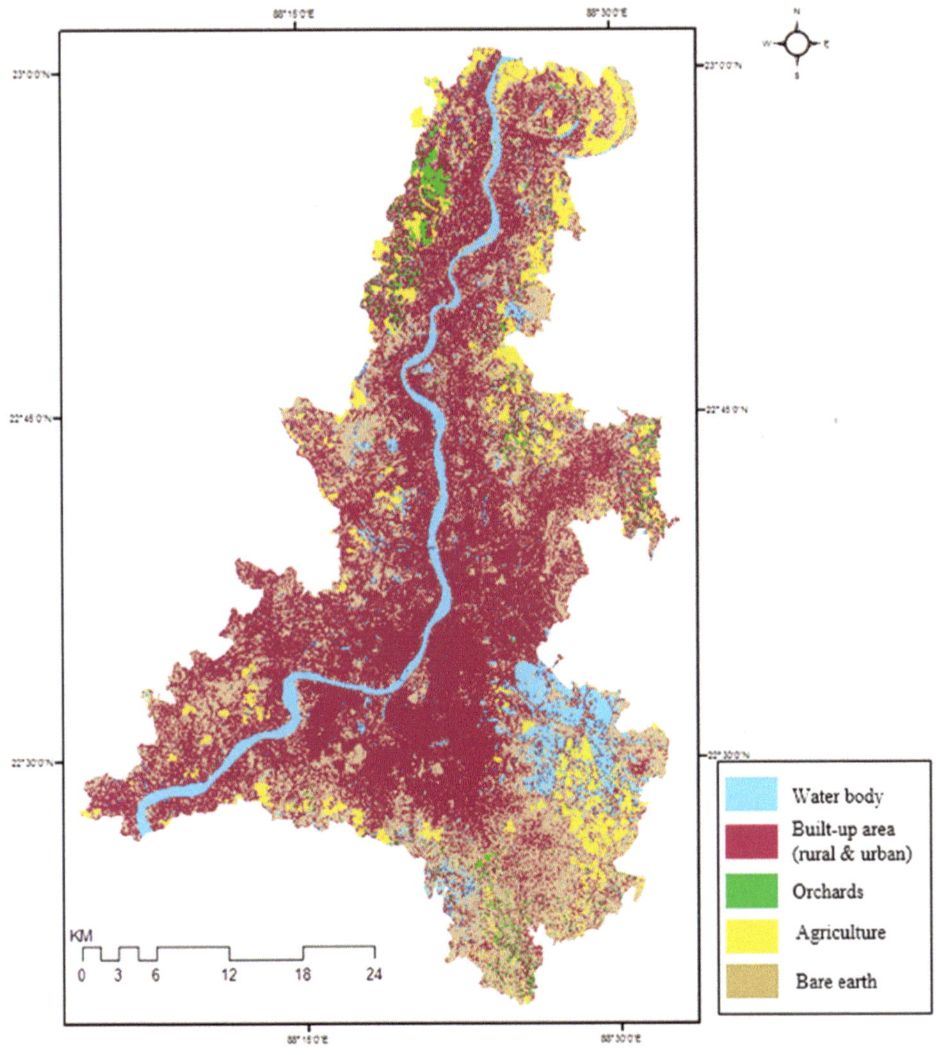

FIGURE 15.5 Land-use land cover map (LULC) of KMA.

TABLE 15.1
Areas under Each LULC Class

LULC Class	Area in Sq. Km	Area in Percentage
Built-up	845.76	51.40
Water body	126.27	7.67
Agriculture with crop	110.77	6.73
Horticulture and other open vegetation	27.63	1.67
Bare earth/ fallow land	534.96	32.51

Source: Author.

areas, including both high-density urban agglomeration and low-density peri-urban settlements, occupy a greater proportion than other land uses. Furthermore, ponds with plantations are represented by the squares or rectangles adjacent to the red-tinted residential areas. Built-up areas are mostly surrounded by agricultural lands which are used for cultivation of rice, jute, oilseed and vegetables such as tomato, onion and cabbage. As the city is moving outwards, farming is seemingly practiced very sparsely in the fragmented lands, occupying only 6.73% of the total area (Table 15.1). Orchards mixed with a variety of shrubs and young trees are seemingly situated along the settlement areas and roads. Spare land with no vegetation is known as bare earth, whereas fallow lands belong to the total land under cultivation, but are temporarily out of cultivation. This category includes areas with sparse vegetation cover like racecourses, golf courses, other open spaces, marshy lands and transactional lands that are likely to be converted to other usages in the near future.

In the Haora district in south-western KMA, UPA areas are small and patchy due to emerging settlements and industries. On the other hand, agricultural lands on both sides of the Durgapur and Andul expressway are being transformed into fallow lands or industries, hotels and the like (Figure 15.6). Landless agricultural labourers and those who are unable to afford land due to skyrocketing prices offered by industrial and housing projects, are gradually advancing towards silent displacements. Furthermore, the cost of living for small-scale their average income, leading to high poverty levels. Nabharanda/Surjyapur, west of Choto Kanthalia, is an example of a fragmented UPA site (Figure 15.7) which is currently fallow mainly due to scarcity of water. The laboratory-based evaluation of surface water suggests that the quality of surface water used for irrigation is deteriorating in many UPA sites with BOD (mg/L) levels being highest at places like Kalikapur, Surjyapur, Bijna and Nihata (Figure 15.8) while escalating pollution levels in the city caused by emissions from cars and industries degrade the soil quality with heavy metal depositions. Soaring demand for land for large-scale infrastructure projects compels farmers to abandon agriculture and take up local small businesses or work as hired laborers. Many farmers are venturing into more profitable agricultural options like horticulture, pisciculture and cash-crop cultivation to cope with land and market competitions. According to Gowda and colleagues (2012), in UPA and rural agriculture, a farmer's per-capita income is at least 50% lower than the per-capita income of an average Indian, whereas shifting to other businesses or employment in the city can earn them at least Rs. 250 to 300 per day. In reality, farmers regard agriculture as no longer profitable, and often sell their land whenever a suitable price is offered.

Farmers from the peri-urban areas face various limitations, including the absence of training support for monitoring soil and water quality; knowledge of the utility and disadvantages of market-introduced seed types as well as fertilizer composition; viable tools and techniques for land and pond reformation, etc. Moreover, the involvement of too many intermediaries within the value-chain processes gives rise to two problems: consumers are compelled to pay higher prices and the farmers are deprived of their rightful remittance as a dominant share of profit goes to the intermediaries.

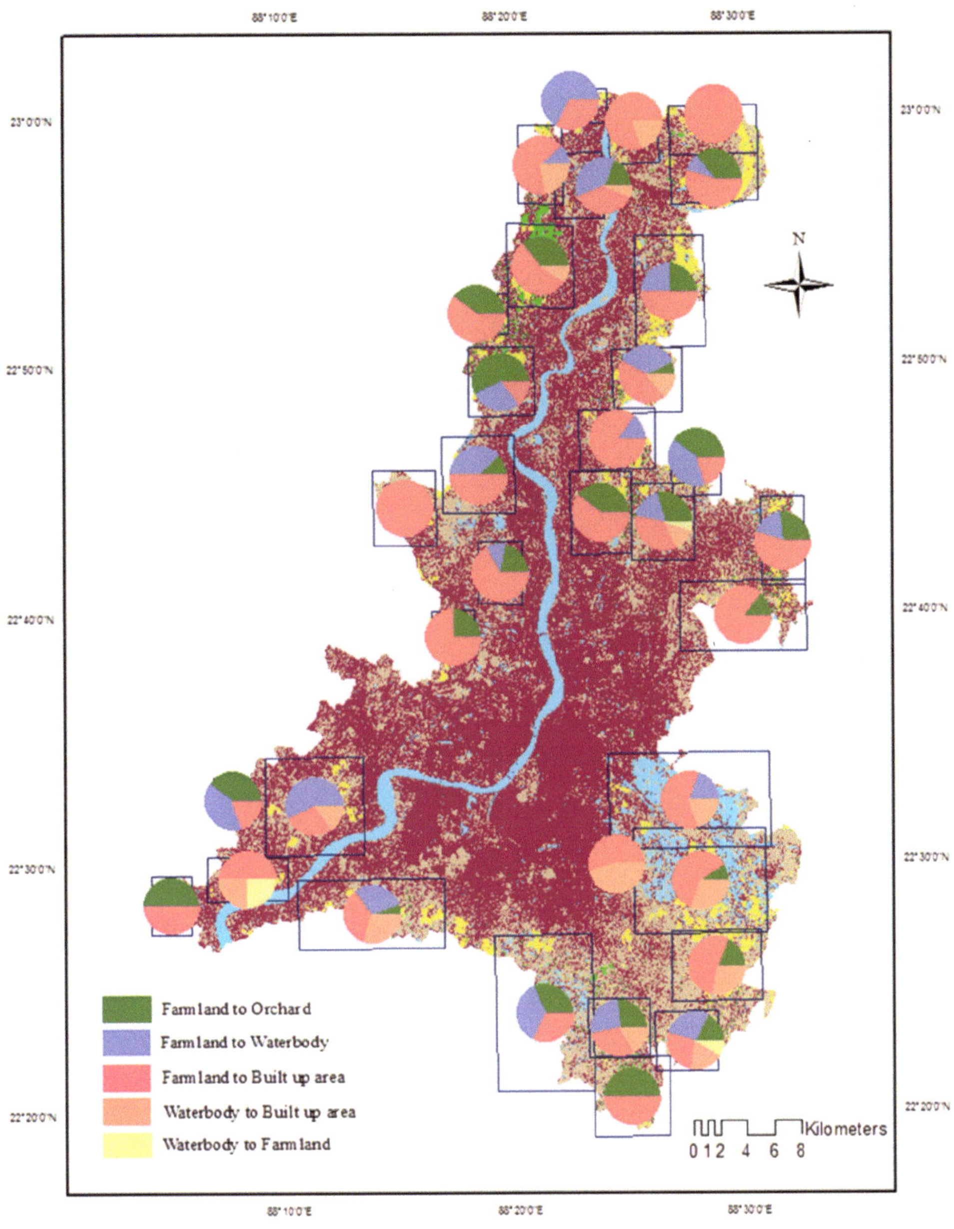

FIGURE 15.6 Changes in land-use practices at KMA based on farmers' perceptions.

In addition to conflicts over land resources, lack of policy supports surrounding farmers' wellbeing, subsidies, property resources, input supplies, poor-quality fish seeds, huge interest rates for credit, and theft of fish produce are some of the key drivers of livelihood challenges to the farmers. The increasing flow of pesticides towards the rivers and canals is also leading to acute eutrophication of ponds (Edwards 1993).

The contexts of UPA relate to modern forms of "capitalist organization and work discipline which are dependent on homogenizing, appropriating and partitioning of urban space" (Alonso 1994). Not only are the urban and peri-urban interfaces fragmented ecologically by the 'hardware' of the city, but marginalization of local communities ahead of their traditional knowledge systems, ecosystem-based practices, values and networks, (re)affirms the evolving nature of urban social

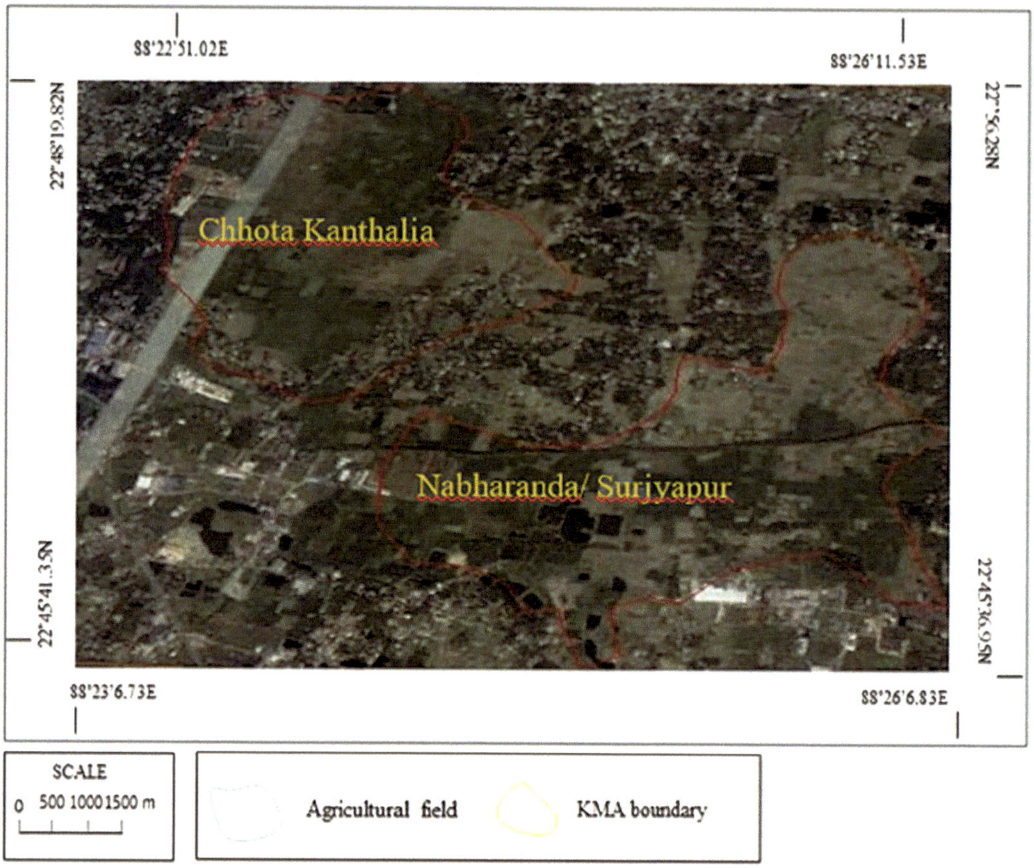

FIGURE 15.7 Chhoto Kanthalia and Surjyapur UPA sites.

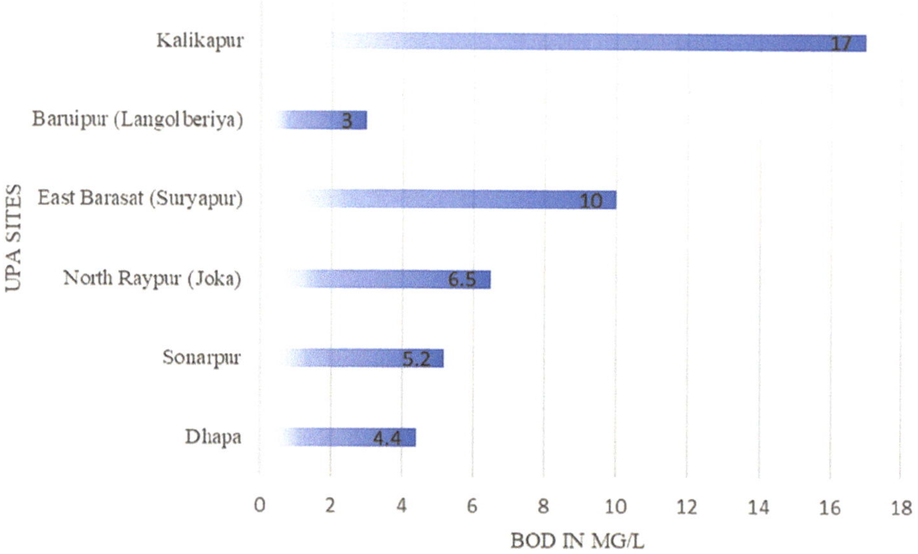

FIGURE 15.8 Biological oxygen demand (BOD in mg/L) in surface water used for irrigation in different UPA sites.

(political) space. Baviskar's (2003) novel idea of "bourgeois environmentalism" emphasizes how the dream of a clean and green city that the "privileged" enjoy remains at variance with the persistent struggles of the peri-urban people who are carrying out ecologically sensible economic exchanges amid proliferating challenges from the powerful sides. Together with fast-fading memories and the refusal to recognize what has been described as "cognitive apartheid" by Kolkata-based ecologist Dhrubajyoti Ghosh in his account of the civilities of the urbanites, downright neoliberal technologies and privatization policies of the governance systems have obscured the stories of peri-urban farmers and fishers in Kolkata.

15.7 CONCLUSION

In light of the present status of UPA, urban places ought to be included under the lexicon of urban resilience, which emphasizes 'urban' as a complex construct of historical and socioecological systems (Mukherjee 2015). Moreover, critical exploration around social hierarchy and networks among multiple actors is required in order to understand policy-making processes and further, gain insights into viable pathways for restoring peri-urban livelihood generation activities. With the propagation of urban squatters and surging poverty levels, it is crucial to shed light on food production avenues and food justice in the face of sharp inequalities across urban resource use and consumption. Notably, UPA has a large scope for contributing to (re)establishing a "conscious metabolic relationship" between human and environment in relation to both intellectual and manual labor. This chapter, thus, examines the scope and challenges of UPA in the context of Kolkata as an unequally urbanizing city, and at the same time strongly emphasizes co-management that accommodates institutional supports, recognition of local knowledge systems and citizens' awareness. By enabling and activating the visions of different stakeholders such as farmers, fishers, researchers, activists, private-sector professionals, citizens and government officials, co-management as a cohesive approach can acknowledge the cooperation between rural-urban functional units and (re)orient the decision-making processes towards efficient and inclusive local-level solutions.

REFERENCES

Alkon, Alison Hope. "The socio-nature of local organic food." *Antipode* 45, no. 3 (2013): 663–680.

Alonso, Ana Maria. "The politics of space, time and substance: State formation, nationalism, and ethnicity." *Annual Review of Anthropology* 23, no. 1 (1994): 379–405.

Awasthi, Pranati. "Urban agriculture in India and its challenges." *International Journal of Environmental Science: Development and Monitoring (IJESDM)* 4, no. 2 (2013): 48–51.

Aubry, Christine, Josélyne Ramamonjisoa, M-H. Dabat, Jacqueline Rakotoarisoa, Josette Rakotondraibe, and Lilla Rabeharisoa. "Urban agriculture and land use in cities: An approach with the multi-functionality and sustainability concepts in the case of Antananarivo (Madagascar)." *Land Use Policy* 29, no. 2 (2012): 429–439.

Baviskar, Amita. "Between violence and desire: Space, power, and identity in the making of metropolitan Delhi." (2003).

Castree, Noel, and Tom MacMillan. "Dissolving dualisms: Actor-networks and the reimagination of nature." *Social nature: Theory, Practice, and Politics* (2001): 208–224.

Corbould, Claire. "Feeding the cities: Is urban agriculture the future of food security." *Future Directions International* (2013). https://www.futuredirections.org.au/wp-content/uploads/2013/11/Urban_Agriculture-_Feeding_the_Cities_1Nov.pdf

De Zeeuw, Henk, René Van Veenhuizen, and Marielle Dubbeling. "The role of urban agriculture in building resilient cities in developing countries." *The Journal of Agricultural Science* 149, no. S1 (2011): 153–163.

Douglas, Mary. *Purity and Danger: An Analysis of Concepts of Pollution and Taboo*. London and New York: Routledge, 2003.

Edwards, Peter. "Environmental issues in integrated agriculture-aquaculture and wastewater-fed fish culture systems." *Environment and Aquaculture in Developing Countries* 31 (1993): 139–170.

Elizabeth Kolady, Deepthi, and Sujana Krishnamoorthy. "9 marketing cooperatives in a case study of HOPCOMS in Karnataka, India." In Biénabe, E., Berdegué, J., Peppelenbos, L., & Belt, J. (eds),

Reconnecting Markets: Innovative Global Practices in Connecting Small-Scale Producers with Dynamic Food Markets London and New York: Routledge (2012): 125.

Forson, Psyche Williams, and Carole Counihan. *Taking Food Public: Redefining Foodways in a Changing World.* London and New York: Routledge, 2013.

Gowda, UC Ramalinge, M. G. Chandrakanth, P. S. Srikanthamurthy, C. G. Yadav, N. Nagaraj, and Channaveer. "Economics of peri-urban agriculture: Case of Magadi Off Bangalore." *Economic and Political Weekly* XLVII (4) (2012): 75–80.

Indraprahasta, Galuh Syahbana. "The potential of urban agriculture development in Jakarta." *Procedia Environmental Sciences* 17 (2013): 11–19.

Lovell, Sarah Taylor. "Multifunctional urban agriculture for sustainable land use planning in the United States." *Sustainability* 2, no. 8 (2010): 2499–2522.

Maxwell, Daniel, Carol Levin, and Joanne Csete. "Does urban agriculture help prevent malnutrition? Evidence from Kampala." *Food Policy* 23, no. 5 (1998): 411–424.

Mazumder, Shrabana, Debajit Datta, Mrinmoyee Naskar, and Raktima Ghosh. "Urban and Peri-urban agriculture in Kolkata metropolitan area: An assessment of status and sustainability". *Contemporary Social Sciences* 207 (2019): 207–222.

McClintock, Nathan. "Assessing soil lead contamination at multiple scales in Oakland, California: Implications for urban agriculture and environmental justice." *Applied Geography* 35, no. 1–2 (2012): 460–473.

Mohapatra, Satya P. "Promoting urban agriculture in India." *Seronica Journal* 3, no. 2 (2012): 17–27.

Mougeot, Luc JA. "Urban agriculture: Definition, presence, potentials and risks." *Growing Cities, Growing Food: Urban Agriculture on the Policy Agenda.* Cities Feeding People Series. Report 31 1 (2000): 42.

M. S. Swaminathan Research Foundation (MSSRF). Twenty-First Annual Report 2010-2011. (2010). Chennai. http://www.mssrf.org

Mukherjee, Jenia, and Gorky Chakraborty. "Commons vs commodity: Urban environmentalisms and the transforming tale of the East Kolkata Wetlands." *Urbanities-Journal of Urban Ethnography* 6, no. 2 (2016): 78–91.

Mukherjee, Jenia. "Sustainable flows between Kolkata and its peri-urban interface." In Allen, A., Lampis, L., & Swilling, M. (eds), *Untamed Urbanisms.* London and New York: Routledge (2015): 33–49.

Mukherjee, Jenia. *Blue Infrastructures: Natural History, Political Ecology and Urban Development in Kolkata.* Singapore: Springer Nature, 2020.

Nandeesha, Mudnakudu Channabasappa. Farmers as scientists: Sewage-fed aquaculture systems of Kolkata: A centuryold innovation of farmers. *Aquaculture Asia* 7, no. 2 (2002): 28–32.

Peng, Jian, Zhicong Liu, Yanxu Liu, Xiaoxu Hu, and An Wang. "Multifunctionality assessment of urban agriculture in Beijing City, China." *Science of the Total Environment* 537 (2015): 343–351.

Sahasranaman, Meera. "Future of urban agriculture in India." *Institute for Resource Analysis and Policy* 2, no. 10 (2016): 1–24.

Shaw, Annapurna. "The informal sector in Kolkata metropolitan area: Appraisal and prospects for local economic development." In *Spatial Diversity and Dynamics in Resources and Urban Development*, pp. 499–516. Dordrecht: Springer, 2016.

Shillington, Laura J. "Right to food, right to the city: Household urban agriculture, and socionatural metabolism in Managua, Nicaragua." *Geoforum* 44 (2013): 103–111.

Smart, Jessie, Etienne Nel, and Tony Binns. "Economic crisis and food security in Africa: Exploring the significance of urban agriculture in Zambia's Copperbelt province." *Geoforum* 65 (2015): 37–45.

Smith, Neil. *Uneven Development: Nature, Capital, and the Production of Space.* Athens and London: University of Georgia Press, 2010.

Swyngedouw, Erik. "Metabolic urbanization: The making of cyborg cities." In *In the Nature of Cities*, pp. 36–55. USA and Canada: Routledge, 2006.

Symoens, Jean-Jacques and Jean-Claude Micha. Management of integrated freshwater agro-piscicultural ecosystems in tropical areas. In *Seminar on The Management of Integrated Freshwater Agro-Piscicultural Ecosystems in Tropical Areas.* Belgium: Royal Academy of Overseas Sciences, 1995.

Tornaghi, Chiara. "Critical geography of urban agriculture." *Progress in Human Geography* 38, no. 4 (2014): 551–567.

Tsuchiya, Kazuaki, Yuji Hara, and Danai Thaitakoo. "Linking food and land systems for sustainable peri-urban agriculture in Bangkok Metropolitan Region." *Landscape and Urban Planning* 143 (2015): 192–204.

Vazhacharickal, Prem Jose, and Andreas Buerkert. "Sustainable cities: An overview of the urban and peri-urban agricultural production in Mumbai Metropolitan Region (MMR)." *Leituras de Economia Política* 19 (2012): 69–87.

16 Degraded Land Management in Urban Areas

The Case of Biodiversity Parks in Delhi

Himanshhu Panwar and Meenakshi Dhote
School of Planning and Architecture, New Delhi, India

CONTENTS

16.1 BACKGROUND

It is often said that agriculture initiated the process of ecological degradation. Widespread agriculture – together with a population increase (Bocquet-Appel 2011) – can be traced back to the Neolithic period, some 10,000–7,500 years ago (Dotterweich 2008; Butzer 2005). The development of agriculture and changes in farming practices led to the conversion of forests and grasslands to farmland. Increased

DOI: 10.1201/9781003131922-19

reliance on domesticated grasses for food production has resulted in soil degradation (Steffen 2005). With the rapid pace of urbanization, these altered ecosystems have contributed to increasing land degradation (IPCC 2019).

About half of the human population resides in urban areas, and according to 2015 UN estimates, this is set to increase to 70% by 2050. This pace of urbanization poses a threat to ecosystem services (Seto 2012). The urban planning process rarely considers the maintenance of ecosystem services or avoidance of land degradation. The focus is more on economic development and creating space for construction. This is resulting in increasing air and water pollution, and the degradation of areas under agriculture and ecosystems – indigenous, natural or semi-natural – both inside and outside urban regions.

The growth of urban areas has modified water cycles, encroached on wetlands and farmlands, and dispersed species, resulting in modified landscapes. About 200 times the city's area is required to provide food, energy, and materials (Cherlet et al. 2018). Seventy per cent of the world's carbon emissions that contribute to global warming are due to production and consumption activities in cities. Also, urban areas are sources of aerosols leading to the alteration of regional precipitation and concentration of particulate matter such a PM2.5 in the atmosphere. The growth of urban areas changes land use, land cover and subsequently, surface temperatures. Cities alter the environment through the utilization of paved surfaces, far and wide, which prevents the infiltration of water into the soil, promoting urban flooding (Cherlet et al. 2018). Despite the scale of these impacts, it is likely that the importance of urbanization in driving biodiversity loss is significantly underestimated and underreported (Bulkeley, Kok and Xie 2021).

In 2016, ISRO prepared a report entitled "Desertification and Land Degradation Atlas of India" (ISRO 2016). The Land Degradation Atlas of India records 29.3% of the country undergoing land degradation between 2011 and 2013. This was 0.57% higher than the figures between 2003 and 2005. The main factors for land degradation during both periods have been water erosion (10.98%), vegetation degradation (8.91%), and wind erosion (5.55%) (ISRO 2016).

The NCT of Delhi ranks third with respect to level of desertification in the country. 6.6% of the state is under desertification and it has also reported the highest increase in land degradation, amounting to 11.03% from 2003–2005 to 2011–2013. In Delhi's case it isn't strictly desertification, but mainly land degradation arising from development that transformed productive areas into non-productive ones (Jain 2018).

This chapter discusses the impacts and causes of land degradation in urban areas, along with techniques used for restoring degraded lands in Delhi. The biodiversity parks that have been designed for this purpose have played an important role in reviving the flora and fauna of the city for the past two decades.

16.2 METHODOLOGY

The chapter begins by explaining the background to and causes of land degradation, and the transformations that have occurred with rising urbanization. It outlines the level of land degradation faced by the city of Delhi over the past two decades. A literature study looks at the reasons for and impacts of land degradation in urban areas and the importance of restoring degraded land. It was observed that one of the successful ways to restore land in urban areas is by developing biodiversity parks. The development of biodiversity parks with their associated ecosystem services is considered, and the seven designated biodiversity parks in the city of Delhi are discussed, with an emphasis on the role they have played in restoring different types of degraded urban ecosystems, especially ridge areas and wetlands. The major restoration techniques used for developing the biodiversity parks are highlighted, with case studies of different sites with similar challenges. The changes that occurred in each park are explored, with particular reference to their impact on the diversity of flora and fauna across sites. The chapter concludes with the need to restore degraded lands, and ways this can be achieved, considering the urban challenges.

16.3 LITERATURE STUDY

The Special Report on Climate Change and Land (SRCCL) defines land as "the terrestrial portion of the biosphere that includes the natural resources, the ecological processes, topography, human settlements and infrastructure that operate within that system". IPCC (2019) defines land degradation as a negative trend in land conditions caused by direct or indirect human-induced processes, anthropogenic climate change, which result in large-scale and long-term impacts of any one of the following: biological productivity, ecological integrity or value to humans. The dynamics of the factors impacting degraded land in urban areas, as well as their direct and indirect repercussions, must be explored to comprehend the restoration measures.

16.3.1 Reasons for Land Degradation in Urban Areas

The pace of land degradation varies from rapid to very slow depending on the causative factors: short and intensive events such as rain storms lasting for 10 minutes, removing topsoil or triggering a gully or landslide (Coppus and Imeson 2002) to slow depletion of nutrients or loss of soil particles (Johnson and Lewis 2007). In India the causative factors of land degradation are mainly deforestation and improper agricultural practices, such as the extension of cultivation to marginal lands, improper crop rotation, improper soil conservation practices, imbalanced use of fertilizers and excessive surface irrigation. Socio-economic reasons include the paucity of land, economic pressure and poverty, leading to over-cultivation. It has been observed that excessive use of fertilizers has resulted in excess nutrients which leach into groundwater, causing its contamination. Improper use and maintenance of canal irrigation, especially in arid and semi-arid areas, has resulted in issues of water logging and salination.

16.3.2 Impacts of Land Degradation

The impacts of land degradation take a number of forms. They may be ecological impacts, such as loss of vegetation, soil erosion, floods, landslides and avalanches, loss of biodiversity, etc. Socio-economic impacts include loss of soil productivity resulting in rural–urban migration. Loss of vegetation leading to soil erosion results in siltation of reservoirs, reducing the life span of hydroelectric dams. Urbanization and associated development activities lead to increasing ecological imbalances, such as a declining groundwater table, land and water pollution, salination, waterlogging, etc.

16.3.3 Need for Restoring Degraded Land

Controlling the social and ecological consequences of changing biodiversity involves preserving what is left by enforcing conservation efforts. Restoration and management of degraded lands are achieved mainly by adopting agronomic measures, soil management methods and mechanical methods. These involve managing vegetation cover, tillage and nutrient supply, and lasting changes to the landscape (IPCC 2019).

Conservation techniques for rescuing flora and fauna are mainly executed in protected areas, such as biosphere reserves, National Parks, sanctuaries and the interiors of forests. While these have helped rescue many species, it does not keep pace with degradation (Koul 2017). Also, while gene banks, and botanical and zoological gardens have helped rescue species they cannot help restore ecosystems in the long term as they do not replicate naturally occurring interaction webs. Since there are no estimates of biodiversity available outside protected areas, we may lose a majority of species. Estimation of species richness should be undertaken to determine the suitability of habitats for biodiversity conservation across the landscapes and in specifically managed lands in urban and suburban areas. There is therefore an urgent need to measure biodiversity while addressing the aspirations and requirements of the urban population.

The concept of biodiversity parks was an outcome of the extensive experience of ecologists and conservation biologists in restoring ecosystems across India. They felt the need to retain some degree of wilderness in city space, which they envisaged as biodiversity parks. These parks would conserve biodiversity in urban and semi-urban areas and would address the three basic objectives of CBD.

16.3.4 Biodiversity Parks

Biodiversity parks are landscapes or riverscapes of wilderness where ecological communities of native species are recreated over marginal/degraded landscapes/riverscapes. Biodiversity parks are based on the ecological restoration principle of establishing self-sustaining ecosystems with biodiversity that generate ecological services contributing to human well-being (Babu 2012).

Biodiversity parks in riverscapes incorporate restored/recreated river ecosystems along degraded stretches of rivers. Biodiversity parks in floodplains include the restoration/recreation of diverse landscape elements such as wetlands, marshes, swamps, lakes, forests and grasslands, besides riparian ecosystems and in-stream communities. It also includes the development of greenways along embankments, forest communities on adjacent uplands, treatment wetlands for cleaning river water and constructed wetlands for treatment of sewage and industrial effluents that enter rivers. The biodiversity park approach is an innovative model for the recreation of lost biodiversity or natural heritage, and it is a conservation approach. It involves the conservation of ecosystems, communities, species and populations, and simulates National Parks/wildlife sanctuaries/nature reserves/wilderness.

Biodiversity parks have a wide range of functions and encompass almost all the four categories of ecosystem services rendered by ecosystems, including: (i) improving the human microbiome, as the parks harbor a rich environmental microbiome, which enhances human health; (ii) serving as habitat for vanishing flora and fauna; (iii) storing flood water and recharging ground water; (iv) preventing soil erosion and stabilizing floodplains; (v) reducing flood water velocity; (vi) serving as a hub for protection, awareness and cultural activities; (vii) promoting ecotourism; (viii) connecting the city and its citizens to nature and biodiversity; (ix) providing livelihoods to local communities; (x) sequestering CO_2 and imparting climate resilience, buffering local weather and even causing local precipitation; (xi) serving as filters for point and nonpoint source of air pollutants (xii); purifying water; (xiii) enhancing biological productivity; (xiv) sustaining river ecosystems; and (xv) rejuvenating rivers.

16.4 BIODIVERSITY PARKS OF DELHI

Delhi city has seven designated biodiversity parks. The Yamuna and Aravalli Biodiversity Parks in NCT Delhi include contain 50 threatened biotic communities of the Yamuna River Basin and Aravalli Hills. These parks are the first of their kind in the country and serve as ecological models for natural heritage conversation in urban areas. The Yamuna and Aravalli Biodiversity Parks serve as nature reserves and habitats for migratory and resident bird species, and have more than 3000 species of flora and fauna. The other five parks are Neela Hauz, Tilpath Valley, Northern Ridge, Tughlaqabad, and Kalindi Biodiversity Park. All the biodiversity parks have unique features and present challenges in terms of their landscapes and the causes of degradation that have determined their present conditions. Aravali, Northern Ridge and Tilpath Valley Biodiversity Parks are located on ridges whereas other biodiversity parks involve reclamation of degraded wetlands (See Table 16.1).

TABLE 16.1
Biodiversity Parks in Delhi and Their Conditions before Restoration

Sno.	Biodiversity Park	Area (ha)	Degraded Land Condition (Before Restoration)	Work Started In	Site Characteristics	
					Wetlands	Ridge
1	Yamuna Biodiversity Park	184	Barren land due to salt accumulation	2005	■	
2	Aravalli Biodiversity Park	282	Degraded land due to past mining and infested with invasive species	2010	■	■
3	Neela Hauz Biodiversity Park	3.90	Dumping ground for construction material and sewage water	2015	■	
4	Tilpath Valley Biodiversity Park	70	Degraded land due to sand/stone/murrum mined quarries	2015		■
5	Northern Ridge (Kamla Nehru Ridge)	122	Infested with invasive species *Prosopis juliflora*	2014		■
6	Tughlaqabad Biodiversity Park	130	Polluted lake due to industrial waste	2016	■	
7	South or Kalindi Biodiversity Park	115	Sewage sludge pond	2020	■	

Source: Compiled from DDA.

16.5 RESTORATION TECHNIQUES

Restoration in the context of biodiveristy parks is best explained by Santurf (2005) as the restoration of a degraded forest back to its undisturbed, natural and historic condition. Urbanization has significantly modified, depleted and chemically changed the soils, exposing them to pollution and compaction, leading to a lack of microbial organisms for plant growth. In the restoration process these soils needed to be cost-effectively cleaned up and decontaminated – in other words, amelioration biophysically, chemically and aesthetically (Elmqvist et al. 2015).

A variety of approaches have been used for restoration in Delhi's biodiversity parks, depending upon the challenges each site faced and the amount of degraded land present (Table 16.2). The revival process starts with a baseline analysis of the degraded site which covers the need for restoration and the probable actions required to restore it (Veblen, Porensky and Young 2019). The causes of degradation need to be analyzed, the likelihood that the degradation can be reversed or lessened considered, and how restoration can be accomplished determined. Some of the restoration techniques implemented in the case of biodiversity parks in Delhi are discussed below:

16.5.1 Soil Treatment

Land degradation is mainly caused by soil erosion which is due to the removal of top soil and its organic content and increases in acidity or alkalinity/salinity, making soil toxic and unfit for agriculture and other uses. Waterlogging due to flooding and excessive irrigation also leads to a rise in toxic subsoil chemicals which are brought to the surface by evaporation

In the case of Yamuna Biodiversity Park, the major challenge was the salinity of the soil, due to which no vegetation was able to thrive. The soils of the Yamuna floodplain have high levels of salinity. Instead of removing the entire soil strata, plants and planting schedules were used to mitigate the effects. Grasses that absorb or retain soil were planted first, such as *Suaeda pluticosa, Bothriochlora, Lepto chloaeptocoa, sporobolusmos*, etc. Apart from phytoremediation, gypsum was also added to the soil to provide stability to the organic matter and aggregates present in the

TABLE 16.2
Restoration Techniques Used at Each Biodiversity Park

Sno.	Aspect/Functions	Techniques of Restoration	Biodiversity Parks A	B	C	D	E	F	G*
1	**Lake rejuvenation and Sewage Treatment**	Wetland construction			■			■	■
		In-situ biological remediation	■						
		Desilting	■		■				■
		Dredging	■		■				■
2	**Soil Treatment**	Farmyard manure	■	■				■	
		Vermi-compost leaf mold	■						
		Phytoremediation: introduction of leguminous species	■						
3	**De-weeding**	Uprooting invasive species		■		■		■	
		Canopy opening		■		■	■		
		Cut rootstock method		■		■	■	■	
4	**Development of Infrastructure**	Construction/strengthening of boundary wall	■	■	■	■		■	■
		Laying of nature trails	■	■	■	■		■	■
5	**Water Conservation**	Series of loose boulder check dams across gullies			■			■	■
		Culverts			■		■		■
		Channels					■	■	
6	**Compost Pit**	On-site compost pit		■				■	
7	**Nursery**	On-site nursery	■	■				■	

Source: Compiled from DDA.
* *Note:* Techniques of restoration at Neela Hauz Biodiversity Park would be replicated at Kalindi kunj (Adak 2019).
A – Yamuna Biodiversity Park; B – Aravalli Biodiversity Park; C – Neela Hauz; D – Tilpath Valley Biodiversity Park; E – Northern Ridge (Kamla Nehru Ridge); F – Tughlaqabad Biodiversity Park; G – Kalindi Biodiversity Park. The restoration work for Kalindi Biodiversity Park is under process and the proposed techniques for revival is similar to the ones used at Neela hauz Park.

soil, help water penetration and promote seed emergence. When the salinity was neutralized, farmyard manure and vermicomposting leaf mold were provided to enrich the nutrients in the soil. In Tughalaqabad Biodiversity Park, where the issues were a polluted water body and invasive species, the soil was prepared only with farmyard manure produced on site.

16.5.2 Wetland Revitalization

A wetland is one of the most productive ecosystems, because it includes a transition between aquatic and terrestrial ecosystems (Ghermandi et al. 2008). Five of the biodiversity parks restored in Delhi have wetland ecosystems that had either dried up or become polluted due to sewage and material deposition.

Neela Hauz Biodiversity Park situated in south Delhi was a small depression to which the Southern Ridge Forest drained. It became filled up with solid waste and untreated sewage from surrounding colonies. In order to restore the lake, the lake was first desilted and a wetland was constructed to treat the water before it entered the lake. This process controlled the raw sewage that used to enter the Neela Hauz and then the Sanjay Van and finally the river Yamuna, using zero energy. The same technique was used in the Tughlaqabad Biodiversity Park, and the recently added Kalindi Biodiversity (or South Biodiversity Park) would also be conserved on the same principles.

For Yamuna Biodiversity Park, where the wetlands completely dried up, locations were dug up to reintroduce wetlands into the site. The locations for digging were identified by looking for– species indicating groundwater saturation: *Phragmities, Tamarix, Vallisneria, Hydrilla, Potemageton, Salvinia and Azolla*. These were introduced to make the wetlands ecologically functional (Jain and Srivastava 2016).

16.5.3 De-Weeding

Weeds are plants that act as competitors with other plants and don't let them thrive. They grow fast in arid conditions, killing any competition, and cause water table depletion. Kikar (*Prosopis julifora*), one of the weeds, was brought to Delhi from Mexico by the British more than a century ago. This exotic plant became invasive and wiped out most of the native plants. It contains toxic chemicals and is an example of a major biological invasion by an alien species. Tughlaqabad, Aravali, Tilpath and Northern Ridge Biodiversity Parks were all infested with it. The menace of invasive weeds such as *Argemone sp*., *Hyptis suvaveolens* and *Parthenium* is tackled by uprooting on a large scale. Weeds like *Lantana camara* and new saplings of *Prosopis Juliflora* are eradicated by using the cut root-stock method (DDA 2016a).

Large tracts of *Vilayati kikar* cannot be removed as they would leave the city with hardly any green cover, an invitation to developers to grab land for construction. The tree can regenerate from the root and would spread again. So, canopies of *Prosopis juliflora* are opened up to allow sunlight to reach the newly planted saplings which are fenced in to protect them from herbivores.

16.5.4 Water Conservation

In Tughlaqabad Biodiversity Park, check dams were made for water conservation and to prevent run-offs. Loose boulder check dams in series across gullies were made, utilizing loose boulders and *Prosopis juliflora* logs (DDA 2016b). Water impounding has many benefits, including replenishment of groundwater, reduction of the run-off into the Kalkaji drain, increase in soil moisture and control of soil erosion. Similarly, to prevent run-off and protect the soil from erosion, site depressions were revived for seasonal water bodies to thrive. In sites with high susceptibility to soil erosion due to water run-offs, check dams should be prepared using channels and culverts to align the course of the water with the prepared wetlands on the site.

16.5.5 Plantation

Native plants provide multiple benefits to people and wildlife, while contributing greatly to healthy soil and water in urban and rural areas. Plants that have evolved in a particular region and occur naturally there are the native plant species of the area. They form the ecological basis to support bird life and humans depend on them. Insects that have co-evolved with these native species are food for the local birds. Any removal of these plants would disrupt the chain and the survival of local birds. Within the Yamuna Biodiversity Park, forest communities such as riparian, tropical dry deciduous, tropical thorn, shrub forest communities and salt bush land have been collected. A total of 30 communities consisting of 850 plant species have been raised ecologically. In Tilpath Valley, the selection of tree species followed a defined process. The topography and terrain dictated the selection of the plant community. Grasslands and broadleaved tree species were selected for plantation in low-lying and plain areas. Tropical dry deciduous forests were developed on gentle slopes and on upper reaches of valleys.

Biodiversity parks have helped restore the lost native biodiversity in urban areas, making them an innovative approach to the control of land degradation. The objective of restoring biodiversity parks is to plant native ethnic species which are suitable for that particular local region, so all the biodiversity parks have collected and planted these species.

16.5.6 Development of Infrastructure

To protect the site from encroachments and grazing, a boundary needs to be constructed, otherwise the restoration process would not be possible. Nurseries for seed germination have been established at Aravali, Yamuna, and Tughlaqabad Biodiversity Parks, and seeds are sent to other biodiversity parks. Citizens play an important role in the implementation and management of biodiversity parks. Site infrastructure such as nature trails, seating, amphitheaters, meeting places, etc. have been constructed to involve the public in the parks. Programs designed to acclimatize people and get them engaged with the wilderness are gaining momentum. Nature walks have become very popular. Nature awareness for school children, and team building and volunteering for corporates are widely appreciated. Volunteering/business teams have also been involved in activities like garbage removal, composting or planting in the nursery.

Table 16.2 summarizes all the restoration techniques used in the seven biodiversity parks.

16.6 RESULTS

In the past ten years, the team of professionals and volunteers at the parks have been able to reintroduce a large number of flora and fauna and recreate representative ecosystems at all the sites. Restorative processes that have taken place from 2002 to 2020 are illustrated in Figures 16.1–16.6.)

16.6.1 Yamuna Biodiversity Park

Since 2005 when the work started, the site has been completely transformed from barren land to a wetland with different ecosystems. Landscaping provides two shallow valleys representing rangelands with 10 mounds representing various ecosystems present from the foothills of the Himalayas (Siwaliks) through the Yamuna basin, till the point of confluence of the Yamuna and Ganga.

16.6.2 Aravali Biodiversity Park

The area, which was highly degraded due to past mining and infested with *Prosopis juliflora* (Figure 16.2a), is now transformed into the nature reserve of Delhi. About 1000 plant species native to Aravalis have been ecologically assembled into 35 communities which have become home to many insects, birds, reptiles and mammals.

16.6.3 Tilpath Valley

Before restoration, the area had a deep and shallow valley, flat lands and steep slopes, hilltops that were degraded and infested with weeds. It has now been converted into different forest, grassland and scrubland communities on each of the land profiles.

16.6.4 Tughlaqabad Biodiversity Park

The historic site had degraded forest communities invaded by *Prosopis juliflora* and highly silted and polluted wetlands due to discharge of untreated sewage. Wetland systems have been constructed and many nullas have been developed into surface waterbodies.

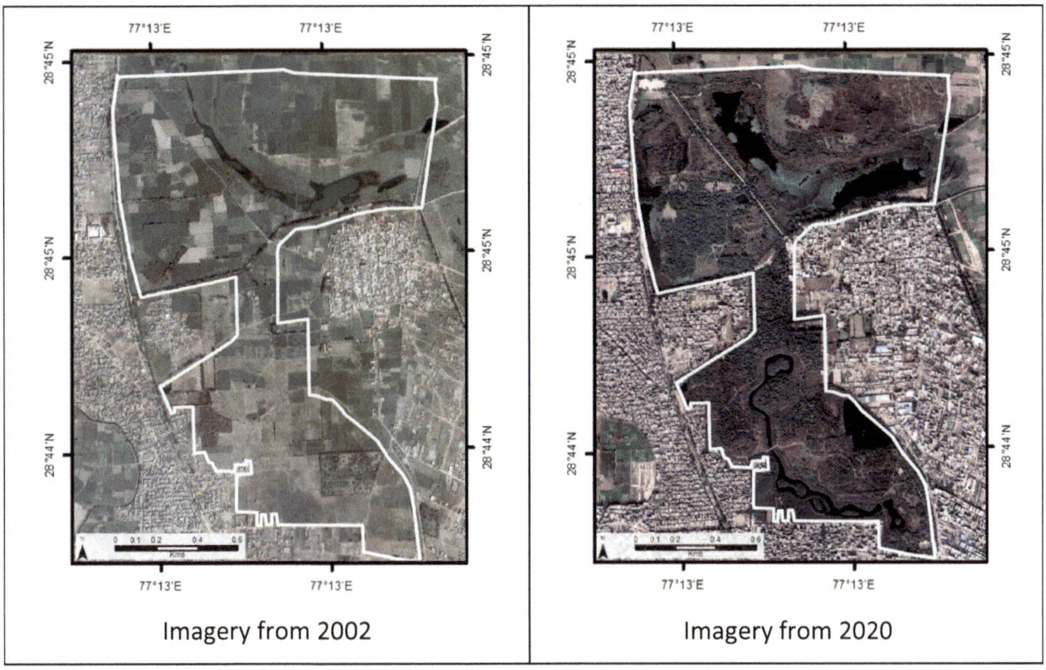

FIGURE 16.1 Changes in Yamuna Biodiversity Park before and after restoration process.

(Source: Google Earth imagery.)

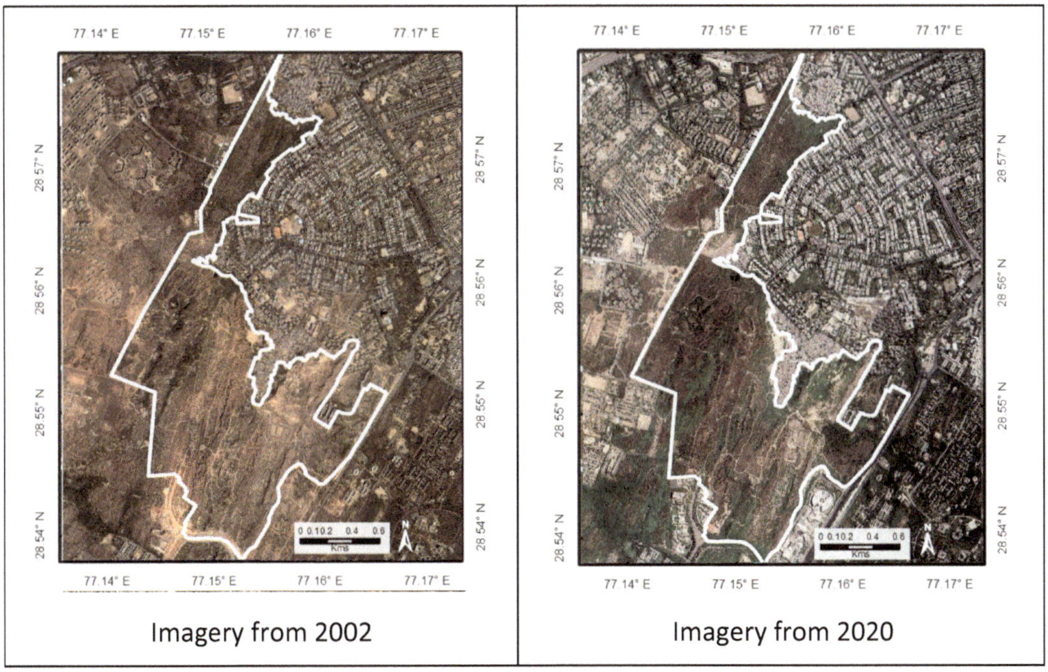

FIGURE 16.2 Changes in Aravali Biodiversity Park before and after restoration process.

(Source: Google Earth imagery.)

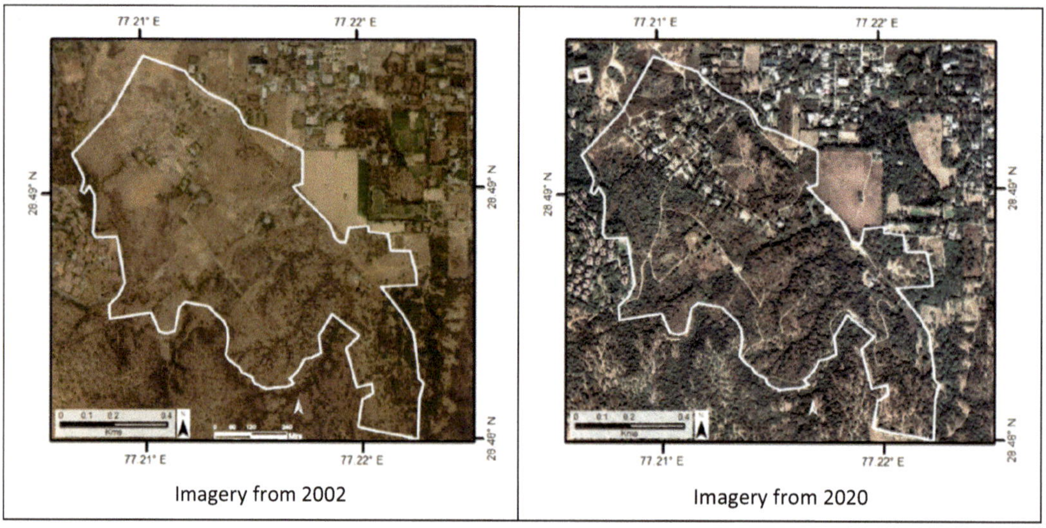

FIGURE 16.3 Changes in Tilpath Valley Biodiversity Park before and after restoration process.

(Source: Google Earth imagery.)

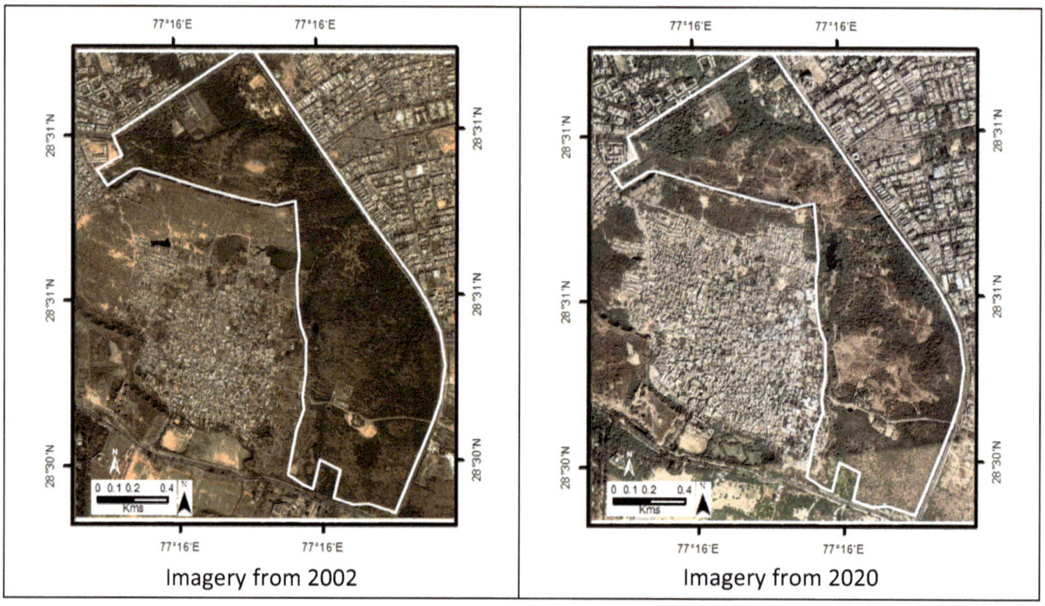

FIGURE 16.4 Changes in Tughlaqabad Biodiversity Park before and after restoration process.

(Source: Google Earth imagery.)

16.6.5 Northern Ridge Biodiversity Park

The ridge is being cleared of invasive species using the canopy opening and cut rootstock methods, and native species have been grown. All the monuments located on the Northern Ridge have been connected through a green corridor consisting of native forest communities of Aravallis.

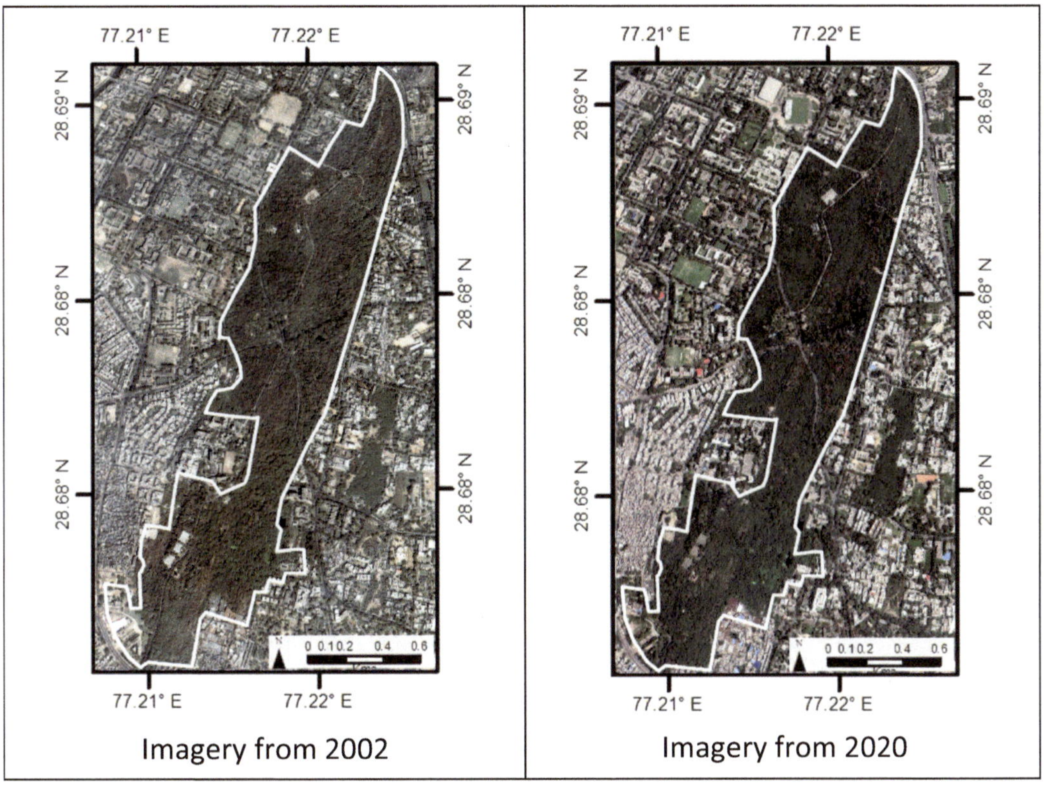

FIGURE 16.5 Changes in Northern Ridge Biodiversity Park before and after restoration process.

(Source: Google Earth imagery.)

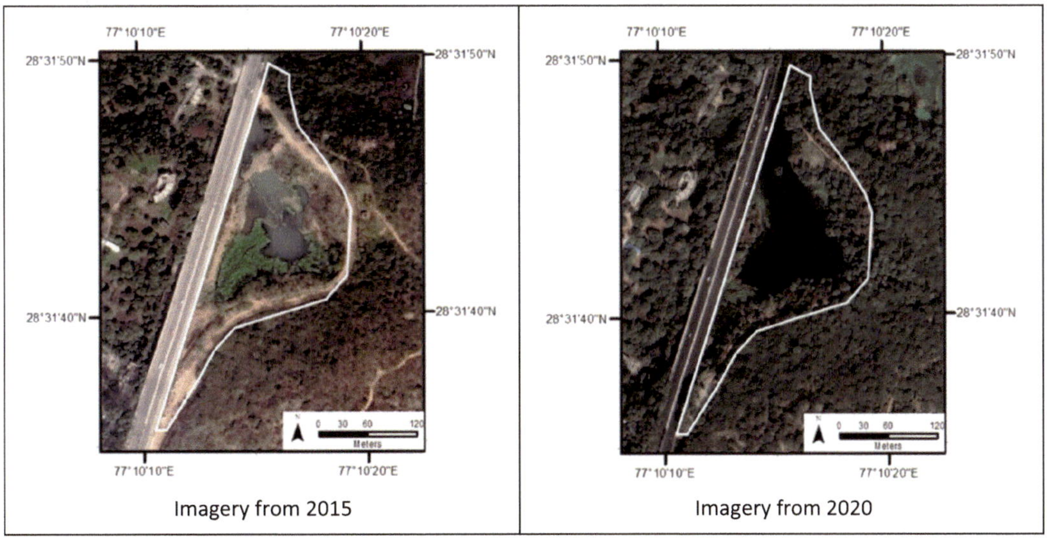

FIGURE 16.6 Changes in Neela Hauz Biodiversity Park before and after restoration process.

(Source: Google Earth imagery.)

TABLE 16.3
Change in Number of Species in Biodiversity (2002 and 2020)

	Yamuna Biodiversity Park (A)			Aravalli Biodiversity Park (B)			Neela Hauz (C)			Tilpath Valley Biodiversity Park (D)			Northern Ridge (Kamla Nehru Ridge) (E)			Tughlaqabad Biodiversity Park (F)			Kalindi Biodiversity Park (G)*
Species group	**2002**	**2020**	**%change**	**2002**	**2020**	**%change**	**2002**	**2020**	**%change**	**2002**	**2020**	**%change**	**2002**	**2020**	**%change**	**2002**	**2020**	**%change**	**2020**
Terrestrial Plants	90	915	**917**	150	950	**533**	90	115	**27.**	156	481	**208**	87	415	**377**	30	155	**416**	38
Birds	37	203	**449**	42	209	**397**	100	131	**31**	81	125	**54**	64	71	**11**	112	112	**0**	80
Butterflies	11	82	**645**	13	113	**769**	15	40	**166**	21	55	**162**	31	46	**48**	35	44	**25**	25
Reptiles and amphibians	3	18	**500**	8	31	**287**	6	7	**16**	14	15	**7**	8	11	**37**	18	28	**55**	4
Mammals	4	22	**450**	5	19	**280**	3	4	**33**	5	7	**40**	7	7	**0**	9	9	**0**	2

Source: DDA (2021).

16.6.6 Neela Hauz Biodiversity Park

Due to urbanization and roads development in 2009, the lake was reduced to 3 hectares in size and filled with solid waste. The restored lake, the green route with a walkway, the manicured lawns, the Phoenix Grove, Butterfly Corner and constructed wetland system (CWS) have made the park a recreational destination for public and tourists.

Table 16.3 shows the change in number of species of terrestrial plants, birds, butterflies, reptiles and amphibians and mammals at all the biodiversity parks. Yamuna Biodiversity Park has a significant percentage change of 917% in plants species and a 450–650% increase in the fauna. The Aravali Biodiversity Park has shown the maximum percentage change in butterflies, up to 770% increase in the number of species.

16.7 CONCLUSION

The concept of biodiversity parks has been by far the most successful initiative for restoring lost plant and animal species. Species that had become extinct can also be seen thriving inside the parks, along with their respective ecosystems involving other species.

With the onset of urbanization, the impacts on urban ecosystems and biodiversity have always been a cause for concern. However, the biodiversity parks mentioned in this chapter can easily be replicated in other urban ecosystems with similar land degradation issues. As an ever-increasing number of individuals live in cities, the revival, safeguarding and management of urban biodiversity have become significant. Concepts related to biodiversity management such as soil treatment, wetland revitalization, de-weeding and in-situ biological remediation can be used to conserve the biodiversity thriving in our cities. Conservation of biodiversity in urban areas often influences the quality of life and education of the population by facilitating the preservation of flora and fauna in their natural habitats.

In order to upscale and mainstream biodiversity conservation projects, which also restore degraded lands, many changes have to be made in policies, principles and legislation. Biodiversity conservation should be understood as an essential infrastructure that impacts resource conservation and production. Policies, along with institutions that would enable effective implementation of conservation, need to be formulated or identified.

REFERENCES

Adak, Baishali. 2019, 12 April. *South Biodiversity Park behind Kalindi Colony coming up soon.* Accessed June 6, 2021. https://www.hindustantimes.com/delhi-news/south-biodiversity-park-behind-kalindi-colony-coming-up-soon/story-QibUaH7q49VxZVKB8sXQ7H.html

Bocquet-Appel, Jean-Pierre. 2011. "When the world's population took off: The springboard of the Neolithic Demographic Transition." *Science* 333(6042): 560–561.

Bulkeley, Harriet, Marcel Kok, and Linjun Xie. 2021. *Realising the Urban Opportunity: Cities and Post-2020 Biodiversity Governance.* The Hague: PBL.

Butzer, K.W. 2005. "Environmental history in the Mediterranean world: Cross-disciplinary investigation of cause-and-effect for degradation and soil erosion." *Journal of Archaeological Science*, 1773–1800.

Babu, C.R.. 2012. *Biodiversity Parks: An Innovative Approach for Conservation of Natural Heritage and Enhancement of Quality of Urban Environment.* Delhi: Centre for Environmental Management of Degraded Ecosystems.

Cherlet, M., C. Hutchinson, J. Reynolds, J. Hill, S. Sommer, and G. von Maltitz (Eds). 2018. *World Atlas of Desertification.* Luxembourg: Publication Office of the European Union.

Coppus, R., and A.C. Imeson. 2002. "Extreme events controlling erosion and sediment transport in a semi-arid sub-andean valley." *Earth Surface Processes and Landforms* 1365–1375.

DDA. 2016a. *Biodiversity Parks.* Accessed June 5, 2021. https://www.delhibiodiversityparks.org/aravalli-biodiversity-park2.html

DDA. 2016b. *Tughlaqabad Biodiversity Park.* Accessed June 2, 2021. https://www.delhibiodiversityparks.org/tughlaqabad-biodiversity-park.html
DDA. 2021. *Biodiversity Parks - Nature Reserves of Delhi.* Delhi: Delhi Development Authority.
Dotterweich, M. 2008. "The history of soil erosion and fluvial deposits in small catchments of central Europe: Deciphering the long-term interaction between humans and the environment – A review." *Geomorphology,* 23(5): 192–208.
Ghermandi, A., J.C. Van den Bergh, L.M. Brander, H.L. De Groot, and P.A. Nunes. 2008. "The economic value of wetland conservation and creation: A meta-analysis." A meta-analysis, Nota di Lavoro, No. 79.2008, Fondazione Eni Enrico Mattei (FEEM), Milano. http://hdl.handle.net/10419/53239
IPCC. 2019. *Special Report on Climate change and Land.* Geneva: IPCC.
ISRO. 2016, June. "Desertification and Land Degradation Atlas of India." *Space Applications Centre.* Accessed June 2, 2021. https://www.sac.gov.in/SACSITE/Desertification_Atlas_2016_SAC_ISRO.pdf
Jain, Neha. 2018, 3 October. *Mongabay.* Accessed May 27, 2021. https://india.mongabay.com/2018/10/why-land-degradation-in-india-has-increased-and-how-to-deal-with-it/
Jain, Pranjal, and D.P. Srivastava. 2016, 23 May. *The Wetlands of Yamuna Biodiversity Park.* Accessed June 5, 2021. https://www.sahapedia.org/the-wetlands-of-yamuna-biodiversity-park
Johnson, D.L., and L.A. Lewis. 2007. *Land degradation: Creation and destruction.* Lanham: Rowman & Littlefield.
Koul, Monika. 2017, June. "Biodiversity Parks: Restoring Degraded Ecosystems." *Science Reporter*: 14–18.
Santurf, J.A. 2005. *What is Forest Restoration? Restoration of Boreal and Temperate Forests?* CRC Press, 3–11.
Seto, K.C. 2012. "Urban land teleconnections and sustainability." *Proceedings of the National Academy of Sciences of the United States of America,* 7687–7692.
Steffen, W.L. 2005. *Global Change and The Earth System: A Planet Under Pressure.* Germany: Springer.
Elmqvist, T., H. Setälä, S.N. Handel, S. van der Ploeg, J. Aronson, J.N. Blignaut, E. Gómez-Baggethun, D.J. Nowak, J. Kronenberg, and R. de Groot. 2015. "Benefits of restoring ecosystem services in urban areas." Current Opinion in Environmental Sustainability 14: 101–108.
Veblen, Kari E., Lauren Porensky, and Truman Young. 2019, 12 July. *Ecological Restoration.* Accessed June 4, 2021. Encyclopedia Britannica. https://www.britannica.com/science/ecological-restoration

17 Geospatial Analysis of Heat Islands in Badulla MC, Sri Lanka

P.L.C.B. Liyanage and R.M.K. Ratnayake
University of Sri Jayewardenepura, Nugegoda, Sri Lanka

CONTENTS

17.1 INTRODUCTION

The presence of warmer air is a common characteristic of urban areas. According to the United Nations in 2018, 55% of the world's population were living in urban areas. However, the proportion increases to 68% by 2050. The main causes are widespread industrialization and the migration of rural population to urban areas. Demand for urban lands has increased, resulting in degradation of environmental conditions.

Urban areas can be identified as specific ecological land areas with growing populations. According to Pickett et al. (2001) there may be a sliding scale from rural to urban when factors such as temperature, CO_2 concentration, nitrogen and length of the growing season are considered. Urbanization has resulted in significant changes in surface and atmospheric properties in most large cities, which can lead to inadvertent local weather and climate changes. Rapid urbanization has both positive and negative repercussions for residents' environmental and social well-being (Senanayake, Welivitiya, & Nadeeka 2013).

According to scholars, a UHI can be defined as an area which is substantially warmer than the surrounding rural areas due to human activity. Usually, the night-time temperature is higher than during the day, and it is most noticeable when the winds are weak. It is a temporary man-made weather condition, which can be solved with improved construction of built-up areas. These conditions are created within the congested built-up areas, due to obstruction of the wind flow.

According to Arifwidodo et al. (2019), a better understanding of the effect of UHI on urban areas can help local governments make better decisions in allocating resources and amenities, as well as being more aware of when to promote or contain growth, while minimizing any externalities that

DOI: 10.1201/9781003131922-20

may occur, particularly in developing countries where resources are scarce. Therefore, the urban heat island is an important area of study for the decision-making process.

The increase in heat within urban areas causes infrastructure demands to increase as well. With increasing demand for electricity and air conditioning, resource allocation needs to serve more and more people. In addition, there are increasing emissions of pollutants, such as sulfur dioxide, carbon monoxide, nitrous oxides, and suspended particles, from power plants. Continuous development of vertical built-up areas changes the wind direction. In this context, natural and man-made heat island sources exist and are increasing the heat of the area. Remote sensing is one of the tools that can be used to measure the heat island effect.

This phenomenon started in Western countries and is now also to be found in Asian developing countries. Especially in Sri Lanka, horizontal development is transformed into vertical congested development where there is high development demand. Colombo, Sri Lanka is one of the locations that can be identified. Yet, most of the area has developed vertically. Further, the principal regional cities are starting to change from horizontal to vertical development, so the urban heat island effect may be found in some of these locations.

Therefore, it is very important to identify local climatic zones within urban zones. Variations in heat conditions directly affect urban residents. The questions are: what are the space-specific reasons for creation of UHIs? What are the root causes behind them? And what are the implications for physical planning in order to minimize the urban heat island effect in urban areas?

17.2 LITERATURE REVIEW

17.2.1 Urban Heat Islands

The concept of UHI as a situation where urban areas have higher atmospheric and surface temperatures than surrounding rural areas was first explored in 1818 (Estoque & Murayama, 2017). When mass development is taking place, demand for land increases and every plot makes the most of its monetary value. As a result, built-up areas are filled with buildings and fire gaps – which also function as wind gaps – are not maintained. Hidden locations through which wind cannot flow are thus created. In the daytime these hidden locations gain energy through long-wave solar radiation. As they have no atmospheric wind flow, energy increases the heat of these particular locations.

Heating, however, is one of the key components that balance atmospheric conditions. Even though the general temperature of the area has a specific value, temperatures vary when heat islands occur, ultimately an effect of urban climatology. Scholars have identified development of a roughness sub-layer creating different buildings of various heights and physical characteristics, separated by trees, canyons and open spaces. This means that high-rise developments do not have a specific locational temperature.

Urban heat islands are a pattern of higher temperature in urban areas than in the surroundings. Voogt (2005) states: "Urban heat island is a name to define the heat parameters of atmosphere and surfaces in town to compare with suburban areas." Most climate-related studies reflect the direct influence of urban heat on people's well-being.

17.2.2 Identification of Urban Heat Islands

The first mention of urban heat islands by Howard (1818) related to an artificial excess of heat in the city of London. Howard's investigation found that urban heat is a result of the collection of vertical surfaces and a lack of moisture available for evaporation due to higher absorption of solar radiation.

According to Memon, Leung, and Chunho (2008), anthropogenic heat from vehicles, air conditioning units and power plants, as well as heat re-radiated and stored in urban structures, is the primary source of urban heat island formation. Elements influencing urban heat islands may be controllable or uncontrollable, according to Memon et al. (2008).

Further, Oke (2006) defines the atmospheric heat creation process as follows. Because the interception role of the canopy is lost when vegetation is removed, evapotranspiration may be reduced, snow cover distribution and duration may be altered, and run-off may be enhanced. By offering a new surface geometry and albedo, removal is likely to disrupt the radiation budget. A new set of heat, moisture, and aerodynamic variables are also likely to change the energy balance.

Urban areas have a higher population density than rural or rural-like locations. As a result, the air in the urban canopy is frequently warmer than the air in the countryside. This urban heat island effect is probably the most visible and well-documented example of unintentional climate change. Due to climatic, locational, and urban features, the exact shape and magnitude of this phenomenon fluctuates across time and space (Oke, 2006). Figure 17.1 shows the characteristic variations in urban heat islands.

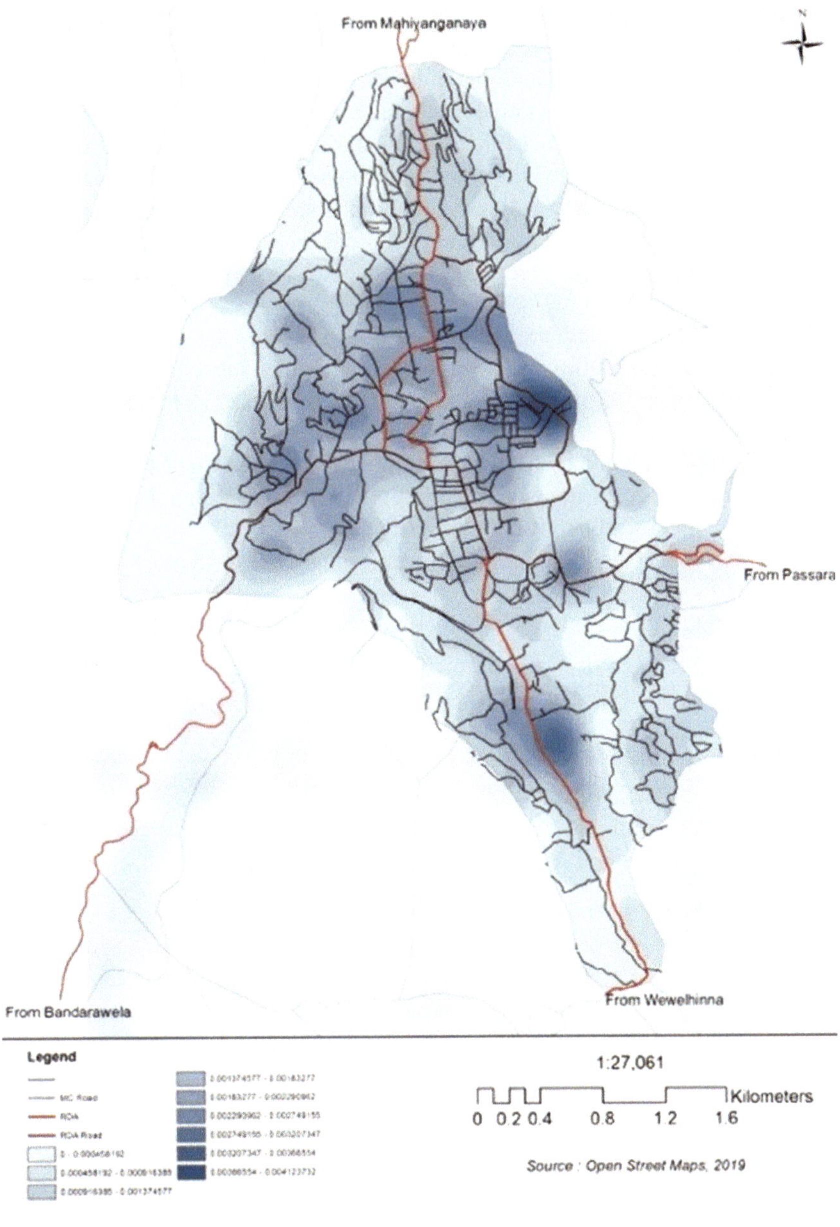

FIGURE 17.1 Badulla MC building density, 2019.

Source: Compiled by the Author, 2020.

Urban heat islands can be regarded as assets in some countries, while other countries consider them a sickness in urban areas. Within the cool climatic areas much energy is needed for heating purposes. Research confirms that the quality of living conditions in urban areas and comfort in indoor and outdoor spaces depend on the variations in climate. Therefore, depending on the climatic conditions, infrastructure supply mechanisms need to change.

Scholars have identified factors affecting the creation of urban heat islands. Voogt (2004) identified six factors: geographic location, time, synoptic weather, city form, city function and city size.

The phenomenon began in Western countries and spread to Asian developing countries. In Sri Lanka in particular, specifically in Colombo, horizontal development changed to vertical congested development, with high development demand.

It is very important to identify local climatic zones within urban zones in order to assess the surface characteristics. Variations in heat conditions directly affect urban residents, giving rise to the following questions. What are the space-specific reasons for creation of UHIs? What are the root causes behind them? And what physical planning efforts are taking urban heat island effects into consideration to minimize their impact on urban areas?

The reciprocal effects of over 11 artificial and natural elements are thought to be one of the mechanisms causing UHI summarized the elements that cause and define the severity of heat islands and divided them into two categories: climatic conditions such as air temperature, wind speed and direction, humidity level and cloud cover; and urban design parameters such as density of urban areas, percentage of built-up area, aspect ratio of urban canyons, sky-view factor, building construction materials and urban form.

According to Voogt (2004), there are three types of urban heat island: surface urban heat islands and atmospheric urban heat islands, with the latter further divided into canopy-layer heat islands and boundary-layer heat islands. Roth (2018) describes these three types as follows:

I. The **surface UHI** is defined by the temperature of the surface that extends over the entire 3D envelope of the surface. It is a surface energy balance phenomenon and involves all urban facets (street, vertical walls, rooves, trees, etc.).
II. The **canopy-layer UHI** is found within the atmosphere below the tops of buildings and trees (in the urban canopy).
III. The urban warmth extends into the urban boundary layer (UBL) through convergence of sensible heat plumes from local areas (bottom up) and the entrainment of warmer air from above the UBL (top down) to create the **boundary-layer UHI**.

The causes and impacts of urban heat islands vary from place to place. Therefore, place-specific urban heat island causes and impacts should be identified as part of the decision-making process.

17.2.3 Technology Used for Analyzing Urban Heat Islands

The character of a region's surface and atmospheric properties changes as a result of urbanization, increasing the temperature of the urban atmosphere. This temperature thus becomes a negative impact on urban human well-being. Generally, there are several manual methods of measuring urban heat island effects. However, with the invention of remote sensing, satellite imagery technology is used in urban heat island studies. Satellite studies of urban heat islands offer easy repeatability and synoptic coverage. Several studies have been carried out using remote-sensing technology.

Nicole (2005) examined the pattern of urban heat islands in the day and at night. He explored the relationship between surface temperature and urban shape at different times of the diurnal temperature cycle, and used remote-sensing technology to interpret these associations in terms of heat island processes. Temperature data were derived from different surveyed and ASTER satellite images. A night-time ASTER satellite sensor thermal image of Hong Kong's western New Territories was compared to a daylight Landsat Enhanced Thematic Mapper Plus (ETM +) thermal image taken

19 days previously. Streutker (2002, 2003) quantified UHIs in Houston, Texas as a two-dimensional Gaussian surface placed on a planer rural background.

Ogashwara and Bastos (2012) also used a quantitative approach to analyze the relationship between urban heat islands and land cover. Brightness temperatures and land-cover types from Landsat TM images of São José dos Campos from 1986, 2001 and 2010 were examined for changes in the spatial distribution of temperature and land cover. A quantitative approach was used to explore the relationships between temperature, land-cover areas and several indices, including the Normalized Difference Vegetation Index (NDVI), Normalized Difference Water Index (NDWI) and Normalized Difference Built-up Index (NDBI).

17.2.4 Development Trends and Factors Affecting Urban Heat Islands

The causes of UHI vary from place to place according to their inherent characteristics. Scholars have identified the main causes of UHI from their experiments. Gartland (2008) found five main reasons: reduced evaporation, increased heat storage, increased net radiation, reduced convection and increased anthropogenic heat.

Asimakopoulos et al. (2001a, 2001b) expressed the root causes of UHIs as:

i. The complex exchange of heat between buildings in urban streets which act as canyons retaining heat.
ii. The thermal properties of building materials that keep the heat within the fabric of urban areas.
iii. Heat released from combustion of fuels, and animal metabolism.
iv. The urban greenhouse effect helps increase the heating effect of the incoming radiation on the polluted and heater urban atmosphere.
v. Reduced evaporative cooling due to impervious surfaces.
vi. Turbulent transfer of heat, by air movement, from within streets is reduced.
vii. Reduced vegetation due to ever increasing demand for land.

Therefore, Nuruzzaman (2015) summarizes the causes as low amounts of evapotranspiration because of less vegetation, absorption of solar radiation due to low albedo, hindrance to the flow of air because of higher rigidity, and a high amount of anthropogenic heat release.

There is ample evidence to suggest that Sri Lanka's climate has already changed (Eriyagama et al., 2010). During 1961–1990, the country's mean air temperature increased by 0.016°C per year, and mean annual precipitation decreased by 144 mm (7%) compared to that of 1931–1960. It is suggested that Sri Lanka's mean temperature may increase by approximately 0.9–4°C above the baseline (1961–1990) by the year 2100, with accompanying changes in the quantity and spatial distribution of rainfall. Scientists attribute this warming trend seen throughout the country to both the enhanced greenhouse (global) effect and the 'local heat island effect' caused by rapid urbanization.

Disanayake et al. (2019) investigate land-use/land-cover changes and their impact on surface urban heat islands, with special reference to Kandy city, Sri Lanka. The study examines the spatio-temporal analysis of Kandy city between 1996 and 2006; the surface UHI effect is discussed mainly through the variation of land-surface temperature (LST) over existing impervious surface and newly added impervious surface. Colombo has already been subject to some outmigration toward cooler areas such as Kandy, due to the increasing trend of thermal discomfort (Emmanuel and Fernando, 2007). Therefore, Kandy also has a high urbanization context with geographically bounded extent.

Little literature exists on contemporary urbanization side-effects in the form of urban heat islands in Sri Lanka, and available literature all focuses on Colombo and its sub-urban area. Other urban centers are experiencing the same urban agglomeration, taking population growth and commuter population into account. Therefore, information not only on Colombo, but also on other main district capitals of Sri Lanka is needed to identify urban heat island effects.

17.3 METHODOLOGY

Since several analyses on UHI have been conducted in Colombo Metropolitan (MC) area the present study omits the Colombo MC and focuses its analysis on selected other areas. In this regard coastal zone cities need to be ignored as they are affected by sea breezes, and the focus of the study is geospatial UHI considerations. It is therefore inland administrative main towns with high levels of urban agglomeration that are considered. The geographic locations were selected based on time, synoptic weather, city size, city form and city function (Voogt 2004).

First the district-wise temperature values need to be examined. According to meteorological data, Kandy, Badulla and Nuwara Eliya have comparatively low average temperatures.

The effect of UHIs in dry-zone flat-terrain town functions can be eliminated by promoting development along the main arteries. Urban-form linear and poly-centric urban forms do not create huge urban agglomeration in one area, which also minimizes development of UHIs. Kandy, Rathnapura and Badulla are therefore appropriate for study. Since Kandy is under the central cultural fund regulations, it high-rise development above the height of the Temple of Tooth Relic is not permitted. Rathnapura has a comparatively larger area for development than Badulla, so its vertical development trend is lower than that of Badulla. Badulla MC area, which extends to 10.5 km^2, was therefore selected for the case study.

From a historical evolutionary point of view, Badulla evolved with the Muthiyanganaya Raja Maha Viharaya, which the Buddha visited on his third visit to Sri Lanka. At that time the town was the major economic center of the region. Currently, it is the capital city of Uva province.

The research has four objectives. First, to identify the historical evolution of Badulla MC, for which purpose historical evolution maps are used. Second, to study the current vertical and horizontal pattern of development. The Urban Development Authority's Survey Department 1:10,000 building data, 1:5000 land-use and roads data are used to analyze the current status of the area.

Third, to recognize the geospatial temperature distribution pattern in Badulla. Two different years' satellite images were used to derive the temperature values. LANDSAT 8 data were used. Band 10 is used to estimate the brightness temperature and bands 4 and 5 are used to calculate NDVI. Onsite temperature values are established by vising places with a thermometer. Three different years' geospatial temperature distribution patterns are also identified.

Fourth, to explore the reasons for the creation of UHIs within the UHI. For this purpose, geospatial patterns of the UHI within the Badulla MC, Sri Lanka are analyzed.

Since the satellite data products are a geometrically corrected data set, dataset projections are limited to the local zone. Other relevant values would be obtained using metadata from satellite images and the land-surface temperature calculated. Landsat 8 satellite images from the following dates were used: (i) 15 March 2016; (ii) 24 March 2019.

Random and space-specific locational temperature values were calculated after LST generation, which were then measured again using a digital thermometer.

17.4 DEVELOPMENT OF URBAN HEAT ISLAND IN BADULLA MC

Badulla is the capital city of Uva province, located at the main junction of the A5 Peradeniya–Chenkalady road. Bandarawela, Mahiyanganaya and Passara are the other main towns located around Badulla.

Badulla's historical evolution goes back for centuries. Archeological evidence shows that Badulla is first mentioned during the Buddha's third visit to Sri Lanka in the 18th Buddhist year. The Buddha stayed at Present Muthiyangana Raja Maha Vihara's residence. This is the first evidence of a built-up area within the Badulla.

Badulla was regarded as the second capital during the Kandy Kingdom. Uva Wansa's story from the Buddhist years 1604–1548 relate how Regional Administrator King Wimaladharmasuriya constructed the Kataragama Devalaya at Badulla. It also mentions that people from 32 races were resident in Kottagoda Village close to city. We can interpret this as showing that during 1604–1548 this was a religious town, which considered Kataragama Devalaya and Muthiyanganaya Raja Maha Viharaya as prominent places. Buddhist people were settled around the town especially near to the Kanupellela and Pinarwa areas.

The socio-economic development of Badulla was affected by the Portuguese and Dutch periods. Christian people resided within the Badulla town limits, particularly along the Mahiyanganaya road and built religious places close to their homes. During the British period there were several changes within Badulla town, including the establishment of regional administrative services. Infrastructure was established, including a rural railway line connecting to the town, which was the main tea transport hub. The Kachcheri Courts Complex and Badulla General Hospital were built and economic activities were promoted. Commercial activities within Badulla town were promoted with the migration of Mo and Moor ethnic groups, who chose to settle closer to the Badulupitiya area.

The built form of Badulla town started with the establishment of the Muthiyanganaya Raja Maha Viharaya and Kataragama Devalya. A royal throne was situated in front of the Kataragama Devalaya facing Lower Street, and overlooking the Muthiyanganaya Perehera. Lower Street is directly connected to the Muthiyanganaya Raja Maha Viharaya. The surrounding lands belong to the Kataragama Devalaya or Muthiyanganaya Raja Viharaya. To establish commercial or residential activities within the town center, people have to lease the Devala or Viahra lands.

Under the National Physical Plan 2030, Badulla is located within the Central Fragile area. It therefore has potential for development in the agricultural, education and tourism sectors. The revised 2050 National Physical Plan designated Badulla as an environmentally sensitive area, further encouraging tourism and agriculture.

Badulla Municipal Council is the first MC within Uva province. The council was first formed as a Town Council in 1933 and upgraded to Municipal Council status in 1963. In 1979 the area was declared an Urban Area under the Urban Development Authority, leading to a change of development directives. In 2010 the Urban Development Authority published the Badulla Development Plan, which concentrated development patterns through a zoning plan, identifying a mixed-development zone in the middle of the town center.

Land-use classification of the Badulla Development Plan, 2010 and 2019–2030, shows that residential land use within Badulla in 2004 was about 54% of the total land area. This increased to 61% in 2016. These variations within the land-use profile show that the town has urbanized rapidly.

The Badulla Development Plan 2019–2030 used satellite images from 2017 and 2018 to calculate the Normalized Difference Built-Up Area Index (NDBI). The analysis shows a 2% increase in the urban portion of the Badulla area in 2018 compared to 2017, highlighting the high demand in terms of population growth.

Badulla has developed as the administrative capital of Uva province. Badulla General Hospital is a key main service in the town. Under the Badulla Development Plan (2010–2030), 100,000 commuters are attracted daily to the town seeking health, administrative and education services. Specific commercial activities are also concentrated in specific locations to support these services. As an example, near Badulla General Hospital, pharmacies and other channeling services are agglomerated.

Further, Badulla Lower Street is one of the commercial streets that were established many years ago. This street has mixed commercial activities. Under the Badulla Development Plan 2010 which is in force to date, buildings surrounding the Muthiyanganaya temple can only contain a ground floor plus three further stories to maintain the height limit of the Muthiyanaganaya Stupa (Figures 17.2 and 17.3).

Land ownership in Badulla town center is a crucial factor in the development process. Since much of the land belongs to the Devalaya and Viharaya, developers are building on leased or rented lands, which brings several limitations, including development of small plots. Unauthorized building encroaches on fire gaps, drainage and river reservations.

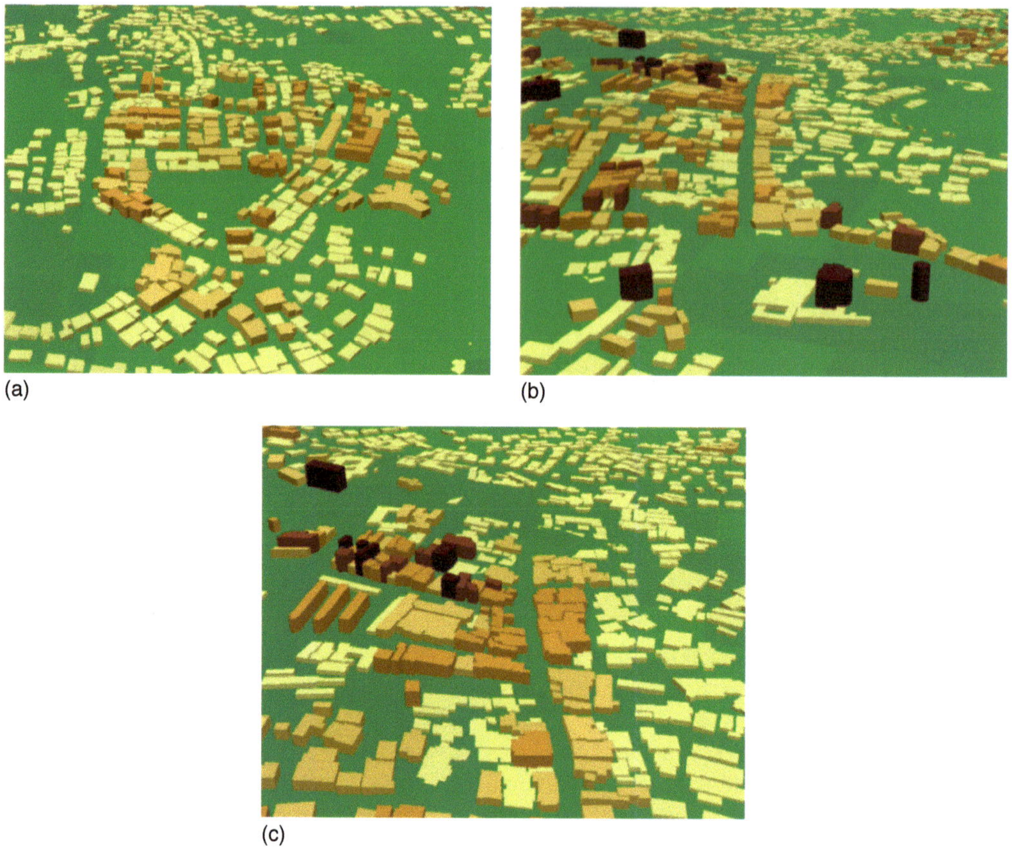

FIGURE 17.2 (a) Badulupitiya, (b) Bazar Street, (c) Lower Street.

Source: Compiled by the Author, 2020.

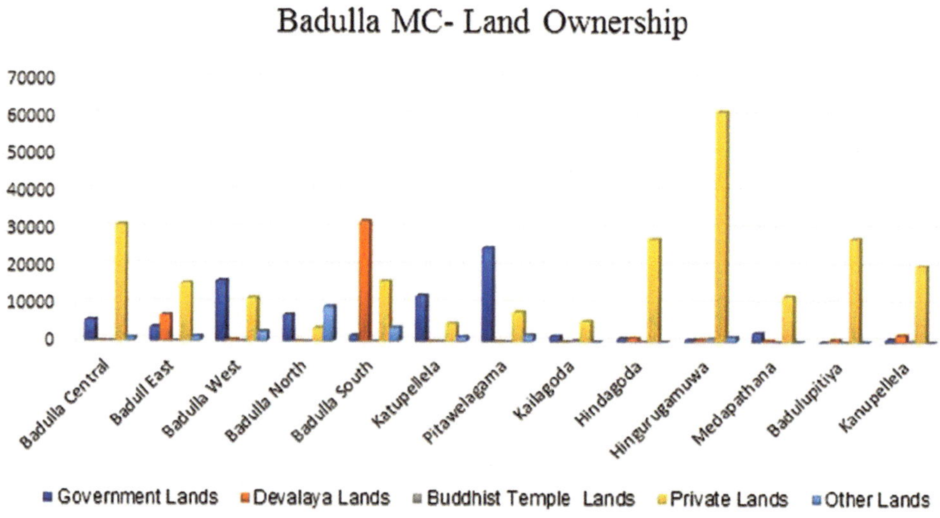

FIGURE 17.3 Badulla MC, land ownership.

Source: Compiled by the Author, 2020.

17.4.1 Distribution Pattern of Urban Heat Islands in Badulla MC

Badulla is a hill-country town with lower elevation than the other towns in the region. Therefore, comparatively, temperature values are higher than in the other towns in the region. The average annual temperature of the area established for the study is 24°C. The average daytime temperature ranges between 24°C and 27°C during the month of July. The area is surrounded by mountain ranges including the Namunukula, so wind flow speed is lower than in other areas.

The urban heat islands were identified by two methods. First, Landsat 8 images from 2016 and 2019 were downloaded from USGS Earth Explorer. Figures 17.4–17.6 show the results.

The distributed temperature values were extracted at 52 selected points, and distributions of the years 2016 and 2019, compared, showing a 2°C increase in temperature between 2016 and 2019. There was also a comparative change in the distribution pattern. Higher temperature values are spread all over within the town limits. It clearly shows that in 2016 higher temperature values were

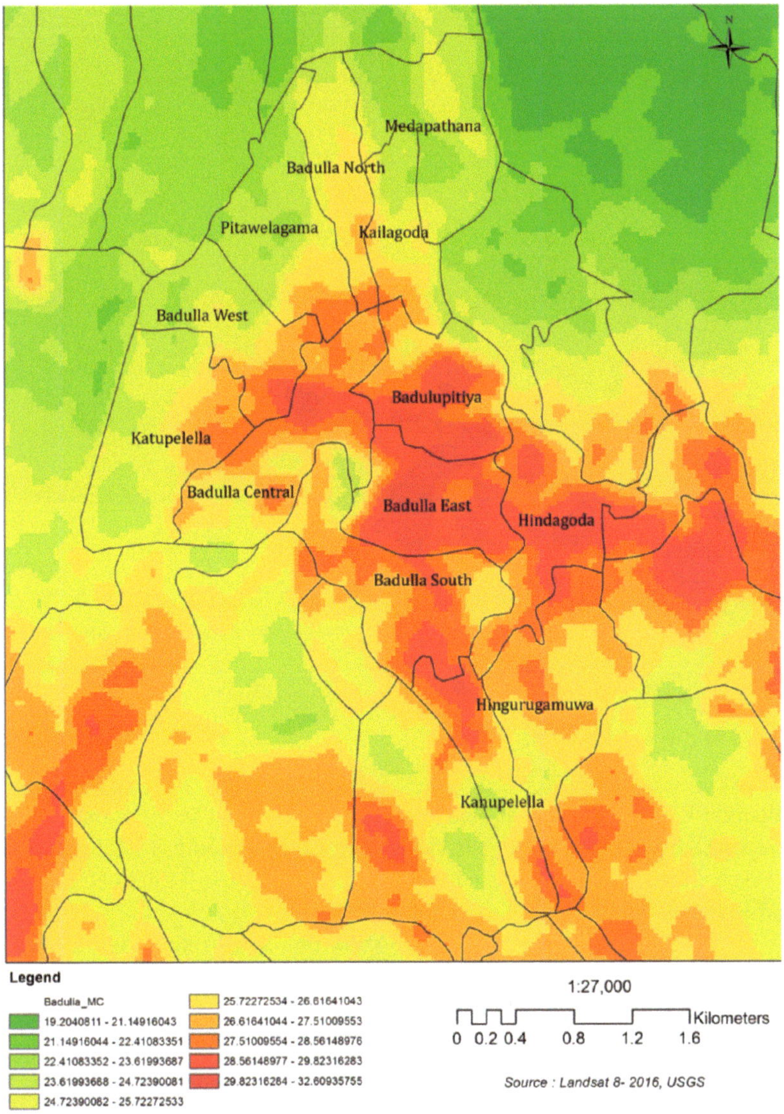

FIGURE 17.4 Badulla MC temperature distribution, 2016.

Source: Compiled by the Author, 2020.

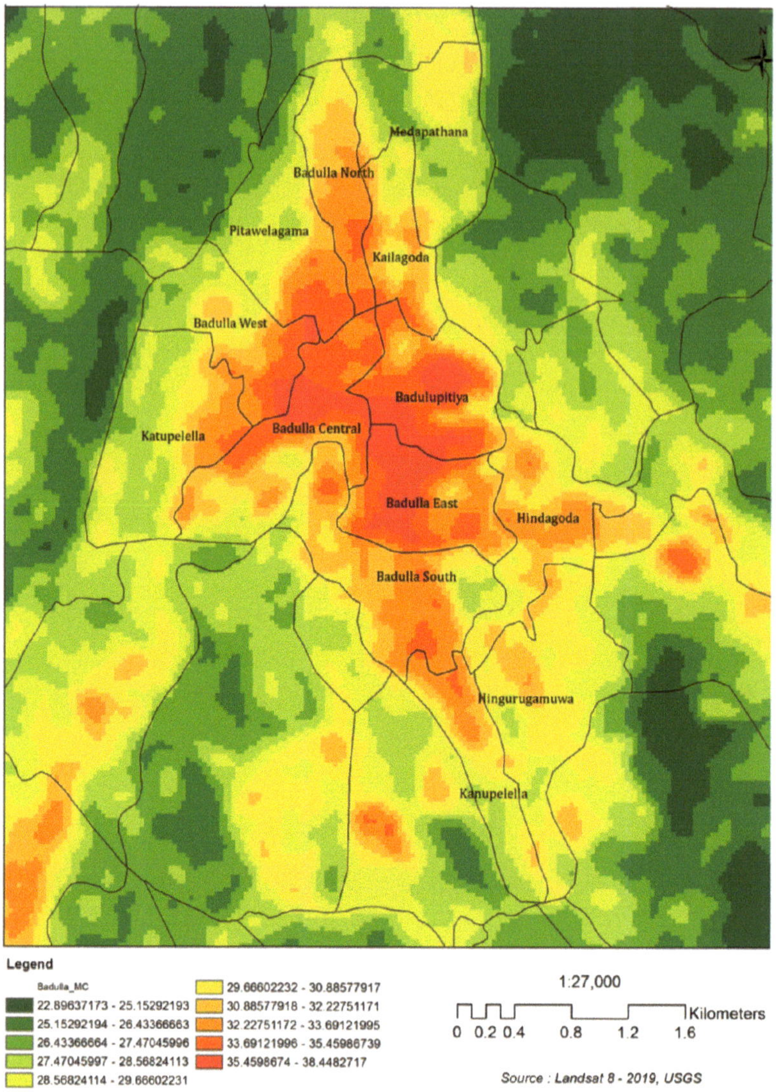

FIGURE 17.5 Badulla MC temperature distribution, 2019.

Source: Compiled by the Author, 2020.

reported in Badulla East, Badulupitiya, and Hindagoda areas. But in 2019 the higher temperature values were reported in Badulla Central, Badulla East, Badulla South, Badulupitiya, Badulla West and Badulla North areas. In particular, the heat has increased in the Badulla North, Badulla Central and Badulla West areas.

The temperature values plotted in Figure 17.6 show that temperature values have not increased by 2°C equally. In some places the values have changed in the range of 0.5–2.5°C. Space-specific reasons have caused the increase in temperature values.

Temperature was also measured manually. The author carried out field visits and collected temperature values in specific locations derived from the Landsat 8 thermal bands. The temperature values are derived from a single day with two different climate conditions. The data were plotted to identify the little gap between these two values, which were checked against the 2019 temperature values and the difference identified (Figure 17.7).

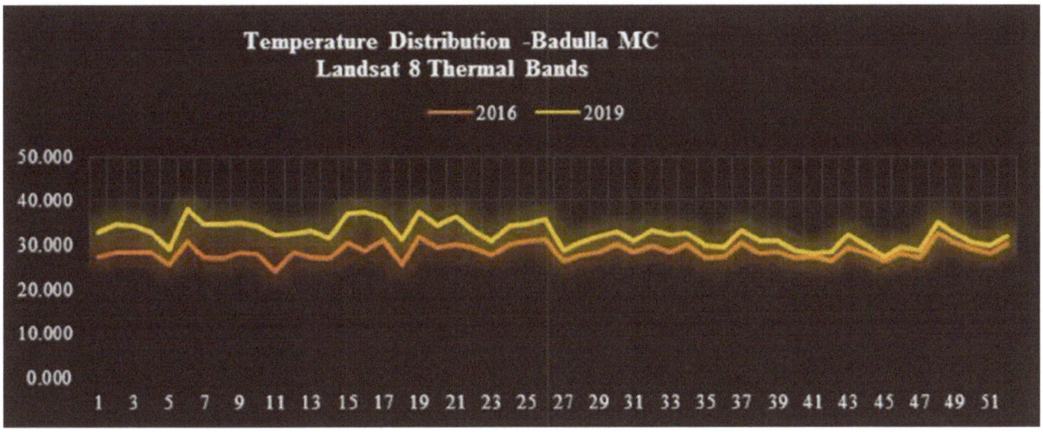

FIGURE 17.6 Temperature distribution, Badulla MC, Landsat 8 thermal bands.

Source: Compiled by the Author, 2020.

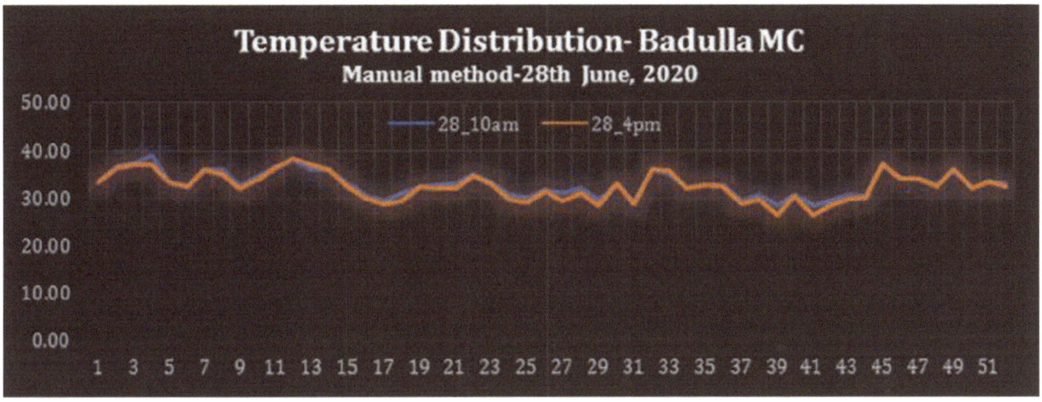

FIGURE 17.7 Temperature distribution, Badulla MC, manual method, 28 June 2020.

Source: Compiled by the Author, 2020.

17.5 ANALYSIS OF DEVELOPMENT OF UHI IN BADULLA MC

The urban heat islands were analyzed to identify the reasons for their creation. First, the land use at each point was identified. According to the Landsat 8 thermal band temperature values, higher values were recorded within the built-up areas in 2016, in particular near the Sri Lanka electricity board grid system, the beginning of Viharagoda Lower Street and Badulla solid waste dump.

The higher temperature values range from 30–32°C. The built form and material collection in these areas are summarized in Table 17.1.

In 2019 the higher heat points have changed from those of 2016. Higher temperature values of 35–38°C are recorded at six points and moderate temperature values of 33–35°C at 10 points. Temperature values are identified by the context and building material, as shown in Table 17.2 (Figures 17.8 and 17.9).

The Landsat 8 data were compared with the onsite temperature values obtained by thermometer on field visits (Figures 17.10 and 17.11).

TABLE 17.1
Temperature Value Points Context and Build Material, 2016

Place	Nature of Built-Up Area	Nature of Material
Sri Lanka electricity board grid substation	Electricity board grid substation is main electricity distribution point of the region.	Includes sharp iron materials
Viharagoda beginning of Lower Street	Attached commercial buildings with 3-story buildings	Concrete and asbestos sheets including buildings and tarmac roads
Badulla solid waste dump	Bare land	Bare land with tin shaded building

Source: Compiled by the author, 2020.

TABLE 17.2
Temperature Value Points Context and Build Material, 2019

		Place	Nature of Built-Up Area	Nature of Material
Higher temperature values	1	Sri Lanka electricity board grid substation	Electricity board grid substation is main electricity distribution point of the region.	Includes sharp iron materials
	2	Viharagoda, beginning of Lower Street	Attached commercial buildings with 3-story buildings	Concrete and asbestos sheets including buildings and tarmac roads
	3	Baulupitiya	Tightly attached residential low-rise building area. Flat terrain.	Predominantly concrete buildings with asbestos sheets
	4	Lower Street, neare Sri Murugesan Kovil	High-rise build form along Lower Street	Concrete buildings with slab areas and tarmac roads
	5	Badulla General Hospital	Tightly build high-rise developments	Concrete buildings with slab area and tarmac roads
	6	Within Badulla Prison	Buildings surrounded by the boundary wall	Concrete boundary wall and buildings
Moderate temperature values	7	Badulla solid waste dump	Bare land	Bare land with tin shaded building
	8	In front of the Bandaranayake Road clock tower	Private tuition institute with closely-packed buildings	Predominantly concrete building with tin sheets and slab area
	9	Badulla Central College	High-rise buildings with playground	Predominantly concrete building with asbestos sheets and bare land
	10	Kanupellela residential units along Wewelhinna main road	Closely-packed residential attached houses within subdivided lands	Predominantly concrete building with tin sheets and asbestos sheets
	11	Main road in front of the District Secretariat building	Main road in front of the high-rise building	Tarmac road
	12	Thisara Florist and its backyard	Closely-packed high-rise buildings and attached houses	Predominantly concrete building with tin sheets and slab area
	13	Pinarawa residential area	Closely-packed attached houses	Predominantly concrete building with tin sheets and slab area
	14	In front of the back gate of the Muthiyanagaya Temple	Closely-packed shanty-type houses	Predominantly concrete building with tin sheets
	15	Kanupellela residential units	Closely-packed residential attached houses within subdivided lands	Predominantly concrete building with tin sheets and asbestos sheets

Source: Compiled by the Author, 2020.

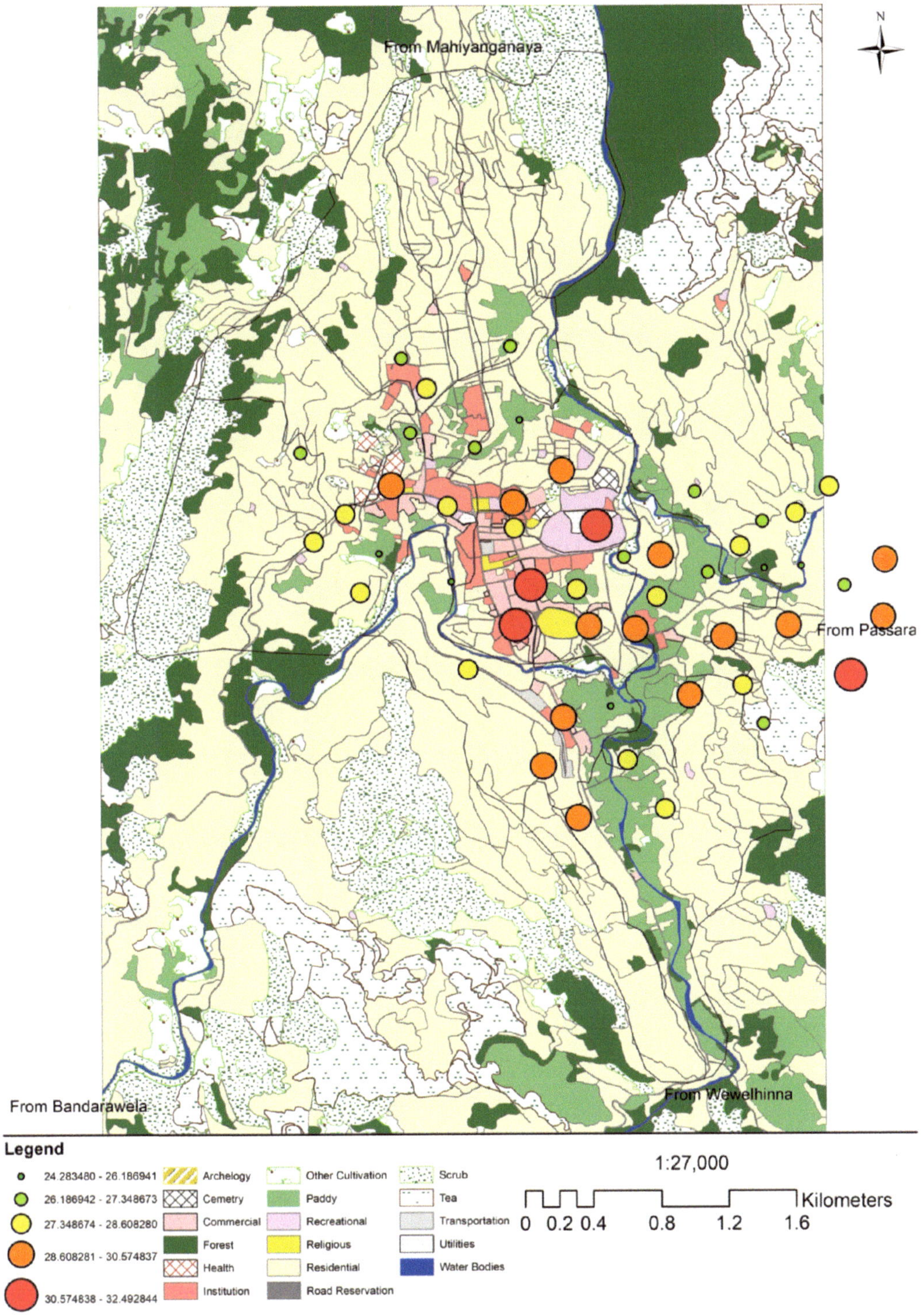

FIGURE 17.8 Land use and temperature distribution, 2016.

Source: Compiled by the Author, 2020.

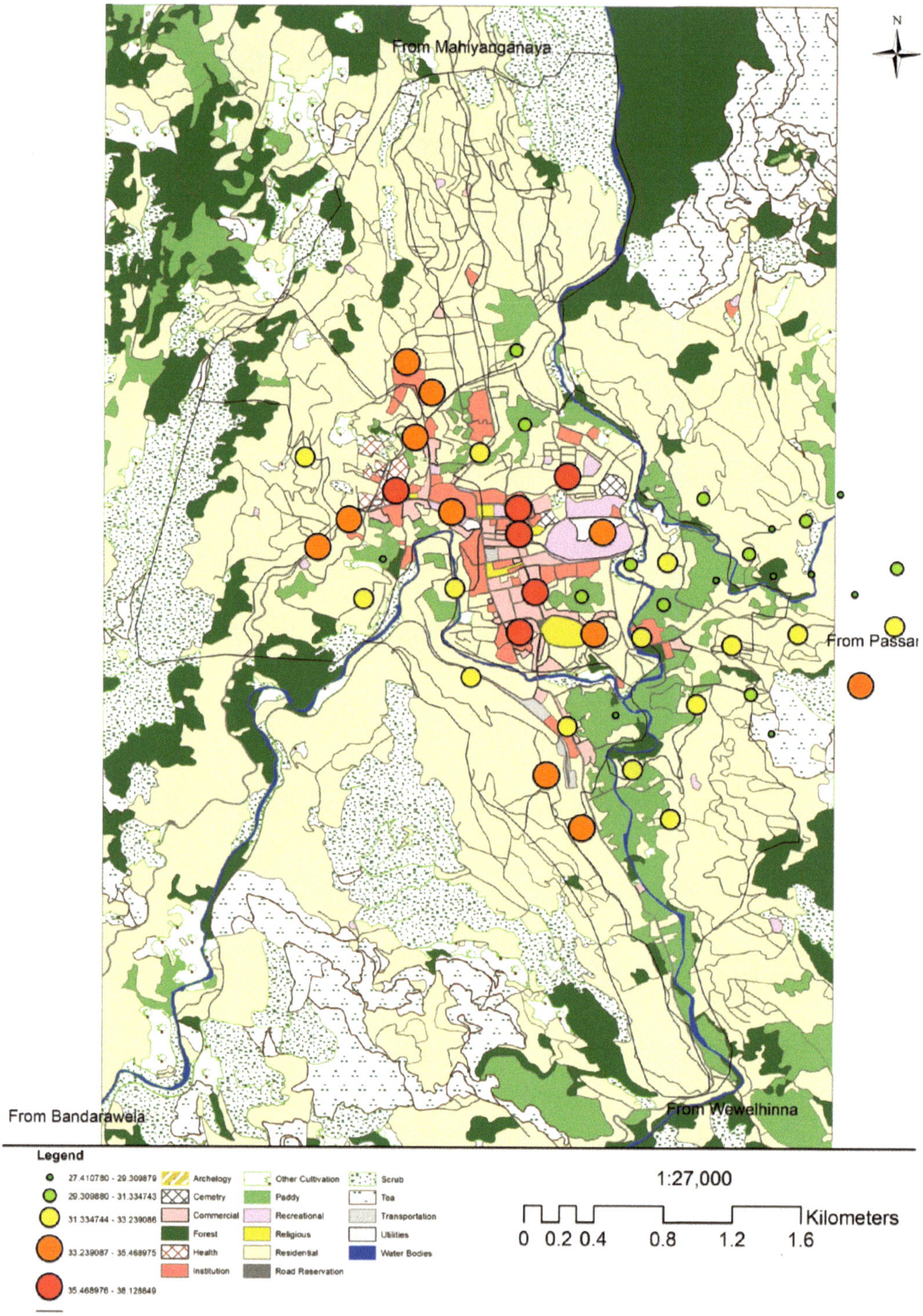

FIGURE 17.9 Land use and temperature distribution, 2019.

Source: Compiled by the Author, 2020.

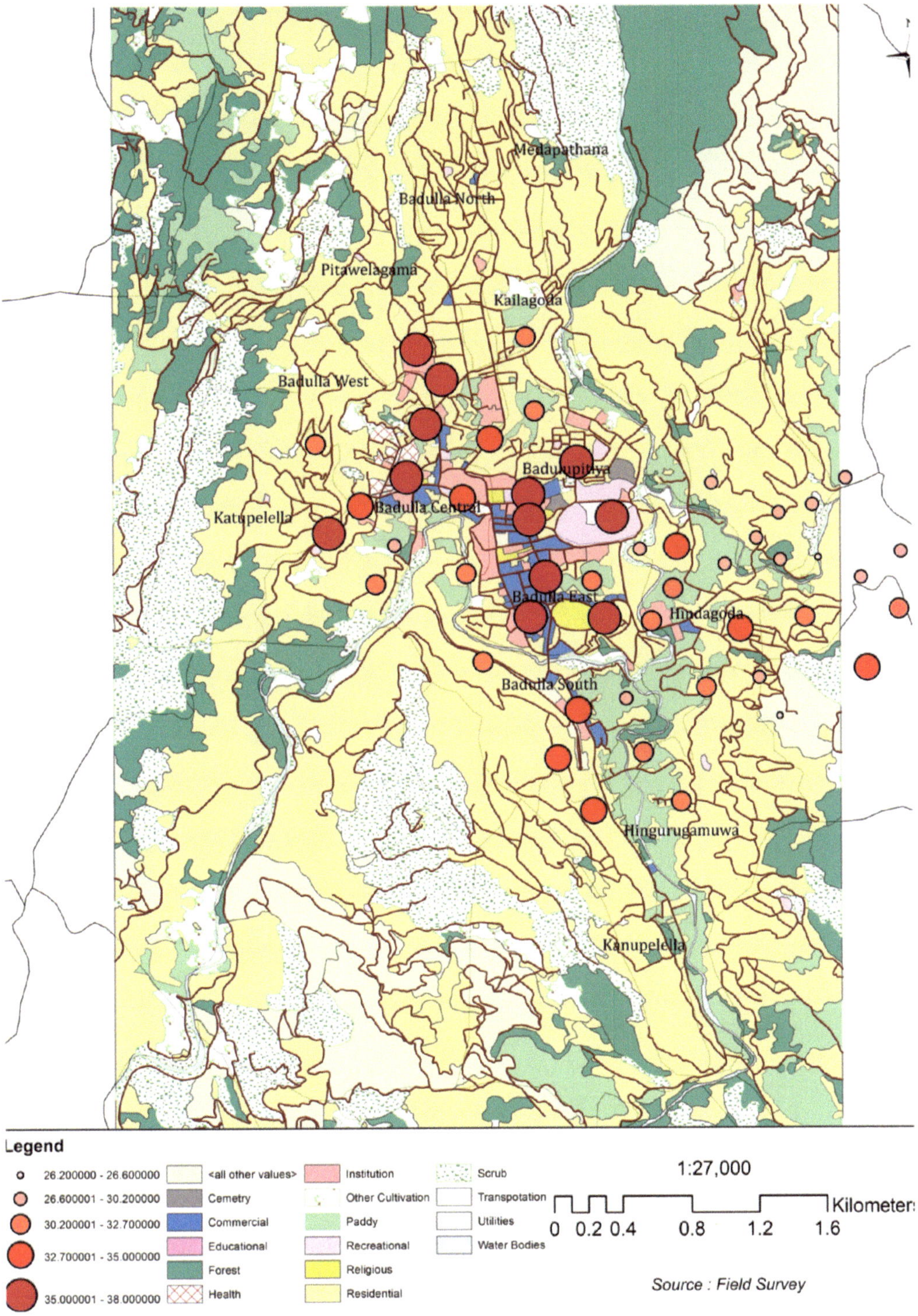

FIGURE 17.10 Onsite temperature values, 2020, 4.00 pm, 28 July 2020.

Source: Compiled by the Author, 2020.

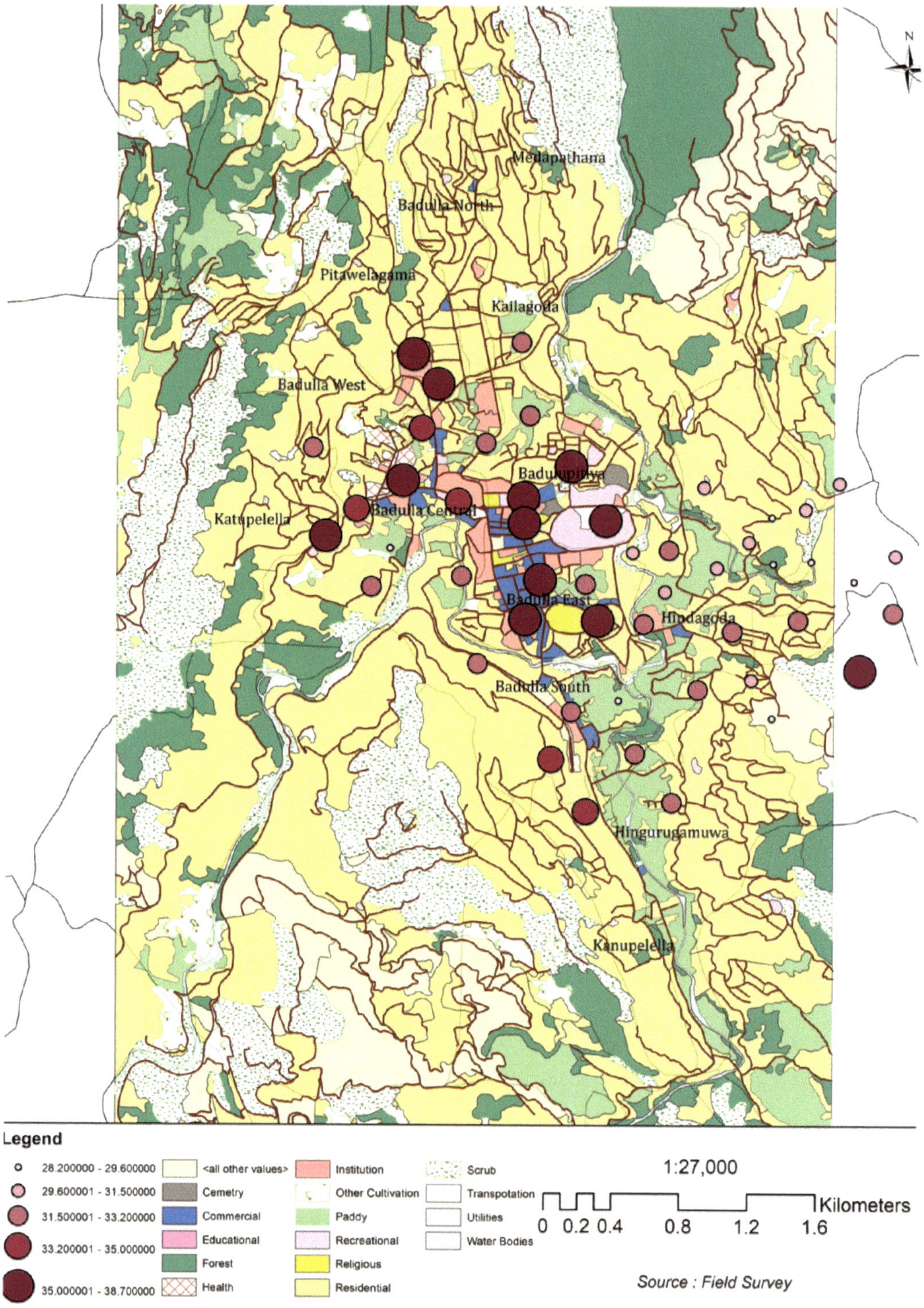

FIGURE 17.11 Onsite temperature values, 2020, 10.00 am, 28 July 2020.

Source: Compiled by the Author, 2020.

Onsite temperature values were derived on sunny days with a cloud-free sky. The values are derived to one decimal point starting from 10.00 am to 11.00 am. There were 11 higher temperature values between 35°C and 38.7°C. All 11 points are within the town center and moderate temperature values are recorded within the residential areas.

The temperature values are derived at 4.00–5.00 pm on the same day at the same geographical points. There were 12 higher temperature values between 35°C and 38°C. These points are all within the town center. The one geographical point which is within the town center residential zone recorded higher temperatures than the 10.00–11.00 am of the day.

Temperature values were then analyzed and compared with the building density. Compared to the 2019 data, the 2020 manual data represent higher temperature values at selected points. The highest building density is recorded within the Badulupitiya, which shows that building density is only one component to determine temperature.

17.6 CONCLUSION

As a regional provincial capital, Badulla has high development demand. Therefore, development cannot be obstructed within the area. But in town planning these precautions can be considered to eliminate future hazardous consequences.

According to the investigation, 2019 temperature values show a 2°C average increase over 2016. It can be considered that the area's temperature is increasing with development, while some places are contributing to the heat island effect. Comparing the 2019 situation with satellite-derived temperature data, the 2020 manual data represent various outputs in selected places that can be considered heat islands. It appears from the investigation that building density does not greatly influence temperature increase, but that surface material, physical formation of the area and physical density are highly influential in the creation of urban heat islands in Badulla MC, Sri Lanka.

When building developments are approved, consideration must be given to the materials to be used, and in many cases this does not appear to be happening. Appropriate provision needs to be made in the macro-level planning panorama.

Second, building density is the main issue that causes UHI effects. It is therefore necessary to consider density distribution when urban forms are being proposed, which could be a crucial aspect of planning attempts. This should be addressed at the micro level according to each development. More open spaces could be created in between physical elements to minimize the hidden places.

Physical elements are the final criterion creating UHIs within Badulla. This should be addressed at both micro and macro levels. At macro level, the wind flow pattern and natural urban form are relevant. If the present urban form results in creation of UHIs, the micro level needs to consider how far buildings are set back, how much space is set aside to the side and behind physical elements. Not only should this be regulated, but it should also be monitored continually and enforced better than is the case at present.

17.7 RECOMMENDATION TO CONTROL UHI IN BADULLA MC

Urban heat islands are a serious problem in Badulla MC, as the analysis shows. Therefore, it is essential to identify planning recommendations to control UHIs. This study has focused on temperature distribution patterns of the area, which could be expanded by overlaying the vegetation pattern of the area.

These are important concepts in urban planning. Maintaining the geographical formation of the area may cause smooth wind flow patterns, otherwise hidden locations may be created without wind flow. Retaining fire gaps in between buildings may be another solution to eliminate these hidden locations. However, the ultimate issue is the comfort of urban living. Nowadays, urban living is a trend in all Sri Lanka's cities. High demand for land may cause discomfort with the climatology of urban areas. Despite the demand for land, there should be regulatory provision to control minimum land sub-divisions that cause high building density.

Further, to eliminate these circumstances it is essential to maintain the wind flow pattern. The use of materials in built-up areas needs to reconsider whether they increase heat emission. Regulatory bodies need to promote green rooves and cooling rooves instead of high heat emission materials and, as an important preventive measure, to introduce regulations to discourage the blocking of wind flows.

BIBLIOGRAPHY

Amed, S. "Assessment of urban heat islands and impact of climate change on socioeconomic over Suez Governorate using remote sensing and GIS techniques", *The Egyptian Journal of Remote Sensing and Space Science*, 21, 15–25, (2018).

Arifwidodo, S., O. Chandrasiri, R. Abdulharis, and T. Kubota. "Exploring the effects of urban heat island: A case study of two cities in Thailand and Indonesia." *APN Science Bulletin*, 9(1), (2019). doi:10.30852/sb.2019.539.

Asimakopoulos, D.N., V.D. Assimakopoulos, N. Chrisomallidou, N. Klitsikas, D. Mangold, P. Michel, M. Santamouris, and A. Tsangrassoulis. "*Energy and Climate in the Urban Built Environment*", M. Santamouris (Ed.) London: James & James Publication, (2001a).

Asimakopoulos D.N., V. D. Assimakopoulos, and N. Chrisomallidou. "*Energy and Climate in the Urban Built Environment*", London, UK: James & James, (2001b).

Bonan, G. "*Ecological Climatology*", United Kingdom: Cambridge University Press, (2002).

Climate Protectiob Partnership Devision (CPPD), "*Reducing Urban Heat Island: Compendium of Strategies*", USA: Urban Heat Islands Basics, (2008).

Disanayake, D.M.S.L.B., T. Morimoto, M. Ranagalage, and Y. Murayama. "Land-Use/Land-Cover Changes and Their Impact on Surface Urban Heat Islands: Case Study of Kandy City, Sri Lanka", *Climate*, 7, (2019).

Disanayake, W.M.P.L., "Study of Spatial Pattern of Urban Heat Island in Colombo District," Faculty of Graduate studies, Msc in GIS & Remote Sensing, University of Sri Jayewardenepura, (2017).

Emmanuel, R., and H.J.S. Fernando. "Urban heat islands in humid and arid climates: Role of urban form and thermal properties in Colombo, Sri Lanka and Phoenix", *Climate Research*, 34, 241–251, (2007). https://doi.org/10.3354/cr00694

Eriyagama, N., V.U. Smakhtin, L. Chandrapala, and K. Fernando. "Impacts of climate change on water resources and agriculture in Sri Lanka: A review and preliminary vulnerability mapping", IWMI Research Reports 94787, International Water Management Institute, (2010).

Estoque, R., and Y. Murayama. "Effects of landscape composition and pattern on land surface temperature: An urban heat island study in the megacities of Southeast Asia." *Science of The Total Environment*, 577, 349–359, (2017). http://dx.doi.org/10.1016/j.scitotenv.2016.10.195

Gartland, L. "*Heat Islands Understanding and Mitigating Heat in Urban Areas in the UK and USA in 2008*", London: Earthscan, (2008).

Grimmond, S. "Urbanization and global environmental change: Local effects of urban warming", *Geographical Journal*, 173(1), 83–88, (2007).

Memon, R.A., D.Y.C. Leung, and L.I.U.A. Chunho. "Review on the generation, Determination and mitigation of Urban heat island", *Journal of Environmental Science-China*, 20(1), 120–128, (2008).

Mobaraki, A. "*Strategies for Mitigating Urban Heat Island Effects in Cities: Case of Shiraz City Center*", USA: Lab Publishing, (2012).

Nuruzzaman, M.d. "Urban Heat Island: Causes, Effects and Mitigation Measures - A Review," *International Journal of Environmental Monitoring and Analysis*, 3(2), 67–73, (2015). doi:10.11648/j.ijema.20150302.15, (2015).

Oke, T.R. "Initial Guidance to Obtain Representative Meteorological Observations at Urban Sites", World Meteorological Organization, Instruments and Observing Methods, IOM Report No. 81, WMO/TD-No. 1250, (2006).

Perera, N.G.R., and R. A. Emmanuel. "Local Climate Zone based approach to urban planning in Colombo, Sri Lanka", *Urban Climate*, 23, 188–203, (2018).

Pickett, S.T.A., M.L. Cadenasso, J.M. Grove, C.H. Nilon, R.V. Pouyat, W.C. Zipperer, and R. Costanza. "Urban ecological systems: Linking terrestrial ecological, physical, and socioeconomic components of metropolitan areas", *Annual Review of Ecological Systematics*, 32, 127–157, (2001). http://dx.doi.org/10.1007/978-0-387-73412-5_7

Ranagalage, M., R.C. Estoque, and Y. Murayama, "An Urban Heat Island Study of the Colombo Metropolitan Area, Sri Lanka, Based on Landsat Data (1997–2017)", *Geo-Information*, (2017).

Roth, M. "Urban Heat islands", *Handbook of Environmental Fluid Dynamics*, 2(11), 143–158, (2018).

Schwarz, N., S. Lautenbach, and R. Seppelt. "Exploring indicators for quantifying surface urban heat islands of European cities with MODIS land surface temperatures", *Remote Sensing of Environment*, 115(12), 3175–3186, (2011).

Senanayake, I.P., W.D.D. Welivitiya, and P. Nadeeka. "Remote sensing based analysis of urban heat islands with vegetation cover in Colombo city, Sri Lanka using Landsat-7 ETM+ data." *Urban Climate*, 5, 19–35, (2013).

Shahmohamadi, P., A.I. Che-Ani, N.A.G. Abdullah, M.M. Tahir, K.N.A. Maulud, and M.F.I. Mohd-Nor. "The Link between Urbanization and Climatic Factors: A Concept on Formation of Urban Heat Island", *WSEAS Transactions on Environment and Development*, 6, (2010).

Unger, J., B. Balazs, Z. Sumeghy, and T. Gal. "Multiple variable model for estimating the maximum UHI using 2 and 3 dimensional surface parameters". In: Lindqvist S., C.S.B. Grimmond, I. Eliasson (Eds.), *Proc 6th Int Conf Urban Clim. (ICUC-6)*, University of Göteborg, pp. 346–349, (2006).

Voogt, J.A. "*Urban Heat Islands: Hotter Cities*", Washington, DC: American Institute of Biological Sciences, (2004).

Weng, Q., D. Lu, and J. Schubring. "Estimation of land surface temperature–vegetation abundance relationship for urban heat island studies", *Remote Sensing of Environment*, 89, 467–483, (2004).

Wu, H., L.P. Ye, W.Z. Shi, and K.C. Clarke. "Assessing the effects of land use spatial structure on urban heat islands using HJ-1B remote sensing imagery in Wuhan, China", *International Journal of Applied Earth Observation and Geoinformation*, 32, 67–78, (2014).

Yamamoto, Y. "Measures to mitigate urban heat islands", *Science and Technology Trends Quarterly Review*, 18(1), 65–83, (2006).

Zhang, H., Z. Qi, X. Ye, Y. Cai, W. Ma, and M. Chen. "Analysis of land use/land cover change, population shift, and their effects on spatiotemporal patterns of urban heat islands in metropolitan Shanghai, China", *Applied Geography*, 44, 121–133, (2013).

18 Analyzing Land-Use/Land-Cover Changes and Its Impact on Land Surface Temperature in Berhampore Municipality, West Bengal, India

Hassan Momin
IDSK-University of Calcutta, Kolkata, India

Chandrakala Tamang
Kazi Nazrul University, Asansol, India

Rubia Biswas
Raja Narendra Lal Khan Women's College (Autonomous), Gope Palace, India

Abdul Odud
Aligarh Muslim University, Aligarh, India

CONTENTS

DOI: 10.1201/9781003131922-21

18.1 INTRODUCTION

Population density, size, amenities or facilities, and socio-economic connectivity are all considered part of the urbanization process (Hugo, Champion, and Lattes 2003; Plane, Henri, and Perry 2005; Montgomery and Balk 2011; Biswas and Momin 2018). Urbanization alters land cover types, resulting in distinct climatic conditions known as 'urban climate' that differ from one urban area to another (Lilly Rose and Devadas, 2009). Urban development drastically alters the biophysical climate by replacing soil and vegetation with changed land use, affecting the land surface's albedo, temperature, and moisture properties (Lo and Quattrochi 2003). The global trend of growing urbanization has contributed to a rapid shift in land-use patterns. The environmental quality (EQ) of the regions is one of the important implications of such an unregulated and unplanned process (Krishnan and Firoz 2021). Rapid urbanization processes trigger conversion of vegetation to built-up areas, resulting in heat islands in urban areas. The temperature in the heart of urban centers ranges from 2 to 5°C, which is relatively high compared to surrounding areas. Temperatures in densely populated settlement areas are significantly greater than those in less densely populated areas (Hua and Ping 2018). The transformation of agricultural land and rural settlements has culminated in urban areas becoming the most prominent category of land use (Roy et al. 2020). It has also been observed that land-surface temperature (LST) has followed a declining trend from urban centers to sub-urban areas and then to rural areas (Tran et al. 2017). Residential zones retain the maximum surface temperature compared to vegetation and water bodies (Pal and Ziaul 2017). Consequently, built-up areas influence the intensity of surface temperature, whereas vegetation mitigates the effect of heat islands in urban areas (Zhou, Huang and Cadenasso 2011).

Each land-use/cover type has distinct traits that influence land surface temperature and vary depending on the season. Hotspots can be found in three types of materials: bare land, rooftops, and marble surfaces (Ding and Shi 2013). Many variables, including industrial factories, barren lands, and dense residential areas, are responsible for heat islands (Stathopoulou and Cartalis 2017). The increased number of mining industries has reduced vegetation cover, resulting in maximum surface temperature. Human activity makes a major contribution to the intensification of urban heat islands (Kayet et al. 2016).

As a result, the nature of land cover has a significant influence on the spatial pattern of the surface temperature distribution. As per LST for various classes of LULC, built-up areas retain the maximum temperature, followed by barren surface, agriculture and vegetation in decreasing order. Climate change and urban heating were observed to have an impact on the study area, as mean LST increased across all categories of land use (Ullah et al. 2019). Sun, Wu and Tan (2012) studied the association between LST and LULC using various methods such as NDVI, NDBI, NDBal, and MNDWI. They discovered that as the concentration of the built-up area increased, so did the temperature of the land surface, and that as vegetation increased, the LST decreased. Chemura et al. (2020) evaluated the association between land-use changes and water balance at the catchment area.

The implications of land-use modifications for temperature trends are determined by the type of LULC and their transformation from one kind to another. Apart from dense forest, all other land cover conversions to agriculture result in cooling, however dense forest conversions lead to significant warming (Nayak and Mandal 2019). Moreover, when studying the changing global environment, LST is an important parameter that is primarily determined by land-use patterns. The surface temperature of land rises as a result of urban renewal, which is exacerbated by modifications in land use (Qiao et al. 2020). According to Gogoi et al. (2019), the Indian subcontinent has significant spatio-temporal variability in climatic phenomena, with decadal, inter-annual and annual changeability in surface temperature. While studying the effect of land-use/cover change on LST in East India, researchers discovered that changing patterns of land use have a substantial effect on the region's climatic conditions in a variety of ways, including land transformation, surface temperature, and rainfall. According to the previous analysis, the rise in land surface temperature in Indian cities is primarily the result of ongoing changes in land use and growing built-up areas (Kant et al. 2009; Mallick, Rahman and Singh 2013; Sharma, Chakraborty and Joshi 2015; Grover and Singh 2016; Singh, Kikon and Verma 2017; Khandelwal et al. 2018). The increase in built-up areas in urban local bodies or municipality areas continuously leads to higher surface temperature (Kotharkar and Surawar 2016; Ivan and Benedek 2017; Ziaul and Pal 2018; Dhar et al. 2019).

In general, the study of urban climatic conditions requires high-resolution images, which may be acquired through satellite imagery (Dousset and Gourmelon 2003). GIS and RS tools are helpful for analyzing LULC change and surface temperature, as well as detecting urban development and studying its effect on LST (Qian, Cui and Chang 2006). The present study examines how land-use change impacts on land surface temperature in Berhampore Municipality. Vegetation changes as well as the association between NDVI and LST for the selected periods (1992, 2005, and 2018) were analyzed. Land surface temperature mapping with a directional profile from 1992 to 2018 was also studied.

18.2 DESCRIPTION OF THE STUDY AREA

Berhampore is a city in the state of West Bengal, situated on the eastern bank of the Bhagirathi river at 24°1′ N latitude and 88°25′ E longitude (Figure 18.1). It is the principal city and administrative center of the Murshidabad District and acts as a connecting link between north and south Bengal.

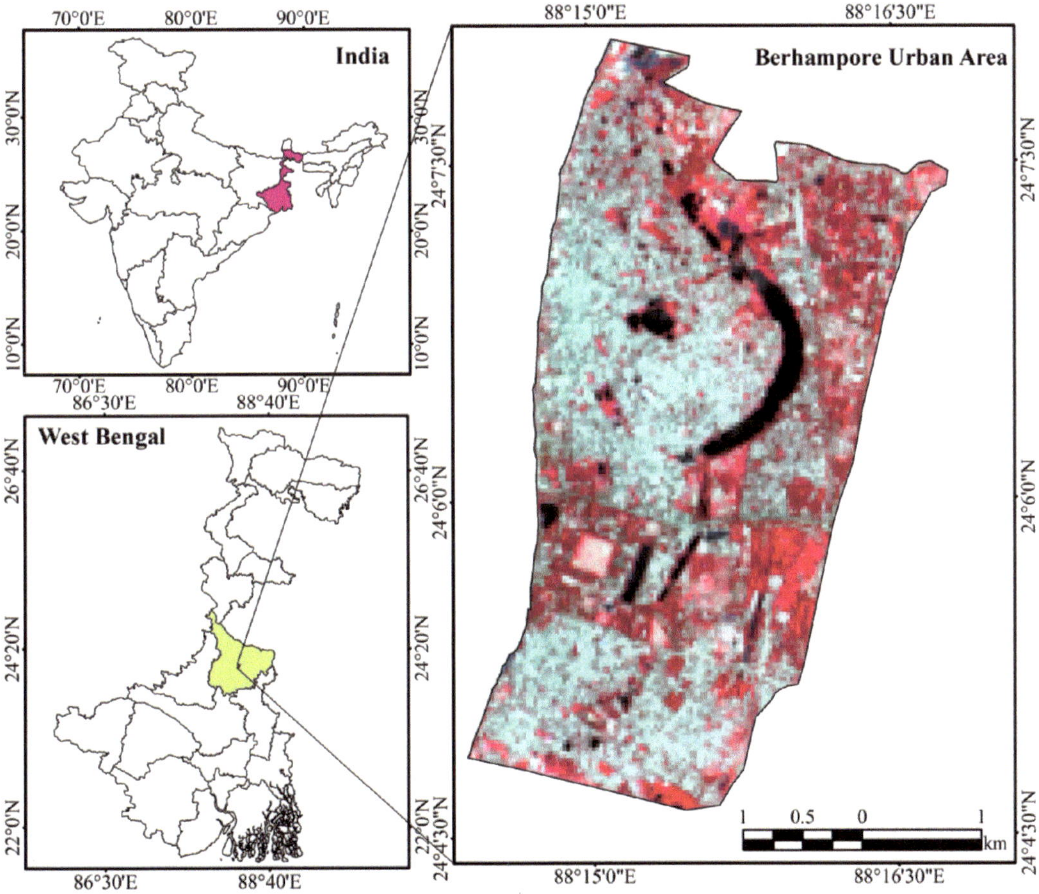

FIGURE 18.1 Map showing the location of the study area.

As per the census data (2011), the city had a total population of 195,223. It is also the seventh most populous city in West Bengal and the only Class I town in Murshidabad District. The land-use pattern of this city has changed drastically over the last few decades. Berhampore municipality and its surrounding region experienced a remarkable growth in settlement between 1990 and 2018 due to the large-scale transformation of agricultural lands and wetlands into built-up areas, causing surface temperature to rise in the city. Even so in the last two decades, seven new census towns have been developed around the Berhampore municipality due to the growing population. The large-scale expansion of residential areas has a significant impact not only on the city but also on the surrounding region (residential, industrial, commercial, etc.). This significant development and growth of settlement expanse reduces the space occupied by water bodies and vegetation.

18.3 DATA SOURCE AND METHODOLOGY

18.3.1 Data Source

The study assessed the Berhampore municipality using satellite images. Landsat 5 TM data from 1992 and 2005 as well as Landsat 8 OLI and TIRS of 2018 were analyzed. Table 18.1 explains in detail the study's data sources, while Figure 18.2 shows the methods adopted for the fulfillment of the study.

TABLE 18.1
Description of Data Used in This Study

Satellite	Sensor	Path/Row	Year	Spatial Resolution	Source
Landsat 5	TM	139/43	1992	30	USGS
Landsat 5	TM	139/43	2005	30	USGS
Landsat 8	OLI/TIRS	139/43	2018	30	USGS

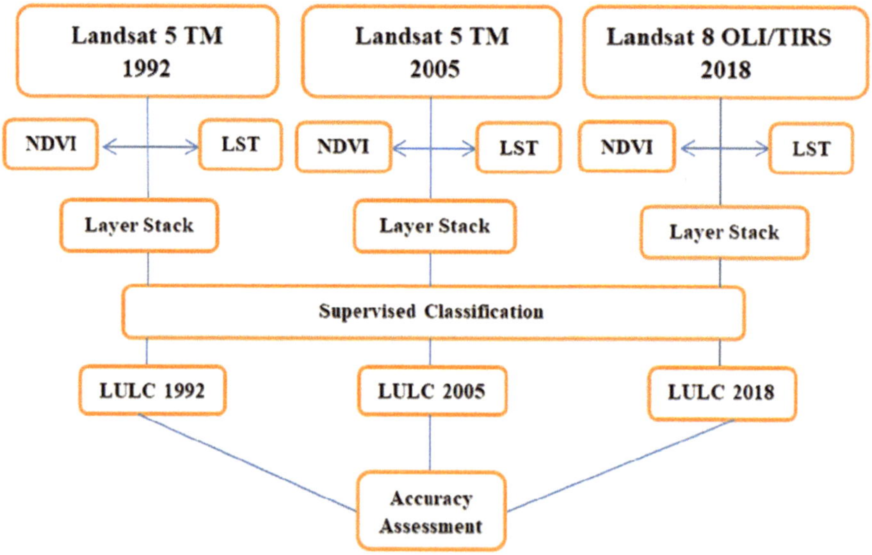

FIGURE 18.2 Methodology adopted for the study.

18.3.2 Change Analysis for LULC

To detect the changes in land-use land cover, remote sensing data were acquired from the USGS. Generally, four classes – vegetation cover, water bodies, fallow/open fields, built-up/settlement areas – were analyzed using supervised classification techniques with a parallelepiped algorithm and Erdas Imagine 2014 software. It is a valuable and widely used tool for classifying LULC. In this analysis, 72 signatures for the entire map were utilized to detect image classification.

In this study, selected years of data were analyzed for categorizing land use and land cover. The cross-tabulation order is then applied to identify each function of the study region. Following that, selected years of maps were evaluated, and the region of transition was determined based on the spatial resolution of the satellite imageries. As a result, the findings demonstrated the actual changes that occurred in the study area between 1992 and 2018.

18.3.3 NDVI

NDVI (Normalized Difference Vegetation Index) is used to calculate the area of vegetation health. NDVI varies from −1 to 1, where the value −1 indicates low vegetation coverage and 1 indicates high vegetation coverage. Rouse et al. (1974) used the following equation:

$$\text{NDVI} = (\text{Near IR} - \text{R})/(\text{Near IR} + \text{R}) \quad (18.1)$$

where Near IR = near infrared band; R = red band.

18.3.4 Process for LST

Nichol and To (2012) used the following formula for the DN (digital number)-to-radiance conversion:

$$\text{L} = \text{ML} * \text{QCAL} + \text{AL} \quad (18.2)$$

where L = radiance; QCAL = thermal band; ML = multiplicative rescaling factor; AL = additive rescaling factor.

After that, conversion of spectral radiance to BT (brightness temperature) is done using Eq. (18.3):

$$\text{BT} = \text{K}_2/\ln(\text{K}_1/\text{L}+1) - 273.15 \quad (18.3)$$

where BT = brightness temperature; K_1 or K_2 = thermal constants.

Thereafter, the PV (proportion of vegetation) was determined by Eq. (18.4):

$$\text{PV} = \{(\text{NDVI} - \text{NDVI}_{\text{min}})/(\text{NDVI}_{\text{max}} - \text{NDVI}_{\text{min}})\}^2 \quad (18.4)$$

Then, emissivity (€) was calculated by Eq. (18.5):

$$€ = 0.004 * \text{PV} + 0.986 \quad (18.5)$$

Finally, the surface temperature value was calculated using Eq. (18.6), given by Artis and Carnahan 1982; Sobrino, Jiménez-Muñoz and Paolini 2004; and Tomlinson et al. 2011.

$$\text{LST} = \text{BT}/1 + (\lambda\sigma\text{BT}/\text{hc}) * \ln(€)\} \quad (18.6)$$

where λ = wavelength of emitted radiance; σ = Boltzmann constant; h = Planck's constant; c = speed of light.

18.3.5 Association between LST and NDVI

After calculating the above techniques, every class of LULC was computed to sum up the values of LST and NDVI (Mukherjee and Singh 2020). Then regression analysis was used to determine the association between these values.

18.4 LULC ACCURACY

Accuracy assessment, an essential part of any thematic mapping task, was applied to each map, taking selective signatures required for identifying the error matrix (Foody 2002). The overall accuracy (OA) was derived by Eq. (18.7):

$$\text{OA} = \left(\text{Overall number of exact values/overall number of values}\right) * 100 \tag{18.7}$$

Then, 149 samples from the ground and Google Earth were chosen for verification. Settlement areas, water bodies, land under fallow and vegetation were chosen as individual classes for accuracy assessment. However, two key methods, user accuracy (UA) and producer accuracy (PA) were utilized for this evaluation. A commission matrix, which is the association between the classified map and the referenced data summarized in an error matrix, is used to compute the controller and explainer's reliability. The following equations were used to measure producer and consumer accuracy:

$$\text{Error of commission} = \text{off diagonal row value/total row value} \tag{18.8}$$

$$\text{Error of omission} = \text{off diagonal column value/total column value} \tag{18.9}$$

$$\text{UA}(\%) = 100\% - \text{commission error}(\%) \tag{18.10}$$

$$\text{PA}(\%) = 100\% - \text{omission error}(\%) \tag{18.11}$$

The computed data was then used to determine the Kappa co-efficient. The value of Kappa co-efficient lies between 0 and 1. The value 1 denotes the ideal agreement with producer value. Eq. (18.12) can be used to calculate the Kappa co-efficient in percentage:

$$k = \frac{N\sum_{i=1}^{r} aii - \sum_{i=1}^{r}(ai + *ai + i)}{N^2 - \sum_{i=1}^{r}(ai + *ai + i)} \tag{18.12}$$

According to Monserud and Leemans (1992), values below 0.4 indicate bad agreement; 0.4–0.55 indicate fair agreement; values between 0.7 and 0.85 indicate quite good agreement, and greater than 0.85 indicate excellent agreement between the images.

18.5 RESULTS

18.5.1 LU/LC Change

Land-use/cover change is crucial for proper and efficient planning, management, maintenance and allocation of resources (Mallupattu, Reddy and Reddy 2013). Four groups (fallow land, vegetation, water body and settlement) were chosen for the classification of LULC change. Figure 18.3 shows

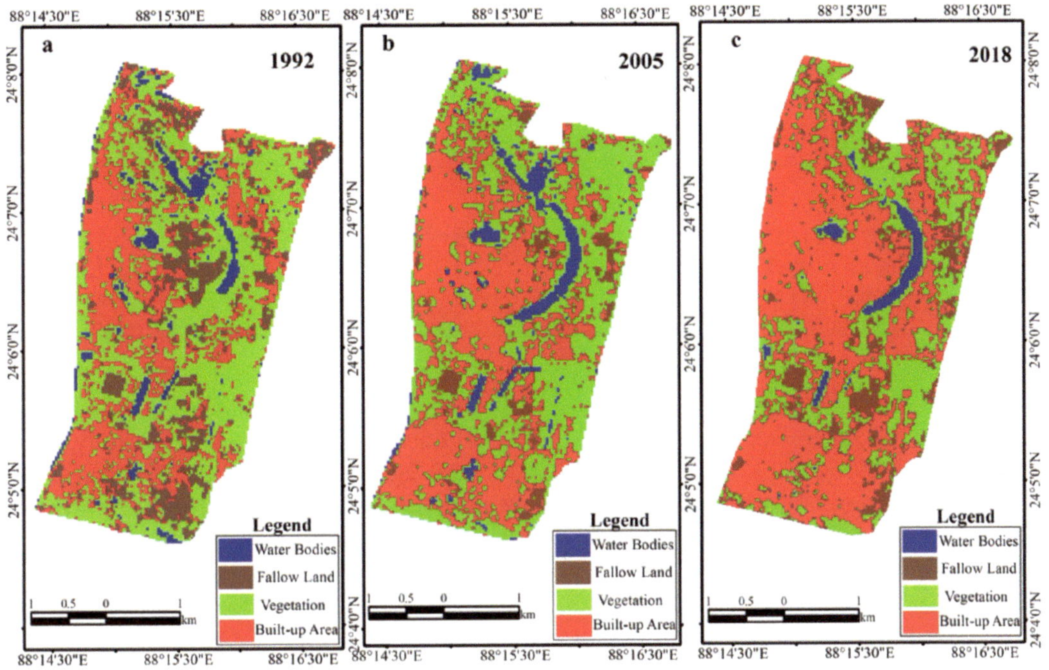

FIGURE 18.3 LULC Classification of Berhampore urban area, (a) 1992; (b) 2005; (c) 2018.

TABLE 18.2
Error Matrix for the Land Cover Study

	Producer Accuracy (%)				User Accuracy (%)					
Year	Water Body	Fallow Land	Vegetation	Built-up Area	Water Body	Fallow Land	Vegetation	Built-up Area	OA (%)	Kappa
1992	100	87.5	86.11	81.82	97.3	75.68	83.78	97.3	88.51	0.847
2005	97.14	83.33	90.63	89.74	91.89	94.59	78.38	94.59	89.86	0.865
2018	94.87	96.67	82.93	97.3	100	78.38	91.89	97.3	91.89	0.892

maps of the different LULC of Berhampore Urban Area in 1992, 2005, and 2018. Table 18.2 shows the accuracy (%) for each land type between 1992 and 2018. The overall accuracy was 88.51% in 1992, 89.86% in 2005, and 91.89% in 2018. The Kappa co-efficient was 0.847 in 1992, 0.865 in 2005, and 0.892 in 2018. Overall, the values denote a very good to excellent agreement of the images, indicating that the classification was completed with the utmost precision.

Table 18.3 shows the calculation of land-use/cover area in selected periods. Between 1992 and 2018, the urban area grew at a faster rate than the rest of the land cover. The total settlement area expanded by 160.8 hectares between 1992 and 2005. Moreover, the total built-up/settled area increased by 322.8 hectares between 1992 and 2018. With the exception of the settlement area, all land-use types decreased in order. The water body increased from 68 to 96 hectares between 1992 and 2005. However, it decreased from 96 hectares to 38 hectares between 2005 and 2018.

Overall, water bodies shrank by 30 hectares over the 26 years. In 1992, there were 216 hectares of fallow land, decreased to 85.5 hectares, but increasing again by 116 hectares between 2005 and 2018. The total fallow land decreased by 5.5 hectares over the entire period (1992–2018). In 1992,

TABLE 18.3
LULC Change in Berhampore for the Years 1992 to 2018

LULC	Areas in Hectares							
	1992	%	2005	%	Change (1992–2005)	2018	%	Change (1992–2018)
Water Body	68	4.56	96	6.45	28	38	2.56	−30
Fallow Land	216	14.52	85.5	5.75	−130.5	210.5	14.14	−5.5
Vegetation	699.3	46.98	641	43.06	−58.3	412	27.68	−287.3
Built-up Area	505.2	33.94	666	44.74	160.8	828	55.62	322.8

vegetation covered 699.3 hectares of the total area. Vegetation cover decreased steadily from 2005 to 2018, accounting for approximately 641 hectares and 412 hectares of total area, respectively.

Between 1992 and 2018, the total vegetation coverage area decreased by 287.3 hectares. The total built-up area in 1992 was 505.2 hectares of the total area. However, by 2018, it had increased to approximately 828 hectares. As a result of the massive urban transformation, the vegetation area has decreased substantially while the built-up area has increased (Figure 18.4).

18.5.2 NDVI Change

The NDVI value ranged between −0.217 and 0.615 in 1992, and it gradually decreased until 2005, when it ranged between −0.153 and 0.605. In 2018, the area covered by vegetation reduced because of the increase in built-up areas. The value of NDVI ranged between 0.468 and 0.0119 in 2018. Vegetation areas decreased significantly as a result of the increased built-up area. Figure 18.5 clearly depicts the decline in vegetation cover as well as the growth in built-up territory.

18.5.3 LST Change

Figure 18.6 displays the variability of surface temperature in Barhampore, Murshidabad District, in 1992, 2005, and 2018. The bright reddish color on the maps of 1992, 2005, and 2018 designates the highest temperature. In 1992, the average land surface temperature was 32.57°C, with a temperature range between 25.83°C and 37.23°C. The minimum land surface temperature increased in 2005

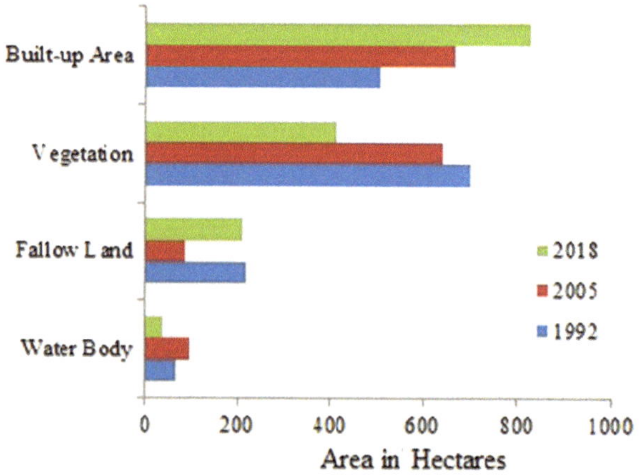

FIGURE 18.4 LULC change analysis of Berhampore urban area.

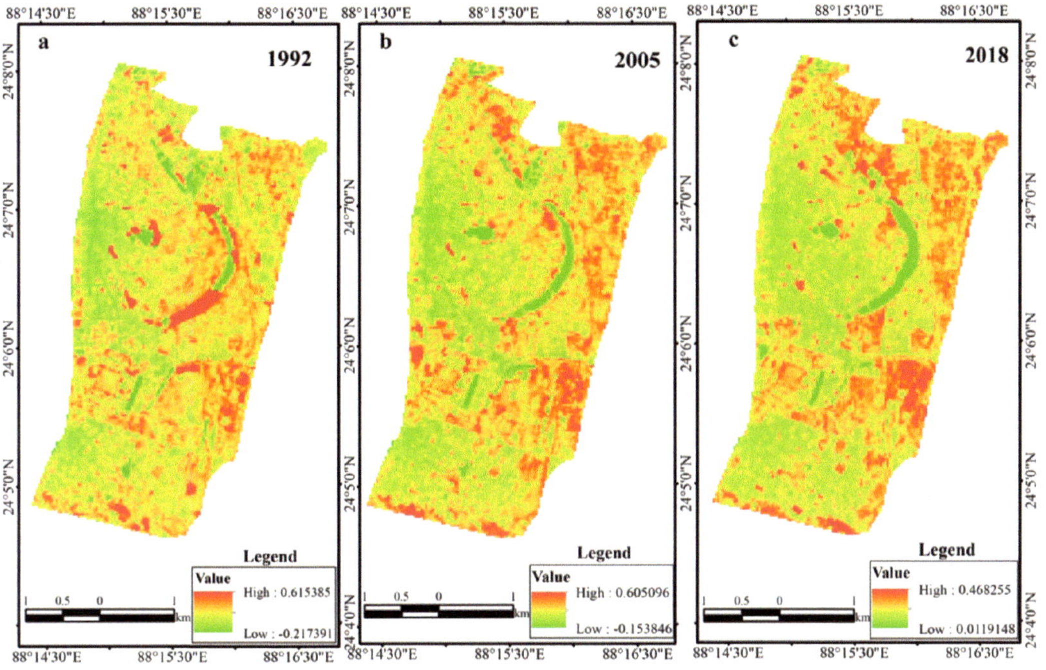

FIGURE 18.5 Spatial distribution of NDVI for Berhampore (a) 1992; (b) 2005; (c) 2018.

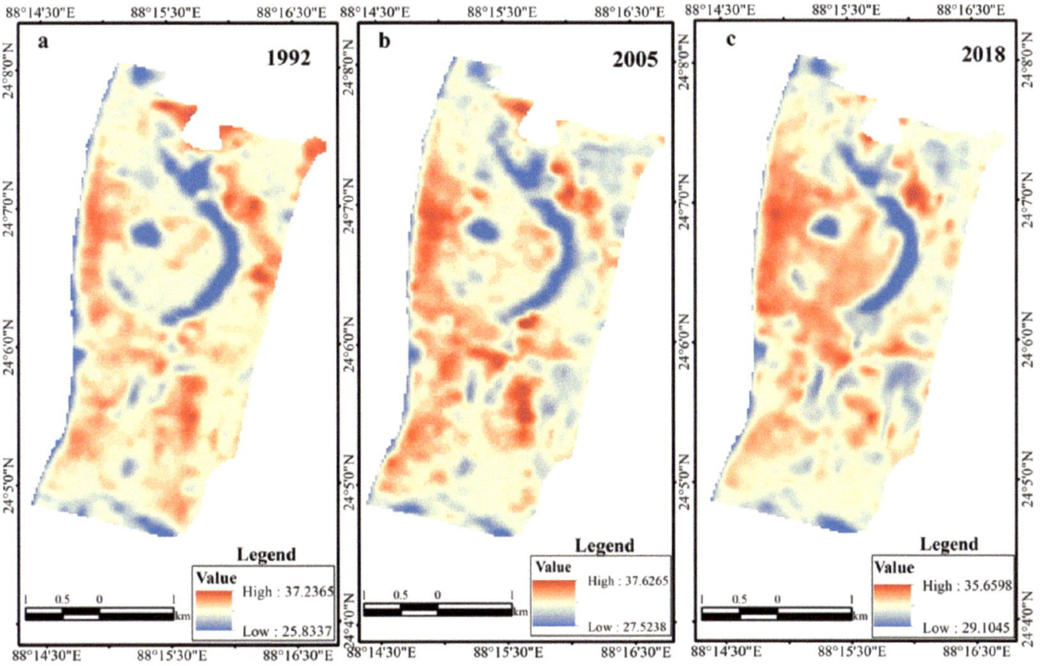

FIGURE 18.6 Spatial distribution of LST for Berhampore (a) 1992; (b) 2005; (c) 2018.

TABLE 18.4
Year-Wise Value of LST (°C) and NDVI

	LST(°C)			NDVI		
Year	**Min**	**Max**	**Mean**	**Min**	**Max**	**Mean**
1992	25.83	37.23	32.57	−0.217	0.615	0.179
2005	27.52	37.62	32.9	−0.153	0.605	0.215
2018	28.96	35.65	32.93	0.011	0.468	0.215

TABLE 18.5
Rise in Temperature (°C) of Different LULC

	Rise in LST(°C)		
LULC	**1992–2005**	**2005–2018**	**1992–2018**
Water Body	1.6901	1.5806	3.27
Fallow Land	0.406	0.7611	1.1671
Vegetation	0.8499	1.0736	1.9235
Built-up Areas	0.4144	0.7203	1.1346

compared to the previous year, the minimum land surface temperature being 27.52°C while the highest temperature was 37.62°C (Table 18.4).

The average land surface temperature was 32.90°C. The minimum range of temperature increased by 1.69°C between 1992 and 2005. The maximum and minimum land surface temperatures in 2018 were 28.96°C and 35.65°C, respectively. The mean value of land surface temperature is 32.93°C. Between 1992 and 2018, the land surface temperatures of the water body increased by 3.27°C, of fallow land by 1.167°C, of vegetation by 1.923°C, and of built-up areas by 1.1346°C (Table 18.5).

18.5.4 LST and NDVI

The correlation between LST and NDVI was found to be significant. Figure 18.7 shows the correlation in the years 1992, 2005, and 2018. The values of LST and NDVI are significantly correlated in 1992, 2005, and 2018 ($R^2 = 0.254$ in 1992, $R^2 = 0.318$ in 2005, $R^2 = 0.088$ in 2018). As a result, there is a significant correlation between these parameters during these years.

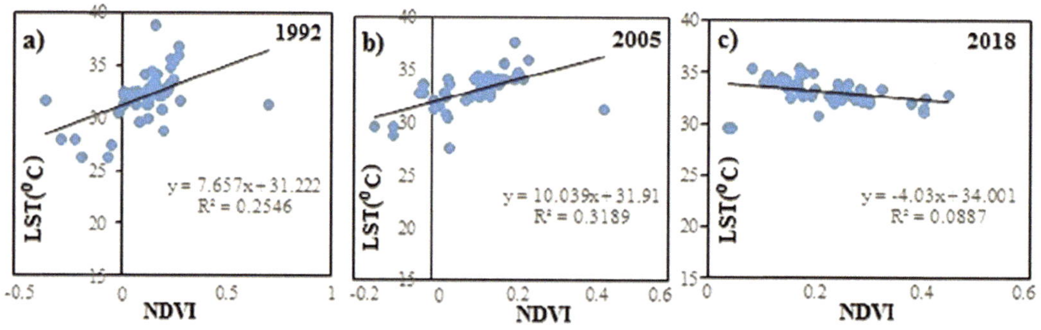

FIGURE 18.7 Regression analysis between NDVI and LST for Berhampore (a) 1992; (b) 2005; (c) 2018.

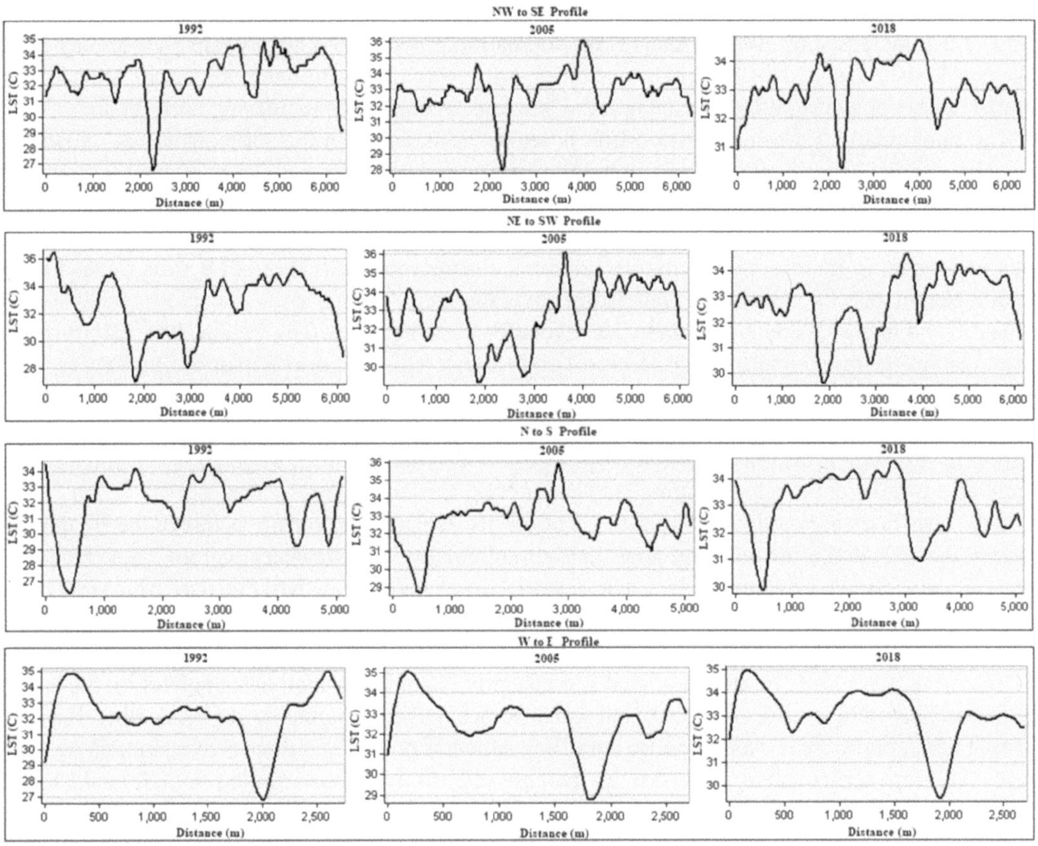

FIGURE 18.8 Directional profiles for Berhampore in 1992, 2005 and 2018.

18.5.5 LST Profile

For the land surface temperature profile, four directional profiles (N to S, W to E, NE to SW, and NW to SE) were taken. The NW to SE profile direction in the study area demonstrated the increase in temperature from 1992 to 2018. Because of the dense vegetation, the temperature was at its lowest in 1992. However, over time, between 2005 and 2018, the profile exhibited an increase. As a result of the growing tendency of built-up areas, the surface temperature showed a few high peaks. Initially, the temperature in the NE to SW profile was significantly lower.

Nonetheless, there are high peaks in 2005 and 2018 due to densely populated areas. In the case of the W to E directional profile, only one high peak was observed, predominantly in the westernmost portion. The westernmost section, the heart of the Berhampore Urban Area, is densely populated, with a highly built-up area. As a result, high surface temperature is most expected in the westernmost part of the area (Figure 18.8). The N to S directional profile, particularly the southernmost part, is consistently the high built-up and low vegetation coverage area, where high temperature peaks were recorded from 1992 to 2018.

18.6 DISCUSSION

Berhampore is a major commercial, educational, and administrative hub of West Bengal. According to Census of India (2011) data, the Berhmapore urban areas have a population of 305,609. The population growth rate changed drastically from 1941 to 2011. The density of the population in

Berhampore is also high, about 6500/km^2. The decadal growth rate in 1961 was +12.05, but by 2011, it had increased to +90.81. As a result, population growth put a huge strain on the study area. The problems associated with rapid urbanization are severe in the Berhampore Urban Area. Urban sprawl has become a significant issue in today's world, owing to the increasing built-up area.

The municipality's westernmost portion is seeing an increase in the settlement area. However, it is also expanding and sprawling in both the western and southern parts of the municipality. The built-up area increased steadily from 505.2 ha to 828 ha between 1992 and 2018 respectively. Being a central hub of the Murshidabad District, Berhampore is a rapidly urbanized and vast settlement area experiencing increasing population pressure. Furthermore, Berhampore's urban growth has been fueled mainly by a decline in vegetation coverage areas.

Scrub and mango gardens dominate the vegetation coverage area. The vegetation zone primarily encompasses the municipality's northernmost and eastern regions. Large expanses of greenery were also visible around the Chatrar Bill and nearby water basins. In 1992, the land-use map showed 699.3 ha of vegetation cover, but in 2018, it showed only 412 ha. Between 1992 and 2018, the vegetation in this region declined from 46.98% to 27.68%. Therefore, the vegetation coverage in the study area is deficient due to changing land cover compared to previous decades.

The water body area of land-use classes is primarily ecologically sensitive, due to the Bhagirathi river in the west and a few wetland areas in the municipality's mid-eastern region. These two wetlands, Chatrar Bill and Bishnupur Bill, also act as a barrier to the growth of the increasing settlement in that area. According to land-use classes, the water body in this region has shrunk from 68 ha to 38 ha., owing primarily to the built-up development.

Therefore, the LULC in the study area is constantly changing, which has an effect on the area's surface temperature. As the vegetation and water bodies in the municipality region decline, the land surface temperature rises. Additionally, the filling of water bodies and their conversion, along with vegetation, into settlement areas lead to an increase in built-up areas, resulting in high land surface temperature in the municipality region.

Furthermore, high temperatures combined with high humidity levels may trigger the incidence of malaria, dengue fever, and other diseases. The rising temperature is primarily associated with an increased risk. As a result, urbanization, along with rising temperatures, has an impact on human health. Moreover, the continuous expansion of industries, factories, air pollution, solid waste, etc., has adverse effects on the human environment.

18.7 CONCLUSION

Urbanization is one of the major driving forces in changing land-use patterns, and it also causes a change in land surface temperature (LST). The massive land-use shifts facilitate development and growth of concrete areas, leading to urban sprawl. This study examined Landsat data between 1992 and 2018, which revealed that the study area, i.e., the Berhampore Urban Area, has experienced significant vegetation loss and increased surface temperature. The study also showed a significant association between LST and NDVI. The surface temperature has risen as a result of the increased settlement, while vegetation has steadily declined. The vegetation, bare soil, fallow land, and water bodies have continued to decline as the built-up area has continued to increase. Berhampore has the fastest-growing urban population in Murshidabad district, which contributes to high built-up areas and ultimately high surface temperatures.

The small and medium urban agglomeration areas in developing countries face tremendous pressure from urbanization. In many places, urban processes are putting their environment in jeopardy. Moreover, climate change poses a challenge to the management of urban areas. Similarly, the area has extensive urban sprawl and urban processes, which has an ongoing impact on the ecological bodies. The growing number of unplanned urban areas can pose a severe threat to the environment. This could lead to urban environmental degradation in the future. Thus, proper planning and policy are necessary for the study area, especially in terms of urban greenery and environment, which require special attention.

BIBLIOGRAPHY

Artis, D. and W. Carnahan. 1982. Survey of Emmissivity Variability in Thermography of Urban Areas. *Remote Sensing of Environment* 12 no. 4: 313-329. https://doi.org/10.1016/0034-4257(82)90043-8

Biswas, R., and H. Momin. 2018. Civic Amenities of English Bazar Municipality of Malda District. In *Social and Economic Development: Issues and Challenges in India*, eds. S. N. Ahmed, A. Ahmed, and N. Khan, pp. 43–52. New Delhi: A.K Publication.

Chemura, A., D. Rwasoka, O. Mutanga, T. Dube, and T. Mushore. 2020. The Impact of Land-Use/Land Cover Changes on Water Balance of the Heterogeneous Buzi Sub-Catchment, Zimbabwe. *Remote Sensing Applications: Society and Environment* 18. https://doi.org/10.1016/j.rsase.2020.100292

Dhar, R. B., S. Chakraborty, R. Chattopadhyay, and P. K. Sikdar. 2019. Impact of Land-Use/Land-Cover Change on Land Surface Temperature Using Satellite Data: A Case Study of Rajarhat Block, North 24-Parganas District, West Bengal. *Journal of the Indian Society of Remote Sensing* 47 no. 2: 331–348. https://doi.org/10.1007/s12524-019-00939-1

Ding, H., and W. Shi. 2013. Land-use/Land-cover Change and its Influence on Surface Temperature: A Case Study in Beijing City. *International Journal of Remote Sensing* 34 no. 15: 5503–5517. https://doi.org/10.1080/01431161.2013.792966

Dousset, B., and F. Gourmelon. 2003. Satellite Multi-Sensor Data Analysis of Urban Surface Temperatures and Land Cover. *ISPRS Journal of Photogrammetry and Remote Sensing* 58: 43–54. https://doi.org/10.1016/S0924-2716(03)00016-9

Foody, G. M. 2002. Status of Land Cover Classification Accuracy Assessment. *Remote Sensing of Environment* 80 no. 1: 185–201. https://doi.org/10.1016/S0034-4257(01)00295-4

Gogoi, P. P., V. Vinoj, D. Swain, G. Roberts, J. Dash, and S. Tripathy. 2019. Land Use and Land Cover Change Effect on Surface Temperature over Eastern India. *Scientific Reports* 9 no. 1. https://doi.org/10.1038/s41598-019-45213-z

Grover, A., and R. B. Singh. 2016. Monitoring Spatial Patterns of Land Surface Temperature and Urban Heat Island for Sustainable Megacity: A Case Study of Mumbai, India, Using Landsat TM Data. *Environment and Urbanization* 7: 38–54. https://doi.org/10.1177/0975425315619722

Guha, S., H. Govil. A. Dey, and N. Gill. 2018. Analytical Study of Land Surface Temperature with NDVI and NDBI Using Landsat 8 OLI and TIRS data in Florence and Naples city, Italy. *European Journal of Remote Sensing* 51 no. 1: 667–678. https://doi.org/10.1080/22797254.2018.1474494

Guha, S., H. Govil, and S. Mukherjee. 2017. Dynamic Analysis and Ecological Evaluation of Urban Heat Islands in Raipur City, India. *Journal of Applied Remote Sensing* 11 no. 3. https://doi.org/10.1117/1.JRS.11.036020

Hua, A. K., and O.W. Ping. 2018. The Influence of Land Use/Land Cover Changes on Land Surface Temperature: A Case Study of Kuala Lumpur Metropolitan City. *European Journal of Remote Sensing* 51 no. 1: 1049–1069. https://doi.org/10.1080/22797254.2018.1542976

Hugo, G., A. Champion, and A. Lattes. 2003. Towards a New Conceptualization of Settlements for Demography. *Population and Development Review* 29: 277–297.

Ivan, K., and J. Benedek. 2017. The Assessment Relationship between Land Surface Temperature (LST) and Built-Up Area in Urban Agglomeration Case Study: Cluj-Napoca, Romania. *Geographia Technica* 12 no. 1: 64–74. https://doi.org/10.21163/GT_2017.121.07

Kant, Y., B. D. Bharath, J. Mallick, C. Atzberger, and N. Kerle. 2009. Satellite-Based Analysis of the Role of Land Use/Land Cover and Vegetation Density on Surface Temperature Regime of Delhi, India. *Journal of the Indian Society of Remote Sensing* 37 no. 2: 201–214. https://doi.org/10.1007/s12524-009-0030-x

Kayet, N., K. Pathak, A. Chakrabarty, and S. Sahoo. 2016. Urban Heat Island Explored by Co-relationship between Land Surface Temperature vs Multiple Vegetation Indices. *Spatial Information Research* 24 no. 5: 515–529. https://doi.org/10.1007/s41324-016-0049-3

Khandelwal, S., R. Goyal, N. Kaul, and A. Mathew. 2018. Assessment of Land Surface Temperature Variation due to Change in Elevation of Area Surrounding Jaipur, India. *The Egyptian Journal of Remote Sensing and Space Science* 21 no. 1: 87–94. https://doi.org/10.1016/j.ejrs.2017.01.005

Kotharkar, R., and M. Surawar. 2016. Land Use, Land Cover, and Population Density Impact on the Formation of Canopy Urban Heat Islands through Traverse Survey in the Nagpur Urban Area, India. *Journal of Urban Planning and Development* 142 no. 1. https://doi.org/10.1061/(ASCE)UP.1943-5444.0000277

Krishnan, S., and M. Firoz. 2021. Impact of Land Use and Land Cover Change on the Environmental Quality of a Region: A Case of Ernakulam District in Kerala, India. *Regional Statistics* 11 no. 2. https://doi.org/10.15196/RS110205

Landsat Project Science Office. Landsat 7 Science Data User's Handbook (2002) https://landsat.gsfc.nasa.gov/wp-content/uploads/2016/08/Landsat7_Handbook.pdf (accessed June 12, 2017).

Lilly Rose, A., and M. D. Devadas. 2009. Analysis of Land Surface Temperature and Land Use/Land Cover types Using Remote Sensing Imagery - A Case in Chennai City, India. In *Proceedings of the 7th International Conference on Urban Climate (ICUC-7), Yokohama, Japan 29.*

Lo, C. P., and D. A. Quattrochi. 2003. Land-Use and Land-Cover Change, Urban Heat Island Phenomenon, and Health Implications. *Photogrammetric Engineering & Remote Sensing* 69 no. 9: 1053–1063. https://doi.org/10.14358/PERS.69.9.1053

Lopez, J. M. R., K. Heider, and J. Scheffran. 2017. Frontier of Urbanization: Identifying and Explaining Urbanization Hot Spots in the South of Mexico City using Human and Remote Sensing. *Applied Geography* 79: 1–10. https://dx.doi.org/10.1016/j.apgeog.2016.12.001

Ma, Y., Y. Kuang, and N. Huang. 2010. Coupling Urbanization Analyses for Studying Urban Thermal Environment and its Interplay with Biophysical Parameters based on TM/ETM+ Imagery. *International Journal of Applied Earth Observation and Geoinformation* 12 no. 2: 110–118. https://doi.org/10.1016/j.jag.2009.12.002

Mallick, J., A. Rahman, and C. K. Singh. 2013. Modelling Urban Heat Islands in Heterogeneous Land Surface and its Correlation with Impervious Surface Area by Using Night Time ASTER Satellite Data in Highly Urbanizing City, Delhi, India. *Advances in Space Research* 52 no. 4: 639–655. https://doi.org/10.1016/j.asr.2013.04.025

Mallupattu, P. K., J. Reddy, and S. Reddy. 2013. Analysis of Land Use/Land Cover Changes Using Remote Sensing Data and GIS at an Urban Area, Tirupati, India. *The Scientific World Journal.* https://dx.doi.org/10.1155/2013/268623

Monserud, R. A., and R. Leemans. 1992. Comparing Global Vegetation Maps with the Kappa Statistic. *Ecological Modelling* 62 no. 4: 275–293. https://doi.org/10.1016/0304-3800(92)90003-W

Montgomery, M. R., and D. Balk. 2011. The urban transition in developing countries: Demography meets geography. In *Global urbanization*, pp. 89–106. University of Pennsylvania Press.

Mukherjee, F., and D. Singh. 2020. Assessing Land Use-Land Cover Change and its Impact on Land Surface Temperature Using Landsat Data: A Comparison of Two Urban Areas in India. *Earth Systems and Environment* 4 no. 2: 385–407. https://doi.org/10.1007/s41748-020-00155-9

Nayak. S., and M. Mandal. 2019. Impact of Land Use and Land Cover Changes on Temperature Trends Over India. *Land Use Policy* 89. https://doi.org/10.1016/j.landusepol.2019.104238

Nichol, J. E. 1994. A GIS-Based Approach to Microclimate Monitoring in Singapore's High-Rise Housing Estates. *Photogrammetric Engineering and Remote Sensing* 60, no. 10: 1225–1232.

Nichol, J. E., and P. H. To. 2012. Temporal Characteristics of Thermal Satellite Images for Urban Heat Stress and Heat Island Mapping. *ISPRS Journal of Photogrammetry and Remote Sensing* 74: 153–162. https://doi.org/10.1016/j.isprsjprs.2012.09.00

Pal, S., and S. Ziaul. 2017. Detection of Land Use and Land Cover Change and Land Surface Temperature in English Bazar Urban Centre. *The Egyptian Journal of Remote Sensing and Space Science* 20 no 1: 125–145. https://doi.org/10.1016/j.ejrs.2016.11.003

Plane, D. A., C. Henri, and M. Perry. 2005. Migration Up and Down the Urban Hierarchy and Across the Life Course. *Proceeding of the National Academy of Sciences* 102 no. 43: 15313–15318.

Qian, L. X., H. S. Cui, and J. Chang. 2006. Impacts of Land Use and Cover Change on Land Surface Temperature in the Zhujiang Delta. *Pedosphere* 16 no. 6: 681–689. https://doi.org/10.1016/S1002-0160(06)60103-3

Qiao, Z., L. Lui, Y. Qin, X. Xu, B. Wang, and Z. Liu. 2020. The Impact of Urban Renewal on Land Surface Temperature Changes: A Case Study in the Main City of Guangzhou, China. *Remote Sensing* 12. https://doi.org/10.3390/rs12050794

Rouse, J. W., R. H. Haas, J. A. Schell, and D. W. Deering. 1974. Monitoring Vegetation Systems in the Great Plains with ERTS. *NASA Special Publication* 351: 309.

Roy, S., S. Pandit, E. A. Evac, M. S. H. Bagmard, M. Papiac, L. Banik, T. Dubeg, F. Rahman, M. A. Razi. 2020. Examining the Nexus between Land Surface Temperature and Urban Growth in Chattogram Metropolitan Area of Bangladesh Using Long Term Landsat Series Data. *Urban Climate* 32. https://doi.org/10.1016/j.uclim.2020.100593

Sharma, R., A. Chakraborty, and P.K. Joshi. 2015. Geospatial Quantification and Analysis of Environmental Changes in Urbanizing City of Kolkata (India). *Environmental Monitoring and Assessment* 187 no. 1. https://doi.org/10.1007/s10661-014-4206-7

Singh, P., N. Kikon, and P. Verma. 2017. Impact of Land Use Change and Urbanization on Urban Heat Island in Lucknow City, Central India. A Remote Sensing Based Estimate. *Sustainable Cities and Society* 32: 100–114. https://doi.org/10.1016/j.scs.2017.02.018

Snyder, W. C., Z. Wan, Y. Zhang, and Y. Z. Feng. 1998. Classification Based Emissivity for Land Surface Temperature Measurement from Space. *International Journal of Remote Sensing* 19 no. 14: 2753–2774. https://doi.org/10.1080/014311698214497

Sobrino, J. A., J. C. Jiménez-Muñoz, and L. Paolini. 2004. Land Surface Temperature Retrieval from Landsat TM 5. *Remote Sensing of Environment* 90 no. 4: 434–440. https://doi.org/10.1016/j.rse.2004.02.003

Stathopoulou, M., and C. Cartalis. 2017. Daytime Urban Heat Islands from Landsat ETM+ and Corine Land Cover Data: An Application to Major Cities in Greece. *Solar Energy* 81 no. 3: 358–368. https://doi.org/10.1016/j.solener.2006.06.014

Sun, Q., Z. Wu, and J. Tan. 2012. The Relationship between Land Surface Temperature and Land Use/Land Cover in Guangzhou, China. *Environmental Earth Sciences* 65 no. 6: 1687–1694. https://doi.org/10.1007/s12665-011-1145-2

Tomlinson, C. J., L. Chapman, J. E. Thornes, and C. Baker. 2011. Remote Sensing Land Surface Temperature for Meteorology and Climatology: A Review: Remote Sensing Land Surface Temperature. *Meteorological Applications* 18 no. 3: 296–306. https://doi.org/10.1002/met.287

Tran, D. X., F. Pla, P. Latorre-Carmona, S. W. Myint, M. Caetano, and H. V. Kieu. 2017. Characterizing the Relationship between Land Use/Land Cover Change and Land Surface Temperature. *ISPRS Journal of Photogrammetry and Remote Sensing* 124: 119–132. https://doi.org/10.1016/j.isprsjprs.2017.01.001

Ullah, S., K. Ahmad, R. U. Sajjad, A. M. Abbasi, A. Nazeer, A. A. Tahir. 2019. Analysis and Simulation of Land Cover Changes and Their Impacts on Land Surface Temperature in a Lower Himalayan Region. *Journal of Environmental Management* 245. https://doi.org/10.1016/j.jenvman.2019.05.063

Zha, Y., J. Gao, and S. Ni. 2003. Use of Normalized Difference Built-Up Index in Automatically Mapping Urban Areas from TM Imagery. *International Journal of Remote Sensing* 24: 583–594. https://doi.org/10.1080/01431160304987

Zhang, J., Y. Wang, and Y. Li. 2006. A C++ Program for Retrieving Land Surface Temperature from the Data of Landsat TM/ETM+ Band 6. *Computers & Geosciences* 32 no. 10: 1796–1805. https://doi.org/10.1016/j.cageo.2006.05.001

Zhang, Y. 2006. Land Surface Temperature Retrieval from CBERS-02IRMSS Thermal Infrared Data and its Applications in Quantitative Analysis of Urban Heat Island Effect. *Remote Sensing* 10: 789–797.

Zhou, W., G. Huang, and M. L. Cadenasso. 2011. Does Spatial Configuration Matter? Understanding the Effects of Land Cover Pattern on Land Surface Temperature in Urban Landscapes. *Landscape and Urban Planning* 102 no. 1: 54–63. https://doi.org/10.1016/j.landurbplan.2011.03.009

Zhou, X., and Y. C. Wang. 2013. Dynamics of Land Surface Temperature in Response to Land-Use/Cover Change: Dynamics of Land Surface Temperature. *Geographical Research* 49 no. 1: 23–36. https://doi.org/10.1111/j.1745-5871.2010.00686.x

Ziaul, S., and S. Pal. 2018. Analyzing Control of Respiratory Particulate Matter on Land Surface Temperature in Local Climate Zones of English Bazar Municipality and Surroundings. *Urban Climate* 24: 34–50. https://doi.org/10.1016/j.uclim.2018.01.006

Part IV

A Multidimensional Approach to Urban Ecology, Space, Society and Sustainability

19 Reflections from East Kolkata Wetlands

A Conceptual Framework for Building Urban Resilience to Climate Change for Sustainable Urbanism in Developing Countries

Sudeshna Kumar and Haimanti Banerji
Indian Institute of Technology Kharagpur, Kharagpur, India

Sudeshna Ghosh
Indiana University of Pennsylvania, Indiana, Pennsylvania, USA

CONTENTS

19.1 INTRODUCTION

Uncontrolled urbanization and its associated uncertainties have compounded the effects of climate change, escalating the intensity and frequency of such risks, making it vital for cities to develop urban resilience (Newell et al. 2016; Leichenko 2018) and compelling the scientific community at large to require the integration of urban resilience into urban management practices (Heinzlef et al. 2020). The apparent increase in the negative impacts of climate change has motivated policy

DOI: 10.1201/9781003131922-23

makers and city leaders to prioritize urban development to be climate resilient; this is regarded as a paradigm shift from sustainable urbanism to climate urbanism (Long and Rice 2019), sustainable urban development to climate-resilient urban development. The concept of urban resilience has been progressively receiving the attention of researchers and policy makers alike in the past two decades (Sharifi and Yamagata 2018a; Sharifi 2020). This has involved formulating sets of indicators developing methods and tools for urban resilience assessment (Sharifi 2016; Sharifi and Yamagata 2017). Until 2009, insubstantial and disintegrated research was conducted on urban resilience, principally focused on risk mitigation and vulnerability assessment (Sharifi 2020). Increased attention in the domain of urban resilience is attributed to the fact that cities are engines of economic growth and the fulcrum of the ever-increasing global population, bringing a myriad of natural and man-made hazards (Klein et al. 2007). Conventionally, urban resilience has been defined as "plan and prepare for, absorb, recover from, and more successfully adapt to adverse events" (National Research Council 2012). Hulme has identified that the concept of resilience is very plastic, similar to the plasticity of climate change. Resilience is loosely associated with enhancing adaptive capacity levels or reducing vulnerability (Leichenko 2018; Hulme 2009). Especially in the last five years, there has been a rapid growth in publications on climate change adaptation and flood resilience. However, these studies are biased towards infrastructural, institutional and environmental dimensions at the cost of social and economic dimensions (Sharifi 2020). Leichenko (2018) has stated that to achieve long-term urban sustainability, it is pertinent to integrate mitigation and adaptation strategies with broad-ranging development plans and policies (Boyd et al. 2008; Sperling et al. 2008; Duxbury and Dickinson 2007; Seto, Sánchez-Rodríguez, and Fragkias 2010), fostering positive social changes alongside long-term urban sustainability. Contemporary cities need to be resilient against a myriad of environmental, economic, and political direct as well as interlinked, overlapped shocks and stresses. The most pivotal characteristics of a resilient city are flexibility, diversity, adaptive governance, and capacity for learning and innovation concerning its population, neighborhoods and systems. These characteristics make a city or an urban ecosystem a unique front-liner of technological innovation that is endeavoring to build up sustainable urban infrastructure (Leichenko 2018).

The term 'resilience', derived from the Latin verb meaning 'rebound or spring back', was conventionally used in the field of ecology and is being increasingly used in the field of social sciences and community development (Adger 2000), engineering designs (White and O'Hare 2014) and urban studies. It implies the ability of an object to withstand shock and trauma and return to its equilibrium state in good time (Sharifi and Yamagata 2016; Sharifi and Yamagata 2018b). In the perspective of an urban system, it refers to the capability of surviving and thriving in adverse, uncertain and both rapidly and incrementally changing situations. Enhancing resilience refers to the essential endeavors during all four phases of a disaster management cycle (Sharifi and Yamagata 2018b). The Intergovernmental Panel on Climate Change (IPCC), in its fifth assessment report, defined resilience as "the capacity of social, economic, and environmental systems to cope with a hazardous event or trend or disturbance, responding or reorganizing in ways that maintain their essential function, identity, and structure, while also maintaining the capacity for adaptation, learning, and transformation". The report proposes the inclusion of climate-resilient pathways to assess the climate change risks that are threatening urban areas (Kim and Lim 2016).

Sharifi (2020) has mapped the changing approach in the last two decades into three sub-periods (1998–2009, 2010–2014, and 2015–2020). A shift was noted from engineering-driven resilience characteristics such as robustness and stability in the first phase to adaptive-based approaches encouraging characteristics such as diversity, modularity, self-organization, connectivity, and adaptive capacity in the second phase, leading to an assessment of urban resilience. Sharifi and Yamagata (2017) have identified five dimensions of urban resilience – environment, economic, social, physical and institutional – that need to be addressed during various stages of a disaster management framework like planning, absorption, recovery and adaptation. The most dominant urban resilience theme has been urban flooding, while the economy and the wellbeing of society are the two most understudied dimensions. Kim and Lim (2016) present a theoretical discussion on different types of resilience. The concept of ecological resilience has been extended to the domain of socio-ecological systems. Engineering resilience is defined by Holling (1973) "as a system's ability to return to a steady-state or equilibrium after a disturbance" (Holling 1985), which includes natural disasters and human-induced social crises. From a socio-ecological perspective, resilience also has a traditional evolutionary resilience approach (Simmie and Martin 2010), which is in contradiction to mechanical operations of the resilience rationale (Kim and Lim 2016). This notion is based on the resilience of a socio-ecological system's capacity to evolve and revamp in response to stress and the system's ability to bounce back to a stable state before an external shock (Davoudi et al. 2012; Berkes, Folke, and Colding 2000), which is a constantly evolving process.

The authors of the current study have focused on developing a conceptual urban resilience framework that would act as a basis for assessing social and economic dimensions alongside the environmental dimension. This study has chosen an important urban ecological system in the eastern part of India, East Kolkata Wetlands (EKW), spreading over an area of 12,500 ha, which is being threatened due to uncontrolled urbanization triggered by unprecedented economic opportunities.

19.2 CONCEPTUAL DIMENSIONAL MODEL OF URBAN RESILIENCE

Despite the huge body of published literature on urban resilience, limited research was found on resilient urban forms (Sharifi and Yamagata 2018b). This study builds upon three important components of an urban area: the natural environment, man-made environment, and an integral human component at the crux of the two. The human component governs the proponent driving forces of the urban ecosystem. Human welfare is dependent on the interaction of the human component with the natural and physical man-made components. This is pivotal for achieving urban sustainability resilience, which is associated with the environmental psychology of the human component.

The study refers to an integrated dimensional model of urban resilience by combining the interlinkage and interconnectedness as explained in the pentagram sustainability dimension model (Kumar, Bhaumik, and Banerji 2020b; Kumar, Santara, and Banerji 2020c), with the three major conceived components of the neighborhood along with the conventional definition of urban resilience (National Research Council 2012). Figure 19.1 illustrates the conceptual development of the model. The study will deal with the understudied aspects of social wellbeing, the economy and man-made urban forms which are lacking in the existing state of the art of urban resilience (Sharifi and Yamagata 2017). The study will also address the issues of the vulnerable and marginalized section of the population residing within the UESA, since to render universally resilient cities, it is of utmost importance to address social inequity issues in their entirety (Meerow et al. 2019).

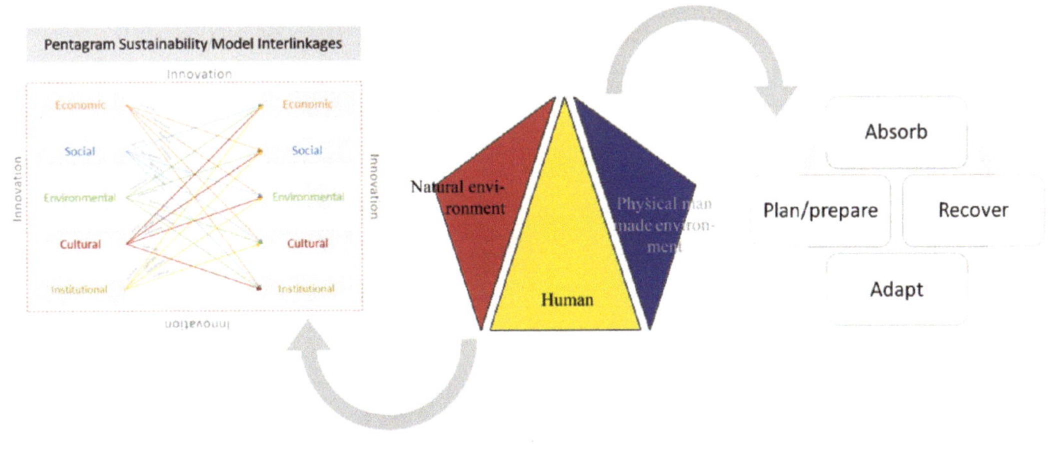

FIGURE 19.1 Conceptualization of the urban resilience dimension model.

In the age of smart cities, human interactions work in conjunction with technologies to enhance the urban level of services and quality of life, which justifies the inclusion of the technological aspect at the core of the dimension model. During the COVID-19 crisis that began in 2020, it was observed that countries like Singapore, South Korea, and Japan performed well in controlling outbreaks of the pandemic and fatality rates, due to their technological advancements and ever-increasing digitization across all generations triggered by high internet and smartphone usage. This included utilizing social media marketing through an expansion of digital payments, working from home and use of social network services, e-commerce shopping, online diagnosis, etc. Singapore has implemented exemplary surveillance, monitoring, and controlling techniques (UN ESCAPE 2020; Euromonitor Consulting 2020). They have conducted extensive testing and developed software such as TraceTogether and SafeEntry for contact tracing which helped in detecting people's proximity to and duration of contact with an infected individual (Singapore Government Agency website 2020; TraceTogether application 2020).

Appropriate use of technology along with proper awareness among people regarding its usage will be essential to develop any futuristic urban resilience model. The proposed urban resilience framework in this study will be mainly based on Sharifi and Yamagata's (2016) urban energy resilience model.

Previous researchers have mentioned the importance and impact of the two-way reaction between human–environmental interactions and climate change that are triggered by technology (Phillips 2016). Figure 19.2 illustrates the dimension model for building urban resilience to the climate change model. In this model, technology, which is vital for bringing about innovation, is at the nexus of climate change and human interactions since it has become indispensable in quantifying environmental changes (*The Lancet* Digital Health 2020). *The Lancet* Digital Health (2020) has emphasized the pertinent role of technological advancements in fighting against climate change

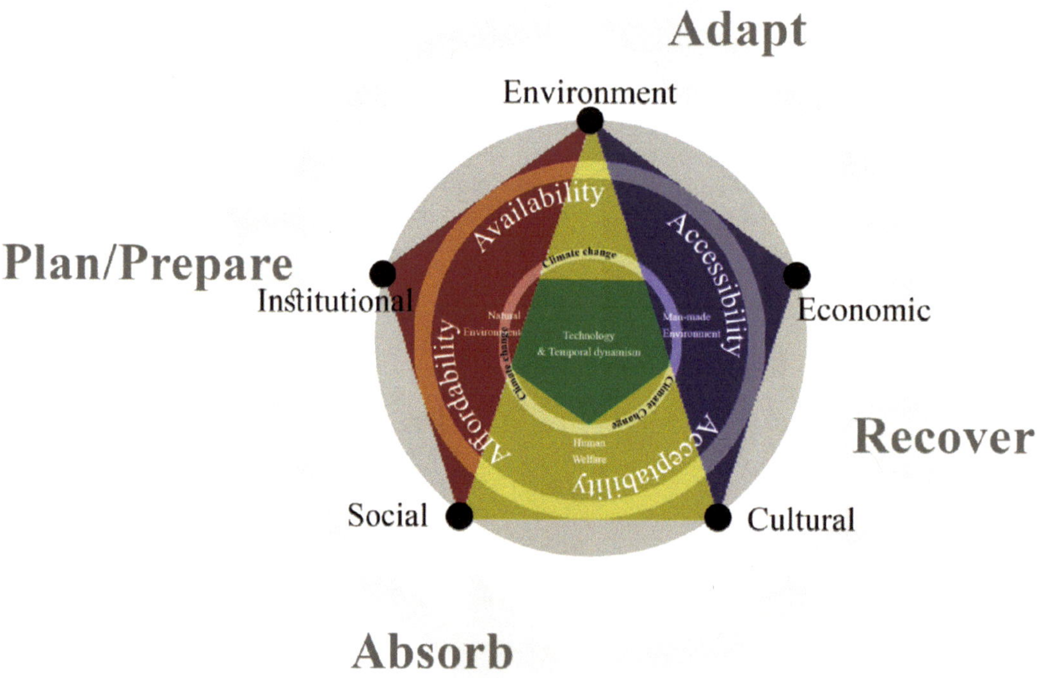

FIGURE 19.2 Dimension model for building urban resilience to climate change.

and COVID-19, equipping planners with real-time information on environmental conditions such as real-time air quality, and revealing linkages between these conditions and disease susceptibility.

Temporal dynamism integrated with technology has been included as one of the key criteria by the evaluation framework adopted in this study for urban resilience assessment. Adaptability, acceptability, affordability and accessibility of a particular resource or activity should be based on human interaction with the natural and man-made environment in conjunction with technology. These aspects are regarded as benefits of sustainability which would enhance resilience abilities such as plan and prepare, absorb, recover and adapt to make a socio-ecological system such as an urban eco-sensitive area (UESA) resilient (Sharifi and Yamagata 2017). The assessment framework will determine the extent of short-term and long-term opportunities and the impact of evolving technology on human interactions affecting urban sustainability in the long run. This study has conceptualized human activity as urban energy and has drawn parallels with the multi-dimensional urban energy resilience assessment model of Sharifi and Yamagata (2017).

Activity interaction triads will be assessed based on adaptability, acceptability, affordability, and accessibility, the four phases of urban resilience as defined by the National Research Council (2012). Planning, absorption, adaptation, and recovery are considered four significant steps to enhance the resilience of a UESA. Additionally, five dimensions of the Pentagram Sustainability Model along with their interlinkages will play a pivotal role within an urban ecosystem; their interactions will be categorized and assessed under each of these dimensions to ensure availability, accessibility, acceptability, and affordability of an activity/interaction within the urban ecosystem. The temporal and dynamic nature of these interactions within the UESA triggered by technological advancement will be considered.

The assessment will be conducted during all four phases of urban resilience. An urban resilient system should be robust in terms of planning and preparation to withstand, absorb and recover from shocks and stresses, and subsequently frame adaptive strategies to mitigate the adverse effects of the events (Sharifi and Yamagata 2016). Sharifi and Yamagata (2015) and Sharifi and Yamagata (2016) have identified 30 underlying principles of urban energy resilience which apply to any resilient urban system. Since resilience is a normative concept, it necessitates the involvement of participatory methods to procure knowledge from the diverse set of stakeholders concerned (Sharifi and Yamagata 2017). One of the key components of building up urban resilience to climate change is developing awareness regarding climate change, its causes, and adverse impacts (Iturriza et al. 2020). This is crucial for encouraging coordination amongst stakeholders. Eventually, the awareness will inculcate pro-environmental attitudes amongst the stakeholders leading to positive behavioral transformation.

19.3 RESEARCH CONTEXT

Urbanization is a human-induced process leading to land-use change which results from a combination of human activities and natural events (Mukherjee, Bebermeier, and Schütt 2018). The growing population, unplanned sprawl and increasing housing densities consequently bring about myriad societal changes (Reveshty 2011) allied with human activities on land, especially in peri-urban and fringe areas, which in turn affects the environment (Mohan, Pathan, Narendrareddy, Kandya, and Pandey 2011; Mukherjee et al. 2018). The EKW lies in a more than three-centuries-old metropolis of eastern India, Kolkata, which was established as capital of the British Indian Empire and later thrived as capital of the state of West Bengal and the major commercial hub of post-independence eastern India (Chowdhuri et al. 2020). Since then the city has experienced various phases of growth and development resulting in massive urban transformation, including some major instances of unplanned urbanization.

Unprecedented and unplanned development in the city due to multiple reasons has led to large-scale consumption of the internationally recognized Ramsar-protected site of EKW, which has experienced significant land-cover changes over the decades. These human-induced perturbations have led to a surge in temperature (Li et al. 2018) adversely impacting its economic, social, and cultural aspects (Kumar, Banerji, and Sengupta 2020a). Agricultural lands have also been utilized for real-estate development, triggered by the unparalleled population explosion in the metropolis. The rapid and haphazard market-driven conversion of land without any consideration for the ecological consequences has led to a massive shrinkage of EKW (Roy-Basu et al. 2020), threatening the entire ecosystem.

In their study Mondal et al. (2017) validated that wetland shrinkage has taken place along the transportation axis outskirts of the old city, leading to extensive expansion of the built-up area and consuming wetlands, croplands, and green spaces. The shrinkage and quality degradation of wetland is strongly correlated with infrastructural development, the influence of new township enclaves, existing developments, and croplands. It is getting worse with each passing day due to the increased human-induced disturbances. The most significant driver of this urban growth is that improved infrastructure accessibility providing the township surrounding the wetlands with access to markets and job opportunities has increased the real-estate demand for housing in this area, all of which can be attributed to economic intervention (Mondal et al. 2017). Baum-Snow (2007) has mentioned that sprawl or low-density spatial expansion mushrooms along newly built transport networks in the periphery rather than the city core, and this phenomenon can be observed along the entire stretch of the Eastern Metropolitan bypass (EMB)adjoining the EKW; Sawyer and Schmid (2015) refer to this as bypass urbanism. The entire stretch along with the Ramsar site is also undergoing landscape transformation.

Landscape transformations are usually referred to as an extensive change in the functions and structure of the landscape and are attributed mainly due to anthropogenic activities. Koffi et al. (2008) state that, in accordance with Bogaert's decision tree model, shrinkage or reduction in the patch is the last of the ten stages of landscape transformation. EKW is in that stage (Roy-Basu et al. 2020) due to the merging development in response to market forces. This new and emerging development along the eastern periphery of KMC adjacent to the EKW lacks basic physical infrastructures such as water supply, electricity, and sanitation (Ramachandra, Aithal, and Sreekantha 2012). According to the report submitted by the World Bank to the Government of West Bengal, Kolkata has topped the list of vulnerable cities of the world, its eastern fringe being the most vulnerable part. The report has forecast a 1.2 to 1.8°C rise in the temperature of Kolkata by 2050, with only a fractional increase in rainfall. Micro-level spatial analysis has predicted that the wards abutting the EMB adjacent to the EKW will be most vulnerable (Basu 2012).

An exploratory study to understand the ecosystem of EKW and the threats it is facing currently from the pressure of urbanization was jointly conducted by the Department of Architecture and Regional Planning, Indian Institute of Technology Kharagpur and the Department of Geography and Regional Planning, Indiana University of Pennsylvania, USA in June 2017. The study methodology included reconnaissance survey, photo-documentation, field testing, personal and group interpretation, and mapping. Questions were raised concerning wastewater treatment and wetland economy to better understand wetland sustainability in a changing environment and global market. The current research refers to some of the inferences drawn from this study in Section 19.4.2.

19.3.1 Aim and Scope of the Current Study

This study presents a dimension model for building urban resilience to climate change and aims to devise an inclusive resilient framework incorporating expert and stakeholder perceptions. The study has focused mainly on the UESA and selected EKW as the testbed. The principal objective of the study is to develop a scientific approach for achieving long-term sustainability for EKW and its adjacent settlements. The study conducts exploratory research and provides an in-depth situational analysis of EKW which paves the way for future data-driven assessment of urban resilience of EKW. The study recommends adaptation and mitigation strategies to enhance resilience by combating climate change through stakeholder participation. Boyd et al. (2008) establish the importance of policies devised for successful livelihood transformation adaptations in developing resilience against the crisis of climate change. The study will discuss enhancing the socio-ecological resilience of UESAs, adopting the evolutionary resilience approach to building a community resilient to climate change induced by urbanization.

19.3.2 Limitations of the Study and Scope of Future Work

This study outlines the aspects that should be considered for enhancing the resilient abilities of UESAs. The conceptual inclusive framework will act as a basis for the scope of future work which will include the assessment of the urban resilience of the UESA with the help of a model or a toolkit. The aspects highlighted in this study could be further operationalized incorporating context specificities as per the case study.

19.3.3 Need for the Protection and Conservation of the Wetlands

Situated along the eastern fringes of the city of Kolkata, EKW is one of the largest clusters of sewage-fed fish ponds in the world, covering an area of 12,500 ha, and comprising the largest wastewater-fed aquaculture area (EKWMA 2021) (see Figures 19.3 and 19.4). A part of the extensive inter-distributary wetland regimes formed by the Gangetic Delta (Kundu and Kumar 2010), EKW provides strong resistance to urban flooding. It is also known as the kidney of Kolkata (Ghosh et al. 2018), since it recycles the city's sewage, utilizing it for pisciculture and agriculture (Roy-Basu et al. 2020).

The post-independence liberation war of the neighboring country of Bangladesh led to a huge influx of refugees to the city of Kolkata, causing a burden for the existing infrastructure (Bose 2007).

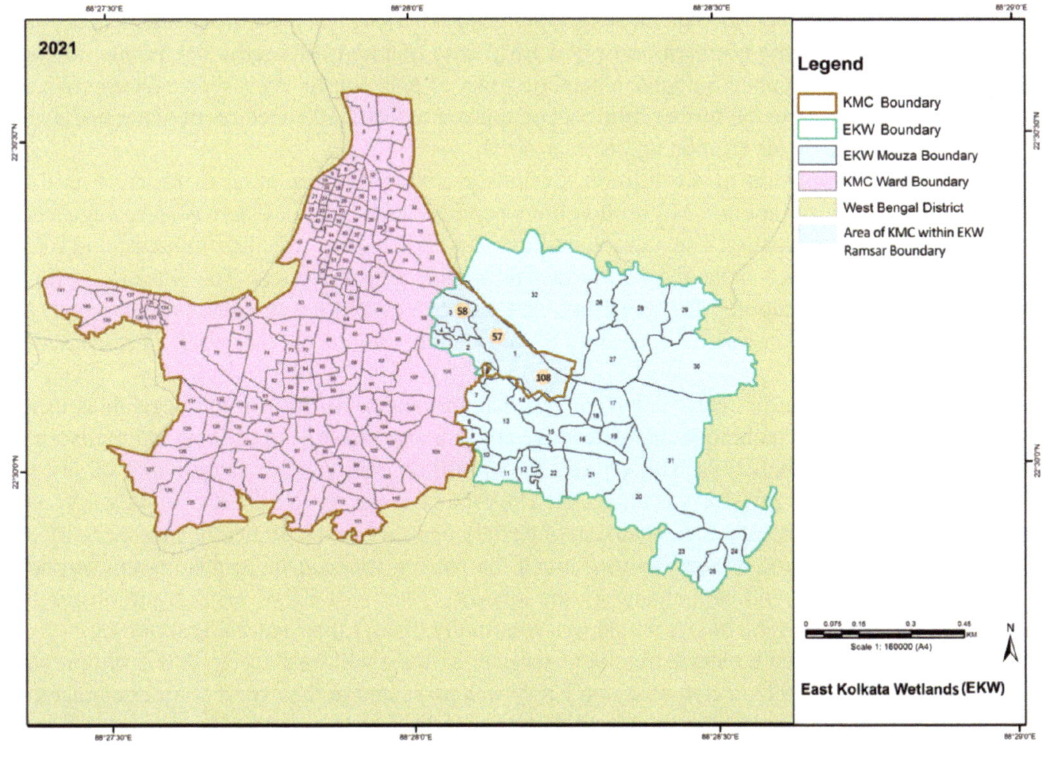

FIGURE 19.3 Map showing the KMC and EKW boundaries.

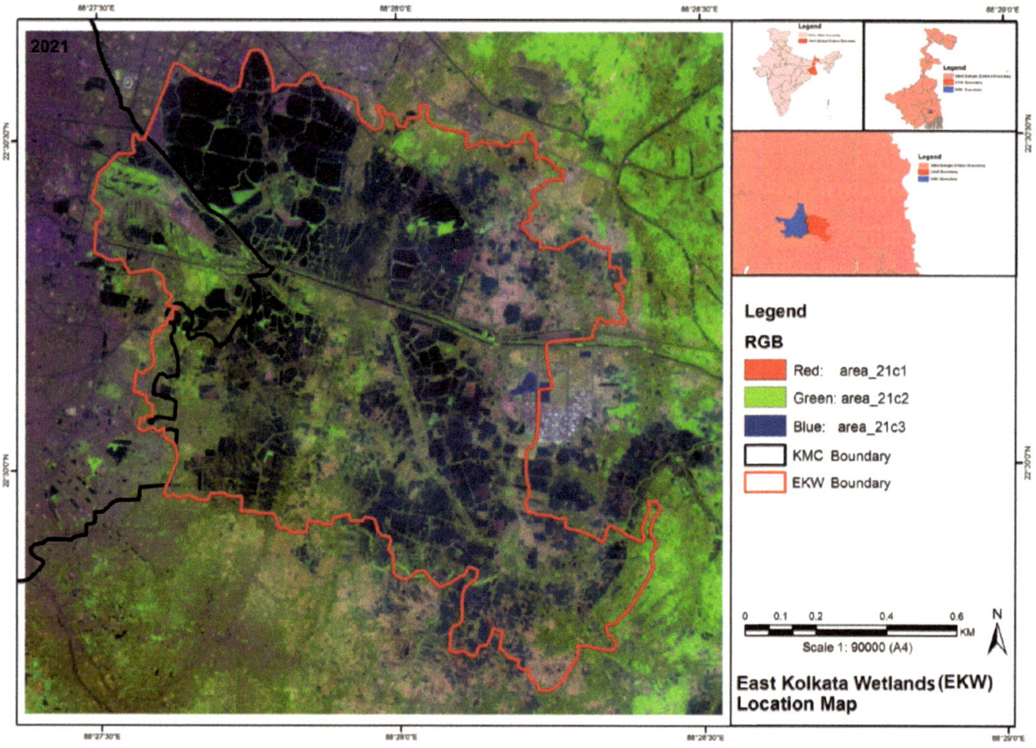

FIGURE 19.4 EKW location map.

To expand the city and accommodate these people (M. Mukherjee 2011), the urban planners encouraged the reclamation of the northern portion of EKW and hundreds of nearby fish ponds, totaling about 1,000 hectares, to create Saltlake, a satellite town of Kolkata. In 1969, redistribution of land through land reforms led to the further filling up of approximately 2,500 hectares of water bodies for conversion into paddy fields (Kundu and Kumar 2010).

Kolkata has a population of 4.6 million, excluding a floating population of nearly 6 million (KMC 2018). It produces about 1,112 million liters per day of sewage, and there is no sewage treatment plant for the municipality. The natural gradient of Kolkata is towards the south-east and EKW is a boon to Kolkata which treats the wastewater through a natural process. The wastewater flows into small ponds where nitrates are converted, and phosphates and heavy metals are absorbed by microbial action. The clean water is then used for agriculture and the organic waste is converted to fish feed (Pal 2017).

The EKW produces more than 10,000 tons of fish and about 40–50% of the vegetables in the market. This indigenous technique of waste recovery is a significantly lower-cost alternative to a sewage treatment plant that could have cost Rs.4,680 million per year, purifying >80% of sewage (Dey and Banerjee 2018; Roy-Basu et al. 2020). The urban growth along the EKW mainly accounts for the increase in city size, and the expansion of tertiary and service-sector activities, especially the IT boom, and the improved transit facilities along the eastern fringe. Due to this urban encroachment, there has been a significant change in the land-use pattern of EKW, resulting in obstructed water flow and siltation in the *bheri*s which will eventually disrupt their natural functioning.

If EKW fails to provide a natural treatment service, Kolkata will lose nearly INR 5 billion/year (Ghosh et al. 2018). In 1992, a case study on EKW was presented in the expert committee meeting of the Ramsar Convention, and the site was declared a wetland of international importance (Kundu and Kumar 2010). To protect, conserve and manage the wetlands, an authority named EKWMA was formed under the state's 2006 EKW Conservation and Management Act.

The next section presents the various physical, environmental and social impacts of urbanization and other forms of human activity on EKW, in order to understand the extent of the threats.

19.4 EKW SITUATIONAL ANALYSIS

19.4.1 LULC Change Detection

The study included an Iso Cluster Unsupervised Classification on multi-temporal Landsat images (2001, 2011, and 2021) obtained from USGS Earth Explorer Portal instrumenting ArcGIS 10.5 version. The details of the satellite images are shown in Table 19.1. The study was designed to detect the changes in land-use land cover (LULC) for an in-depth understanding of the existing EKW scenario, especially the major land-use conversions and the extent of shrinkage of the wetlands. The LULC map was prepared using seven classes as shown in Table 19.2.

The study observed a significant shrinkage in the area of the wetlands. Almost 19km^2 of wetlands have been consumed over the two decades since 2001, which is reflected in the drastic increase in the percentage of the built-up area and decrease in both water bodies (wetlands and wastewater-fed aquaculture) and agriculture farming lands. Most of the built-up area has mushroomed along the eastern peripheral regions of the KMC, which is located within the EKW, especially in wards 57, 58 and 108 (see Figure 19.3). Figure 19.8 shows the multi-temporal (2001, 2011, and 2021) LULC

TABLE 19.1
Details of the Satellite Images

Year	Date	Landsat Sensor	Path/Row	Sensor	Band Information	Resolution (m)
2001	4 Jan	Landsat 7	138/44	ETM+C1 Level-1	1	30
					2	30
					3	30
					4	30
					5	30
					6	30
					7	30
2011	24 Jan	Landsat 4–5	138/44	TM C1 Level-1	1	30
					2	30
					3	30
					4	30
					5	30
					6	30
					7	30
2021	19 Jan	Landsat 8	138/44	OLI TIRS C1 Level-1	1	30
					2	30
					3	30
					4	30
					5	30
					6	30
					7	30

TABLE 19.2
Land-Use Categories Considered for LULC Classification

Sl.No	Land-Use Category	Elements
1	Wetlands	Wetlands, marshy lands
2	Wastewater-Fed Aquaculture	Fisheries, artificial water body
3	Built-up/settlements	Roads, urban settlements including residential and commercial settlements
4	Natural vegetation/parks	Trees, grass cover, parks
5	Barren land	Open ground
6	Agriculture farming	Cultivated lands for farming
7	Waste dumping ground	Solid waste disposal

classification of EKW and Figure 19.6 shows the area for each of the land-use categories of EKW. Existing wetlands within EKW are subjected to high fragmentation, leading to the degradation of the EKW (Ghosh and Das 2020). It has been observed that a significant percentage of wetland has been converted to wastewater-fed aquaculture, as is evident from Figures 19.7 and 19.8. The fishponds which constitute the wastewater-fed aquaculture in EKW are regarded as a wise way of maintaining and sustaining the EKW, since this is compatible with the ecosystems' natural properties (Bunting, Pretty, and Edwards 2010; Finlayson et al. 2011). Ghosh and Das (2019) found that 28.307% of the wetland has been converted to agriculture; 7.27% of wetland was converted to urban area from 2007 till 2017, while only 2.156% of the urban area was converted to wetland (Figures 19.5 and 19.6).

This conversion can be attributed to the increase in artificial water bodies included within the wastewater-fed aquaculture which began to be practiced in the 20th century, as shown in Figure 19.8. Wastewater-fed aquaculture has been proved beneficial in protecting the wetlands. It has been observed that the land-use percentage of built-up settlements increased drastically, signifying land-use change compelled by development pressure which is leading to the degradation of the EKW. Figure 19.7 shows that the wetlands are subjected to high fragmentation, especially along the eastern and south-eastern fringe areas, and depletion in the central portion, signifying alarming shrinkage. The major reason for the shrinkage of the EKW is the land-use conversion which can be seen in the increase in a) fragmentation of the wetlands and b) urban area settlements evident in the multi-temporal data as shown in Figures 19.6, 19.7, and 19.8. Observations from the time-series stellate images by authors in the current study correspond with an earlier study by Mondal et al. (2017), which states that most of the wetland conversion has taken place along the north-eastern and western parts of EKW, which is contiguous with the eastern periphery of KMC. The reason can be attributed to drastic growth along the eastern periphery of KMC, due to the phenomenon of 'bypass urbanism' arising from the development of the major transportation axis of the EMB and the New Town development along the north-eastern part of EKW. Mondal's study also predicts that by 2025, 9 km^2 of EKW is at serious risk and almost 46% of the 38 km^2 wetland will be converted to other land uses.

19.4.2 Dimension-Specific Situational Analysis

19.4.2.1 Environmental Dimension

As part of the 2017 IIT-IUP study (Section 19.3.1), wastewater testing was conducted at sites representative of different stages of the operation, including wastewater treatment and/or utilization within the overall wetland system. The essence of the study was to understand the progressive change in the water quality conditions of the wetlands on exposure to pollutants and natural treatment processes (e.g., canal entry into a wetland area, inflow and outflow points of the pond). Finally,

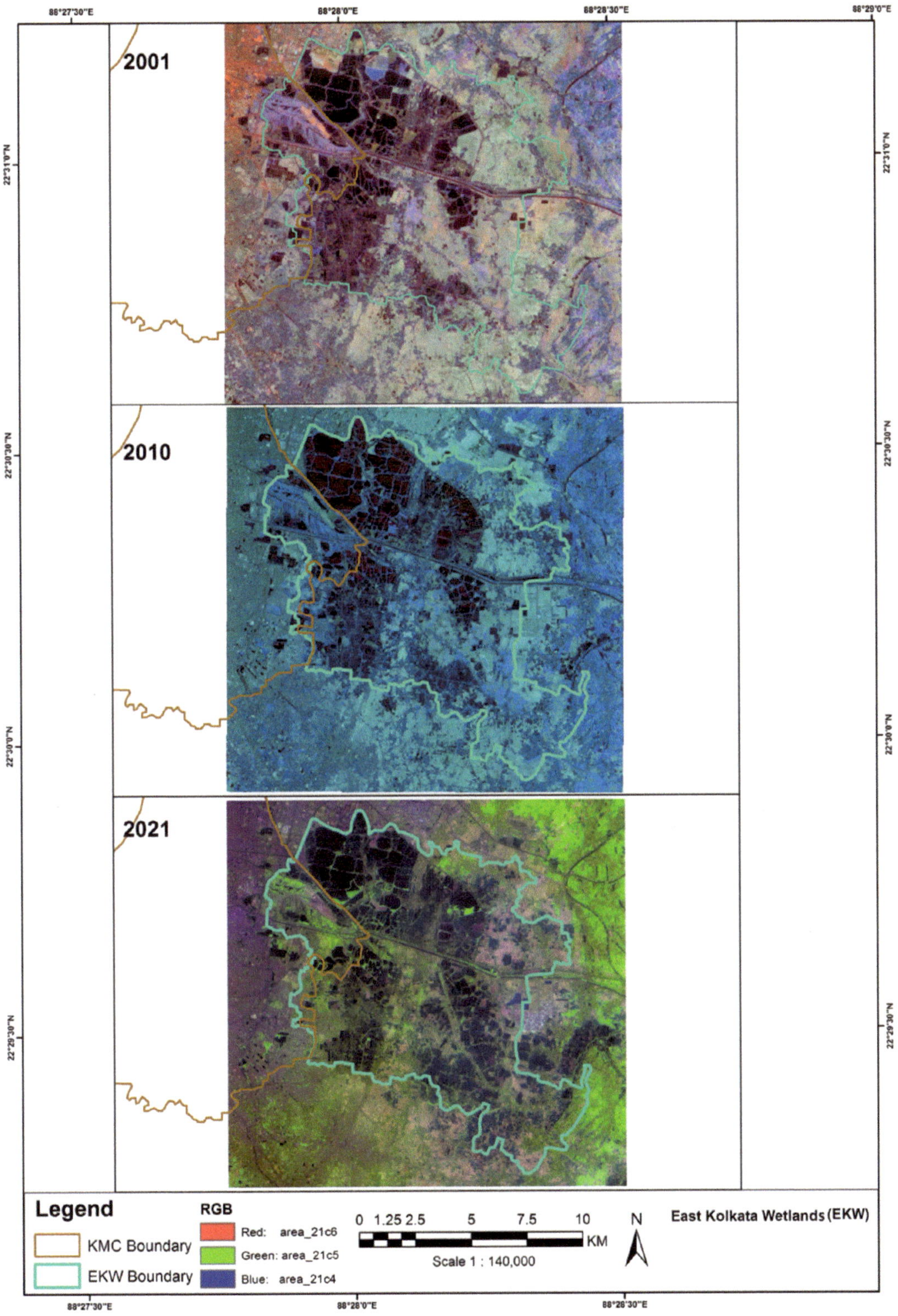

FIGURE 19.5 Multi-temporal landsat satellite images of EKW obtained from USGS earth explorer portal.

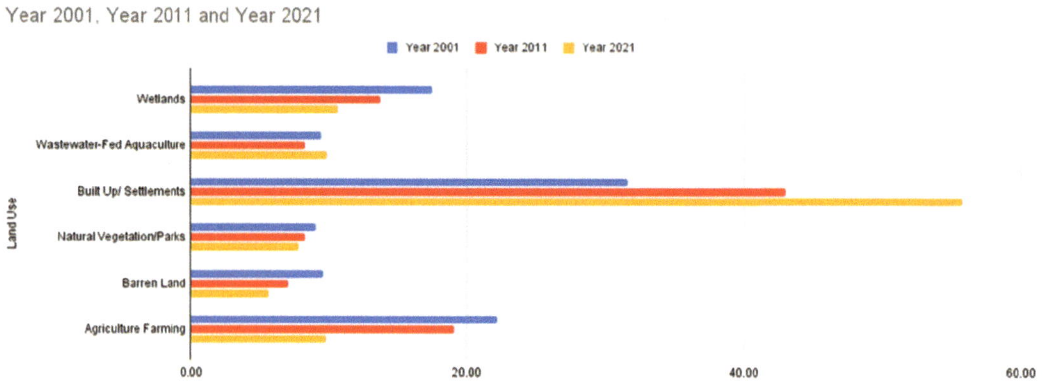

FIGURE 19.6 Percentage changes in EKW area land use

eight sample point locations were purposively selected, based on accessibility conditions and effective coordination with local stakeholders (see Figure 19.9).

The study observed a higher concentration of phosphate levels at sites above the wetland ponds, i.e., before being subjected to biological treatment. A reduction in phosphate levels was observed from inflow to outflow points. This implies that the biological treatment had a substantial effect on water quality, since the nutrients were utilized by the wetland for biological production. The lowest dissolved oxygen levels were also observed above the wetland pond sites, implying the biological processes are responsible for maintaining suitable conditions for fish. In addition to the high phosphate levels, low dissolved oxygen levels and the growth of algae also contributed to increased alkaline pH values. These sites also reported high conductivity levels, attributed to the concentration of solids due to evaporation. This indicates that organic treatment of water by algae–bacteria symbiosis happens here simultaneously with pisciculture (wastewater-fed aquaculture). The organic compounds of the sewage and wastewater biodegrade via the plankton population after a few days. The wastewater-fed aquaculture constituting the fish ponds also acts as a solar reactor, solar energy being trapped by a dense population of plankton. These fish ponds have an organic loading rate ranging between 20 and 70 kg per hectare per day in the form of biochemical oxygen demand (BOD). The cumulative efficiency of reducing the BOD of the sewage wastewater is above 80% and for coliform bacteria 99.99% on average.

19.4.2.2 Economic Dimension

EKW supports three types of economic activities: (i) fish harvesting, (ii) farming, and (iii) trash digging. The wastewater-fed aquaculture and sewage treatment at EKW share a unique symbiotic relationship. They operate simultaneously and are independent of each other. The fish are harvested using nets from the edges of the embankments of the fish ponds or from boats. Once collected, the fish are transported live to the markets or vendors. Vegetables are grown utilizing the garbage and waste found in the Dhapa, the huge official landfill site of KMC within EKW, which is nearly 30 meters high and spread over an area of 21.47 ha. The Dhapa vegetable market contributes almost 40% of the green vegetables of the Kolkata market. The structure of the fish and vegetable market is sustainable and the time gap between the two markets is well coordinated, which allows the fish to be cleaned and delivered by 8.30 a.m., followed by the setting up of the vegetable market, which helps the market vendors to avoid the unpleasant scorching sun at midday.

The rag pickers of EKW remove tons of waste for the municipality but they still belong to the unorganized labor sector. Despite providing an indispensable service free of charge to the municipality, which reaps huge economic and environmental benefits, rag pickers are considered to be the

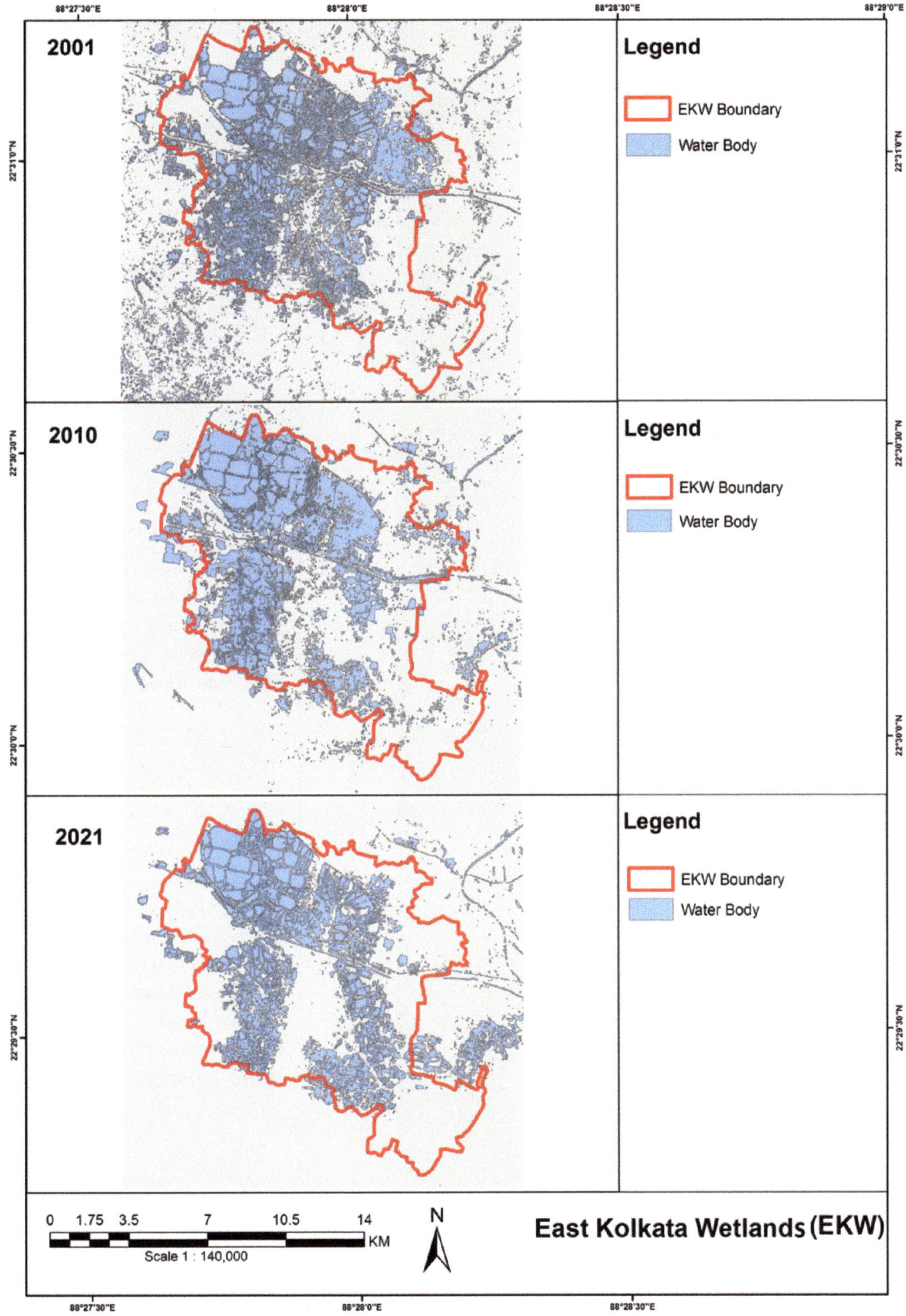

FIGURE 19.7 Multi-temporal map of water bodies (wetlands and wastewater-fed aquaculture) of EKW.

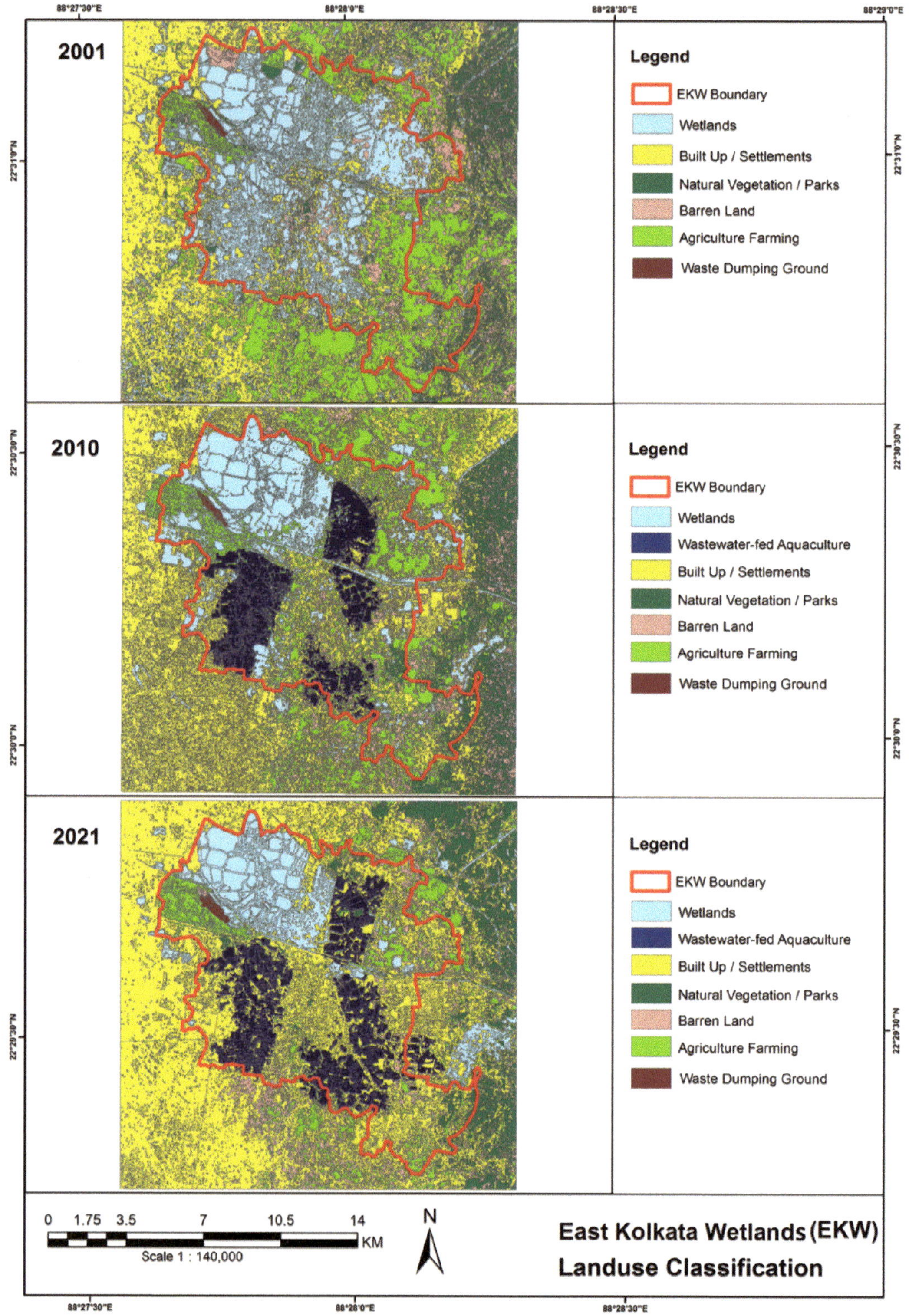

FIGURE 19.8 Multi-temporal LULC classification of EKW.

FIGURE 19.9 Study sites along the wastewater-carrying canal in EKW.

lowest in the social pecking order (Ghosh 2017). Garbage picking leads to recycling much of the waste generated, which provides a decent income for rag pickers.

19.4.2.3 Social and Cultural Dimensions

Development pressures leading to uncontrolled land speculation associated with stealthy land-use conversion happened mainly in the 1980s and 1990s, and was accelerated by the absence of a proper land-use policy and wetland protection rights (Bunting, Pretty, and Edwards 2010). Another major obstacle that has been identified by Bunting, Pretty, and Edwards (2010) is that EKW is inhabited by heterogeneous wetland communities and user groups, each with different development priorities, resulting in a lack of coordination amongst themselves and a hurdle to the successful implementation of the EKW conservation and protection policies.

Previous EKW risk and vulnerability assessments have ignored the social aspects of the EKW land-use conversion that have resulted in its degradation. People residing within the EKW share a common socio-economic status, being mostly the poor and migrants. Since there is no distinct difference in the socio-economic status of the residents, social conditions have not been regarded as an important factor during the decision-making process (Ghosh and Das 2019). Apart from the increase in population density and housing demand, the social needs and priorities have not been considered much in previous research, even though these do evolve over time (Mukherjee, Bebermeier, and Schütt 2018). This study recommends that social conditions should be integrated within the EKW urban resilience assessment framework and in the resilience and vulnerability assessment. Moreover, since the resident population is predominantly poor, they are compelled to engage in precarious livelihood activities, including rag pickers and sex workers (Bunting, Pretty, and Edwards 2010).

The increasing migrant population in EKW is also aggravating urban expansion, leading to further encroachment and degradation. The socio-environmental, socio-economic, socio-cultural and socio-institutional aspects should be well represented in the assessment framework. The government should be aware of the constantly evolving social conditions of the EKW and come up with policies aligned with their livelihood and cultural interests. They should also devise policies to curb further settlement of migrants in these areas and should adopt relocation/rehabilitation policies for the new migrants without disrupting their livelihoods. Proper awareness programs should be conducted regularly to inform the resident population about wetland loss and the associated threats and opportunities (Dey and Banerjee 2013). This would eventually strengthen biodiversity conservation and help to build resilience. The shrinkage of the wetland and degradation of its quality are worsening with each passing day due to the increased human-induced disturbances, and are strongly correlated with the infrastructural development, the influence of new township enclaves, existing developments, and croplands.

19.4.2.4 Institutional Dimension

The wetland comes under the jurisdiction of several institutional bodies (see Table 19.3), but as Mondal et al. (2017) point out, none of these institutions' fundamental policies deal with development plans. Neither do they adopt an integrated approach to assimilating policies from different institutions (Pablo Shiladitya Bose 2015), nor they have framed management proposals to address the issues related to urban growth in the area concerned (Rajashekariah 2011). Conflicting policies and lack of inter-institutional coordination further complicate the situation (Hettiarachchi, Morrison, and McAlpine 2015).

The West Bengal legislature passed the East Kolkata Wetlands (Conservation and Management) Act 2006 for the conservation and management of EKW, which facilitated the formation of the East Kolkata Wetlands Management Authority (EKWMA). The East Kolkata Wetlands (Conservation and Management) Act was amended in the year 2017,leading to the reconstitution of EKWMA, under the jurisdiction of the Ministry of Environment, Forest and Climate Change, Government of India.

TABLE 19.3
List of Institutions Looking after the Jurisdiction of EKW

Institutional Body	Deals With
Kolkata Metropolitan Development Authority (KMDA)	Urban development
West Bengal Housing and Infrastructure Development Corporation (WBHIDCO)	Urban development
West Bengal Pollution Control Board (WBPCB)	Environmental monitoring, waste management, Automobile pollution regulation
East Kolkata Wetlands Management Authority (EKWMA)	Demarcation of boundary, detection of wetland character, measures pollution levels, conserves biodiversity and conducts awareness program.

The EKWMA is guided thoroughly by rules and regulations formulated by the East Kolkata Wetlands (Conservation and Management) Act, concerning the conservation, preservation, maintenance, restoration and inclusion of any area which is deemed suitable to be within EKW. It also includes the granting of sanctions for land-use change and imposes penalties for non-compliance with the rules and regulations (Environmental Law Research and Guidance Foundation India 2013; EKWMA 2021).

19.5 RECOMMENDATIONS

The study has explored the dimension-specific human interactions in EKW and identified the conflicting issues hindering the conservation of the EKW and protecting it from further encroachment. The study recommends that to make a UESA like the wetlands resilient to the ill-effects of climate change, a stakeholder participation approach should be adopted. To enhance the resilience of the EKW communities, enabling user groups and stakeholders to withstand adverse effects of climate change, the study has developed an Expert-Stakeholder Matrix (ESM) framework to obtain expert knowledge and stakeholders' feedback at every stage of urban resilience:1) planning and preparation, 2) absorption, 3) recovery, and 4) adaptation (see Table 19.4).

The framework is based on the dimension model discussed in Section 19.2. This framework will also provide an understanding of the perceptual extent of adaptive capacities of the locals to an adverse event. Consequently, it will help raise awareness of the effects of human interactions with the natural and the physical environment on different dimensional aspects of the sustainability of UESA.

The study has developed a 5-phase strategy to make a UESA resilient to climate change.

1) First, the environmental, economic, social, cultural and institutional aspects of the socio-ecological system concerned will be studied, which will constitute the situational analysis of the UESA. Qualitative and quantitative assessments of the threats and opportunities of the UESA will also be encouraged during this phase. This will be followed by an expert opinion survey. The authors have prepared an Expert-Stakeholder Matrix (ESM).
2) A group of experts will be formed to address the persisting issues. An expert opinion survey adopting a Delphi technique can be conducted to obtain a pool of strategies and policies for addressing the issues. Two to three rounds of Delphi will be conducted depending on consensus on responses (Muhuri 2017; Kumar, Bhaumik, et al. 2020).
3) After finalizing the list of strategies that will be carried out in all four stages (planning/preparation, absorption, recovery and adaptation) to make a UESA resilient to climate change, a feedback questionnaire will be circulated to the stakeholders. Data-driven assessment of the respondents' feedback will be carried out followed by revision of strategies and policies as required. Stakeholder feedback will be beneficial for incorporating context specificities and will aid in preparing a tailor-made comprehensive urban resilient framework.
4) The final list of policy frameworks will undergo mandatory administrative proceedings, when target-oriented, pragmatic action plans will be drafted for a phased implementation.
5) The final stage will involve initiation of implementation programs and execution of the plans at ground level with an inbuilt mechanism for monitoring which would involve group discussions with locals, awareness programs, and incentivizing programs.

The study recommends that ESM should be developed for each of the four stages of resilience, highlighting the experts' and stakeholders' perceptions of 1) availability, 2) accessibility, 3) acceptability, and 4) affordability, catering to each of the human interactions within the UESA (EKW) concerning climate change. All the different categories of wetland user groups should be identified. Focus group discussions should be held regularly to devise policies to mitigate the direct and indirect impacts of climate change on the EKW, and how to enhance the socio-ecological resilience aligned by adopting the evolutionary resilience approach. Coordination amongst the various institutions governing the

TABLE 19.4
Expert-Stakeholder Matrix (ESM), Exploring the Human Interactions concerned in the Environmental Interlinkages for the Planning and Preparation Stage

				Expert Framework (*Policymaker, Scientists*)							
				Likert Scale Scoring 1(Low)-5(High)				Climate Change Impacts			
Planning & Prepare	**Interlinkages**	**Interactions**	**Strategies**	**Availability**	**Accessibility**	**Acceptability**	**Affordability**	**Direct**	**Indirect**	**Technology to Be Adopted**	**Temporal Dynamism**
	Environ-Socio										
	Environ-Eco										
	Environ-Cultural										
	Environ-Institutional										

			Stakeholder Feedback (*Residents, Users*)					
			Likert Scale Scoring 1(Low)-5(High)					
Interlinkages	**Interactions**	**Selected Strategies**	**Availability**	**Accessibility**	**Acceptability**	**Affordability**	**Technology to Be Adopted**	**Suggestions**
Environ-Socio								
Environ-Eco								
Environ-Cultural								
Environ-Institutional								

EKW, adopting good governance practices, and ensuring inclusion through community participation and stakeholder involvement are the priority areas.

The local governing body of EKW should devise policies and strategies in the best interests of the healthy livelihood and socio-cultural realm of EKW. These may include giving incentives to local people to abide by the existing rules and regulations and setting up a vigilance committee involving local people to report any inappropriate development and malpractice within the EKW. Given the ongoing COVID-19 crisis, and with EKW located on the outskirts of Kolkata, the authors propose the establishment of an exclusive awareness center, equipped with medical consulting rooms, which could double up as a healthcare center to deal with disaster-related health emergencies for the EKW community. The awareness center would also be instrumental in local capacity building through the use of technology and provision of internet access to the socio-economically vulnerable groups of the wetland community. The center would hold interactive exhibitions and training workshops to raise awareness regarding the wetlands, and drilling sessions to enhance the disaster preparedness of the resident community to natural disasters or pandemics. The center would also act as a homeless shelter or an isolation center, depending on what type of disaster (flood, pandemic) struck the UESA.

19.6 CONCLUSIONS

This chapter has outlined the importance of building resilience to climate change for UESAs. It has studied the emergence of the concept of resilience and explored its different dimensions The environmental, economic, socio-cultural and institutional perspectives of the EKW have been highlighted through a situational analysis. The adverse effects of urbanization on the EKW have also been underlined. The chapter has emphasized the importance of stakeholder participation for boosting resilience to climate change. It has come up with a proposed framework that takes all dimensions of sustainability into account, with a special focus on achieving the long-term sustainability of the UESA by making it resilient to climate change and any other adverse natural or man-made shocks. The authors believe there should be stringent rules and regulations imposed by the government to curb speculation-driven real-estate investments, and consequently protect UESAs. Since the adjacent stretch of bypass poses huge potential for development, the government should integrate sustainability strategies and stakeholder interests in the formulation of real-estate development plans. The authors recommend that subaltern knowledge be incorporated within the participatory framework for enhancing the adaptation of climate change (Olazabal et al. 2021). Future researchers should further operationalize the ESM and develop a robust indicator framework for assessing UESA climate resilience.

Furthermore, since India is a major South Asian country, and Kolkata Global South mega-city, the analysis and practical implications devised in this study for a Ramsar-protected site have the potential to be generalized for application in other developing countries.

ACKNOWLEDGMENT

The IIT-IUP study referred to in this chapter was funded through program fees with the assistance of the IUP Offices of International Education and Accounts Payable, and scholarships provided by the Honors College at IUP.

REFERENCES

Adger, W Neil. 2000. "Social and Ecological Resilience: Are They Related?" *Progress in Human Geography* 24 (3). Sage Publications, Thousand Oaks, CA: 347–364.

Basu, J. 2012. "East Kolkata Most Vulnerable to Climate Change_World Bank." *Down To Earth*. https://www.downtoearth.org.in/news/east-kolkata-most-vulnerable-to-climate-change-world-bank-35882

Baum-Snow, Nathaniel. 2007. "Did Highways Cause Suburbanization?" *The Quarterly Journal of Economics* 122 (2). MIT Press: 775–805.

Berkes, Fikret, Carl Folke, and Johan Colding. 2000. *Linking Social and Ecological Systems: Management Practices and Social Mechanisms for Building Resilience*. Edited by Fikret Berkes and Carl Folke. Cambridge: Cambridge University Press. https://www.cambridge.org/vi/academic/subjects/life-sciences/ecology-and-conservation/linking-social-and-ecological-systems-management-practices-and-social-mechanisms-building-resilience?format=PB#bookPeople

Bose, Pablo S. 2007. "Dreaming of Diasporas: Urban Developments and Transnational Identities in Contemporary Kolkata." *Topia* 17: 111–130.

Bose, Pablo Shiladitya. 2015. *Urban Development in India: Global Indians in the Remaking of Kolkata*. Routledge.

Boyd, Emily, Henny Osbahr, Polly J Ericksen, Emma L Tompkins, Maria Carmen Lemos, and Fiona Miller. 2008. "Resilience and 'Climatizing' Development: Examples and Policy Implications." *Development* 51 (June 2014): 390–396. doi:10.1057/dev.2008.32.

Bunting, Stuart W, Jules Pretty, and Peter Edwards. 2010. "Wastewater-Fed Aquaculture in the East Kolkata Wetlands, India: Anachronism or Archetype for Resilient Ecocultures?" *Reviwes in Aquaculture*, no. February 2018. doi: 10.1111/j.1753-5131.2010.01031.x.

Chowdhuri, Indrajit, Subodh Chandra, Asish Saha, Rabin Chakrabortty, Manoranjan Ghosh, and Paramita Roy. 2020. "Significant Decrease of Lightning Activities during COVID-19 Lockdown Period over Kolkata Megacity in India." *Science of the Total Environment* 747: 141321.

Davoudi, Simin, Keith Shaw, L Jamila Haider, Allyson E Quinlan, Garry D Peterson, Cathy Wilkinson, Hartmut Fünfgeld, Darryn McEvoy, Libby Porter, and Simin Davoudi. 2012. "Resilience: A Bridging Concept or a Dead End? 'Reframing' Resilience: Challenges for Planning Theory and Practice Interacting Traps: Resilience Assessment of a Pasture Management System in Northern Afghanistan Urban Resilience: What Does It Mean in Plannin." *Planning Theory & Practice* 13 (2). Taylor & Francis: 299–333.

Dey, Debanjana, and Sarmila Banerjee. 2013. "Ecosystem and Livelihood Support: The Story of East Kolkata Wetlands." *Environment and Urbanization Asia* 4 (2). SAGE Publications Sage India: New Delhi, India: 325–337.

Dey, Debanjana, and Sarmila Banerjee. 2018. "How Expensive Is the Decay of East Kolkata Wetlands? An Estimation of Opportunity Cost for Kolkata." In *Sustainable Urbanization in India*, edited by Jenia Mukherjee, 181–205. Singapore: Springer. doi:10.1007/978-981-10-4932-3_10.

Duxbury, Jane, and Sarah Dickinson. 2007. "Principles for Sustainable Governance of the Coastal Zone: In the Context of Coastal Disasters." *Ecological Economics* 63 (2–3). Elsevier: 319–330.

EKWMA. 2021. "East Kolkata Wetlands Management Authority (EKWMA)." *EKWMA*. http://ekwma.in/ek/.

Environmental Law Research and Guidance Foundation India. 2013. "The East Kolkata Wetlands (Conservation And Management) Act, 2006 - Elaw." *Elaw.In*. https://www.elaw.in/wetland/ecwact.html

Euromonitor Consulting. 2020. "Future Watch Update on Consumer Trends and Lifestyles in Japan and South Korea in Light of the Covid-19 Pandemic."

Finlayson, C Max, Nick Davidson, Dave Pritchard, G Randy Milton, and Heather MacKay. 2011. "The Ramsar Convention and Ecosystem-Based Approaches to the Wise Use and Sustainable Development of Wetlands." *Journal of International Wildlife Law & Policy* 14 (3–4). Taylor & Francis: 176–198.

Ghosh, D. (2017). *The Trash Diggers: Rethinking Solid Waste Management in Urban India* (1st ed.). New Delhi, India: Oxford University Press.

Ghosh, Sasanka, and Arijit Das. 2019. "Urban Expansion Induced Vulnerability Assessment of East Kolkata Wetland Using Fuzzy MCDM Method." *Remote Sensing Applications: Society and Environment* 13 (October 2018). Elsevier B.V.: 191–203. doi:10.1016/j.rsase.2018.10.014.

Ghosh, Sasanka, and Arijit Das. 2020. "Wetland Conversion Risk Assessment of East Kolkata Wetland : A Ramsar Site Using Random Forest and Support Vector Machine Model." *Journal of Cleaner Production* 275. Elsevier Ltd: 123475. doi:10.1016/j.jclepro.2020.123475.

Ghosh, Subrata, Santanu Dinda, Nilanjana Das Chatterjee, and Kousik Das. 2018. "Analyzing Risk Factors for Shrinkage and Transformation of East Kolkata Wetland, India." *Spatial Information Research*, no. August. Springer Singapore. doi:10.1007/s41324-018-0212-0.

Heinzlef, Charlotte, Benoît Robert, Yannick Hémond, and Damien Serre. 2020. "Operating Urban Resilience Strategies to Face Climate Change and Associated Risks : Some Advances from Theory to Application in Canada and France." *Cities* 104 (May). Elsevier: 102762. doi:10.1016/j.cities.2020.102762.

Hettiarachchi, Missaka, T H Morrison, and Clive McAlpine. 2015. "Forty-Three Years of Ramsar and Urban Wetlands." *Global Environmental Change* 32. Elsevier: 57–66.

Holling, C. 1985. "Resilience of Ecosystems: Local Surprise and Global Change." In: *Global Change*. (Eds. Roederer, J.G. & Malone, T.F.), pp. 228–269, Cambridge: Cambridge University Press.

Holling, Crawford S. 1973. "Resilience and Stability of Ecological Systems." *Annual Review of Ecology and Systematics* 4 (1). Annual Reviews 4139 El Camino Way, PO Box 10139, Palo Alto, CA, USA: 1–23.

Hulme, Mike. 2009. *Why We Disagree about Climate Change: Understanding Controversy, Inaction and Opportunity*. Cambridge University Press.

Iturriza, M, L Labaka, M Ormazabal, and M Borges. 2020. "Urban Climate Awareness-Development in the Context of Climate Change Resilience." *Urban Climate* 32 (February). Elsevier: 100613. doi:10.1016/j.uclim.2020.100613.

Kim, Donghyun, and Up Lim. 2016. "Urban Resilience in Climate Change Adaptation: A Conceptual Framework." *Sustainability*, 1–17. doi:10.3390/su8040405.

Klein, Richard J T, Saleemul Huq, Fatima Denton, Thomas E Downing, Richard G Richels, John B Robinson, and Ferenc L Toth. 2007. *Inter-Relationships between Adaptation and Mitigation*. Cambridge: Cambridge University Press.

KMC. 2018. "Official Website of Kolkata Municipal Corporation." *Kolkata Municipal Corporation*. https://www.kmcgov.in/KMCPortal/jsp/Mayor'sDesk.jsp

Koffi, Kouao Jean, Vincent Deblauwe, Serge Sibomana, Danho Fursy Rodelec Neuba, Dominique Champluvier, Charles De Cannière, N Barbier, D Traore, B Habonimana, and Elmar Robbrecht. 2008. "Spatial Pattern Analysis as a Focus of Landscape Ecology to Support Evaluation of Human Impact on Landscapes and Diversity." In *Landscape Ecological Applications in Man-Influenced Areas* (eds S-K Hong, N Nakagoshi, B Fu, Y Morimoto, 7–32. Dordrecht: Springer.

Kumar, Sudeshna, Haimanti Banerji, and Biplab Kanti Sengupta. 2020a. "Methodology for Sustainable Development along an Eco-Fragile Zone for Land Use Management and Mitigating Climate Change : Evidence of East Kolkata Wetlands." In *EGU General Assembly 2020*, 0–1. Creative Commons Attribution 4.0 License. doi: 10.5194/egusphere-egu2020-2080.

Kumar, Sudeshna, Sumitro Bhaumik, and Haimanti Banerji. 2020b. "Methodology for Framing Indicators for Assessing Economic-Socio-Cultural Sustainability of the Neighbourhood Level Urban Communities in Indian Megacities: Evidence from Kolkata." *Social Indicators Research* 152 (3). Springer Netherlands. doi:10.1007/s11205-020-02559-6.

Kumar, Sudeshna, Anirban Santara, and Haimanti Banerji. 2020c. "Pentagram Sustainability Model." *International Review for Spatial Planning and Sustainable Development* 8 (2). International Community of Spatial Planning and Sustainable Development: 100–117.

Leichenko, Robin M. 2018. "Climate Change and Urban Resilience." *Current Opinion in Environmental Sustainability*, no. July. doi:10.1016/j.cosust.2010.12.014.

Li, Xia, Chandana Mitra, Li Dong, and Qichun Yang. 2018. "Understanding Land Use Change Impacts on Microclimate Using Weather Research and Forecasting (WRF) Model." *Physics and Chemistry of the Earth* 103. Elsevier Ltd: 115–126. doi:10.1016/j.pce.2017.01.017.

Long, Joshua, and Jennifer L Rice. 2019. "From Sustainable Urbanism to Climate Urbanism." *Urban Studies* 56 (5): 992–1008. doi:10.1177/0042098018770846.

Maiti, Sutapa, and Praween K Agrawal. 2005. "Environmental Degradation in the Context of Growing Urbanization: A Focus on the Metropolitan Cities of India." *Journal of Human Ecology* 17 (4). Taylor & Francis: 277–287. 10.1080/09709274.2005.11905793

Meerow, Sara, Pani Pajouhesh, Thaddeus R Miller, Sara Meerow, Pani Pajouhesh, Thaddeus R Miller, Sara Meerow, and Thaddeus R Miller. 2019. "Social Equity in Urban Resilience Planning Social Equity in Urban Resilience Planning." *Local Environment The International Journal of Justice and Sustainability* ISSN: 9839. Taylor & Francis. doi:10.1080/13549839.2019.1645103.

Mohan, Manju, Subhan K Pathan, Kolli Narendrareddy, Anurag Kandya, and Sucheta Pandey. 2011. "Dynamics of Urbanization and Its Impact on Land-Use/Land-Cover: A Case Study of Megacity Delhi." *Journal of Environmental Protection* 2 (9). Scientific Research Publishing: 1274.

Mondal, Biswajit, Gour Dolui, Malay Pramanik, Santu Maity, and Sumantra Sarathi. 2017. "Urban Expansion and Wetland Shrinkage Estimation Using a GIS-Based Model in the East Kolkata Wetland, India Urban Expansion and Wetland Shrinkage Estimation Using a GIS-Based Model in the East Kolkata Wetland, India." *Ecological Indicators* 83 (December). Elsevier: 62–73. doi:10.1016/j.ecolind.2017.07.037.

Muhuri, Soumi. 2017. "For High-Rise Group Housing Complexes in India." *Social Indicators Research*, no. Bernard 1999. Springer Netherlands. doi: 10.1007/s11205-017-1633-1.

Mukherjee, Mala. 2011. "Urban Growth and Spatial Transformation of Kolkata Metropolis: A Continuation of Colonial Legacy." *ARPN Journal of Science and Technology* 2: 365–380.

Mukherjee, Subham, Wiebke Bebermeier, and Brigitta Schütt. 2018. "An Overview of the Impacts of Land Use Land." *Land* 7 (91). doi: 10.3390/land7030091.

National Research Council. 2012. *Disaster Resilience: A National Imperative*. Washington, DC: The National Academies Press.

Newell, Joshua Peter, Sara Meerow, Joshua P Newell, and Melissa Stults. 2016. "Defining Urban Resilience: A Review Landscape and Urban Planning Defining Urban Resilience : A Review." *Landscape and Urban Planning* 147 (March). Elsevier B.V.: 38–49. doi: 10.1016/j.landurbplan.2015.11.011.

Olazabal, Marta, Eric Chu, Vanesa Castán Broto, and James Patterson. 2021. "Subaltern Forms of Knowledge Are Required to Boost Local Adaptation." *One Earth* 4 (6). Elsevier: 828–838.

Pal, Sanchari. 2017. "This Wetland Is the World's Largest Organic Sewage Management System." *The Better India*. https://www.thebetterindia.com/84746/east-kolkata-westland-dhrubajyoti-ghosh-organic-sewage-management/

Phillips, Jason. 2016. "Climate Change and Surface Mining: A Review of Environment-Human Interactions & Their Spatial Dynamics." *Applied Geography* 74. Elsevier: 95–108.

Rajashekariah, Kiran. 2011. *Impact of Urbanisation on Biodiversity: Case Studies From India*. WWF-India Report, New Delhi: World Wide Fund for Nature-India.

Ramachandra, T V, Bharath H Aithal, and S Sreekantha. 2012. "Spatial Metrics Based Landscape Structure and Dynamics Assessment for an Emerging Indian Megalopolis." *Facilities* 1 (1).

Reveshty, Mohsen Ahadnejad. 2011. "The Assessment and Predicting of Land Use Changes to Urban Area Using Multi-Temporal Satellite Imagery and GIS: A Case Study on Zanjan, IRAN (1984–2011)." *Journal of Geographic Information System* 3 (4). Scientific Research Publishing: 298.

Roy-Basu, Avanti, Girija K Bharat, Paromita Chakraborty, and S K Sarkar. 2020. "Science of the Total Environment Adaptive Co-Management Model for the East Kolkata Wetlands: A Sustainable Solution to Manage the Rapid Ecological Transformation of a Peri-Urban Landscape." *Science of the Total Environment* 698 (August 2019): 134203.

Sawyer, Lindsay, and Christian Schmid. 2015. "Bypass Urbanism." In *Jahrbuch Yearbook, ETH Zurich*, 213–215. doi:10.3929/ethz-b-000308980.

Seto, Karen C, Roberto Sánchez-Rodríguez, and Michail Fragkias. 2010. "The New Geography of Contemporary Urbanization and the Environment." *Annual Review of Environment and Resources* 35: 167–194.

Sharifi, Ayyoob. 2016. "A Critical Review of Selected Tools for Assessing Community Resilience." *Ecological Indicatorsx* 35: 629–647. doi: 10.1016/j.ecolind.2016.05.023.

Sharifi, Ayyoob. 2020. "Urban Resilience Assessment : Mapping Knowledge Structure and Trends." *Sustainability* 12 (15): 5918.

Sharifi, Ayyoob, and Yoshiki Yamagata. 2015. "A Conceptual Framework for Assessment of Urban Energy Resilience." *Energy Procedia* 75. Elsevier B.V.: 2904–2909. doi:10.1016/j.egypro.2015.07.586.

Sharifi, Ayyoob, and Yoshiki Yamagata. 2016. "Principles and Criteria for Assessing Urban Energy Resilience: A Literature Review." *Renewable and Sustainable Energy Reviews*, no. July. Elsevier. doi:10.1016/j.rser.2016.03.028.

Sharifi, Ayyoob, and Yoshiki Yamagata. 2017. "Towards an Integrated Approach to Urban Resilience Assessment." *APN Science Bulletin*, *7*(1). doi:10.30852/sb.2017.182.

Sharifi, Ayyoob, and Yoshiki Yamagata. 2018a. "Resilience-Oriented Urban Planning: Theoretical and Empirical Insights." In *Resilience-Oriented Urban Planning* (eds Ayyoob Sharifi and Yoshiki Yamagata), 3–27. Cham: Springer.

Sharifi, Ayyoob, and Yoshiki Yamagata. 2018b. "Resilient Urban Form: A Conceptual Framework Chapter 9 Resilient Urban Form: A Conceptual Framework." In *Lecture Notes in Energy*. doi:10.1007/978-3-319-75798-8.

Simmie, James, and Ron Martin. 2010. "The Economic Resilience of Regions: Towards an Evolutionary Approach." *Cambridge Journal of Regions, Economy and Society* 3 (1). Oxford University Press: 27–43.

Sperling, Frank, Corinne Validivia, Roberto Quiroz, Roberto Valdivia, Lenkiza Angulo, Anton Seimon, and Ian Noble. 2008. *Transitioning to Climate Resilient Development: Perspectives from Communities in Peru*. Washington, DC: World Bank.

Sudhira, H S, T V Ramachandra, and K S Jagadish. 2004. "Urban Sprawl: Metrics, Dynamics and Modelling Using GIS." *International Journal of Applied Earth Observation and Geoinformation* 5 (1). Elsevier: 29–39. doi:10.1016/j.jag.2003.08.002.

The Lancet Digital Health. 2020. "Technology : The Nexus of Climate Change and COVID-19?" *The Lancet Digital Health* 3 (2). The Author(s). Published by Elsevier Ltd. This is an Open Access article under the CC BY 4.0 license: e67. doi:10.1016/S2589-7500(21)00007-8.

UN ESCAPE. 2020. "Managing the COVID-19 Pandemic in China, Japan and the Republic of Korea _ ESCAP." © *United Nations ESCAP*. https://www.unescap.org/news/managing-covid-19-pandemic-china-japan-and-republic-korea

White, Iain, and Paul O'Hare. 2014. "From Rhetoric to Reality: Which Resilience, Why Resilience, and Whose Resilience in Spatial Planning?" *Environment and Planning C: Government and Policy* 32 (5). SAGE Publications Sage UK: London, England: 934–950.

20 Urban Societal Needs and Sustainability in Zambia

A Case of Public Land and Housing

Anthony Mushinge, Ephraim Kabunda Munshifwa, Roy Alexander Chileshe, and Niraj Jain
Copperbelt University, Kitwe, Zambia

CONTENTS

20.1 INTRODUCTION

Urban land is crucial for human shelter and other economic activities like commercial and industrial activities. It is a platform around which human endeavors revolve, and accounts for what is globally referred to as a principal necessity of life (Oladokun et al., 2010). Recent statistics show an ever-increasing demand for urban land, bringing the issues of sustainability to the fore. This is because the pace of urbanization and the number of people living in urban settings has rapidly increased, with statistics indicating that more than half of the world's population now resides in urban areas (USAID, 2014). This form of urbanization is driven by the concentration of investment and employment opportunities in urban areas (USAID, 2014). Estimates further show that half of Africa's population will also live in cities by 2030 (UN-Habitat, 2010), signifying future challenges which African cities will have to contend with. Although there is increased demand for all other urban uses, the demand for land for shelter often outstrips these others. Thus, equitable access to public land is critical in urban areas because it enables many people to have access to decent housing as a societal need.

In Zambia, the Constitution (Amendment) Act of 2016 Part xix, Section 253 provides for equitable access to public land for all citizens (Government of the Republic of Zambia [GRZ], 2016).

DOI: 10.1201/9781003131922-24

Furthermore, the country's Vision 2030 envisages a Zambia that ensures equitable access to public urban land to all citizens for sustainable socio-economic development (GRZ, 2006). In practice, however, the situation is far from the vision. In many towns and cities, public urban land for housing has become a commodity that is difficult for low-income earners to access (Civic Forum on Housing and Habitat, 2015; GRZ, 2017b, 2017c; Lupale and Hampwaye, 2019). Current statistics show that the majority of low-income earners are facing serious challenges in accessing public urban land for housing, resulting in construction in environmentally vulnerable areas, increased squatting and the growth of unplanned settlements. Further consequences are that towns and cities resemble amoeboid shapes and grow unsustainably without regard to the provision and vision of development plans, thus adversely affecting the urban environment.

Sustainability in this chapter is viewed within the context of the World Bank's Sustainable Development Goal (SDG) 11 on Sustainable Cities and Communities, which aims at ensuring access for all to adequate, safe and affordable housing by 2030, while providing basic services and upgrading slums. With the foregoing scenario in mind, this chapter answers the following questions: first, what is the process of accessing public land for urban housing? Second, what are the barriers to accessing public land and how can these be minimized?

20.2 HOUSING, ACCESS AND SUSTAINABILITY

Access to basic housing is one of the key societal needs, and yet studies show a serious deficit in most developing countries. In addition, studies such as Gilbert (2004), Rondinelli (1990) and others show that provision of land and housing by government has also failed in many African countries. For instance, as early as 1990, studies (such as Rondinelli, 1990) argued that public housing, sites and services, slum upgrading and government-assisted self-help programs had failed and yet 25 million households needed to be accommodated by the year 2000. Fast forward, current estimates show that 881 million of the urban population reside in slums in developing countries, an increase from 689 million in 1990 (Habitat for Humanity, 2021); this represents an increase of 192 million over 30 years. This rapid growth rate of informal settlements is partly indicative of failure by the public sector to provide land and housing for all.

Informal settlements are defined differently by different scholars. However, in the context of this chapter, informal settlements include (UN-Habitat, 1996): (a) residential areas where a group of housing units have been constructed on land which the occupants have no legal rights to, and (b) unplanned settlements and areas where housing is not in compliance with the current planning and building regulations. They are thus characterized by lack of or inadequate access to basic infrastructure (proper roads, drainage, safe water, and sanitation), inadequate access to health facilities, overcrowding, substandard housing and, most importantly, lack of land tenure security (UN-Habitat, 2003).

The growth of informal settlements is argued to be one unsustainable way of building towns and cities. A step back to understand the key components of a housing sector explains why informal settlements are seen as an unsustainable development. Broadly, five interdependent components interact in any housing sector: land, infrastructure, finance, labor and building materials (UN-Habitat, 2011a). How these components are organized and shaped is determined by the country's legal and institutional frameworks (UN-Habitat, 2011a). Access to affordable land is a core component for the provision of affordable housing. It thus goes without saying that provision of land is vital to achieving efficient and sustainable housing delivery in towns and cities (UN-Habitat, 2011a; Arjjumend and Seid, 2018). As noted earlier, governments in developing counties have used various means to provide housing for their citizens, such as public housing, sites and services, and upgrading and government-assisted self-help programs. As well as the direct provision of housing, the other mechanisms employed presuppose that citizens have access to bare land (plots). Thus, a policy on planning, servicing and allocation of public land becomes critical for ensuring that development costs are reduced, opportunities increased for all and needs met for the lower-income earners of society (Hickey and Sturtevant, 2015). A policy on planning, servicing and allocation of public land is based on the key objective of ensuring that all vulnerable members of society are able to afford decent housing (Munshifwa et al., 2017). A key difference between most African, and American and Western countries is the way public land is allocated. While in Western countries the market is more dominant in the allocation of land, in the majority of African countries, such as Zambia, the whole system of accessing land is through grants by the state (Munshifwa et al., 2017). It is argued that such a system is able to provide the cheapest means of public housing (Munshifwa et al., 2017).

Furthermore, a policy on planning, servicing and allocation of public land is supposed to enable peoples' legal access to urban land to, among other things, build houses. Land access means an opportunity for temporary or permanent access to shelter (UN-Habitat, 2008). In addition, UN-Habitat (2003) notes that legal access to land is a strategic prerequisite for the provision of adequate shelter for all and for the development of sustainable human settlement affecting urban areas. Thus, providing access to land for the urban populace is one effective way of contributing to alleviating poverty (Arjjumend and Seid, 2018).

Literature reveals problems of access to land for housing development in most parts of Sub Saharan Africa (Oladokun et al., 2010; UN-Habitat, 2010; Eglin, 2017). This can be attributed to challenges in urban land governance related to lack of transparency, unresponsive institutions, and corruption (Transparency International Zambia, 2011). It further points to challenges related to cumbersome land transaction administration, excessive delays and unfairness (UN-Habitat, 2010). The result of lack of access to land in most parts of Sub Saharan Africa can be seen in the rapid growth of informal settlements (UN-Habitat, 2011b). Therefore, to build more sustainable and livable cities, access to land and housing becomes of paramount importance.

20.3 RESEARCH METHODOLOGY

Both secondary and primary data were used for this chapter, with secondary data comprising both published and grey literature. Different types of documents were reviewed, including peer-reviewed journal articles, books, theses, Zambian land laws, technical reports, and conference proceedings. Secondary data was supplemented by primary data obtained from three purposively selected public land institutions – Kitwe City Council (one employee was interviewed), Ndola City Council (one employee was interviewed), and Ministry of Lands and Natural Resources (two employees were interviewed). Telephone interviews were used to obtain information from these institutions. The study was conducted between January 2021 and March 2021.

20.4 PUBLIC URBAN LAND AND HOUSING IN ZAMBIA

This section discusses the process of accessing public land for urban housing and barriers to accessing it.

20.4.1 Process of Accessing Public Land for Urban Housing

Zambia's land governance system is based on the grant of land rights through a public land allocation structure which runs from the President (by virtue of vestment) down to local authorities (as agents). The primary responsibility, though, falls on the Ministry of Lands and Natural Resources (MLNR), which holds delegated powers from the President. This system largely depends on detailed state rules and regulations formulated at central government level. Table 20.1 illustrates the various land administration functions performed under the various departments within the MLNR.

While organizational structures exist at national and regional level, these are missing at district level. For instance, MLNR has no offices at district level, further delegating responsibilities to local authorities through land agency agreements. Under these agency agreements, local authorities are

TABLE 20.1
MLNR Departments and Their Functions

Department	Head	Functions
Lands	Commissioner of Lands	Identify land for allocation Facilitate planning of the identified land Issue offer letters Prepare lease agreements Oversee land agency agreements with local authorities
Lands and Deeds	Chief Registrar	Issue certificate of titles Register any rights on land, i.e., mortgages
Survey	Surveyor General	Undertake surveys, i.e., cadastral Maintain survey records Compile property index maps Examine and process survey records Maintain and improve National Survey Control Network In charge of national planning

Source: Mushinge (2017).

then empowered to undertake certain land administration functions on behalf of MLNR, which include identification of land for development, preparation of plans, initial processing of applications and selecting suitable candidates. In Zambia, local authorities are generally referred to as councils and comprise 5 city councils, 14 municipal councils and 98 district/town councils (Ministry of Local Government and Housing, 2020). The designation of status depends on a number of considerations; for instance, city councils are located in urban districts with large populations and diversified economic activities, while municipal status is granted to councils which cover the subregions (Commonwealth Local Government Forum, 2018). District council status is mostly given to rural districts with low population and heavy reliance on agriculture as the main economic activity (Commonwealth Local Government Forum, 2018). Although this partnership between MLNR and local authorities helps as a stop-gap measure, in practice land governance at this level is littered with several challenges, including corruption (Mushinge et al., 2020), creating further barriers to accessing land for low-income earners. The process of accessing public land for urban housing is captured in Table 20.2.

20.4.2 Barriers to Accessing Public Land for Housing in Urban Areas

Access to urban land is considered a critical element in the improvement of living conditions for urbanites, thus improving overall development of towns and cities (Arjjumend and Seid, 2018). In addition, adequate supply of land is considered a prerequisite to sustainable housing delivery, notwithstanding its role in the socio-economic life of people (Arjjumend and Seid, 2018). Evidence, however, reveals that provision of adequate development land in urban areas is a challenge in Zambia (Civic Forum on Housing and Habitat, 2015; Chitengi, 2015). This is especially severe amongst low-income earners who lack economic and political power (Adedeji and Olotuah, 2012). These barriers to accessing public land for urban housing in Zambia are discussed below.

20.4.3 High Plot Fees

Zambia is classified amongst the poorest countries in the world, with World Bank statistics showing that 58% of the population lives on US$1.90 or less per day, the international poverty line. This

TABLE 20.2
Process of Accessing Public Land for Urban Housing

Steps	Description
Stage 1	Involves the acquisition of land, starting with the local authority in which it is situated. The local authority first identifies land for development, prepares layout plan and requests the Lands Department at MLNR to check whether it is available.
Stage 2	Lands Department then requests Survey Department to prepare a general plan including numbering of the plots. The numbered plan is then sent back to Lands Department for further processing.
Stage 3	Copies of the numbered plans from Survey Department are then further sent to the relevant local authorities.
Stage 4	Local authority is then able to advertise the parcels of land available (plots) and invite the general public to apply.
Stage 5	Applicants pay a non-refundable application fee and then apply.
Stage 6	Local authority then conducts interviews and makes assessments to determine who should be allocated a particular piece of land before issuing recommendation letters. Letters of recommendation submitted to MLNR include key details such as applicant's name, property number (or description), location (low, medium or high cost), council minutes (as proof of allocation by the full council meeting), and process applications to the stage of issuing recommendation letters.
Stage 7	Commissioner of Lands, at Lands Department, is the first stop at MLNR. An 'invitation to treat' is first issued, requesting the allotee to pay the stipulated consideration to government. Once this is paid, the Commissioner of Lands issues the final offer letters.
Stage 8	Physical allocation of the land is undertaken.
Stage 9	In a number of cases, land is allocated unsurveyed and unserviced; in such instances it becomes the responsibility of the allotee to engage a private land surveyor to prepare survey diagrams. These are then submitted to the Surveyor General at MLNR for examination and approval. On the appendage of the Surveyor General's signature, these become legal documents.
Stage 10	The approved and signed survey diagram is then submitted to the Lands Department for further processing.
Stage 11	With the lease agreement prepared and signed by the Commissioner of Lands and with the approved and signed survey diagrams from Survey Department, the documents are then submitted to the Lands and Deeds Department (still at MLNR) for the preparation of the 99-year Certificate of Title.
Stage 12	Verification of the lease agreement and survey diagram is the done at Lands and Deeds Departments.
Stage 13	Registration then follows at Lands and Deeds Department, concluded by the preparation and signing of the Certificate of Title by the Chief Registrar/Registrar.
Stage 14	Applicant is then notified by Lands Department.
Stage 15	Collection is the final stage in this process of land acquisition.
Stage 16	Upon obtaining a Certificate of Title, the land owner has to apply for a building permit from the local authority. The building plans will only be scrutinized upon payment of a plan scrutiny fee. Once a building permit is obtained, then the land owner can start building a house.

Source: Munshifwa et al. (2017), Mushinge (2017).

means that the majority of Zambian citizens live on US$57 per month[1] or less. The implication is that the majority will not be able to pay for plots if they are highly priced.

Recent advertisements for sale of plots by councils around the country show that most of these sales are indeed priced beyond the means of those living on US$1.90 or less per day. For example, for a low-cost plot in a city, a person needs to pay an application fee of about ZMW 250 (US$11.90) (interview with municipality employee, February 2021). This fee is non-refundable even if one has not been offered the land. Furthermore, for one to actually own the public land after successful interviews, one needs to pay a further amount of about ZMW8, 000 (US$380.95) for council fees and service charges as well as plan scrutiny fee of about ZMW600 (US$ 28.57) for scrutinizing building drawings (interview with municipality employee, February 2021). These fees are paid to local authorities.

Since the councils are not the final authority on land, a recommendation is then made to the MLNR so that the land is formally offered to the applicant. At the Ministry, the applicant will have to pay more fees, which include annual ground rent (ZMW 166 or US$7.90), registration fee (ZMW 333.60 or US$15.89), consideration fee (ZMW 2000 or US$95.24), and preparation fee (ZMW 166.20 or US$7.91) (interview with MLNR employee, February 2021). Since most land in towns is allocated unsurveyed, the responsibility of ensuring that land is cadastrally surveyed is then shifted to the allotee who has to engage the services of a private land surveyor at a cost of ZMW3, 500 or US$166.67 for a survey fee. The total cost of accessing public land in towns and cities, then, is beyond the earnings of the 58% of the population who live on US$1.90 or less per day. The end result is that low-income earners are forced to use other means to acquire land, such as squatting or forcible occupation (Chitengi, 2015; Lupale and Hampwaye, 2019). This inequitable access to urban land is therefore exacerbated by the many fees to be paid before embarking on the construction of a house. Thus, pricing of land services such as surveying, registration and other related fees and charges become a prohibitive factor in the acquisition of urban land (GRZ, 2017a).

20.4.4 Inadequate Supply of Public Urban Land

Land is a finite resource, meaning its provision for development cannot be infinite. This is compounded by the fact that development land (often defined as land with services and ready for development) is further constrained by a number of institutional and organizational challenges. For instance, serviced public land is in short supply, coupled with increased demand in most towns and cities, thus exerting pressure on MLNR and all affected councils (interview with MLNR employee, February 2021). Those institutions responsible for the provision of public land have an increased workload, contributing to underperformance in the provision of serviced land (GRZ, 2014). As a way of ameliorating these challenges, the government, under Section 16(1) of the Lands Act of 1995, made provision for the establishment of a Land Development Fund (LDF). The LDF is administered by the MLNR for the purpose of assisting city, municipal and district councils to open up new areas for development. In particular, the fund is intended for the preparation of layout plans, demarcation and surveying of land, construction of roads, water reticulation, and electricity provision in the subject areas (Ministry of Lands and Natural Resources, 2021). Thus, any council in need of funds for opening up new areas for development may apply to the MLNR for some of these funds (MLNR, 2021). Although the fund is available, the finance from the Treasury (Ministry of Finance) is often low, resulting in relatively low impact (GRZ, 2017a). The result is that councils are still unable to provide sufficient serviced public urban land for all.

20.4.5 Cumbersome Public Land Allocation Process

Low-income earners, who are often semi-literate, face the further challenges of following long, cumbersome and costly processes. This is the case for the process of acquiring public land, which involves a number of steps and different point of decision making (see Table 20.2). The result is delays in the processing of applications for land. Ironically, the MLNR Service Charter is very explicit in terms of timelines for the processing of various requests. For instance, the Charter prescribes that MLNR pledges to process offer letters within thirty working days and prepare lease agreements within ten working days after submission of approved survey diagrams (Mushinge, 2017). The reality is that very few of these timelines are actually met, resulting in long waiting periods for offer letters and lease agreements, often extending to several months or years (interview with MLNR employee, February 2021). This delay in the processing of documents is usually attributed to shortage of staff and inertia amongst the few that are given these responsibilities (Sikazwe, 2005).

20.4.6 Corruption in Land Sector

A number of studies point to the land sector as one of the most corrupt sections of government in Zambia, hindering access to land for most low-income earners (GRZ, 2012, 2014; Mushinge et al., 2020). Corruption in the land sector takes many forms, from low-level bribes and fraud (such as administrative corruption) to high-level abuse of government powers and political positions (such as political corruption) (Arjjumend and Seid, 2018). According to GRZ (2012, 2014, 2018), corruption in public land institutions was widespread. Table 20.3 shows some common types of corruption in the land sector in Zambia.

Corruption, whether administrative or political, adversely affects people's access to public urban land. The consequence of corruption is that it is very difficult for the great majority of people, especially those in the low-income group, to acquire land for development.

TABLE 20.3
Categorizing Corruption in the Land Sector

Land Administration Activity	Type of Corruption
Land Use Planning	Fraud mostly happens at identification of land. Some politicians (for example, councilors) in connivance with some land surveyors and planners could privately identify land, prepare layout plans, survey the land and share and/or sell the land for private gain.
Cadastral Surveying	Private land surveyors sometimes paid bribes to MLNR officials for the survey records to be examined and approved quickly. This was due to a backlog of survey records to be examined and approved. The backlog was caused by the lack of manpower and the centralized nature of the examination and approval offices – offices are located in Lusaka and Ndola districts only, meaning that all surveys done in all districts of Zambia had to be lodged, examined and approved in Lusaka or Ndola.
Land Allocation	On receipt of the applications, the councils will interview and select the most suitable applicants for the parcels of land and make recommendations in writing to the MLNR, giving reasons supporting the recommendations. Evaluation sheets were used to rate suitable candidates during interviews. However, these evaluation sheets showed that there was no clear basis for selecting one candidate over another even when they all provided the necessary requirements such as bank statement or pay slip. Therefore, selection of potential developers was susceptible to corruption (fraud, bribery and favoritism). For example, some of the public urban land available at a particular point in time could sometimes be allocated to some politicians (for example, councilors) and political cadres (belonging to the ruling party) as well as close associates such as family and friends.
Land Registration and Titling	The Lands and Deeds Department (under MLNR) registers property rights and issues certificates of title based on the approved lease terms and survey diagrams. The period for the lease is often 99 years. Because of long delays in processing of these documents, potential land users often believe that an 'incentive' in the form of a bride would speed up the process. As a result, people end up paying a bribe to register their property rights and obtain certificates of title quickly.

Source: Mushinge et al. (2020).

20.4.7 Rapid Increase in Population

Natural growth in population and urbanization exerts serious pressure on available services in cities. For instance, Zambia's population was estimated at 16,405,299 in 2017. With an annual growth rate of 2.8%, the Zambian population is expected to increase to 17,885,422 and 23,576,214 by 2020 and 2030, respectively (GRZ, 2020). Furthermore, with the current rate of urbanization at 4.3%, Zambia's urban population is now estimated at 41.8% (GRZ, 2020). This makes Zambia one of Sub-Saharan Africa's most urbanized countries (UN, 2014 cited in Tembo and Sommerville, 2018). The unprecedented growth of the urban population in the country is resulting in an exceptionally rapid increase in demand for public urban land (Civic Forum on Housing and Habitat, 2015). A further result is that public land institutions are unable to meet the demand for land (interview with municipality employee, February 2021). In a number of cases, often reported in the media, people end up invading vacant public and private land or acquiring land illegally which has led to the mushrooming of informal housing settlements. Invasion of vacant land or illegal acquisition of land by squatters leads to scarcity of land available for other legal uses such as commercial, industrial, and civic and community (for example, clinics and schools) (GRZ, 2017a). Statistics further show that 70% of Zambia's urban population is accommodated in informal housing (GRZ, 2020). These settlements are characterized by poor accessibility, congestion, low housing standards, overcrowding, high crime rates, urban delinquency, higher health risks and poor solid waste management (GRZ, 2020). In addition, households in these settlements depend on unprotected wells for water, septic pits for sewerage and forests for energy sources, threatening the sustainability of human settlements (GRZ, 2020), breeding conditions for unsustainable development. If not managed properly, urbanization can create stress on urban land that is detrimental to development (Arjjumend and Seid, 2018).

20.4.8 Land Hoarding and Speculation

Land hoarding and speculation is very common in urban areas in Zambia (Mushinge et al., 2019). Land hoarding refers to a situation where a person keeps many parcels of land despite having no intention of developing or selling them (Mark and Mpandikizi, 2013). Land speculation, on the other hand, refers to a situation where someone keeps on acquiring many parcels of land with the intention of selling them at a high price (Mark and Mpandikizi, 2013). The speculators wait to benefit from increase in land values due to market forces of demand and supply (Mark and Mpandikizi, 2013). Land hoarding and speculation leads to artificial shortages of land (Civic Forum on Housing and Habitat, 2015) and in many cases makes it difficult for those with less money and power to access this land. In other words, the rich and powerful get the land while the low-income earners squat on the remaining spaces, often in the periphery of cities or environmentally weak sites.

20.5 CONCLUSION

This chapter has linked access to urban land and housing to sustainability, arguing that lack of equitable access gives rise to land invasion (of both public and private land) and illegal acquisition of land, especially by low-income earners. Barriers to accessing urban public land for housing construction by low-income earners was also discussed, including high plot fees, inadequate supply of public urban land, cumbersome public land allocation process, corruption in the land sector, urbanization, and land hoarding and speculation. The result is that there are high incidences of land invasion and illegal acquisition of land. The foregoing scenario has led to scarcity of land for other legal uses such as commercial, industrial, and civic and community, as land is taken up by informal housing settlements. Consequently, urban development takes place without regard to the provisions and vision of development plans, adversely affecting the urban environment.

The policy implication for this chapter is that cities in the developing world will continue seeing more rapid growth of unplanned settlements if governments do not seriously address the various impediments to housing for the poor. Ignoring this sector of society would breed conditions for unsustainability and civil discontent with government. Thus, government should provide deliberate policies for the provision of low-priced or subsidized land and housing to this sector of society, which is critical for sustainable development. The reality of most developing countries is that the private property market cannot economically provide housing below a certain cost threshold; below this point housing becomes a social good with the responsibility falling squarely on government.

NOTE

1 Approximately ZMW1,254.00

REFERENCES

Adedeji Y. M. D., and O. A. Olotuah. "An evaluation of accessibility of low income earners to housing finance in Nigeria". *European Scientific Journal* 8, no. 12 (2012): 80–95.

Arjjumend, H., and N. M. Seid. "Challenges of access to land for urban housing in Sub-Saharan Africa". *Journal of Global Resources* 4, no. 1 (2018): 1–11.

Civic Forum on Housing and Habitat. "Housing at the Centre of Urbanisation". Zambia Housing Symposium Report. Lusaka. Civic Forum on Housing and Habitat (2015).

Chitengi, H. S. "Deriving Lessons for Urban Planning and Housing Delivery from the Resilience of Informal Housing Systems in Zambia". PhD Thesis. University of Dundee, Dundee, (2015).

Commonwealth Local Government Forum. "The local government system in Zambia". Available at: http://www.clgf.org.uk/default/assets/File/Country_profiles/Zambia.pdf (2018, March)

Eglin, R. "Equitable access to land". Available at: http://afesis.org.za/wp-content/uploads/2017/12/12-Equitable-Access-to-Land_final_RE_NN.pdf (2017, July)

Gilbert, A. "Helping the poor through housing subsidies: lessons from Chile, Colombia and South Africa". *Habitat International* 28, no 1 (2004): 13–40.

GRZ. "Vision 2030: a prosperous middle-income nation by 2030". Available at: https://www.mohe.gov.zm/download/policies/2030-Vision.pdf (2006, November)

GRZ. "Report of the Committee on Lands, Environment and Tourism for the Second Session of the Eleventh National Assembly". Lusaka: National Assembly of Zambia (2012).

GRZ. "Strategic Plan 2014-2016". Lusaka: Ministry of Lands, Natural Resources and Environmental Protection (2014).

GRZ. "Constitution of Zambia (Amendment) No. 2 of 2016". Lusaka: Government Printers (2016).

GRZ. "Draft National Land Policy". Lusaka: Ministry of Lands and Natural Resources (2017a).

GRZ. "Report of the Committee on Agriculture, Lands and Natural Resources on the Report of the Auditor General on Government's efforts to ensure access to land in an effective manner". Lusaka: National Assembly of Zambia (2017b).

GRZ. "Seventh National Development Plan 2017-2021". Lusaka: Ministry of National Development Planning (2017c).

GRZ. "2019-2021 Strategic Plan and Balanced Scorecard". Lusaka: Ministry of Lands and Natural Resources (2018).

GRZ. "National Housing Policy 2020-2024". Lusaka: Ministry of Housing and Infrastructure Development (2020).

Habitat for Humanity. "Affordable Housing in Developing Countries". Available at: https://www.habitat.org/emea/about/what-we-do/affordable-housing (2021, March)

Hickey, R., and L. Sturtevant. "Public Land and Affordable Housing in the Washington DC Region: Best practices and recommendations". Available at: https://nhc.org/wp-content/uploads/2017/10/Public-Land-Affordable-Housing.pdf (2015, April)

Lupale, M., and G. Hampwaye. "Inclusiveness of urban land administration in the city of Lusaka, Zambia". *Bulletin of Geography Socio-economic Series* 46, no. 46 (2019): 53–70.

Mark, R., and M. Mpandikizi. "A study on affordability and access to planned urban land by the ordinary Tanzanian citizens". *Report prepared by Christian Social Services Commission (CSSC) on behalf of Tanzania Ecumenical Dialogue Group (TEDG)* (2013).

Ministry of Local Government and Housing. "Local Authorities in Zambia". Available at: https://lgazambia.org.zm/municipalities/ (2020, December)

Ministry of Lands and Natural Resources. "Land Development Programme". Ministry of Lands and Natural Resources. Available at: https://www.mlnr.gov.zm/?page_id=4872 (2021, March)

Munshifwa, E. K., W. Ngoma, and I. Makenja. "Major determinant of physical development on urban residential land: The case of Kalulushi Municipality in Zambia". *International Journal of Social Science Studies* 5, no. 6 (2017): 79–89.

Mushinge, A. "Role of Land Governance in Improving Tenure Security in Zambia: Towards a Strategic Framework for Preventing Land Conflicts". PhD Thesis. Technical University of Munich, Munich, Germany (2017).

Mushinge, A., E. K. Munshifwa, and C. M. Mwenya. "An investigation into the causes of land invasion in Zambian cities: Evidence from the City of Lusaka". Paper presented at the conference on Infrastructure Development and Investment Strategies for Africa: Innovation in Infrastructure Development and Investment-Is Africa Ready for Change? Held in Livingstone, Zambia from 23–24 July (2019).

Mushinge, A., S. Mwando, A. Issa, and J. M. Tshitaka. "Mitigating land corruption through computerisation of statutory land governance activities: Evidence from DRC, Tanzania and Zambia". *African Journal of Land Policy and Geospatial Sciences* 3, no. 3 (2020): 71–80.

Oladokun, T. T., J. T. Gbadegesin, and A. Odebode. "Enhancing access to land by the urban poor: exploring viable alternatives". *Proceedings of The International Conference of the School of Environmental Technology*, Federal University of Technology, Akure, Nigeria, held between 25–27, October (2010).

Rondinelli, D. A. "Housing the urban poor in developing countries: The magnitude of housing deficiencies and the failure of conventional strategies are world-wide problems". *American Journal of Economics and Sociology* 49, no. 2 (1990): 153–166.

Sikazwe, F. M. "Land Administration in Zambia: An Appraisal of the Efficacy of the Lands Act No. 20 of 1995". Master of Law Thesis. Lusaka: University of Zambia (2005).

Tembo, E., and M. Sommerville. "Land tenure dynamics in peri-urban Zambia". Policy brief. Available at: https://www.land-links.org/wp-content/uploads/2018/04/USAID_Land_Tenure_Peri-urban_Zambia-Brief_-1.pdf (2018, January)

Transparency International Zambia. "Land Administration in Zambia: A guide to a corruption free acquisition of land". Lusaka: Transparency International Zambia (2011).

UN-Habitat. "Indicators of sustainable development: Framework and methodologies". New York: United Nations (1996).

UN-Habitat. "The Challenge of Slums". Global Report on Human Settlements. Nairobi: United Nations Human Settlements Programme (2003).

UN-Habitat. "Secure Land Rights for All". Nairobi: United Nations Human Settlements Programme (2008).

UN-Habitat. "The State of African Cities 2010: Governance, Inequality and Urban land markets". Nairobi: United Nations Human Settlements Programme (2010).

UN-Habitat. "Affordable Land and Housing in Africa". Volume 3. Nairobi: United Nations Human Settlements Programme (2011a).

UN-Habitat. "Cities and Climate Change: Global Report on Human Settlements 2011". London: Earthscan (2011b).

USAID. Land tenure in urban environments. Issue Brief. Available at: https://www.land-links.org/wp-content/uploads/2016/09/USAID_Land_Tenure_Urban_Brief_061214-1.pdf (2014, April)

21 Impact of Noise Pollution on Human Health in Barasat Urban Area, West Bengal

Kasturi Mukherjee
Adamas University, Kolkata, India

Nandita Deb
Adamas University, Kolkata, India

Abira Dutta Roy
Bankura Zilla Saradamani Mahila Mahavidyapith (affiliated with Bankura University), Bankura, India

Pratik Dash
Khejuri College (affiliated with Vidyasagar University), Purba Medinipur, India

CONTENTS

21.1 INTRODUCTION

Sound is defined by acousticians as the sensation in the ear created by pressure variations or vibrations in the air (Corbisier, 2003), whereas noise is considered an unwanted sound which can create impact on human psychological or cardiovascular processes, and disrupt the natural balance of human life (Singh and Davar, 2004). From a physical point of view there is no difference between the concepts of sound and noise, although it is an important distinction for the listener. Noise are a class of sounds that are considered unnecessary, are composed of many frequencies and can be produced from various sources like machines, construction activities, transportation and musical performance, etc. (Cabrera and Lee, 2000).

DOI: 10.1201/9781003131922-25

In urban areas, the problem of noise is becoming increasingly pronounced due to rapid population growth, heavy traffic movement, economic activities and other functions. This problem has become aggravated in fast-growing cities of developing countries like India. Noticeably, small cities and towns that are adjacent to megacities have acute noise pollution, like Barasat city which is adjacent to Kolkata city. In this chapter, GIS-based mapping was used to study the problem of noise in an urban context of Barasat to understand the spatial distribution of noise pollution levels and their impact on human health. The study demonstrates a causal relationship between noise sources and the built environment that is necessary for an evidence-based planning approach. This study is significant from the point of view of the Indian context of urbanization, as the results of the survey corroborate the impacts of noise on the health of people who are exposed to the outdoor environment.

Urbanization facilitates the provision of a comfortable and easy urban life for citizens. But with the growth of population and infrastructure development, sources of noises are also increasing rapidly and are more impactful in creating noise pollution. Moreover, the removal of noise barriers like vegetation, playgrounds, open spaces, etc. is aggravating noise levels in the urban centers (Ow and Ghosh, 2017; Shen et al., 2011). The magnitude of noise pollution in terms of duration, frequency, and intensity continues to increase due to technological advances, population growth and urbanization (Halonen et al., 2017; Wang et al., 2016). In Indian cities, noise from different sources including increasing population, vehicular density, traffic congestion, etc. is increasing (Hunashal and Patil, 2012; Das et al., 2019). Traffic congestion due to insufficient street capacity, heavy traffic load, unwillingness to follow traffic rules, etc. is one of the major problems in Indian cities such as Delhi, English Bazar and Kolhapur that has led to increasing noise pollution (Firdaus and Ahmad, 2010; Das et al., 2019; Hunashal an Patil, 2012). At low traffic speeds, vehicle engines, transmissions, exhausts and brakes cause "bulk traffic flow noise" (Nagi et al., 1999). Approximately, 60–80% of the noise in cities comes from this bulk traffic flow noise, which is a combination of all noises from vehicles traveling along a road (Mishra et al., 2010). On the other hand, noise level also increases as vehicles accelerate. Rough surfaces, poorly maintained roads with potholes, and older vehicles produce more noise (Bazaras et al., 2008). In India, where urbanization is mostly haphazard, the roads are in a bad condition and poorly maintained, and carry a large number of old vehicles with outdated technology, all of which contribute to noise pollution near the roads (Mishra et al., 2010).

Construction is another aspect of development that is necessary to meet the demands of an increasing population. Construction of highways, roads and buildings involves large machinery, such as bulldozers, loaders, dump trucks and drilling machines, that are good sources of outdoor noise (Geetha and Ambika, 2015; Kantová, 2017). The use of household equipment like generators, air conditioners, vacuum cleaners, mixers, and various kitchen appliances make urban life easy but create indoor noise (Suter, 1991). The crowds in markets, roads and public gatherings also contribute to urban noise pollution. In India, festivals are mostly celebrated with songs and dance in large groups using musical instruments. On many festivals, marriage ceremonies and other occasions, loudspeakers are used that disturb the ambience of the locality (Bhargawa, 2001).

Noise can be treated as harmful, in the same way as other pollutants like toxic chemicals present in air and water. Unlike other pollutants, however, noise pollution does not change the chemical composition and structure of the environment, and is not impactful for long-term natural environmental change; still it has a significant impact on human physical and mental health (Stansfeld et al., 2005; Firdaus and Ahmad, 2010). The World Health Organization (WHO) has documented seven categories of adverse health effects of noise pollution on humans: hearing impairment, interference in spoken communication, sleep disturbance, cardiovascular evidence, disturbance in mental health, the effects of noise on performance, effects of noise on residential behavior and annoyance (Berglund and Lindvall, 1995). Urbanization increases exposure to noise pollution, which has profound public health implications (Sygna et al., 2014; Lercher and Schulte-Fortkamp, 2003; Begou et al., 2020). Urban areas where administrative, commercial, educational, health and cultural activities are carried out are heavily affected by noise pollution (Yuan et al., 2019; Bouzir and Zemmouri, 2017).

Noise levels greater than 55 dB are deemed disturbing for comfortable hearing (WHO, 1995). Past studies have identified noise to be a common cause of sleep disturbance, nervousness, emotional instability, psychological distress, and so on (Dzhambov and Dimitrova, 2014, Sygna et al., 2014; Lercher and Schulte-Fortkamp, 2003; Begou et al., 2020, Recio et al., 2016). This subjective perception of noise and its impacts varies from person to person, whether it impacts their physical or mental condition, adaptability to or awareness of noise pollution (Vienneau et al., 2015; Roswall et al., 2017). It has been found that people working in and around noisy areas and residing in buildings situated in the central urban area and beside main roads, railway lines, airports and factories are directly exposed to excessive noise (Morel et al., 2016; Flores and Gagliardi, 2017; Bazaras et al., 2008). Studies carried out among these people have established that exposure to noise at <70 dB does not appear to affect hearing loss, regardless of the duration of exposure. But noise exposure at >85 dB for more than eight hours is potentially hazardous for hearing impairment, more so for children than adults (Berglund and Lindvall, 1995). It was found that people exposed to noise at a range of 90–115 db may have many physiological and psychological problems like sleep disturbances, increased heart rate, nausea, hypertension, hearing loss, annoyance, etc. (Goiner and Hagler, 2007). Moreover, stress and hypertension resulting from noise may lead to other health problems. A study conducted by Singh (1984) found that workers exposed to high noise levels are more prone to hypertension and neurosensory, circulatory problems, cardiac disease, peptic ulcers, and motor impairment. Goiner and Hagler (2007) in their study assessed that exposure of people to noise at sufficient intensity and duration may increase heart rate and peripheral resistance, high blood pressure, blood viscosity and blood lipids. They may also increase the levels of epinephrine, cortisol and norepinephrine. Various studies have shown a close relationship between noise exposure and subsequent cardiovascular disease (Berglund and Lindvall, 1995; Begou et al., 2020; Vienneau et al., 2015). Apart from contributing to cardiovascular problems, noise may also increase the level of coronary artery disease (Hoffmann et al., 2006). It has also been found that exposure to high-intensity noise in industry may boost the level of non-adrenaline and adrenaline (Stansfeld and Matheson, 2003).

Noise pollution may also lead to a number of personality problems and behavioral changes like lack of confidence, concentration fatigue, uncertainty, irritation, decreased work capacity, stress reactions, anxiety, nervousness, nausea, headache and emotional instability (Berglund and Lindvall, 1995). Noise affects performance and can increase the number of mistakes made at work; it may also cause accidents due to lack of concentration and poor performance (Lercher and Schulte-Fortkamp, 2003; Recio et al., 2016; Licitra et al., 2016). Thus, noise pollution is a serious issue of public concern, yet little importance or in-depth research has been accorded to it (Singal, 2005).

The problems of noise pollution have been treated seriously by some developed countries. For example, the Netherlands restricted building of houses in areas where daily average noise levels exceed 50dB (Granneman, 2013). The USA has taken the initiative to create zones where human-caused noise pollution is not allowed (Geary, 1996). But in India, the movement against noise pollution has been weak, although Indian cities are considered the noisiest in the world (Goiner and Hagler, 2007). However, the Noise Pollution (Regulation and Control) Rules 2000 were drafted under the Environment (Protection) Act (1986), recognizing noise as a pollutant. The guidelines for regulation and control of noise have limited the maximum tolerable noise level for different land-use zones as shown in Table 21.1.

A few studies like that of Kumar et al. (2018), after analyzing Central Pollution Control Board (CPCB) data from 2006–2011, found that the average noise level in the three megacities of Delhi, Mumbai, and Kolkata varies from 60–64, 67–75, and 64–72 dB, respectively, which is much higher than the permissible limit of 55 dB for residential areas during the day. A study of Asansol found maximum day-time noise around the silence zone was 73.5 dB, while maximum night-time noise was in traffic intersection areas (71.18 dB) and the lowest was in the industrial areas (Banerjee et al., 2008).

In addition to noise from traffic, transportation, construction activities, and industry, the community is also greatly affected by noise caused by festivals, public address systems, loudspeakers, and so on. Loudspeaker noise from various community festivals like Durga Puja and Diwali, late-night

TABLE 21.1
Guidelines for Regulation and Control of Noise by Noise Pollution (Regulation and Control) Rules, 2000

	Noise Limit in dB	
Category of Area	**Day Time**	**Night Time**
Industrial area	75	70
Commercial area	65	55
Residential area	55	45
Silence zone	50	40

parties, marriage processions, and election campaigns have caused a lot of noise that affect the quality of life (Bhargawa, 2001). The burning of fire-crackers is another source of noise pollution in urban areas, not only during festivals, but also during weddings, games, announcements of election results, etc. For example, noise level in a residential area of Delhi during Diwali increased to an average range of 76–80 dB (1.2–1.3 times higher than a normal day) due to fireworks (Mandal et al., 2012).

The Government of India, along with the state governments, is taking a number of steps to control noise pollution, especially from community festivals. The West Bengal Pollution Control Board (WBPCB), like those in other states, has initiated steps to control the sale of fire-crackers exceeding the specified noise limits, in consultation with their respective local administrations. However, restrictions and vigilance are more or less limited to the metropolitan cities. In West Bengal, attention has been paid to noise pollution control mostly within Kolkata city, but not in other parts of Kolkata Metropolitan Development Authority (KMDA). Hence, an in-depth study of the spatial variation of noise pollution and its impact on human health, as well as its relationship with vegetation as a barrier to noise pollution is very relevant in the present era of rapid urbanization. The major objectives of this chapter are: (1) measurement and spatial mapping of the sound levels in Barasat municipality, (2) evaluation of the role of vegetation cover as a noise barrier, and (3) understanding the impacts of noise pollution on human health in Barasat municipality. The findings of this chapter may help administrators, planners and policy-makers make strategic decisions in urban growth planning and smart city development.

21.2 MATERIALS AND METHODS

In this chapter, Barasat has been used as the case study because it is one of the fastest-growing cities in West Bengal, as well as a district headquarters town with diverse activities. Therefore, it can be considered a model town representing various aspects of other growing towns in India, such as increased population, immigration, deforestation, urban sprawl, high vehicular frequency, unplanned growth, etc. The study measured noise levels enabled with GPS location at various places including near major roads and along the selected peripheral areas, and with varying land cover conditions, during peak traffic hours. The measured data was analyzed in a GIS environment to obtain the spatial variation of noise levels and their association with different land-use zones. A perception survey with questionnaire was conducted at various locations and analyzed to understand the impact of noise pollution on human health.

21.2.1 STUDY AREA

Barasat, an important city and district headquarters of North 24 Parganas district in West Bengal, is one of the fastest-growing urban areas in greater Kolkata, as well as in West Bengal. Barasat municipality is the administrative headquarters of the North 24 Parganas district, extending between

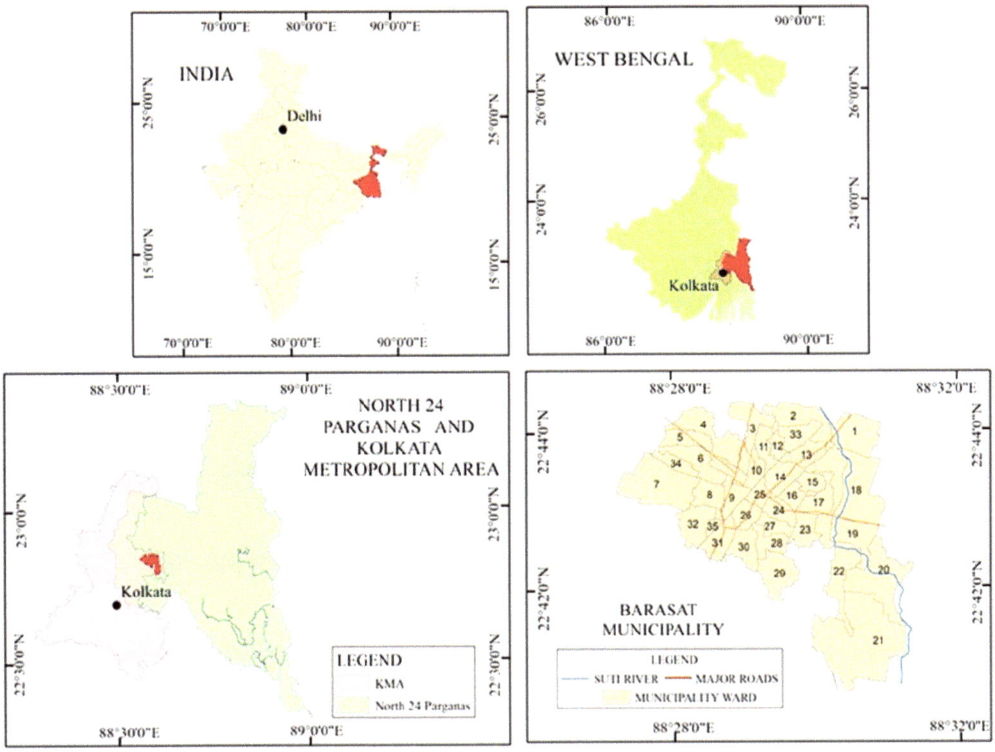

FIGURE 21.1 Location map of the study area.

22°40′58″ and22°44′04″ N latitude and 88°27′00″ to 88°31′30″ E longitude (Figure 21.1). Barasat town consisted of 15 wards in 1986 covering an area of 20.77 km^2, and increased to 32 wards with an area of 28.62 km^2 in 2010 with the incorporation of the nearby panchayat areas within its jurisdiction. Before the partition, the town was located on the eastern side of the Sealdah–Bongaon railway line, though in the post-independence period, the western side of the town grew particularly with the influx of Hindu refugees from erstwhile Pakistan (Chakraborty, 2000).

In Barasat municipality, population growth, an indicator of urbanization, has increased exponentially (Figure 21.2). According to the Census of India, the total population of the town increased from 231,521 in 2001 to 283,443 in 2011. Population density is also uneven in the city, being

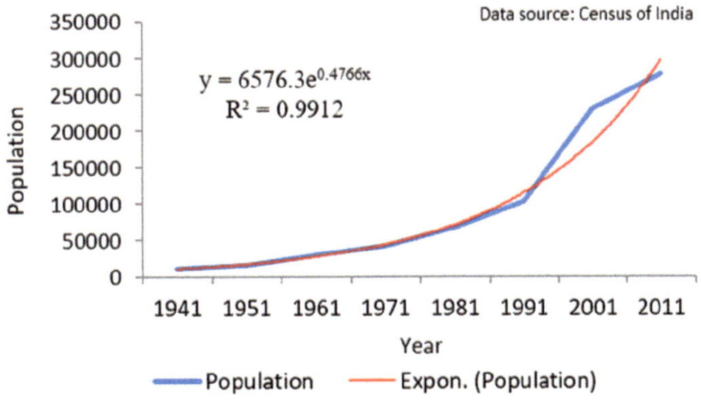

FIGURE 21.2 Population growth in Barasat municipality, 1941–2011.

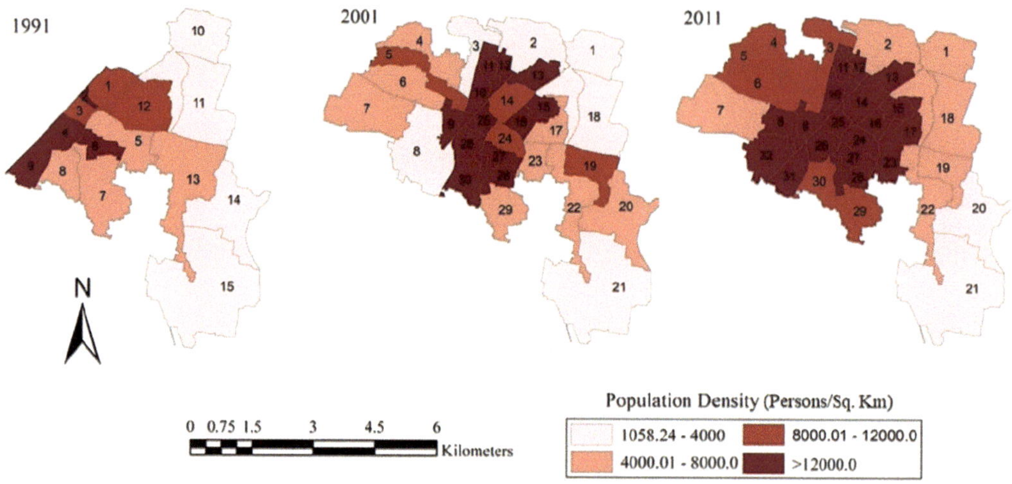

FIGURE 21.3 Population density growth in Barasat over the last three decades.

higher near the core and in the western part of the city (Figure 21.3) on account of the proximity to Madhyamgram and Kolkata. The wards further away from the center have lower density.

The town attracts a large number of people from the surrounding area because of its locational advantage, nearness to the Bangladesh border and linkages, as well as the availability of agricultural land. It has been reported that 90% of the population of the wards are migrants (Chakraborty, 2000). A large number of people have settled for working in brickfields or as builders, while some have settled for an easy commute to Kolkata and some have set up businesses. After the creation of Bangladesh, thousands of homeless Hindu refugees crossed the border and settled in the town because of its closeness to the border. Owing to the rapid population growth, the share of the residential area has also been increasing. Refugee colonies in Kazipara and Sangmura are proofs of immigration (Chakraborty, 2000).

There is no planned development of industrial sectors within the municipality limits. The major industries are brick kilns, cotton mills, sawmills and pitch board factories, and there are some small-scale manufacturing industries for metal and alloy, polyethylene and plastic, and steel products, as well as oil mills situated within the congested residential area along the transportation links. Moreover, Barasat town is a major roadway junction (see Figure 21.4). It is connected by road to Kolkata and other towns of West Bengal and to Bangladesh. NH 34 and NH 35 and SH 02 run through the city. The NH 34 (originally known as Krishnanagar Road) goes towards North Bengal and is a major link between Kolkata and NE India. The NH 35 (also known as Jessore Road) leads to Bongaon and ultimately to the Bangladesh border at Petrapole. The SH02 (Taki Road) leads to Barsirhat, Hasnabad and Taki (Figure 21.4). The town is well connected to other major municipalities of North 24 Parganas by rail. Barasat Junction is a major railway station in the northern section of Sealdah, connecting Bongaon, Bashirhat and Hasnadad. Approximately 187 passenger trains help transport commuters from different parts of Kolkata every day. An extension of the Kolkata metro is under construction at present. The Netaji Subhas Chandra Bose International Airport (CCU) is just 11 km away from Barasat city, and flights from its international terminal go all over the world. With such locational advantages, Barasat has become a pull center for a huge immigrant as well as floating population.

As a consequence of urban growth, increasing population, traffic load, infrastructure development and commercial activities, levels of various types of pollutions like air, water, soil and noise have increased. The Barasat municipality website states that the most serious of all the problems in

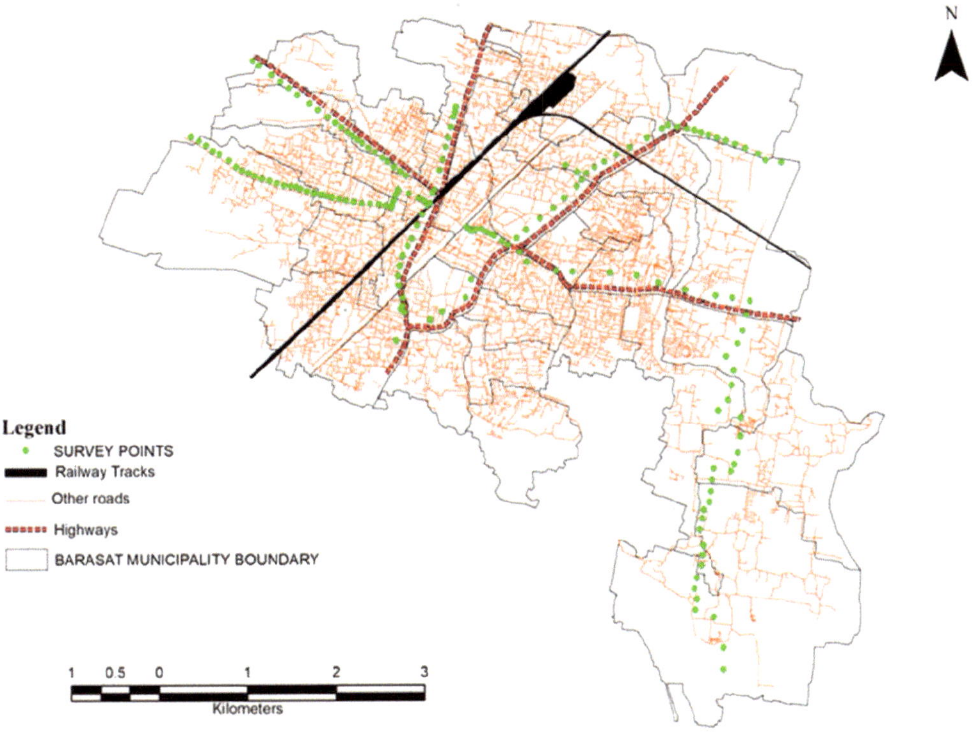

FIGURE 21.4 Road network map of Barasat.

Barasat is noise pollution. Hence, the present chapter carries out a detailed investigation to measure sound pollution levels at various places associated with different land-use zones, and their impact on human health through a perception study.

21.2.2 Methodology

The methodological flow chart is shown in Figure 21.5. The administrative boundary map (ward-level) and road network map were collected from Barasat Municipal Corporation (BMC) and vectorized in GIS environment to prepare a plan for survey and data collection. Locations of industries, schools, markets, hospitals, and commercial complexes were surveyed with the help of GPS and mapped to identify locations that are prone to high vehicular frequency and noise pollution as well as those which should be within a 'no-honking' zone. Silence zones were also identified after mapping and a buffer of 100 meters was drawn around the identified hospitals and schools in the GIS platform, following the Noise Pollution (Regulation & Control) Rules, 2000 guidelines.

The land-use land cover (LULC) map, generated from LANDSAT 8/OLI-TIRS satellite imagery, was designed for noise sample collection points for each LULC condition so that the relationship between LULC and noise propagation and intensity could be assessed. The Landsat-8/OLI-TIRS image (Path 138, Row 44) for 22 April 2016 was obtained from the U.S. Geological Survey (USGS) web portal (https://earthexplorer.usgs.gov/). The LULC map of the area was prepared using supervised classification techniques through a maximum likelihood algorithm on the TntMips platform. The accuracy of the classified map has been estimated with the ground truth data obtained from a field survey. The area was divided into four LULC classes (Table 21.2).

The noise-level measurement survey was conducted from 6 March 2018 to 29 March 2018. All measurements were taken twice a day during rush hour: the first from 9 am to 10 am and the second

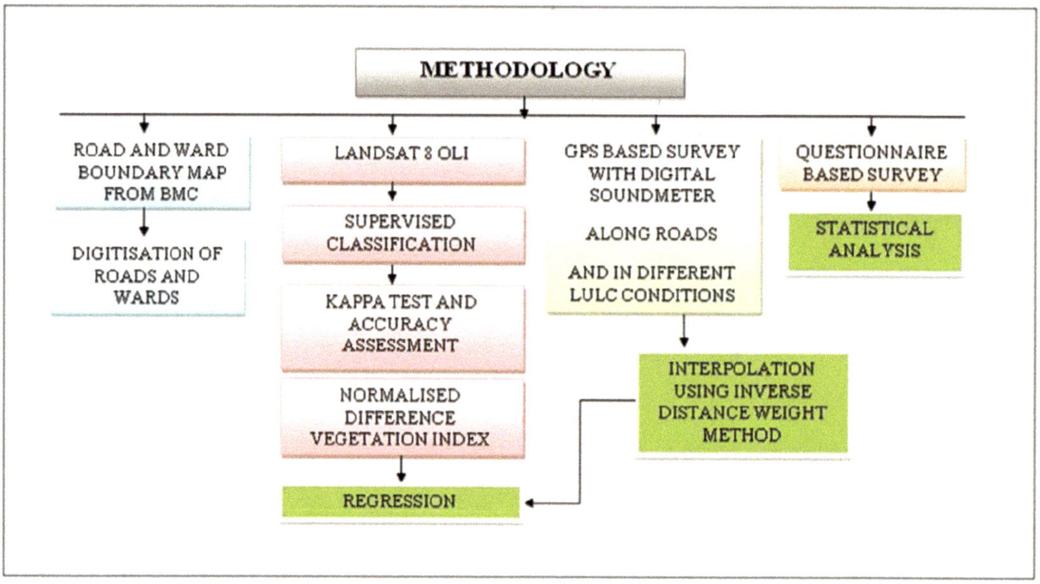

FIGURE 21.5 Research methodology flowchart.

TABLE 21.2
Description of Land-Use Land Cover Categories for Barasat

Land-Use Land Cover Type	Description
Dense settlement	All infrastructure-residential, commercial, mixed use and industrial areas, settlements, construction sites, road network, pavements and man-made structures
Semi-dense settlement	Scattered settlement with vegetation
Vegetation	Natural vegetation and plantation, gardens, parks and playgrounds, grassland, vegetated lands
Agricultural fallow	Agricultural lands, crop fields
Open space	Fallow land, earth and sand in-fills, open space, bare soil
Water bodies	Ponds, tanks, marshy land and swamps

from 5 pm to 6 pm. All measurements were carried out on working days and under calm weather conditions. Noise measurements were standardized for each point by considering the average value of observations taken for one minute, because noise levels fluctuated while vehicles approached and crossed the observation point. A handheld GPS device with a digital sound level monitor (model SL-4030) was used along the major roads, crowded zones, residential areas, agricultural land, fallow land and open spaces of Barasat municipality. The intensity and type of human activities associated with each land-use class were considered during the survey. The total noise samples collected from 194 points (Figure 21.4) across the different land-use types are summarized in Table 21.3.

In order to map the spatial distribution and obtain the unknown values using the observed points with known values, interpolation in QGIS was executed. Research papers by Aumond et al., 2018; De Mesnard, 2011; Harman et al., 2016 showed that other than modeling, kriging is one of the best interpolation techniques used for mapping noise pollution. The kriging interpolation method was chosen over IDW, Spline and other interpolation techniques (Shepard, 1968; Watson and Philip, 1985). Kriging assumes that the distance or direction between sample points reflects a spatial correlation that can be used to explain variation in the surface. Kriging is similar to IDW, in that it weights

TABLE 21.3
Distribution of Noise Samples Collected across Various Land-Use Zones

Zone	No. of Points
Urban area (including the silence zone and industrial plots)	88
Areas of informal settlement	24
Agricultural fallow/barren land	63
Vegetated area	19
Total	194

the surrounding measured values to derive a prediction for an unmeasured location. The general formula for both interpolators is formed as a weighted sum of the data: in IDW, the weight, λi, depends solely on the distance from the prediction location. However, with the kriging method, the weights are based not only on the distance between the measured points and the prediction location but also on the overall spatial arrangement of the measured points. To use the spatial arrangement in the weights, the spatial autocorrelation must be quantified. Thus, in ordinary kriging, the weight, λ_i, depends on a fitted model to the measured points, the distance to the prediction location, and the spatial relationships among the measured values around the prediction location. Kriging uses the following formula (Eq. 21.1)

$$\hat{Z}(s_0) = \sum_{i=1}^{N} \lambda_i Z(s_i) \tag{21.1}$$

where:

$Z(s_i)$ = the measured value at the ith location
λ_i = an unknown weight for the measured value at the ith location
s_0 = the prediction location
N = the number of measured values

The interpolated raster map was divided into very high (>74 dB), high (71–74 dB) and moderately high (<71 dB) noise zones.

In order to understand the importance of vegetation as a noise barrier (Cohen et al., 2014; Pathak et al., 2011; Islam et al., 2012), a Normalized Difference Vegetation Index (NDVI) was computed and correlated with noise level by random sampling. NDVI is the most frequently used vegetation index for determining the health and spatial extension of vegetation cover by remote sensing (Weier and Herring, 2000). The near infrared (NIR) band at the wavelength range of 700–1100 nm is strongly reflected by plant cell structures in green leaves. The high reflectance of the NIR band as compared to the red band (RED) from healthy vegetation (Gates, 1980) is a crucial property used in remote sensing for analyzing vegetation cover by calculating the NDVI. Equation 21.2 is used for computing NDVI (Cracknell, 1997).

$$\text{NDVI} = \frac{NIR - RED}{NIR + RED} \tag{21.2}$$

The value of NDVI ranges from −1 to 1, where positive values indicate vegetated areas and negative values signify non-vegetated surface features such as water, barren land, clouds, etc. Water typically has an NDVI value less than 0, bare soil ranges between 0 and 0.1. Moderate values, around 0.2 and 0.3, correspond to shrub and grasslands and high values like 0.5 and above typically correspond to dense vegetation like rainforest. In this chapter, NDVI is calculated for 22 April 2016.

From the continuous data of both noise map and NDVI, values were extracted through randomly selected feature points in the GIS platform. This spatially associated data was used for regression analysis. Considering vegetation as independent variable and noise as dependent, a bi-variate regression analysis was performed. The regression scatter plots of noise versus NDVI values was created to analyze the influence of vegetation on noise. Regression analysis using Equation 21.3 between two variables is also tested using Pearson's correlation co-efficient as given in Equation 21.4.

$$Y_i = \int(X_i,\beta) + e_i \tag{21.3}$$

where Y_i = *dependent variable*; X_i = *independent variable*; β = *unknown parameters*; and e_i = *error terms*

$$r = \frac{\Sigma(x_i - \overline{x})(y_i - \overline{y})}{\sqrt{\Sigma(x_i - \overline{x})^2 \, \Sigma(y_i - \overline{y})^2}} \tag{21.4}$$

where r = correlation co-efficient; x_i = values of the x – variable in a sample; $\overline{x}$ = mean of the values of the x – variable; y_i = values of the y – variable in a sample; $\overline{y}$ = mean of the values of the y – variable

Finally, a questionnaire survey was also conducted among 265 people from different occupations and exposure time, to understand people's perception of noise pollution and its impact on their health. The data were collected through household survey and spot survey. The participants were randomly selected and were usually adults. Depending on the duration of exposure, the impact of noise on their health was evaluated.

21.3 RESULTS

21.3.1 Land-Use Land Cover Classification

The land-use land cover (LULC) of the study area was classified using supervised classification techniques and merged into four broad classes (Figure 21.6). The overall accuracy and kappa coefficients were found to be 85.6% and 82.8%, respectively (Table 21.4). The classified map shows that the urban area/settlement area is the dominant land cover with a share of about 76% of the total land. Agriculture occupies 12% of the municipality, 8% is vegetation and the other LULCs occupy 4 % of the area.

All the major roads, like NH 34, NH 35 and SH 02, which connect the city with Kolkata, the surrounding urban area and Bangladesh, run through the central part of the Barasat municipality (Figure 21.4). The urban area has a contiguous area of informal settlements which incorporates slums, small houses with gardens and water bodies. Peripheral areas are characterized by the presence of vegetation and agricultural fallows where the presence of both road and settlements are especially low.

21.3.2 Noise Level Analysis

Noise levels measured with a handheld GPS device at 186 locations were used to interpolate. The interpolated map was classified into very high (>74 dB), high (71–74 dB) and moderately high (<71 dB) zones (Figure 21.7). Though noise level varied from place to place, it remained high in the core of the city. The noise level was found to be very high near the main roads, especially at Halabottola, Colony More, Dakbunglow More and Chapadali, compared to the peripheral areas. These are crossroads between the NH35, Jessore Road, Taki Road, K.B. Basu Road and Barrackpore Main Road. This is mainly attributed to the condition of national and state highways and the heavy traffic volumes. The lowest noise level (66dB) was found in the vegetation zone, which is still higher

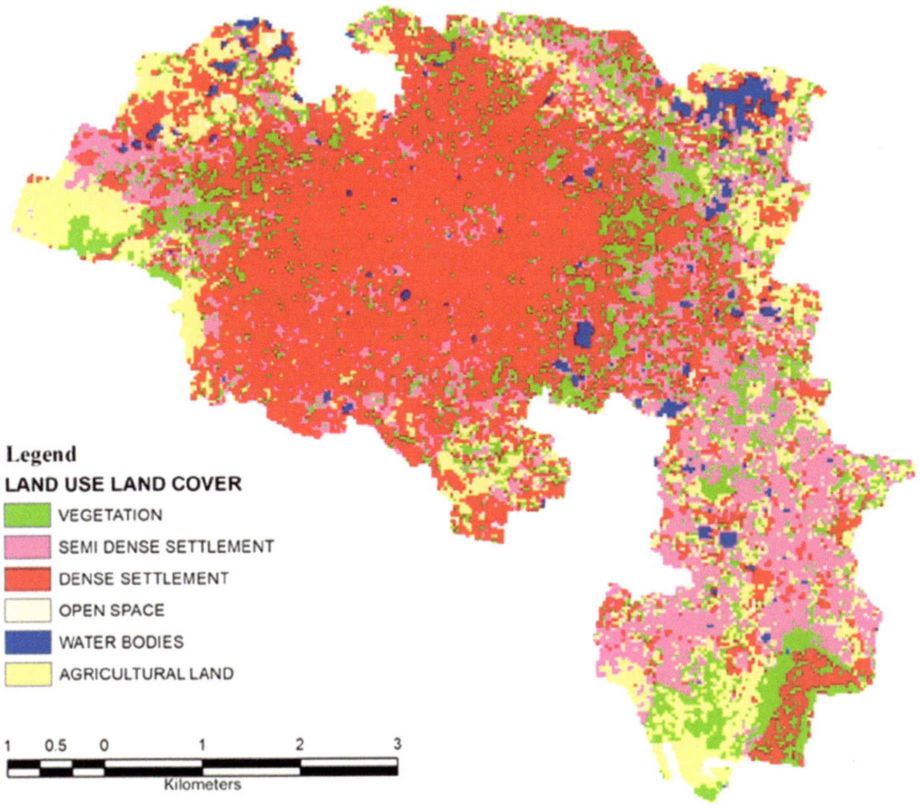

FIGURE 21.6 Classified land-use land cover map of Barasat, 2016.

TABLE 21.4
Confusion Matrix and Accuracy Measures of Classified Land Use Map, 2016

Semi-Dense Settlement		Water Body	Agricultural Land	Dense Settlements	Vegetation	Open Space	Total	User Accuracy
Semi-Dense Settlements	39	0	5	0	4	2	50	78.0
Water Body	0	49	0	1	0	0	50	98.0
Agricultural Land	3	0	43	0	2	2	50	86.0
Built-up Area	3	0	2	41	0	4	50	82.0
Vegetation	4	0	3	0	43	0	50	86.0
Open Space	2	0	1	3	2	42	50	84.0
Total	51	49	54	45	51	50	300	
Producer Accuracy (%)	76.5	100.0	79.6	91.1	84.3	84		

Overall Accuracy: 85.7%; Kappa Co-efficient: 82.8%.

than recommended noise level according to the Noise Pollution (Regulation and Control) Rules, 2000. Nilganj bus stop, Ashok nagar, and Dutta pukur are situated towards the periphery and have less noise pollution. The highest noise levels were recorded as 80dB in the urban built-up areas near Hallabotala and Chapadali. It was evident from the findings that the whole of Barasat faced the problem of high levels of noise pollution.

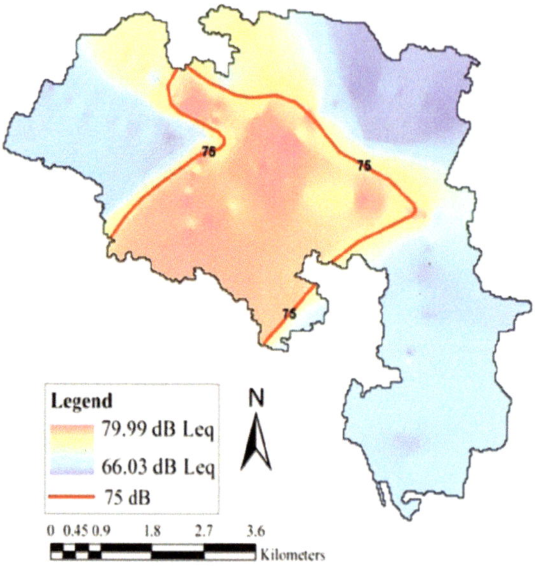

FIGURE 21.7 Spatial distribution of noise levels in Barasat municipality.

Being an unplanned city, Barasat does not have distinct zones for industry and commercial activities. Five industries have been located within the municipality boundary near the congested residential plots (Figure 21.8c). All the hospitals and schools (Figure 21.8a and 21.8b) are also situated within the city center along the main roads. The Noise Pollution (Regulation & Control) Rules, 2000 limit the noise level within 100 meters of schools and hospitals to 50 dB. Schools and hospitals have to be located in the silence zone where honking, use of fire crackers and loudspeakers are strictly prohibited, but on the ground the situation was completely contradictory. According to the regulations, schools, colleges, universities and hospitals are supposed to put up notices such as 'Silence Zone' or 'No Honking Zone', but during field observations it was noted that these regulations were being violated. The study shows that almost all hospitals and schools are built in the high noise pollution areas of Barasat where the sound exceeds the permissible level (Figures 21.8a and 21.8b).

Figure 21.7 demonstrates that the isoline of 75dB passes through the core dense settlement area of Barasat. The noise pollution limit has exceeded the 55dB day-time and 45dB night-time regulatory norms in residential areas (Table 21.1). It is seen that even the norms for industrial areas have been exceeded. This is an indication of the grim situation of noise pollution in Barasat. Only the peripheral areas have less noise pollution, which is still beyond the permissible limit. These conditions ought to have a health impact, which is assessed in the following section.

21.3.3 Health Impacts of Noise

To understand the direct impact of high noise levels on human heath, 265 people were surveyed who are directly exposed to the zone within and above 75dB of noise, based on the duration of exposure. The respondents were asked about their occupations, hours of exposure to noise, health impacts pertaining to noise exposure and information related to consultations with doctors for these health problems. It was found that they are exposed to sound pollution either due to their occupation or residence in high noise area. The surveyed respondents' duration of exposure to direct noise varied from 2 hours to 18 hours (Figure 21.9). They were categorized into four groups, those exposed for < 6 hours, 6–10 hours, 10–16 hours and >16 hours.

People in jobs such as vehicle drivers, shop keepers and ATM guards who work for 6–10 hours on or near roads are exposed to high noise zones and suffer from many diseases like headaches,

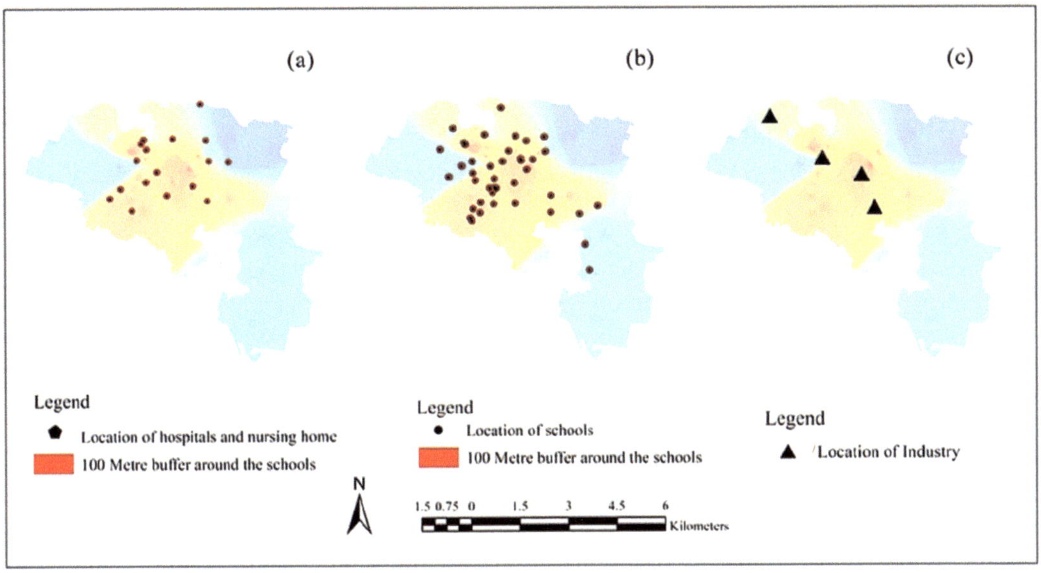

FIGURE 21.8 Locations of (a) hospitals and nursing homes (b) schools, and (c) heavy industries in Barasat.

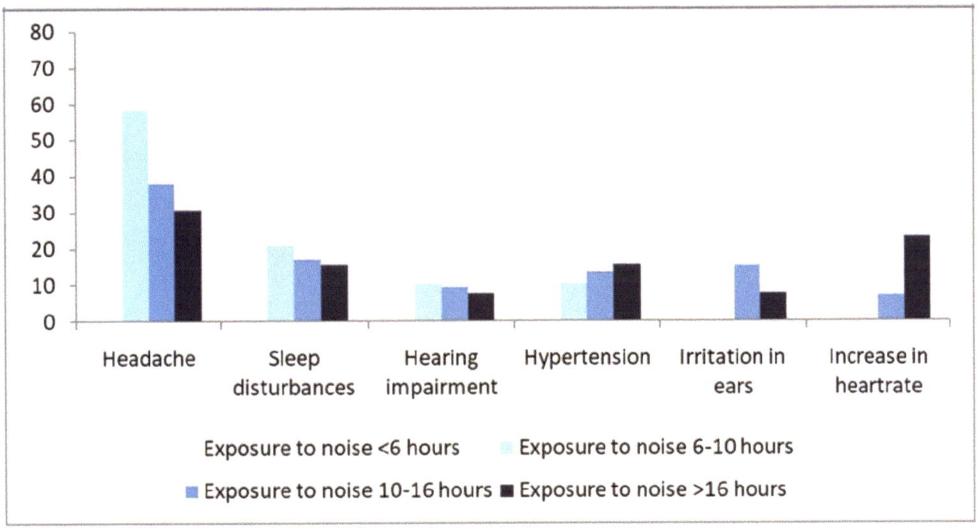

FIGURE 21.9 Impact of noise pollution on human health for various exposure times among the sample respondents.

hypertension, sleep disturbance, irritation in the ears, and increase in heart rate. The people most exposed to noise are traffic police, car drivers, and taxi drivers working for 10–16 hours a day. These kinds of working groups face many health issues like hearing impairment, anxiety, headaches, hypertension, and sleep disturbance. From Figure 21.9 it is evident that of the 265 respondents, around 160 people suffered from headaches, mostly those who were exposed for 6–10 hours on a daily basis. Around 70 people suffered from sleep disturbance though they were exposed for less than 6 hours to noise pollution. 10% of the surveyed respondents complained of hearing impairment and irritation in the ears. Hypertension and increased heart rate were reported by people mostly with exposure time to noise pollution of over 16 hours. In order to mitigate the adverse health implications of noise pollution, tree buffers have proved a good alternative, hence their impact on noise pollution was assessed for Barasat.

21.3.4 Relationship between Noise Level and Vegetation

The NDVI map is shown in Figure 21.10. It is evident that the core region of Barasat lack dense vegetation patches or lines of trees as well as tall trees. The outskirts of the city have vegetated land and agricultural plots. The green patches in Figure 21.10 show vegetated land with value below 0.02, which is very low and hence could not have any impact on controlling noise pollution.

A scatter plot and regression analysis (Figure 21.11) ascertained that noise has a negative relation with higher vegetation cover levels. The trend line represents the decreasing sound level with increasing NDVI values. The b (slope) has a negative value (–32.95). The co-efficient of determination (R^2) value is 0.647. It implies that the model can explain only 65% of the total relationship between the variables. But correlation analysis has shown a more substantial relation between vegetation and noise level with the value of Pearson's product moment –0.82.

However, these statistical measures have provided evidence that vegetation can play a role in combating noise pollution in a city. There have been many examples of plunging noise levels in places of higher vegetation cover. With urban expansion, the vegetation cover starts plummeting. The city center of the study site is devoid of any vegetative surface. Some healthy vegetation is scattered around the periphery of the municipality, and the rest of the area has been covered by sparse vegetation. But this sparse, scattered vegetation has proved effective for reducing the noise pollution.

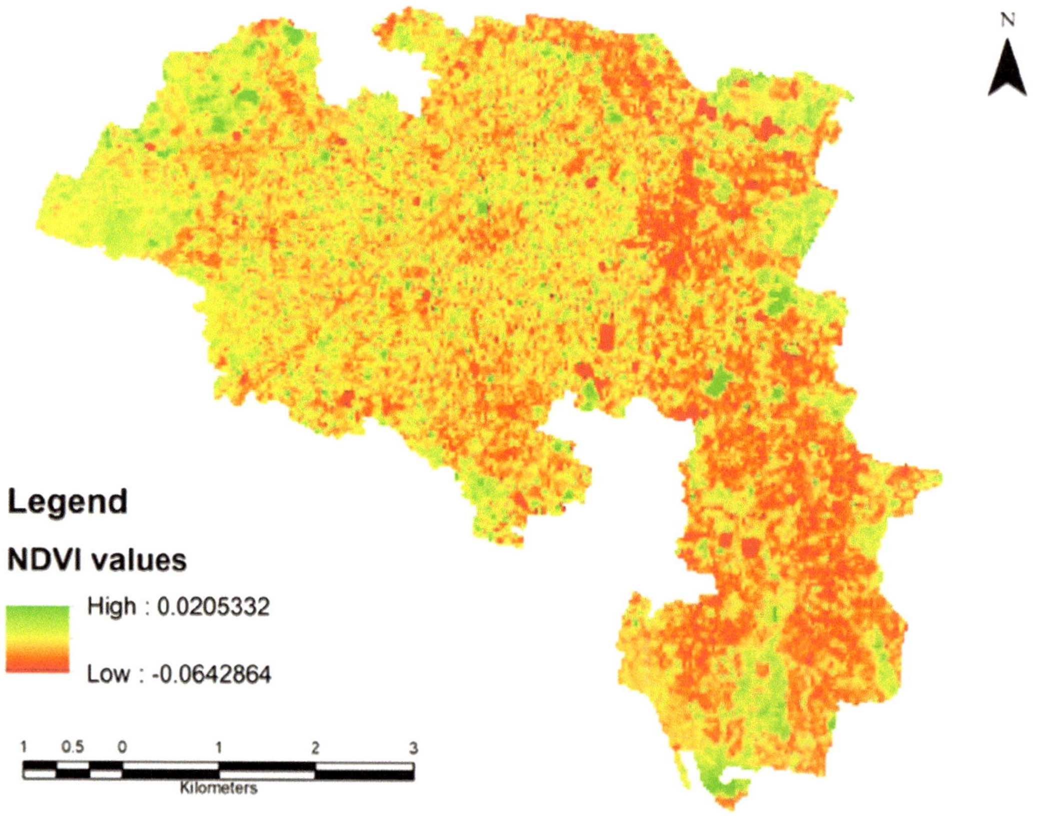

FIGURE 21.10 Spatial distribution of NDVI values in 2016.

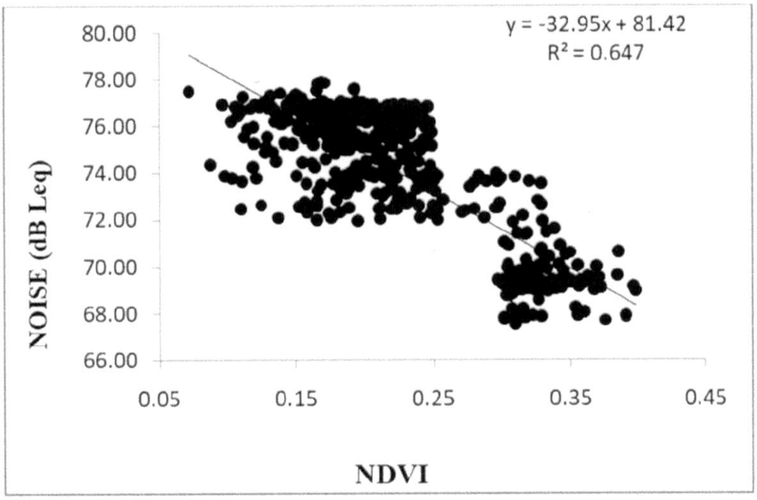

FIGURE 21.11 Scatter plot showing relationship between noise level and NDVI.

21.4 DISCUSSION

In last two decades, many studies have investigated the source, spatial variation and impacts of noise pollution in many cities in various countries, including Italy, UK, Brazil, Turkey, India, and Bangladesh (Zannin et al., 2003; Stansfeld and Matheson, 2003; Pathak et al., 2008; Ozer et al., 2009; Haq et al., 2012). Most of these studies found that cities are seriously affected by noise, mostly from traffic. Like previous studies, this chapter found that noise levels in most of the sampling points were higher than the CPCB standards limits for different area categories (Pathak et al., 2008; Kumar et al., 2018). Throughout the Barasat municipality, the measured noise level at sampling points was found to be high, with maximum noise intensity in the city core. The city core is characterized by dense residential areas crossed by major roads like NH35, Jessore Road, Taki Road, K. B. Basu Road and Barrackpore Main Road. The heavy volume of traffic along with congested narrow passages and poor road maintenance are the causes of noise in this compact residential zone, as noticed in developing countries (Haq et al., 2012). The noise level in many noise-sensitive areas – 'Silence Zone' or 'No Honking Zone' like schools, colleges, universities, hospitals, clinics, etc. – is very high compared to the permissible limits. Similar findings were observed by Fiedler and Zannin (2015) in Curitiba city (southern Brazil). Unfortunately, educational institutions in the rapidly growing cities of India are often built in high-noise areas, as observed in Barasat. This is due to their good connectivity and proximity to residential areas, which attracts investors to construct new educational institutions. However, the municipality administration and state government should implement strict regulations to minimize noise pollution in the silence zone, either by traffic control, traffic diversification, or by imposing restrictions to the establishment of any institution sensitive to noise like schools, hospitals, etc. Additionally, creating noise barriers by planting trees around the silence zone and residential area may reduce noise levels, as observed in this study and suggested by previous researchers (Nagi et al., 1999; Pathak et al., 2011; Shen et al., 2011; Ow and Ghosh, 2017).

From the literature review and the findings of this chapter it is evident that noise pollution is a serious issue in the urban environment, potentially linked to many non-communicable health problems (Zannin et al., 2003; Pathak et al., 2008; Recio et al., 2016). The analysis of the data obtained through the questionnaire confirms that noise is a serious cause of a number of health problems, and the magnitude and criticality of the problems tends to rise with exposure and noise intensity (Hsu et al., 2012). People who work near roads are often exposed to high noise for a duration of

6–10 hours and suffer from many diseases like headaches, hypertension, sleep disturbance, irritation of the ears, and increase in heart rate. On the other hand, people who are exposed to noise for less duration (<6 hours) suffered from sleep disturbance. With the increase of exposure time, the criticality of problems increased, as evidenced by the result that people who are exposed to noise for more than 16 hours in moderate to high noise zones often suffer from hypertension and increased heart rate. The previous study shows that cardiovascular response is related to noise exposure, and episodes of anxiety and arrhythmia are frequently observed in noisy conditions (Hsu et al., 2012). As well as the duration of exposure, people's age is related to the impact of noise. Previous studies by Koushki et al. (1993) and Pathak et al. (2008) have shown that the level of annoyance increases with age.

21.5 CONCLUSION

The unplanned growth of urban areas creates different environmental problems, of which noise pollution is a big concern for Indian cities that is often neglected due to a lack of political will, good governance and social awareness. This chapter has investigated the various aspects of noise pollution, including its impact on human health, in Barasat city. The study measured sample noise levels at various places such as near major roads and residential areas, schools, hospitals and other land-use types during peak traffic hours. The collection and processing of data and the spatial variation of noise levels was analyzed in a GIS environment, focusing on the association of noise levels with different land-use zones. Thereafter, a questionnaire-based perception survey was conducted at various locations to explore the potential impact of noise pollution on human health.

The study found that overall, the city is exposed to high noise levels exceeding the permissible limits suggested by the Central Pollution Control Board (CPCB). The Barasat city core, which contains dense settlement areas, has a noise level of more than 75dB, which is very much higher than the permissible limits of 55dB and 45dB day-time and night-time, respectively, in residential areas. The noise levels near hospitals and schools are far beyond the recommended level.

The high noise level is putting the urban population into a high-risk category for numerous noise-related effects, including hearing impairment, psychological stress, sleep disturbance, and behavioral disorders. However, these diseases are largely preventable. Stringent enforcement of regulations, proper maintenance of vehicles, a ban on old vehicles, effective infrastructure materials, and careful design of roads, considering their proximity to human communities, have all been recommended as preventive and control steps.

The use of GIS technology facilitates noise pollution study in a number of ways. It enables various spatial databases to be created, encompassing essential information such as settlements, roads, locations of schools and hospitals, and spatial coverage of land-use types on the one hand, while providing opportunities for data processing, spatial analysis and visualization on the other. The collection, creation and integration of spatial data has become essential in various research domains and is now easy and cost-effective in GIS platforms. However, sophistication in GIS technology provides new opportunities to explore the various aspects of GIS modeling pertain to noise pollution for better decision making in urban planning.

Noise attenuation through vegetation has been discussed in this study. Implementation of a scheme for tree planting, especially near schools and hospitals, is also suggested. Moreover, above all, awareness of the community and stakeholders is critical in the prevention and control of community noise pollution in cities like Barasat. People have every opportunity to learn about noise emission through common sources of noise pollution the adverse health effects, and the public preventive measures and conditions punishable by law by reading newspapers, magazines, listening to the radio, watching television, and seeing cinema advertisements. The study is significant from the perspective of urban planning, and can help planners, administrators and stakeholders to minimize noise pollution to mitigate consequent impacts on human health.

REFERENCES

Aumond, P., A. Can, V. Mallet, B. De Coensel, C. Ribeiro, D. Botteldooren, and C. Lavandier. "Kriging-based spatial interpolation from measurements for sound level mapping in urban areas." *The Journal of the Acoustical Society of America* 143, no. 5 (2018): 2847–2857.

Banerjee, D., S. K. Chakraborty, S. Bhattacharyya, and A. Gangopadhyay. "Evaluation and analysis of road traffic noise in Asansol: an industrial town of eastern India." *International Journal of Environmental Research and Public Health* 5, no. 3 (2008): 165–171.

Bazaras, J., J. Jablonskyte, and E. Jotautiene. "Interdependence of noise and traffic flow." *Transport* 23, no. 1 (2008): 67–72.

Begou, P., P. Kassomenos, and A. Kelessis. "Effects of road traffic noise on the prevalence of cardiovascular diseases: The case of Thessaloniki, Greece." *Science of The Total Environment* 703 (2020): 134477.

Berglund, B., and T. Lindvall (Eds.). *Community Noise*. Stockholm: Center for Sensory Research, Stockholm University and Karolinska Institute (1995).

Bhargawa, G. *Development of India's Urban and Regional Planning in 21st Century*. Glan Plushing House: New Delhi, India (2001).

Bouzir, T. A. K., and N. Zemmouri. "Effect of urban morphology on road noise distribution." *Energy Procedia* 119 (2017): 376–385.

Cabrera, I. N., and M. H. Lee. "Reducing noise pollution in the hospital setting by establishing a department of sound: A survey of recent research on the effects of noise and music in health care." *Preventive Medicine* 30, no. 4 (2000): 339–345.

Chakraborty, M. T. (2000). *Study of Barasat town: Its problems and planning for future development* (Doctoral dissertation, University of North Bengal).

Cohen, P., O. Potchter, and I. Schnell. "The impact of an urban park on air pollution and noise levels in the Mediterranean city of Tel-Aviv, Israel." *Environmental Pollution* 195 (2014): 73–83.

Corbisier, C. "Living with noise." *Public Roads* 67, no. 1 (2003): 22–22.

Cracknell, A. P. *Advanced Very High Resolution Radiometer AVHRR*. CRC Press (1997).

Das, P., S. Talukdar, S. Ziaul, S. Das, and S. Pal. "Noise mapping and assessing vulnerability in meso level urban environment of Eastern India." *Sustainable Cities and Society* 46, (2019): 101416.

De Mesnard, L. "Pollution models and inverse distance weighting: Some critical remarks." *Computers & Geosciences* 52 (2011): 459–469.

Dzhambov, A. M., and D. D. Dimitrova. "Urban green spaces' effectiveness as a psychological buffer for the negative health impact of noise pollution: a systematic review." *Noise and Health* 16, no. 70 (2014): 157.

Fiedler, P. E. K., and P. H. T. Zannin. "Evaluation of noise pollution in urban traffic hubs—Noise maps and measurements." *Environmental Impact Assessment Review* 51 (2015): 1–9.

Firdaus, G., and A. Ahmad. "Noise pollution and human health: A case study of Municipal Corporation of Delhi." *Indoor and Built Environment* 19, no. 6 (2010): 648–656.

Flores, R., and P. Gagliardi. "A case study of the influence of urban morphology on aircraft noise." *Acoustics Australia* 45, no. 2 (2017): 389–401.

Gates, D. M. *Biophysical Ecology*. Springer Verlag: New York (1980): 611.

Geary, J. "Saving the sounds of silence." *New Scientist* (1996, April): 45.

Geetha, M., and D. Ambika. "Study on noise pollution at construction site." *International Journal of Research in Engineering and Technology* 4 (2015): 2319–2321.

Goiner, L., and L. Hagler. "Noise pollution: A modern plague." *Southern Medical Journal* 3 (2007): 287–294.

Granneman, J. H. "Construction noise: Overview of regulations of different countries." In *INTER-NOISE and NOISE-CON Congress and Conference Proceedings*, Institute of Noise Control Engineering 247, no. 3 (2013): 4892–4900.

Halonen, J. I., H. M. Dehbi, A. L. Hansell, J. Gulliver, D. Fecht, M. Blangiardo, ... C. Tonne. "Associations of night-time road traffic noise with carotid intima-media thickness and blood pressure: The Whitehall II and SABRE study cohorts." *Environment International* 98 (2017): 54–61.

Haq, M. A., M. M. Islam, M. S. Ali, M. F. Haque, and M. M. R. Akhand. "Status of noise pollution in mixed areas of Dhaka city: a GIS approach." *Journal of Environmental Science and Natural Resources* 5, no. 1 (2012): 9–17.

Harman, B. I., H. Koseoglu, and C. O. Yigit. "Performance evaluation of IDW, Kriging and multiquadric interpolation methods in producing noise mapping: A case study at the city of Isparta, Turkey." *Applied Acoustics* 112 (2016): 147–157.

Hoffmann, B., S. Moebus, A. Stang, E. M. Beck, N. Dragano, S. Möhlenkamp, ... K. H. Jöckel. "Residence close to high traffic and prevalence of coronary heart disease." *European Heart Journal* 27, no. 22 (2006): 2696–2702.

Hsu, T., E. Ryherd, K. P. Waye, and J. Ackerman. "Noise pollution in hospitals: impact on patients." *JCOM* 19, no. 7 (2012): 301–309.
Hunashal, R. B., and Y. B. Patil. "Assessment of noise pollution indices in the city of Kolhapur, India." *Procedia-Social and Behavioral Sciences* 37 (2012): 448–457.
Islam, M. N., K. S. Rahman, M. M. Bahar, M. A. Habib, K. Ando, and N. Hattori. "Pollution attenuation by roadside greenbelt in and around urban areas." *Urban Forestry & Urban Greening*, 11, no. 4 (2012): 460–464.
Kantová, R. "Construction Machines as a Source of Construction noise." *Procedia Engineering* 190 (2017): 92–99.
Koushki, P. A., L. F. Cohn, and A. A. Felimban. "Urban traffic noise in Riyadh, Saudi Arabia: Perceptions and attitudes." *Journal of Transportation Engineering* 119, no. 5 (1993): 751–762.
Kumar, A., P. Kumar, R. K. Mishra, and A. Shukla. Study of air and noise pollution in mega cities of India. In: *Environmental Pollution*. Springer: Singapore (2018): 77–84.
Lercher, P., and B. Schulte-Fortkamp. "The relevance of soundscape research to the assessment of noise annoyance at the community level." In *Proceedings of the Eighth International Congress on Noise as a Public Health Problem*, (2003): 225–231.
Licitra, G., L. Fredianelli, D. Petri, and M. A. Vigotti. "Annoyance evaluation due to overall railway noise and vibration in Pisa urban areas." *Science of the Total Environment* 568 (2016): 1315–1325.
Mandal, P., M. Prakash, and J. K. Bassin. "Impact of Diwali celebrations on urban air and noise quality in Delhi City, India." *Environmental Monitoring and Assessment* 184, no. 1 (2012): 209–215.
Mishra, R. K., M. Parida, and S. Rangnekar, "Evaluation and analysis of traffic noise along bus rapid transit system corridor." *International Journal of Environmental Science & Technology* 7, no. 4 (2010): 737–750.
Morel, J., C. Marquis-Favre, and L. A. Gille. "Noise annoyance assessment of various urban road vehicle pass-by noises in isolation and combined with industrial noise: A laboratory study." *Applied Acoustics* 101 (2016): 47–57.
Nagi, G. K., M. K. Dhillon, and G. S. Dhaliwal. *Noise Pollution*. Commonwealth Publishers: New Delhi (1999).
Ow, L. F., and S. Ghosh. "Urban cities and road traffic noise: Reduction through vegetation." *Applied Acoustics* 120 (2017): 15–20.
Ozer, S., H. Yilmaz, M. Yeşil, and P. Yeşil. "Evaluation of noise pollution caused by vehicles in the city of Tokat, Turkey." *Scientific Research and Essays* 4, no. 11 (2009): 1205–1212.
Pathak, V., B. D. Tripathi, and V. K. Mishra. "Evaluation of traffic noise pollution and attitudes of exposed individuals in working place." *Atmospheric Environment* 42, no. 16 (2008): 3892–3898.
Pathak, V., B. D. Tripathi, and V. K. Mishra. "Evaluation of anticipated performance index of some tree species for green belt development to mitigate traffic generated noise." *Urban Forestry & Urban Greening* 10, no. 1 (2011): 61–66.
Recio, A., C. Linares, J. R. Banegas, and J. Díaz. "Road traffic noise effects on cardiovascular, respiratory, and metabolic health: An integrative model of biological mechanisms." *Environmental Research* 146 (2016): 359–370.
Roswall, N., O. Raaschou-Nielsen, M. Ketzel, A. Gammelmark, K. Overvad, A. Olsen, and M. Sørensen. "Long-term residential road traffic noise and NO^2 exposure in relation to risk of incident myocardial infarction–A Danish cohort study." *Environmental Research* 156 (2017): 80–86.
Shen, L. Y., J. J. Ochoa, M. N. Shah, and X. Zhang. "The application of urban sustainability indicators–A comparison between various practices." *Habitat International* 35, no. 1 (2011): 17–29.
Shepard, D. "A two-dimensional interpolation function for irregularly-spaced data." In *Proceedings of the 1968 23rd ACM national conference* (1968): 517–524.
Singal, S. P. *Noise Pollution and Control Strategy*. Narosa Publication: New Delhi (2005).
Singh, N., and S. C. Davar. "Noise pollution-sources, effects and control." *Journal of Human Ecology* 16, no. 3 (2004): 181–187.
Singh, P. "Noise pollution." *Every Man's Science* 25, no. 1&2 (1984): 231–235.
Stansfeld, S. A., B. Berglund, C. Clark, I. Lopez-Barrio, P. Fischer, E. Öhrström, ... RANCH Study Team. "Aircraft and road traffic noise and children's cognition and health: a cross-national study." *The Lancet* 365, no. 9475 (2005): 1942–1949.
Stansfeld, S. A., and M. P. Matheson. "Noise pollution: non-auditory effects on health." *British Medical Bulletin* 68, no. 1 (2003): 243–257.
Suter, A. H. "Noise and its effects." *In Administrative conference of the United States* (1991): 1–47.
Sygna, K., G. M. Aasvang, G. Aamodt, B. Oftedal, and N. H. Krog. "Road traffic noise, sleep and mental health." *Environmental Research* 131 (2014): 17–24.

Vienneau, D., C. Schindler, L. Perez, N. Probst-Hensch, and M. Röösli. "The relationship between transportation noise exposure and ischemic heart disease: A meta-analysis." *Environmental Research* 138 (2015): 372–380.

Wang, V. S., E. W. Lo, C. H. Liang, K. P. Chao, B. Y. Bao, and T. Y. Chang. "Temporal and spatial variations in road traffic noise for different frequency components in metropolitan Taichung, Taiwan." *Environmental Pollution* 219 (2016): 174–181.

Watson, D. F., and G. M. Philip. "A refinement of inverse distance weighted interpolation." *Geo-Processing* 2, no. 4 (1985): 315–327.

Weier, J., and D. Herring. "Measuring Vegetation (NDVI & EVI)." http://earthobservatory.nasa.gov/Features/MeasuringVegetation/measuring_vegetation_1.php (2000)

World Health Organization (WHO). *Physical Status: The Use of and Interpretation of Anthropometry, Report of a WHO Expert Committee*. World Health Organization (1995).

Yuan, M., C. Yin, Y. Sun, and W. Chen. "Examining the associations between urban built environment and noise pollution in high-density high-rise urban areas: A case study in Wuhan, China." *Sustainable Cities and Society* 50 (2019): 101678.

Zannin, P. H., A. Calixto, F. B. Diniz, and J. A. Ferreira. "A survey of urban noise annoyance in a large Brazilian city: the importance of a subjective analysis in conjunction with an objective analysis." *Environmental Impact Assessment Review* 23, no. 2 (2003): 245–255.

22 Water Vendors of Peri-Urban Kolkata

Life, Livelihood and Lived Experiences

Srabanti Bhattacharya
Saheed Anurup Chandra Mahavidyalaya, South 24 Parganas, India

CONTENTS

22.1 INTRODUCTION

Water is a unique resource. It not only sustains life; it is also absolutely necessary for public health and hygiene. The urge to cater to this basic human need for the expanding global population has resulted in immense pressure on available fresh water resources. Lowered water tables, reduced natural flows, complex pollution and quality problems are affecting people's access to water for sustenance as well as for agriculture and ecosystems. Demographic trends of increasing population density, migration and urbanization further complicate the situation, leading to an ever-changing pattern of competition for fresh water (Gronwall, 2008).

In 2000 the member states of the United Nations signed the Millennium Development Goals (MGDs). Goal 7 included a target that challenged the global community to halve the proportion of people without sustainable access to safe drinking water by the year 2015. The global MDG target was met in the year 2010, yet in 2015, it was estimated that 663 million people worldwide still used unimproved drinking water. Nearly half of all people using unimproved drinking water live in

DOI: 10.1201/9781003131922-26

Sub-Saharan Africa, while one-fifth live in South Asia (UNICEF and WHO, 2015). By 2025, it is predicted that half of the world population will be living in water-stressed areas.

It is essential at this point to conceptualize the notions of access to water and water stress. There is no absolute shortage or scarcity of fresh water on earth in relation to its population. The resources, however, are not evenly distributed across the globe, creating varied actual availability. This is because absolute availability of the required quantity of water is not a guarantee of its accessibility.

Physical accessibility broadly implies availability of "water, and adequate water facilities and services" that "must be within safe physical reach for all sections of the population" (UNCESC 2002, clause 12/c/i). The notion of "safe physical reach" is commonly interpreted as "within a certain distance from the household" (Roaf et al., 2005). For instance, in urban areas a person is assumed to have access to safe water if there is water within 200m of their dwelling through a tap or a public standpipe (Zerah, 2000 cited in Gronwall, 2008). Howard and Bartram (2003) noted the direct health benefits of accessibility to water, referring to a source more than 100 m from a dwelling as "no access" with "very high" health concerns and to multiple connections in a dwelling as "optimal access" with "very low" health concerns (cited in Gronwall, 2008). However, Roaf et al. (2005) developed a proxy indicator for ascertaining physical accessibility which, for practical purposes, emerges as a better indicator. It is the time that is taken to collect water. It proposes that the amount of time it takes to collect water is the deciding factor on how much water is collected.

This notion of physical accessibility is particularly effective in understanding access issues in urban areas. The question of economic accessibility is a much more complex dimension, related to the basic conceptualization of water as a 'free good' or an 'economic good'. This itself is an intensely debated issue. However, keeping in mind the cost of collection and distribution, a larger consensus is on pricing water to a certain extent. As WHO (2005) states, "[p]roviding water is never free; the water needs to be collected, stored, treated and distributed".

Water stress, which more often than not indicates a situation where the available resources are not being accessed by the population according to their need, is defined by Schulte as "the ability, or lack thereof, to meet human and ecological demand for water". Compared to scarcity, 'water stress' is a more inclusive and broader concept. It considers "several physical aspects related to water resources, including water scarcity, but also water quality, environmental flows, and the accessibility of water" (Schulte, 2014). Water stress is a global phenomenon, especially in the warm tropics and equatorial nations which have abundant water resources (World Bank 2011). Water stress is a function of physical and economic inaccessibility of water, even when water is actually available.

In urban areas this problem is further multiplied and the issue of access is further complicated by ever-changing demographic profiles, extending city limits and sub-urban layouts. Most urban centers of the developing world suffer from peri-urbanization and unclear boundaries; both are decisive for the possibility of claiming right of access to safe water and for issues of legal regulation and control. Thus, zones of water stress are created in these fringe areas where formal supply is inadequate or fractured. Water stress primarily affects the poor who are vulnerable due to their location in zones beyond the formal supply areas. The gap between demand and supply to these areas is systematically ignored by the urban service providers, and consequently is filled by informal suppliers. These informal suppliers are broadly defined as individuals or groups operating outside the legal framework of water management in an area. The informal suppliers, ranging from water vendors and resellers to private water tankers, are commonly known as small-scale water providers (SSWPs), and operate widely in poor urban areas of Asia, Latin America and Africa, providing services to millions of urban poor (Angueletou-Marteau, 2008). They are an important component of the water sector of the developing world, often in the face of strong official discouragement.

The current research originates from the need to interrogate the lives and livelihood practices of such informal water suppliers in the south-eastern fringes of Kolkata. Informal water providers are not new to Kolkata. The *bheestis* and *bharis* or water carriers were part of the city's water scene even before the advent of piped water. Both the *bheestis* and *bharis*, the predecessors of the present-day SSWPs, were essential community workers who transported water from distant sources to the doorstep. They survived on the need for convenient water access and thus ceased to exist, or became rarer, as the city's network expanded and improved plumbing became available. But water access is not only about convenience. Other than price, the four significant aspects of water supply that determine ideal water access and customer satisfaction are quantity, quality, reliability and convenience (Burt and Ray, 2014). The present-day independent water providers of Kolkata are a response to these requirements, in particular acceptable quality and adequate quantity of potable water. This paper looks at how the present-day informal systems operate and how effective they are in meeting the needs of the residents of the fringe area. It tries to understand if such services can make a positive contribution to the existing urban water infrastructure. It is a micro-level study focusing on one ward (109) of the Kolkata Municipal Corporation area where these suppliers operate.

22.2 STUDY AREA

Kolkata, the largest urban agglomeration of eastern India, is geographically placed in the humid tropical region, and endowed with abundant rainfall (1647 mm/year). The location of the city on the banks of a perennial and large river like the Hugli adds to its situational advantage from the perspective of water availability.

The city started supplying filtered surface water to its citizens way back in 1890. But the ever-expanding population, especially with the huge refugee influx since the 1970s, posed a major challenge to the existing water supply infrastructure. As a crisis response, the state took to supplying groundwater alongside filtered water to the old town in areas of crisis and exclusively groundwater to the fringe areas of the south, south-east and south-west which the distribution network was yet to reach.

The supply of groundwater, which started as an interim measure, continues to be an integral part of the city's formal supply system, in spite of the significant expansion of the filtered water supply network.

At present the city with its 144 wards and a population of 6.2 million (2011) is served by five water treatment plants at Palta Water Works (260 million gallon per day(mgd)), Garden Reach Water Works (120 mgd), Watgunge Water Works (5 mgd), Jorabagan Water Works (8 mgd) and Jai Hind Water Works (30 mgd). Groundwater is supplied from 439 large-diameter (300 mm) tube wells and 1050 small-diameter (40 mm) tube wells. The intermittent supply of 8 hours/day provides 271 mgd of filtered water and 25 mgd of groundwater to 245,019 house connections and 17,000 standpipes. Domestic water is supplied without charge in Kolkata, but commercial connections for bulk supply are chargeable. The city boasts a supply of 160 liters per capita per day (lpcd), which is much higher than most metropolitan areas of the country.

This elaborate network of formal water supply system, however, does not reflect the realities on the ground in their entirety. As in all urban areas across the developing world, water supply statistics do not match the water access situation on the ground. This is even truer for the marginal or fringe areas of the urban centers which merge with peri-urban layouts. In Kolkata, such zones bordering the south-eastern limits of the city's boundary are the worst affected by fragmented, inconsistent and poor quality of supply. These distal areas of ward 108,109 and 110, of borough XII (Figure 22.1), though within the perimeter of Kolkata Municipal Corporation (KMC) area and facing large-scale urbanization, are outside the surface-water network of KMC. They are supplied largely by groundwater. It is in these areas that a city otherwise well-endowed with water is facing acute shortage of acceptable quality of water. The result is a steady growth of informal water providers who are sourcing water from corporate surface-water supply to adjoining wards and supplying it to these pockets, mainly to the urban poor.

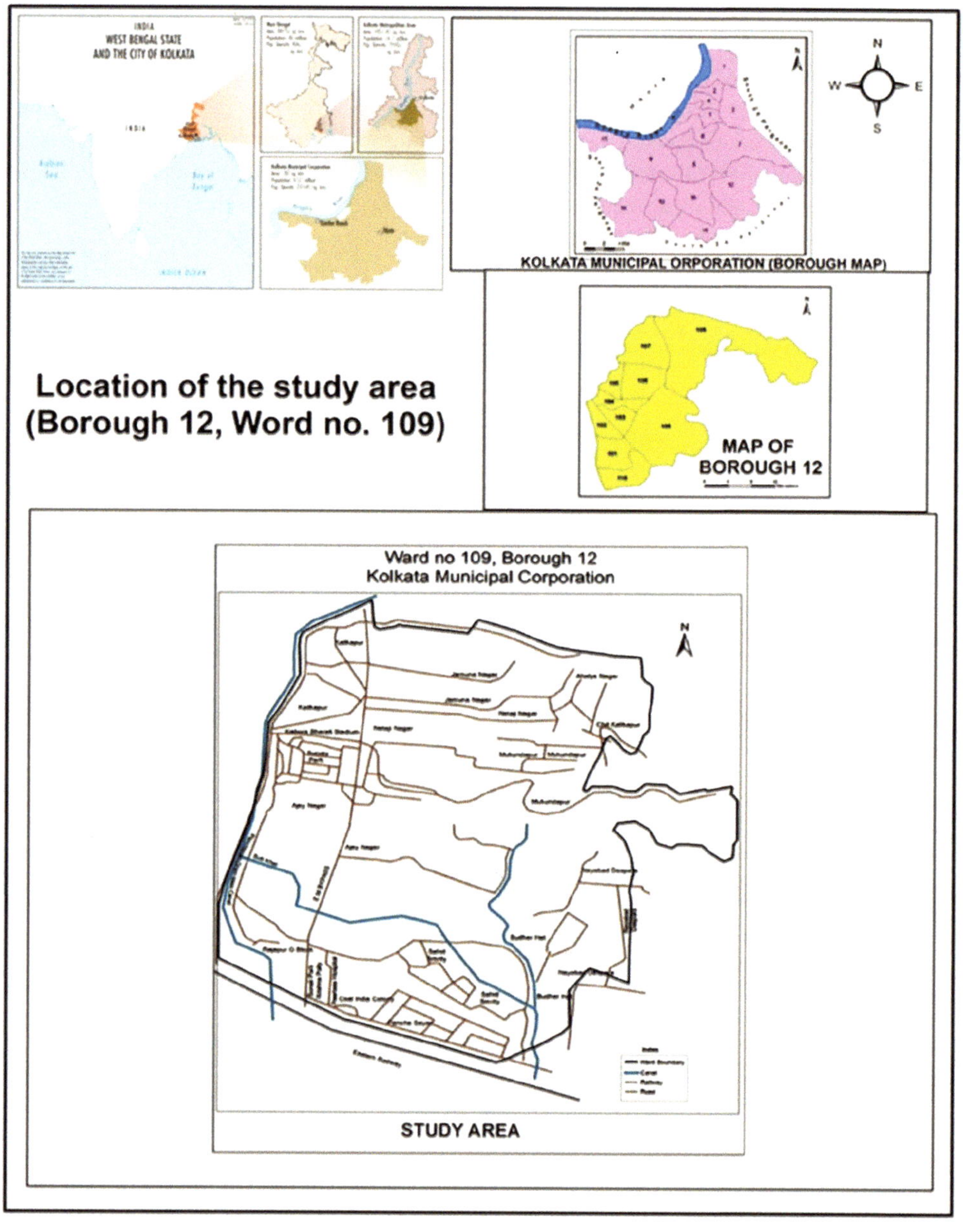

FIGURE 22.1 Location of the study area.

The current paper is an exploratory, micro-level study, restricted to the fringe areas of ward 109 of KMC, and does not claim to be statistically representative. The ward lying in the extreme south-eastern corner of the city spreads over both sides of the EM bypass. The greater part of the ward lies to the east of the bypass. It is interesting to note that the piped filtered water network covers the part of the ward lying on the western side of the bypass, which houses upscale apartment blocks and upper middle-class residential areas. The larger eastern part of the ward, extending into the rural fringes, is not connected to the formal supply network. It is in this part where the informal suppliers are mainly functioning. The unique disposition of the ward in terms of its water accessibility was the reason for selecting this ward for the current study.

22.3 MATERIALS AND METHODS

22.3.1 Research Strategy and Methods

The current research is an exploratory study primarily aimed at understanding the life of the informal water vendor, the modalities of informal water provision practices, and its underlying social and economic implications. The study involved a questionnaire survey of 45 distributing vendors[1] and five wholesale vendors working in and around ward 109. The interviews were carried out in ward 109 of Kolkata Municipal Corporation.

22.3.2 Sampling

The respondents of the vendor survey were selected at will and there was no formal sampling frame. The reason behind this was that the vendors are not licensed or registered. One could have gone out, as the author did, and just listed the vendors. But to actually find all vendors functioning in the area appeared impossible, partly because they work at different times and partly because they were not under any one forum or listed under any organization. As Kjellén (2006) puts it, "one can also imagine sampling from different geographical points from which the first, second or third vendor that appeared would be approached for an interview". In a sense, this was done, but in a non-formalized manner.

The most common way of selecting a vendor for interview would be to spot a couple of vendors while walking through a neighborhood known to be supplied by them, then approach them and ask for an interview. At times the 'snowball' technique was also employed when one vendor introduced us to another. In many instances people would come to listen to the interviews, and at times these would be other vendors. If they were not present for the entire course of the previous interview, these curious bystanders were also interviewed. In some areas, while interviews were being conducted some vendors came up voluntarily and wanted to be interviewed.

22.3.3 The Questionnaire

The interviews were carried out on the basis of a structured questionnaire. The questionnaire was in English and every question was translated into Bengali during the course of the interview. The questionnaire consisted of three sections. The first section contained general information about the social standing of the vendor; the second section was on their occupational history. In this section the vendors were questioned about the reasons for taking up their occupation, their previous jobs and the modalities of getting introduced to the trade. The last section was an elaborate one dealing with occupational details. The focus of this section was to understand the modalities of operation, income, challenges and future plans. The survey instrument was first tested for compliance. It was amended to some extent after a short pilot run.

22.3.4 The Interviews

The interviews were conducted by the researcher. Each interview took 30 to 45 minutes. On some occasions the interviews were recorded (after obtaining permission from respondents), but as the interviews were structured, the recordings were not very significant in the study. It was mainly done to collect potential side comments. The interviews were generally carried out in public spaces like roadsides, water filling points and delivery points. All the respondents were a little apprehensive at the beginning, but with the progress of the interview, as we made it clear that this was not a secret mission and would not affect the vendor's business at all, they gradually opened up. In general, no compensation was given to the vendors who were interviewed. For the record, illustrative and documentary photos were taken during some interviews.

22.3.5 Bias and Errors

There are a few biases and errors in the study. First, it was not always easy to assess whether the respondent had clearly understood the question. Second, for questions on income some inconsistencies were noted as people in general try to avoid such questions. The lack of a proper sampling frame for the given study is an obvious source of bias. But as explained earlier, proper care was taken to include as many respondents as possible in as random a way as possible.

22.4 RESULTS AND DISCUSSIONS

The study of the vendors in the south-eastern fringe of Kolkata provides an insight into the life, livelihood and working conditions of the informal sector. The following sections elaborate the major facets revealed by this study.

22.4.1 Who Are the Vendors?

The vendors were found to be young men, mostly (58%) below 30 years of age. Eighty-three percent of them were married, with 71% being the sole earners of the family. They supported families with an average family size of three members (34%). Fifty per cent belonged to the scheduled caste category while equal numbers were from the general category. Most of them (75%) were local residents, while a few (25%) were migrants from different villages of South 24 Parganas.

With very poor educational backgrounds, many (21%) were found to be illiterate, while most others had only completed primary education (63%). When asked about their aspirations and why they had taken up this trade, the vendors mostly referred to their poor educational background and lack of other employment opportunities. Most of them worked from early morning until late in the evening with a few hours' break during the day between the morning and evening shifts. They started their day early, at around 6 a.m. as they queued up to collect water at different sources. This was followed by distribution to consumers. The same routine was followed in the afternoon when they started queuing up at water points around 3 p.m. and finished delivery by 7.30 p.m.

Most of them worked for 7–9 hours/day (29%), and some (25%) working three shifts worked for more than 9 hours/day. But some vendors (25%), mostly the ones above 40 years old, worked for around 4–5 hours/day and made only one trip. They said that their health did not permit any more work.

The vendors not only worked for long hours, they also worked for 7 days a week (67%) as water is a daily need for most consumers. Only 8% took a day off during the week and some worked for 5 days (12%)

22.4.2 Occupational History

In Kolkata the water vending trade has flourished in the past 7–10 years, when filtered water from Garden Reach Water Works began to be supplied to parts of 104 and 103 wards in the Garfa and Santoshpur areas. Access was easy, which led to the mushrooming of the vendors. It was found that 21% of the vendors had taken to this trade in the last three years and 37% had been engaged in it for between 3 and 6 years, 25% for years and 17% for the last 10 years or more.

For most (83%), this was not their first employment. They had experience in various employments, mostly in the unorganized sector, such as laborer, carpenter, van driver, rickshaw puller, hawker, mason, pizza delivery boy, etc. A significant number (30%) were van rickshaw pullers who started ferrying water in their vans as demand for this trade grew. 38% of them were introduced to the trade by friends and family, while 68% took to the trade themselves, realizing the income potential and noticing the demand for filtered water. Many (29%) vendors told us that they were in the trade with members of their families.

22.4.3 Modalities of Functioning

Of the total respondents interviewed, 59% were independent distributing vendors or water resellers and 33% were distributing vendors working for the wholesale vendors for a monthly salary or on the basis of commission. The remaining 8% were wholesale vendors.

The functioning of the independent distributing vendors was a little different from the others. The independent vendors mainly collected water in 20-liter jerry cans and ferried them to the consumers in their rickshaw vans. They mostly carried 15–20 cans per trip and made two trips per day. Most of them (68%) owned the cycle vans, which cost them Rs. 5000–7000/, and the cans which were bought for Rs. 150/- each. The working hours also ranged from 5–9 hours per day to more than 10 hours per day.

The income of the vendors also varied. Twenty-five percent of the vendors earned less than Rs. 6000/-, 34% earned Rs. 6000–9000/-; 25% earned Rs. 9000–Rs. 12000/-, whereas only 16% earned above Rs. 12000/- per month.

The wholesale vendors were bigger players than the distributing vendors, and there were fewer of them than the independent providers. They employed a number of vendors as well as water collectors and truck drivers. They typically owned three to four mini trucks which people employed by them used to collect water from various sources. They also owned a number of cycle vans and jerry cans which, along with water cans, were given to distributing vendors working for them. These distributing vendors who collected water from wholesale vendors only distributed it to consumers. The wholesale vendors owned around 400–500 water cans and they charged Rs. 8/–9/- per can from the vendors who took water from them. Each can was then sold at prices ranging from Rs. 15–17/- depending on the area where it was being sold and to which floor the cans were to be taken. In some cases, the distributing vendors worked for the wholesale vendors for a monthly salary of Rs. 9000–10000/-.

The selling price of the cans for independent vendors commonly ranged between Rs. 12/- and Rs. 15/- per can. But in some cases, they sold water at Rs. 10/- per can to commercial establishments like hotels and guesthouses who took large amounts of water daily.

It is to be noted that the distributing vendors who collected water themselves sold it in areas near the Eastern Metropolitan Bypass which lies just on the fringes of the filtered water supply area. They did not go inside as they had to take the heavy load of water-filled vans from the collecting points to the consumer's doorstep.

The price they charged was also marginally lower as there were many vendors supplying these areas. The interior areas of Nayabad-Daspara, Budher Hat, Ahalla Nagar and Chitkalikapur were served by the distributing vendors who bought water from the wholesale vendors who mainly brought water in trucks to various distributing points located on the eastern side of the EM bypass.

The vendors collect their water from connected households (23%) or from standpipes (45%) in connected areas of Garfa, Santoshpur and Dhakuria or ward numbers 103,104 and 105. Those vendors collecting water from connected houses mostly paid Rs. 700–1000/- per month to the house owner for collecting water. At times they also paid in kind by doing odd jobs like running errands or cleaning the premises for the house owner. For those collecting water from standpipes, there was no such fixed payment to be made, but they had to pay subscriptions for religious festivals like various pujas and Eid celebrations to local clubs or the neighborhood strongmen. There was no clear answer as to who they actually paid, but all of them indicated that they were the decision makers in the localities, and that they had alliances with the ruling political party. These subscriptions varied from Rs. 200/- to Rs. 10,000/-, depending on the status of the vendor. The wholesale vendors commonly paid higher amounts to maintain their water collection.

The vendors supplied water to low-income households (living in *kuccha* or semi-*pucca* houses), standalone houses, apartment blocks and commercial establishments of ward 109[2] where there is no filtered water supply.

22.4.4 Regulation and Control

Our enquiry into the domain of regulation and control on the modalities of vending in the study area obtained three significant findings. First, the study revealed that the trade was completely unorganized and informal in nature with no umbrella organization or union for the vendors. All respondents denied any political affiliation, but some hinted that they were operating with the permission of the local political class.

Second, it was noted that most of them thought they were not doing anything illegal by selling water that was provided free of charge by the authorities. They emphasized that the price they were charging was for transporting good-quality water to the doorstep of the consumer, who would otherwise be forced to consume poor-quality water. They reiterated that the price they were charging was the cost of physical labor, and so it could not be illegal. To make their claim stronger, they added that the police had full information on their trade and they were never told that they were doing anything illegal. Remarkably, they pointed out that the police had never stopped them from selling water, but had been booked on several occasions for crossing the Eastern Metropolitan Bypass with their cycle vans, the bypass being out of bounds for rickshaw vans. To gain permission for these crossings, almost all of them paid the local police station in cash and also with occasional water supply to the police station, which was also in need of filtered water. Money was given either directly to police personnel or traffic guards, or to some middle man locally known as *dalal* of the local police. The frequency of payment was erratic. According to some, payment was made at regular intervals like once a week or once a month. But many pointed out that they did pay, but only on occasions when they were caught by the police while crossing the bypass. They made a one-time payment to release their vans and cars, but they insisted it was not a regular affair. One could not clearly understand these varying dynamics between the police and the vendors, as it was not possible to talk to the police about this. However, one thing was clear – the underlying alliance with law enforcers provided the vendors with a sense of protection in running this trade.

Finally, since there was no organization of the vendors, there was no regulation or control over their trade. The price of each can of water thus varied according to the demand, season and location of the consumer. Nor was there any palpable competition between the vendors, as they univocally voiced that there was enough work for all vendors working in the area. I could not get an exact number of vendors functioning in the area, but my rough estimate is that about 150–200 of them are supplying water to large parts of ward 109 in Nayabad-Daspara, Panchasayar, Mukundapur, Budher Hat, Ajoynagar, Ahalla Nagar and Chitkalikapur areas.

22.4.5 Issues and Challenges

The major problem that the vendors face, especially while being initiated into the trade, is the physical toughness of the job. Rickshaw vans loaded with water-filled jerry cans, are difficult to ply, especially to areas with bad roads or no paved roads at all. Their transport requires considerable physical strength. Many vendors complained of pain in the chest and joints and recurrent bouts of breathlessness. The job is also tough for vans which have repeated breakdowns due to ferrying such large loads on bad roads.

The vendors failed to point out any other problems that they faced, and yet when asked if they wanted any help from the government, they all answered in the affirmative. Interestingly, though, they could not specify the areas in which they needed assistance from the authorities.

The greatest challenge that the vendors faced was the impending supply of filtered water to their supply areas from the water treatment plant under construction at Dhapa. They were aware that filtered water from the Dhapa plant would be supplied to these areas in near future and that once that happens their trade will take a major hit. Most of them had no backup plan and for some the only option would be to go back to their old occupations. Such is the uncertainty of their trade that it has prepared them to take on the inevitable as and when required.

22.4.6 Discussion

The key points that have emerged from this study are as follows:

i. In Kolkata the mushrooming of informal water vending is a response to dissatisfaction with quality of water served through open-access sources like stand pipes or tube wells. The areas that are catered to by informal suppliers are not completely beyond the city's supply network. These areas are supplied with groundwater extracted from deep tube wells. It is the poor quality of this groundwater compared to the perceived superior quality of filtered surface water supplied to other parts of the city and promised to be brought to the entire population that has created a demand for better-quality surface water. As the surface-water network is being extended at a painstakingly slow pace, the demand for surface water is being met by informal providers.
ii. Since these areas are not devoid of formal water supply, the services of the informal suppliers are not universally accessed. It thus creates economic and spatial inequality at three levels. First, between the residents of the areas supplied with surface water in the city and those beyond such network. Second, it creates inequality between the wealthy households of the fringe areas and the poor ones. The wealthy consumer of poor-quality groundwater purifies the available water using various domestic water purifiers available in the market. This is a costly option because the purifiers are not only expensive, they also incur recurring maintenance costs (like changing the filter candle, etc.) along with the cost of the electricity to run them. Third, it discriminates against the poorest who are left out of this provision as it is beyond their budget. The poorest of the poor can hardly afford the prices and thus they rely on available open-access sources.
iii. The success of this phenomenon lies in the nature of the service. It is sustainable as it reaches pockets where no formal alternative is providing the required service. They are stop-gap arrangements linking the formal and the informal.
iv. The services are consumer specific, responding to the physical and social characteristics of the clients. The supply is in small required quantities, which is affordable as payment is a daily sum of Rs. 12–15/-, and not a substantive monthly amount. This is especially beneficial for the poor, as most of them are daily wage earners and spend on a daily basis. For them it is less convenient to pay consolidated sums at the end of the month. There is no threat of discontinuation of service, as vendors will provide water whenever they are asked to, and the consumer has the choice of not buying water when funds are not available. Such an arrangement is not possible in a formal supply system.
v. The trade is limited to a niche market, with low initial investment cost. Most vendors had to spend Rs. 5000–7000/- on a cycle van and a few jerry cans. Those working for wholesale vendors in most cases made no investments at all as they used the wholesale vendors van and jerry cans.
vi. Although quality of water is the main cause of the inception and growth of this phenomenon, the informal suppliers are not conscious of the quality control and maintenance of hygiene standards required for the supply of potable water. Most of the cans are never cleaned properly and collection points like public standpipes with unpaved or damaged base and stagnant water can be sources of potential contamination. Since there is no regulation of their services, and they are completely unrecognized and informal, there is no quality check in place to standardize the water quality.
vii. The trade is completely unregulated with no umbrella organization or union for the vendors. This leads to varying prices in different areas.
viii. Vending is a competitive activity in which market forces like demand for water and the distance the vendor has to travel to deliver the water, determine the price of water. The vendors have a strong mutual collaboration which controls the interaction among themselves. Each vendor

respects the service territory of other vendors and will always ask for permission from the local vendor when their service is requested by a new client in that provider's supply area. In some cases, the vendors may substitute for other vendors in the absence of regular vendors.

ix. There is no institutional relationship between the vendors and the public authorities. The Municipal Corporation does not formally acknowledge the presence of such providers and the police allow them to supply water as long as they pay for their passage (in the form of a bribe) across the Eastern Metropolitan Bypass with their rickshaw vans. At times they themselves are consumers of such providers. For the local authorities the informal water vendor sector is too small, diverse, localized in marginal pockets and dispersed to regulate. As long as this service is available there is no widespread discontent among the electorate. The promise of extension of the surface-water network and the availability of canned surface water from informal providers thus maintains a status quo which the authorities do not want to disrupt.

x. In India, the law forbids the selling or subleasing of services that anyone acquires from the government. The failure of the government to enforce such existing formal rules, or overlooking the illegal activities related to government urban services, leads to displacement of formal rule by informal ones. This has helped informal service providers to flourish, ignoring or manipulating the existing formal rules, which are not responsive to local realities (Mahiteme, 2008).

xi. The informal water sector is a major job guarantor in the fringe areas. The unorganized workforce residing in these areas, and previously engaged in various blue-collar activities, has taken up this trade as it is reliable and ensures greater income. The uncertainty of the trade in case of the extension of surface water to these areas has not deterred people from taking up this job. Most of them were satisfied with their present status and did not want to consider such possibilities at present.

xii. Physical labor is a major challenge of this trade, as most complained of breathlessness, and pain in the chest and joints. This is also the main cause of a young workforce, who are mostly under the age of 35.

22.5 CONCLUSION

22.5.1 Major Observations

All across the developing nations of the Global South, neither the state nor formal privatized suppliers nor community-based non-profit water organizations have been able to keep pace with the burgeoning population and its ever-increasing demand for acceptable quality and required quantity of potable water. The standard response to this perpetual crisis has traditionally been a simplistic effort to extend the city-wide network under the control of a single operator. But this monolithic solution does not provide an adequate response to the wide variation in demand from the consumers living in different environments and requiring different amounts of water at different times of the day and at different seasons (Collignon and Vézina, 2000).

This becomes even more difficult when it comes to serving the poor living in unplanned settlements located at the edge of the cities. These marginal locations are very difficult to serve through formal distribution networks. The system is thus highly fragmented with differentiated and inequitable services (Kjellén, 2006). In such pockets of deprivation, the poor are the worst affected. With little reserves to bail them out of the situation, they rely on the informal water providers for their water. Their crude and informal services are the only alternative for them to ameliorate the apathy of an inequitable supply system.

In Kolkata this informal arrangement of supplementing the shortage of acceptable-quality water in isolated pockets of the city by drawing from filtered water supplied elsewhere in the city is a way of life or "territorial practise" (Roy and AlSayyad, 2004) under the formal supply set-up of the city.

Informal water vending is a reality of the urban centers of the developing world. Informal water vendors are unique examples of a private entity with profit motive delivering a public service (Pangare and Pangare, 2008). The Kolkata story is no different. The metropolis is unable to provide water to all and keep pace with the ever-expanding limits of the city. The resultant service failure is a reality which needs immediate redressal. But for a utility under the state government jurisdiction, and controlled by the municipal corporation, the available funds are inadequate to change the scenario at an acceptable pace. The long wait for the service to reach every consumer in the city, for a resource as essential as water, is unacceptable. In such a scenario one is inclined to acknowledge the services of the informal vendors, as they not only serve as a significant link between the consumer and the utility, but their service also brings about an appreciable change in the wellbeing of the citizens of the fringe area.

In conclusion, some consideration should be given to ways and means of making this service of greater benefit until such time as formal services are available.

First, it is essential to recognize the role played by the vendors. It is 'informal' and 'illegal' according to the law of the land, but it is the only option available to a section of the citizens. This brings us to the second consideration, which is to accept the vendors as an integral part of the water system, just as the concept of 'last mile connectivity' in the transport sector encourages the existence of unique transports like the 'bike taxi'. By giving them formal recognition, the water utilities can also co-opt the vendors as 'peripheral providers'. Third, formal recognition must come with the direct supply of water from points ascertained by the utility, and from where the vendors can purchase or take water and then supply it to the consumers. This supply at a subsidized nominal cost can in turn reduce the cost of water paid by the consumer and also allow the utility to earn some dividends from the sale of water they are filtering at a cost. The vendor can be a registered service provider working as an official partner providing water to the fringe areas. Fourth, the registered vendor can be trained in hygiene practices and provided with cans and motorized vehicles to improve the quality of water and reduce the physical strain involved in vending. This in turn will increase the efficiency and reliability of the service. Finally, formation of associations could be considered, as such organizations could take responsibility for training a potential vendor. They could also monitor the day-to-day functioning of the vendors and negotiate with the parent utility in addressing various issues and challenges. The association could also control the price of water and prevent variations in various localities and seasons. This would benefit consumers, especially those living in distant parts of the interior.

These proposals, if implemented, could bring about a considerable change in the functioning of the vendors. It is utopian to wish away the 'informal' from our urban lives, which are confounded by the challenges of the changing times. Whether in the realm of water, transport or sanitation, these players will exist and are an integral part of our fractured systems. This paper thus argues that it is only prudent to accept and co-opt these players and maximize the benefits from their services. Having said that, the researcher also recognizes the various challenges in such formal recognition of the informal, and understands the various negotiations required at different levels for such an acceptance. However, the study submits that it is a challenge worth exploring, bearing in mind the benefits to the marginally located and the poor.

NOTES

1 The various forms of water vending practiced across nations have been aggregated into three sub-types by Whittington et al. (1989):

i) **Wholesale Vendors**: obtaining water from a source and selling it to distributing vendors.

ii) **Distributing Vendors**: obtaining water from a source or from wholesale vendors and selling it to consumers door to door.

iii) **Direct Vending**: selling water to consumers coming to the source to purchase.

2 The vendors also supplied water to some areas of ward 110, however my study focuses only on their trade in ward 109 of KMC.

REFERENCES

Angueletou-Marteau, A. "Informal Water Suppliers Meeting Water Needs in the Peri-Urban Territories of Mumbai, an Indian Perspective". Global Changes and Water Resources: Confronting the Expanding and Diversifying Pressures: XIIIth World Water Congress, September 2008, France. http://iwra.org/member/congress/resource/abs208_article.pdf Assessed on 24.06.2021

Bakker, K. 2003. "Archipelagos and Networks: Urbanization and Water Privatization in the South". *The Geographical Journal* (Blackwell Publishing and The Royal Geographical Society), Vol. 169, No. 4 (2003): 328–341.

Burt, Z. and Ray, I. "Storage and Non-Payment: Persistent Informalities within the Formal Water Supply of Hubli-Dhrawad, India". *Water Alternatives*, Vol. 7, No. 1 (2014): 106–120.

Collignon, B. and Vézina, M. *Independent Water and Sanitation Providers in African Cities: Full Report of a Ten Country Study*. UNDP-World Bank Water and Sanitation Program, Washington, DC (2000).

Gronwall, J. T. *Access to Water: Rights, Obligations and the Bangalore Situation*, No. 439. Linkoping University, Linkoping (2008).

Howard G., and Bartram J. *Domestic Water Quantity, Service Level and Health*. World Health Organisation, Geneva (2003).

Kjellén, M. *From Public Pipes to Private Hands: Water Access and Distribution in Dar es Salaam Tanzania*. Department of Human Geography, Stockholm University, Sweden (2006).

Mahiteme, Y. "The Informalization of Urban Service Provisions and Management in Peri-Urban Addis Ababa." *Journal of Ethiopian Studies*, Vol. 41, No. 1/2 (2008): 61–82.

Pangare, G. and Pangare, V. "Informal Water Vendors and Service Providers in Uganda: The Ground Reality". *The Water Dialogues: Informal Water Vendors, Uganda* (2008).

Roaf, V., Khalfan, A., and Langford, M. *Monitoring Implementation of the Right to Water: A Framework for Developing Indicators*, Global Issue Papers, No. 14. Heinrich Böll Foundation, Berlin (2005).

Roy, A. and Alsayyad, N. "Urban Informality: Crossing Borders", In Roy, A. and Alsayyad, N., (eds.) *Urban Informality: Transnational Perspectives from the Middle East, Latin America and South Asia*. Lexington Books, Lanham, Maryland (2004).

Schulte, P. "Defining Water Scarcity, Water Stress and Water Risk, Pacific Institute", (2014). https://pacinst.org/water-definitions/ Assessed on 24.06.2021

UNCESCR. "General Comment No.15, The Right to Water" (2002). UN Doc E/C.12/2002/11.

UNICEF, and WHO. "Progress on Sanitation and Drinking Water – 2015 update and MDG Assessment", (2015).

Whittington, D., Lauria, D. T., Okun, D. A., and Mu, X. "Water Vending Activities In Developing Countries: A Case Study of Ukunda, Kenya". *Water Resources Development*, Vol. 5, No. 3 (1989): 158–168.

WHO. *Technical Notes for Emergencies*, No. 9. World Health Organization, Geneva (2005).

World Bank. "India Vulnerability of Kolkata Metropolitan Area to Increased Precipitation in a Changing Climate", Report No. 53282-IN, Environment, Climate change and Water Resources Department South Asia Region (2011).

Zerah, M. H. "Household Strategies for Coping with Unreliable Water Supplies: The Case of Delhi". *Habitat International*, Vol. 24, No. 3 (2000): 295–307.

23 'Sacred Land Capture'
An Introduction to a New Phenomenon of Encroachment in Urban India

Somenath Halder
Kaliachak College, Malda, India

CONTENTS

23.1 INTRODUCTION: FACTS AND FACETS OF URBAN INDIA

According to the 2011 census, the declining population growth trend observed in India's urban population in the 1980s and 1990s has been reversed, and the pace of urbanization has increased faster than expected (Bhagat 2011). While the broad phenomenon of urbanization changes demography and cultural landscape, it is the vital and dramatic transformation of land that is irreversible. Urbanization is manifested globally, and especially in India, in an unprecedented and uncontrolled rate of growth, and its scope remains beyond our understanding, or ability to measure and monitor it (Taubenböck et al. 2009; Bettin and Wollni 2020). As an important nation (in the Global South), India is increasingly urban, and a healthy urban environment as well as the quality of life of urban dwellers and commuters are growing concerns. Between 2001 and 2011 the urban population of India increased by 92 million (Kundu 2013; Shaw 2018), and the coming decade is expected to see a turning point towards a growing 'urban future' where the net addition to the urban population surpasses that of the rural population.

In simple terms, the nature of urban India's transformation is characterized by uninterrupted concentration of population and activities additional to primary activity in large settlements. Pragmatically, the pace of urbanization in India is slower than in other developing nations in the Global South. The term "over-urbanization" (Davies and Golden 1954) has sometimes been used to emphasize the presence of urban misery, a modified form of rural poverty which may reasonably pull back a city to become a dynamic one. In over-urbanized habitable settlements, the features of an inefficient, less productive informal sector (Kundu and Basu 1998) are becoming increasingly apparent. Another scholar (Breese 1969) depicts India's urbanization as "pseudo urbanization",

DOI: 10.1201/9781003131922-27

where immigrants reach cities due not to urban pull but to rural push. Moreover, Raza and Kundu (1978) rightly describe the dysfunctional urbanization and urban accretion that results in a concentration of population in a few large cities without a corresponding increase in their economic base. Moreover, Prasad (2019) argues that the key feature of urban phenomena in several countries, especially in the Global South, is that they are an "urban laboratory for the world". Prasad (2019) asserts that India, with its "plurality in cultures, languages, climate zones and landscapes", together with institutional efforts, is setting a "unique global benchmark" for sustainable urbanization.

Out of this chaos theory, the key features of urban India, or Indian urbanization, are the growth of Class-I cities as lopsided urbanization; urbanization without a strong economic base (or viable industrialization); demographic explosion generating (rural-to-urban) immigration urbanization; increase in slums and unofficial squatter settlements; the decline in the quality of urban life due to distress migration; and other specific problems such as urban poverty, unemployment, exploitation, inequalities and so forth (Dutta 2006; Bhagat 2011). The UN (2015) has set 17 sustainable development goals (SDGs) as an agenda for 2030. These are the eradication of poverty, hunger and starvation, good quality of health and well-being, value-added education, gender equality and sensitization, safe drinkable water and sanitation, affordable and clean energy, decent work and economic growth, industry and infrastructure, reduced inequalities, sustainable cities and communities, accountable production and consumption, climate action, life under water, life on land, workable institutions for peace and justice, and partnership for the above targets. At present, to reach the SDGs, especially in Indian cities, NITI Aayog[1] (2015) has formulated seven mission programs for all-round development. The programs are 100 Smart Cities Mission, Atal Mission for Rejuvenation and Urban Transformation (AMRUT), National Heritage City Development and Augmentation Yojana (HRIDAY), Sardar Patel National Urban Housing Mission (SPNUHM), National Mission on Sustainable Habitat (NMSH), Clean India Mission (CIM), and National Urban Information System (NUIS).

Nevertheless, there is something missing in the current "subaltern urbanization" phenomenon (Denis et al. 2012; Denis and Zérah 2017) in a country like India. One can easily observe that without any effective industrialization and provision of skill-oriented employment, the towns and cities are becoming overcrowded and congested and malfunction due to several socio-political and socio-cultural factors. The current study should be regarded as a preliminary attempt to focus on an unidentified phenomenon: small-scale land capture especially in "reluctant urban" areas (Tiwari et al. 2015). This chapter looks at the progress of this phenomenon and how its spatial outlook corresponds with its overall causes and consequences. For better understanding, pictographic models and ground-level images are included. The chapter concludes with policy suggestions and further research proposals.

The design of this chapter is as follows. Following the introduction, the second section considers the basis of the concept, including proposed definition, process and optimal spatial dimension. The third section discusses methodology. The fourth section gives an overview of how sacred land capture (SLC) in the urban environment has emerged and developed in recent decades. The coexistence of causes and consequences of SLC among sustainable towns and cities, ground-level evidence with metadata and views on the essential theoretical basics are indicated in the following subsections. The fifth section concludes the main findings and provides policy suggestions.

23.2 SLC: WHAT, HOW AND WHERE THE PHENOMENON TAKES PLACE

Beyond mainstream urbanization, in India "reluctant urbanization" (Tiwari et al. 2015) and/or "subaltern urbanization" (Denis and Zérah 2017) has been taking place in recent decades. What the two concepts have in common is the large amount of settlement agglomeration beyond previously established urban regions and a huge influx of rural migrants who are trying to cope with the changing urban environment. This brings us to the meaning and description of the concept *sacred land capture*. India has unique features of heritage, culture, linguistics, and rural economy linked with its prehistory. Consequently, in the Western world, India is regarded as a land of pilgrimage. It is also true that among all the aspects of peoples' lives and lifestyles, the impact of religion is ever dominant. In the face of urban overcrowding, the majority of settlers in the urban areas depend on non-agricultural activities. Within the ever-expanding urban spaces there are vacant lands or small pieces of unused land (sometimes incorporating 'sacred groves'). In this respect it is not slum areas or squatter settlements that are highlighted but also the core parts of the city or town (known as the CBD). Newcomers or immigrants (who have migrated from rural to urban areas not only due to the *pull factor* of urban areas but also due to *push factor* of rural areas) have automatically been attracted towards these vacant or unused lands in search of a sustainable way of earning a living. In the initial stages, they engage in temporary selling activities with tiny sitting or standing arrangements. Moreover, if these places are (partially) free from any objections or legislation, these kinds of marginal selling activities are going on and other immigrants and the urban poor replicate them in other corners of towns and cities. After a decade or two a new phase begins when a (temporary) religious establishment develops in the form of a temple or mosque (*Majār* or *Eid-Gaha* or a sacred place of community prayer) and if there are any sacred groves this type of establishment become more prominent. Thereafter a number of temporary shops or small-scale commercial activities gradually start to flourish and in this way sacred land capture (SLC) becomes a reality. In short, based on religious and sacred behavioral grounds, small pieces of vacant or unused patches of land in towns and cities, are being gradually and illegally (without proper legal documents of ownership and taxation/levies) captured by the urban poor for subsistence-type commercial activities known as 'sacred land capture'. SLC is most common among the census towns[2] and peri-urban areas[3] of India, but limited in the big cities or mega-cities due to strict legislation, urban renewal initiatives by local authorities, and so forth.

23.3 METHODOLOGY: FROM UNACCOUNTED REALITY ON THE GROUND TO THEORETICAL CONTEXTUALIZATION

Urban social science tends to concentrate its searching lens on the largest metropolitan cities, when half of the globe's city dwellers actually live in small and medium-sized towns (UN 2018). By reconsidering the importance of small and medium-sized towns under the larger phenomenon of contemporary urbanization in India, this chapter seeks to discuss an unrecognized silent scenario of small-scale land encroachment in urbanized localities, based on religious or sacred sentiments. Being an introductory concept, this study is based on a simple research methodology that involves two main components: meticulous literature review and long-term ground observations. References for the literature study appear at the end of the chapter, while the related terminology is shown as notes. Discussions and interviews with various experts, including town planners and senior geographers, have proven valuable for the standardization and development of the concept.

The study focuses on understanding the process of gradual sacred land capture, the key reasons for illegal encroachment of vacant or unused small patches of street-side land, and the unenviable consequences for city dwellers and commuters. Simple pictographic model(s) and a Venn diagram representing the theoretical base of the whole scenario are incorporated to provide context. In support of the narratives, several ground images (with their respective geo-reference and address tags) have also been included with this volume. It may not be possible to transpose the lessons from diversified spatial dimensions directly to a single set of models, but they may be adapted to suit the concept and further design of the reframed policy. While the study is limited to preliminary findings and recommendations, it may encourage younger researchers to conduct detailed research for applied purposes.

23.4 SMALL-SCALE ENCROACHMENT (SACRED) IN URBAN AREAS

From the first pivotal city, the city itself has played a significant role in the multi-phase centric change of a surviving society, which in itself is an impact of overall socio-economic and socio-political transformation. Objectively, 'urbanization' can be described as a vital, intertwined social and economic phenomenon that is being seen across the globe at an unparalled rate (Liu et al. 2010; Shi et al. 2012; Sun et al. 2013). Within a short period of time, urban areas as a whole are capable of significantly transforming the existing landscape with a variety of environmental and social impacts. To accommodate the ever-growing number of urban immigrants (especially rural migrants), towns and cities are encroaching on the peripheral hinterland beyond the fixed administrative urban map (Atu et al. 2013; Mosammam et al. 2016; Sahana et al. 2018). Again, due to the acceleration of phenomena like 'urbanization' and 'subaltern urbanization', there is growing interest among urban geographers in considering *horizontal* and *vertical* (spatial and temporal dimension) urban development. Also relevant for academic analysis are urbanization and its impact on a broad set of environmental factors, including loss of arable land and change of natural vegetation cover at micro and macro level (Grimm et al. 2000; Lopez et al. 2001). Chadchan and Shankar (2012) point out that unplanned and uncontrolled rapid urban growth adversely affect good living conditions and a healthy environment for urban dwellers.

It is a fact that the emerging census towns[2] as well as peri-urban areas[3] of India have suffered from basic problems like overcrowding, shortage of open space, deterioration of quality urban life, and so on (Dutta 2006). In parallel, sacred land capture repeatedly casts its shadow over these problems, frequently enhancing their magnitude. This sub-section focuses on how sacred land capture has become established in the urban environment. Traveling through the emerging small urban areas of India, nothing seems unusual to an urban ecologist. However, the situation on the ground becomes increasingly clear when analyzed in a temporal paradigm, with even some specific spatial variations manifesting themselves. For overall generalization, a model-based approach has been included here. Figure 23.1 also helps us understand the process of development of SLC in its non-rigid temporal dimension. In the first phase (Figure 23.1(a)), small vacant or unclaimed areas of land in any town and city are usually seen along disputed lands like railway junctions and national highways, where major highways meet local roads, peripheral areas of any governmental institutions/offices, footpaths alongside urban streets, and so on. These kinds of unused (very) small patches of land in small towns and cities are the starting point for SLC. These kinds of disputed or unclaimed (vacant) small pieces of land initially remain covered by shrubs and small trees from common Indian genuses. After a while, in the next phase, some of the existing trees are transformed into sacred groves. In this phase (Figure 23.1(b)), within a decade new kinds of temporary selling activities (with tiny sitting or standing arrangement for sellers) flourish, meaning that preliminary encroachment of small strips of urban land (disputed or unclaimed) begins. At first this kind of encroachment is overlooked and ignored by local government and residents. With the passage of time, over a decade or so, those (informal) small selling activities with minimum sitting or standing arrangements along the unclaimed street sides or junction points become semi-permanent and acquire concrete shop-like structures (Figure 23.1(c)). At the same time sacred spots develop as concretized temples or *mazar* (a kind of sacred place for the Muslim community) – the beginning of sacred land capture. Thus it would not be wrong to state that both sacred place and sacred land capture boost as well as supporting each other to become a common phenomenon in small urban areas. In the fourth phase (Figure 23.1(d)), many changes occur gradually over the next decade. The established SLC becomes more prominent through the steady infrastructural development of concretization of shops and sacred spots, electrification, construction of footpaths alongside captured small strips of land, buying and selling interactions, and so on. The SLC sites become overcrowded through multi-selling activities and even begin to recapture the footpaths used by pedestrians. One surprising fact is the ever-changing ownership of the tiny area occupied by an individual through illegal person-to-person contact. During the entire process of development of SLC (model-based discussion), from the (hypothecated) beginning to the end, this kind of unauthenticated capture of vacant or unclaimed disputed land is replicated in many corners of a particular town or city and in neighboring small urban areas. The subsequent paragraphs discuss the causes and consequences of SLC development and present metadata as evidence of the phenomenon.

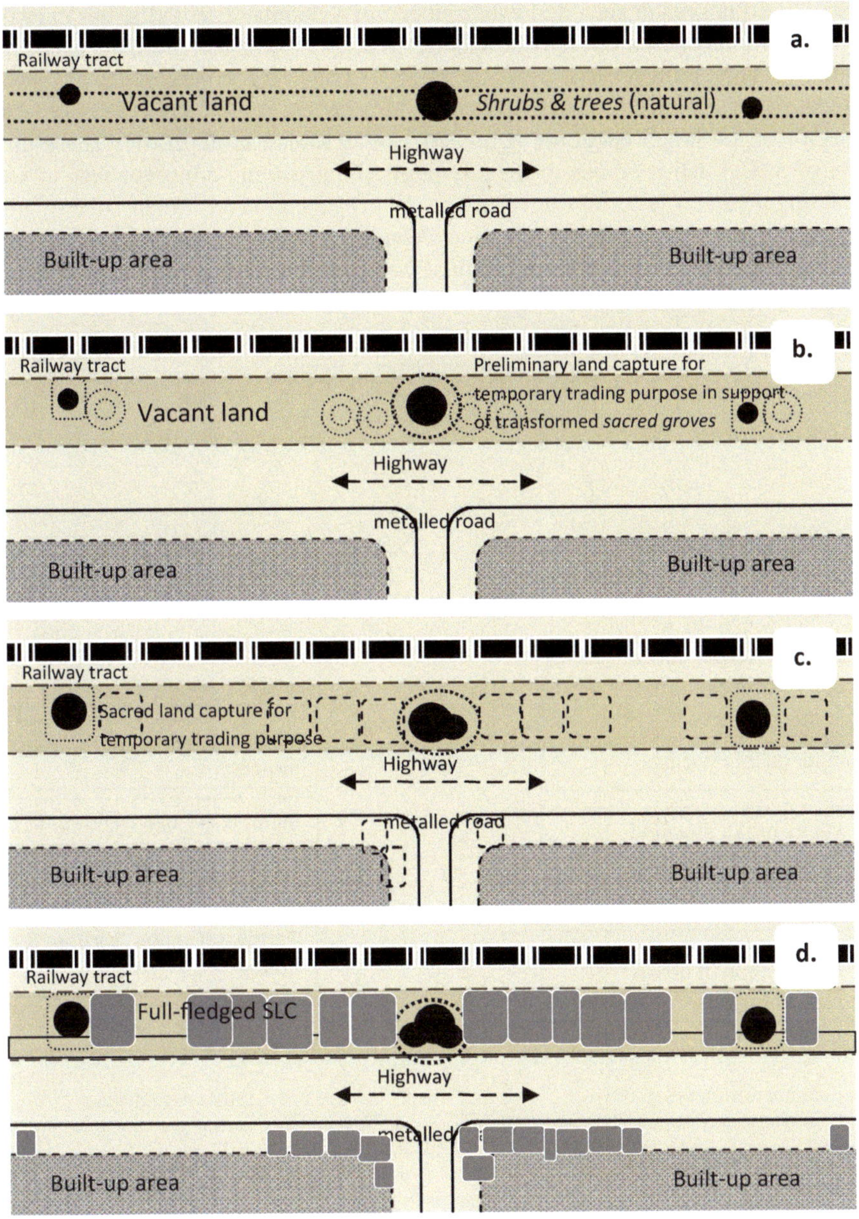

FIGURE 23.1 Phase-wise schematic development of SLC in the urban environment in India.

(Source: Developed by author (based on long-term ground observation and research).)

23.4.1 Causes and Consequences of SLC

It has been observed that in a country like India globalization, liberalization, and privatization mutually or individually downbeat the urbanization process (Kundu and Gupta 2000; Dutta 2006). Liberalization alone allows cheap import of goods without incurring and utilizing domestic resources, which in due course adversely influences the maintenance of the rural economy, handicrafts and household industry on which rural peasants depend. The desired positive benefits of liberalization do not reach each and every citizen of a country like India, but only those who are able to acquire new skills (Despande and Despande 1998). Over and over again, globalization and privatization cause retrenchment for rural peasants. The combination of these negative factors affects poverty-induced (push factor) rural-to-urban migration and creates pressure on the informal sector of the urban economy (Mukherji 1993; Kundu et al. 2001). These are the pre-phenomena of SLC when it comes to the key point of causes and consequences of sacred land capture (see Figure 23.2). The unwanted effects of urbanization on Indian society are regional inequality, traffic jams and blocked streets, slums/squatter settlements and occupied land conflicts, deteriorating ambient air quality, poor or irregular water supply, garbage and sewage management, and the growing reluctance on the behavioral side of both stakeholders and urban local bodies[6]

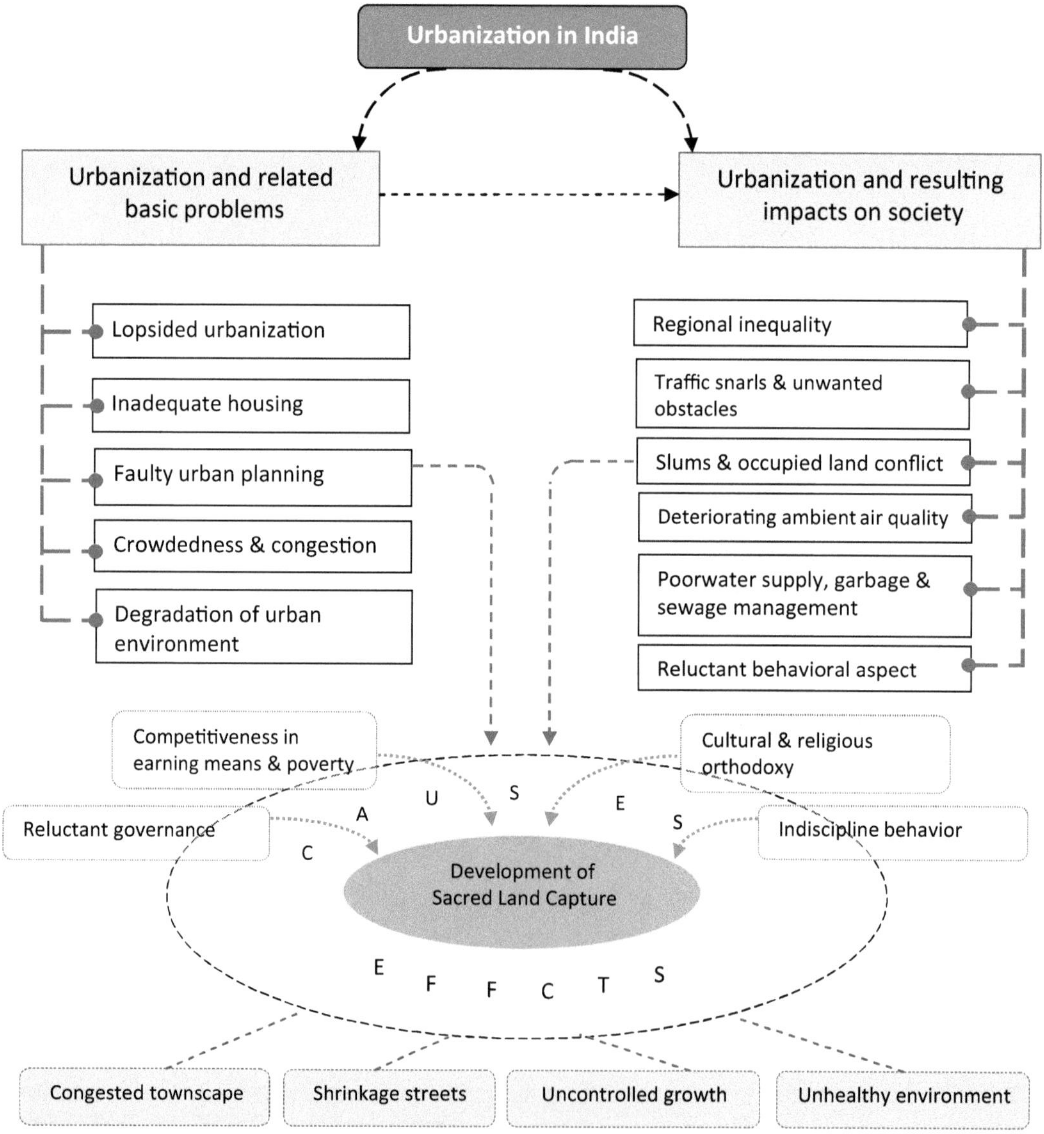

FIGURE 23.2 Schematic outline of causes and consequences of SLC.

(Source: Developed by author, after Jaysawal and Saha (2014).)

(Jaysawal and Saha 2014). Thus, it is pre-existing factors, especially faulty urban planning, slums and occupied land conflicts, that predominantly give rise to the phenomenon of sacred land capture. Behind this phenomenon lies the religious and cultural mindset of Indian citizens, with its deep-rooted linkages in any kind of spatial transformation (e.g., the establishment and development of former Indian cities is religious place centric). Contemporarily, due to a huge influx of rural immigrants into newly formed urban centers (census towns and peri-urban areas) there is big pressure on as well as competition for urban land-use management and the unorganized sector of the sustainable urban economy. In parallel, the slack institutional set-up or reluctant governance of the newly emerged urban bodies are unwittingly providing opportunities to capture or encroach on vacant, unclaimed or disputed tiny piece of land. Slowly but surely, urban dwellers and commuters are accepting and becoming accustomed to this somewhat filthy, unclean, unsystematic, noisy urban environment, all of which contributes to a mindset that encourages undisciplined behavior by residents of small urban areas. The World Bank (2011) has also commented on the problematic features and challenges of urban areas and has emphasized the legal, regulatory and institutional basis for land management. Chandran (2018) has rightly raised the question of the fastest-growing Indian urban provinces with unclear (disputed) land ownership, which would trigger legal disputes among the rightful owners and capturers; the problems become intrinsic when legal group(s) report violent evictions of poorer communities. This is a serious problem for the *smart cities* vision of a South Asian country like India. At the same time, the inevitable consequences of sacred land capture are congested townscapes, shrinking streets, uncontrolled or haphazard growth, scant sources of earnings for ULBs[4], and an unhealthy overall environment in newborn urban centers. The next section endeavors to capture a reality check on the ground and focus on spatial aspects of the phenomenon (Figures 23.3 and 23.4).

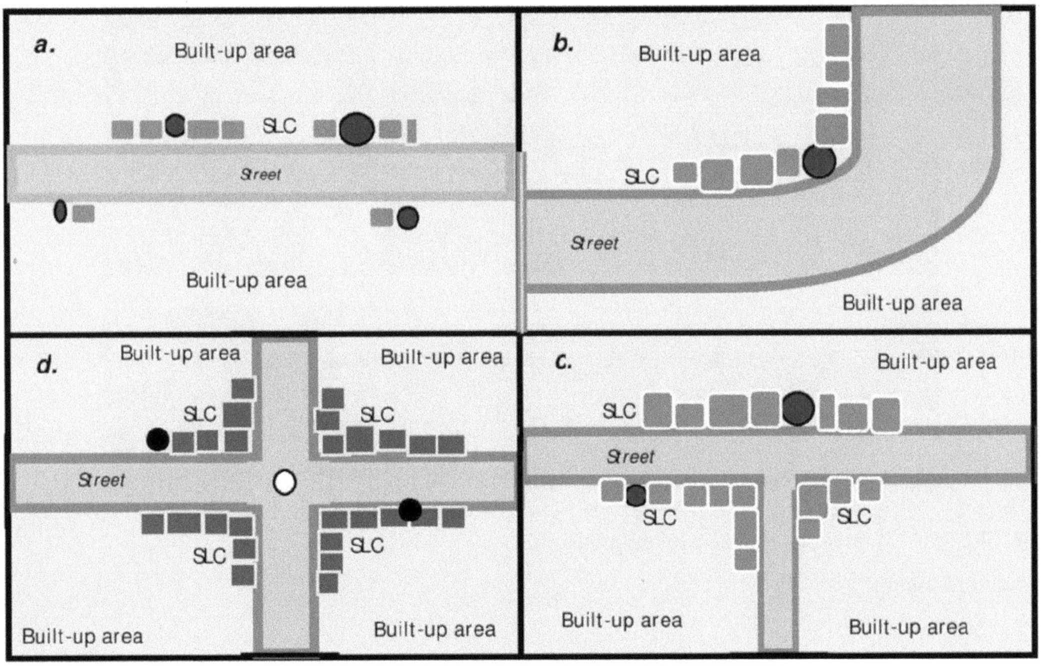

FIGURE 23.3 Major geometric patterns of sacred land capture a. Linear; b. 'L' shaped; c. 'T' shaped; d. 'X' shaped).

(Source: Developed by author (based on ground observation).)

FIGURE 23.4 Collage of ground-level examples of SLC.

(Source: Author.)

23.4.2 Exemplifying Undocumented Spatial Features

The inevitable outcomes of the rapid urbanization or subaltern urbanization are the spatial expansion of towns and cities beyond their juridical limits and into their hinterlands and peripheries in order to accommodate the growing urban population, as a result of natural demographic changes (Mosammam et al. 2016; Sahana et al. 2018). In the meantime, under the broad umbrella of urbanization, a phenomenon like sacred land capture (SLC) silently challenges the illegal encroachment of small urban lands with religious petitions and recreated beliefs, taking various areal forms and dimensions.

23.4.3 Observation of Inner Layout of SLC

The SLC phenomenon is undoubtedly not the simple outcome of human–nature interactions, rather it is a complex and long-term consequence of the interactions between multiple factors in the urban environment. This sub-section includes the crux of the philosophical pillars of the development of SLC, with a detailed discussion. Figure 23.5 illustrates that the three major pillars on which the phenomenon of SLC in mainstream urbanization rests are political economy, political ecology and the religious behavioral aspect. At the beginning when a small urban area is going through its initial formation and growth stage, the rural immigrants and native urban poor are allowed to grab tiny pieces of unused land for temporary trading, making the most of sympathy for their ultra-poor status. This is the political economy factor.[5] In the later stages when any ULBs[4] enforce any judicial initiative against unlawful land capture, opposing political groups raise their voices in support of the ultra-poor and their illegal encroachment. These situations merely represent the politicization of the issue (playing the 'victim' card) and a way of capturing the votes of marginalized people (the urban poor). Second, there is growing environmental and ecological awareness among ordinary people, and establishing sacred groves is a reflection of environmental sensitivity. SLC flourishes with the support of sacred groves and it would be near impossible for any ULBs[4] to reclaim captured urban lands for planned development. Local opposition political parties and NGOs may proactively prohibit the local administration from amending lawful systematic land-use processes. The political ecology[6] aspect of the situation is when those groups radicalize the issue as environmental ethics and morality. The third standpoint, the religious behavioral aspect, is the most important, being an inherent feature of Indian society and culture, and can be regarded as solely responsible for the development of the entire phenomenon of SLC. It is therefore a phenomenon capable of penetrating the entire societal set-up, thus to eradicate or mitigate the problem a new kind of well thought-out policy is needed.

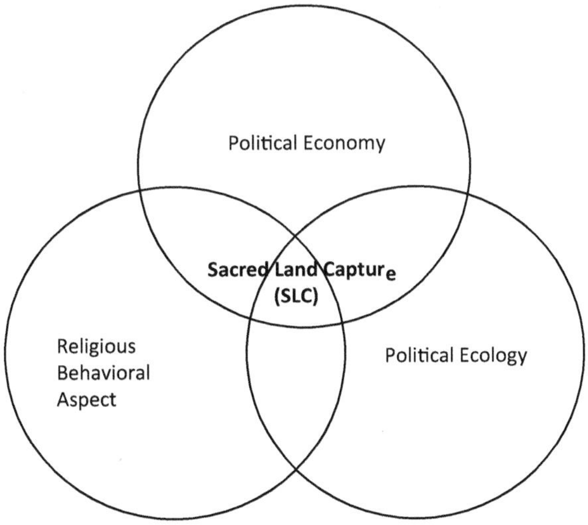

FIGURE 23.5 Conceptual pillars of SLC.

(Source: Developed by author.)

23.5 CONCLUSIONS AND POLICY SUGGESTIONS

This chapter has given comprehensive witness to the unaccounted gradual phenomenon of unlawful capture and small-scale alteration of urban land use and landscape, supported by cultural and religious practices. The creation of this kind of land-based problem is strongly evident among emerging small urban centers and their unmarked peripheries in India and is causing problems for smart city development agendas and further infrastructural development. Land-capture issues are also present in India's mega-cities but their magnitude is less. The factors responsible for SLC are political, economic, socio-cultural, religious, ecological and behavioral. The undesirable outcomes are shortage of open space, conflicts between legal owners and capturers, reduction in streets' carrying capacity, intolerable obstacles and traffic jams, reduced-width footpaths for pedestrians, and greater obstacles to the futuristic development of small urban centers.

Due to shortages of time and funding, this work has many limitations, but it does reopen a (possibly) new avenue of urban study (for example, micro-level land-use management, quantitative analysis of SLC with diversified spatial dimensions, SLC and the micro-economics aspects, congestion-related policy making, etc.). A number of academics and urban thinkers have been occupied with the progressive change of land use and land cover in Indian towns and cities, and have advocated land-based financing and land-value capture. There is no doubt that such initiatives and steps are needed for census towns and peri-urban areas (Gupta 2014; Pethe et al. 2014; Gandhi and Pathak 2016). It is almost impossible to eradicate these problems completely in a country like India. However, this study has the following policy suggestions: systematic collection of taxes/revenues from SLC by ULBs, step-wise cleaning and rehabilitation of street-side small/marginal traders, creating eco-friendly footpaths for pedestrians, controlling and monitoring further illegal land capture, roadside plantation and beautification of newly growing census towns and peri-urban areas. Thus, a win–win situation would be achieved for both urban dwellers (including the urban poor) and the urban environment.

NOTES

1 The term 'NITI Aayog' stands for National Institution for Transforming India, formed through a resolution of the Union Cabinet in 2015. This constitutional body, established with a vision of reform agenda, replaced the former Planning Commission. The NITI Aayog acts as the quintessential platform of the Indian government to bring states to act together in the national interest and thereby encourage cooperative federalism.
2 A Census Town is a comparatively more densely inhabited place in India, but not statutorily notified and administratively recognized as a town. Though those settlement sites already having urban characteristics like population exceeding 5,000, with a minimum population density of 400 persons per square kilometer, and out of total inhabitants at least 75% of main workers (especially the male population) engaged other than in agricultural activities (Pradhan 2013; Guin and Das 2015a, 2015b; Sircar 2017).
3 Peri-urban areas are mainly newly emerging habitable places known previously as outskirts. This/these area(s) are zones of transition from rural to urban land uses. The boundaries of peri-urban areas are porous and transitory as urban development extends into rural and commercial/industrial land. Thus, it would be viewed as a contemporary landscape type in its own right (UNESCO 2014).
4 The term stands for Urban Local Body(s), a grass-roots-level administration called a municipality. Municipal corporations or 'nagar panchayat' in India have socio-cultural, democratic political and financial functions.
5 Political economy is a sub-discipline of social science, which studies the intra-relationship or interrelationship between humans (as individuals) and society, and between dominant markets and the country or state. This branch of social science uses a diverse set of tools and methods broadly drawn from economics, political science and sociology (Veseth and Balaam 2020).
6 Political ecology is a multidisciplinary umbrella in which various social actors with different levels of authority challenge each other to control natural resources. It emphasizes the interlinkages between the perspectives of ecology and society by comparing ecological and societal hierarchical backgrounds. It also contributes to their intellectual communication and the creation of social landscapes (Halder 2018).

REFERENCES

Atu, J. E., Ayama, O. R., and Eja, E. I. 2013. "Urban sprawl effects on biodiversity in peripheral agricultural lands in Calabar, Nigeria." *Journal of Environment & Earth Science* 7(3): 219–231.

Berling-Wolff, S., and Wu, J. 2004. "Modeling urban landscape dynamics: A case study in Phoenix, USA." *Urban Ecosystem* 7(3): 215–240. doi:10.1023/B:UECO.0000044037.23965.45.

Bettin, J., and Wollni, M. 2020. "Environmental concern and urbanization in India: Towards psychological complexity." *Sustainability 12*: 1–25. doi:10.3390/su122410402.

Bhagat, R. B. 2011. "Emerging pattern of urbanisation in India." *Economic & Political Weekly 46*(34): 1–8.

Breese, G. 1969. *Urbanisation in newly developing countries*. New Delhi: Prentice Hall.

Census. 2011. Census of India Website: Office of the Registrar General & Census Commissioner, India (ORGI). https://censusindia.gov.in/digitallibrary/Tables.aspx (accessed May 11, 2020)

Chadchan, J., and Shankar, R. 2012. "An analysis of urban growth trends in the post-economic reforms period in India." *International Journal of Sustainable Built Environment 1*(1): 36–49. https://doi.org/10.1016/j.ijsbe.2012.05.001

Chandran, R. 20 February 2018. *Urbanization triggers land disputes in Indian cities*. Thomson Reuters Foundation. https://www.reuters.com/article/us-india-landrights-tech-idUSKCN1G416D (accessed January 11, 2015)

Davies, K., and Golden, H. H. 1954. "Urbanisation and development in pre-industrial areas." *Economic Development and Cultural Change 3*(1): 6–26.

Denis, E., and Zérah, M. H. 2017. *Subaltern urbanisation in India: An introduction to the dynamics of ordinary towns*. India: Springer, 1–614. doi:10.1007/978-81-322-3616-0.

Denis, E., Mukhopadhyay, P., and Zérah, M. H. 2012. "Subaltern urbanisation in India." *Economic & Political Weekly 47*(30): 52–62.

Despande, S., and Despande, L. 1998. "Impact of liberalisation of labour market in India: What do facts from NSSO's 50th round show." *Economic & Political Weekly 33*(22): L21–L31.

Dutta, P. 2006. Urbanisation in India—Regional and sub-regional population dynamic population process in urban areas, European Population Conference, 21–24 June, 2006. https://casi.sas.upenn.edu/sites/default/files/iit/Urbanisation%20in%20India.pdf (accessed October 21, 2020)

Gandhi, S., and Pathak, V. K. 2016. "Land-based financing in metropolitan cities in India: The case of Hyderabad and Mumbai." *Urbanisation 1*(1): 31–52. doi:10.1177/2455747116644106.

Grimm, N. B., Grove, J. M., Picket, S. T. A., et al. 2000. "Integrated approaches to long-term studies of urban ecological systems." *Bioscience 50*(7): 571–584. doi:10.1641/0006-3568(2000)050.

Guin, D., and Das, D. N. 2015a. "New census towns in West Bengal: 'Census activism' or sectoral diversification?" *Economic and Political Weekly 50*(14): 68–72.

Guin, D., and Das, D. N. 2015b. "Spatial perspectives of the new census towns, 2011: A case study of West Bengal." *Environment and Urbanization Asia 6*(2): 109–124. doi:10.1177/0975425315589155.

Gupta, R. 2014. "The pattern of urban land-use changes: A case study of the Indian cities." *Environment and Urbanization Asia 5*(1): 83–104. doi:10.1177/0975425314521539.

Halder, S. 2018. "Political ecology of snake charming: Evidence from West Bengal, India." *South Asian Survey 24*(1): 54–87. doi:10.1177/0971523118783373.

Jaysawal, N., and Saha, S. 2014. "Urbanization in India: An impact assessment." *International Journal of Applied Sociology 4*(2): 60–65. doi:10.5923/j.ijas.20140402.04.

Kundu, A. 2013. Processes of urbanization in India: The exclusionary trends. *International Affairs Forum 4*(1): 83–90. doi:10.1080/23258020.2013.824238.

Kundu, A., and Basu, S. 1998. "Informal manufacturing sector in urban areas–An analysis of recent trends." *Manpower Journal 34*(1), 41–52.

Kundu, A., and Gupta, S. 2000. "Declining population mobility, liberalisation and growing regional imbalances–the Indian case." In *Inequality, mobility and urbanization*, ed. A. Kundu. Indian Council of Social Science Research, New Delhi: Manak Publications.

Kundu, A., Lalitha, N., and Arora, S. L. 2001. "Growth dynamics of informal manufacturing sector in urban India: An analysis of interdependence." In *Informal sector in India*, ed. A. Kundu, and A. N. Sharma. New Delhi: Institute of Human Development.

Liu, X., Li, X., Chen, Y., et al. 2010. "A new landscape index for quantifying urban expansion using multi-temporal remotely sensed data." *Landscape Ecology* 25: 671–682. doi:10.1007/s10980-010-9454-5.

Lopez, E., Bocco, G., Mendoza, M., et al. 2001. "Predicting land-cover and land-use change in the urban fringe: A case in Morelia city, Mexico." *Landscape Urban Planning 55*(4): 271–285. doi:10.1016/S0169-2046(01)00160-8.

Mosammam, H. M., Nia, J. T., Khani, H., et al. 2016. "Monitoring land use change and measuring urban sprawl based on its spatial forms: The case of Qom city." *Egyptian Journal of Remote Sensing & Space Science 20*(1): 103–116. doi:10.1016/j.ejrs.2016.08.002.

Mukherji, S. 1993. *Poverty induced migration and urban involution in India: Cause and consequences.* International Institute for Population Sciences (IIPS), pp. 1–91.

Mundia, C. N., and Murayama, Y. 2010. "Modeling spatial processes of urban growth in African cities: A case study of Nairobi City." *Urban Geography 31*(2): 259–272. doi:10.2747/0272-3638.31.2.259.

NITI Aayog. 2015. Official website, Government of India. Seven-Mission Program. http://www.niti.gov.in (accessed June 26, 2019)

Pethe, A., Nallathiga, R., Gandhi, S., et al. 2014. "Re-thinking urban planning in India: Learning from the wedge between the de jure and de facto development in Mumbai." *Cities 39*: 120–132. doi:10.1016/j.cities.2014.02.006.

Pickett, S. T. A., Cadenasso, M. L., Grove, J. M., et al. 2001. "Urban ecological systems: Linking terrestrial ecological, physical, and socioeconomic components of metropolitan areas." *Annual Review of Ecology and Systematics 32*: 127–157. doi:10.1146/annurev.ecolsys.32.081501.114012.

Pradhan, K. C. 2013. "Unacknowledged urbanisation: New census towns of India." *Economic & Political Weekly 48*(36): 43–51.

Prasad, S. 18 January 2019. Why should world watching the India's fast-growing cities. World Economic Forum (Part of World Economic Forum Annual Meeting). https://www.weforum.org/agenda/2019/01/india-urbanization-why-the-world-should-watch/ (accessed March 3, 2020)

Raza, M., and Kundu, A. 1978. *Some aspects of disfunctional characteristics of urbanisation. Socio-economic development problems in South and South East Asia.* Bombay: Popular Prakashan.

Sahana, M., Hong, H., and Sajjad, H. 2018. "Analyzing urban spatial patterns and trend of urban growth using urban sprawl matrix: A study on Kolkata agglomeration, India." *Science of the Total Environment 628-629*: 1557–1566. doi:10.1016/j.scitotenv.2018.02.170.

Shaw, A. 2018. "Towards sustainable cities in India." In *Sustainable urbanization in India, exploring urban change in South Asia*, ed., J. Mukherjee, 23–37. Singapore: Springer Nature Singapore Pvt. Ltd. doi:10.1007/978-981-10-4932-3_2.

Shi, Y., Sun, X., Zhu, X., et al. 2012. "Characterizing growth types and analyzing growth density distribution in response to urban growth patterns in peri-urban areas of Lianyungang City." *Landscape and Urban Planning 105*(4): 425–433. doi:10.1016/j.landurbplan.2012.01.017.

Sircar, S. 2017. "'Census Towns' in India and what it means to be 'urban': Competing epistemologies and potential new approaches." *Singapore Journal of Tropical Geography*: 1–16. doi:10.1111/sjtg.12193.

Sun, C., Zhi-Feng, W., Lv, Z., et al. 2013. "Quantifying different types of urban growth and the change dynamic in Guangzhou using multi-temporal remote sensing data." *International Journal of Applied Earth Observation and Geoinformation 21*: 409–417. doi:10.1016/j.jag.2011.12.012.

Taubenböck, H., Wegmann, M., Roth, A., et al. 2009. "Urbanization in India–Spatiotemporal analysis using remote sensing data." *Computers, Environment and Urban Systems 33*: 179–188. doi:10.1016/j.compenvurbsys.2008.09.003.

Tiwari, P., Nair, R., Ankinapalli, P., et al. 2015. *India's reluctant urbanization: Thinking beyond.* New York, US: Palgrave Macmillan. doi:10.1057/9781137339751.

UN. 2015. Sustainable development goals—United Nations *United Nations.* www.un.org/sustainabledevelopment/sustainabledevelopment-goals/ (accessed April 17, 2017).

UN. 2018. World urbanization prospects: The 2018 revision. Economic & Social Affairs. https://esa.un.org/unpd/wup/Publications/Files/WUP2018-KeyFacts.pdf (accessed December 11, 2018)

UNESCO. 2014. Peri-urban landscapes; water, food and environment security. https://en.unesco.org/events/peri-urban-landscapes-water-food-and-environmental-security (accessed April 17, 2017)

Veseth, M. A., and Balaam, D. N. 2020. "Political economy." *Encyclopedia Britannica* Updated May 22, 2020. https://www.britannica.com/topic/political-economy (accessed April 11, 2013).

Weber, C., and Puissant, A. 2003. "Urbanization pressure and modeling of urban growth: Example of the Tunis metropolitan area." *Remote Sensing of Environment 86*(3): 341–352. doi:10.1016/S0034-4257(03)00077-4.

World Bank. 2011. Urbanization in India. https://www.worldbank.org/en/news (accessed February 11, 2020).

24 Aging in Urban India
Needs and Policies for Sustainable Development

Aakanksha S. Bhargava and Barnali Gogoi
Cotton University, Guwahati, India

CONTENTS

24.1 INTRODUCTION

The 2030 Agenda of the United Nations, adopted in the year 2015, aims at transforming the development process in a way that promotes the cause of each and every thing on the earth. This program, which consists of 17 Sustainable Development Goals (SDGs), also tries to incorporate within it the cause of cities and their populations under Goal 11, 'sustainable cities and communities'. The objective is to ensure a livable, inclusive and resilient environment in the cities for all. Accomplishment of this goal depends on a city's ability to hold all its communities. This is extremely relevant for the most vulnerable residents: women, children, differently abled and older persons.

Accessibility and inclusiveness, green public spaces and security are key concerns for Goal 11 with reference to women and children, persons with disabilities and older persons (United Nations 2020, 46–48). The present chapter is therefore an attempt to recognize the older population as an important constituent of Indian cities. The aim is to showcase the Indian aging scenario in terms of changing demography, the associated needs and city plans, especially the Smart Cities Mission which is an element of Goal 11 in the Indian Sustainable Development Plan (Niti Aayog 2018). In general, the chapter examines how, as a member state of the United Nations, India is fulfilling the measures needed to achieve Goal 11, with reference to the old-age population.

24.1.1 Evolving Prospects of Aging: India

Aging in India has always been a matter of pride: Indian culture teaches the generations to be humble and dutiful towards older persons. Members of a family in India, especially the older members, have always filled a position of supremacy and authority (Pastithil 2009, 10, 1). This is because of

DOI: 10.1201/9781003131922-28

the predominance of a joint family system where all the members have a sense of security and are cared for by each other, with the elders of the house placed in the highest positions. Hence, care giving and supporting elders have never been a challenge in Indian societies. However, changes in living patterns, adapting to modern ways of life and a shift from a simple and passive lifestyle to a more complex and fast-paced life have generated a wide range of complications arising from changing family structures, complex work cultures, financial stress, lack of time, and so on. One of the major changes to affect elders in India is the shift from an extended family system to smaller, nuclear families; again this type of family system is more common in the urban than the rural environment (Shah 1999, 1–2). Most of the people living in the cities have migrated there in search of employment, and hence time they invest in work and living is exceptionally high compared to the rural areas, resulting in less time to care for their elders, so that older people are forced to live in less dignified conditions. The issue also needs to be addressed because of the phenomenon of the aging population, which has been following a global trend, at first in European countries and latterly in India. The country thus needs to be prepared to support this huge surge in the elderly population. It is important to study this community in Indian cities to see how they are embraced or marginalized in the planning process for sustainable development. Lifestyle and demographic changes make it necessary to treat older communities as a separate and vulnerable constituent of Indian cities and identify their needs so they can be incorporated within smart city and master plans.

24.1.2 Growth of Older Population: India

Planning for the elderly in India is now of crucial concern because of the continuous growth in the older population with improved life expectancy and an increase in the total percentage of the aged population from 5.6% in 1961 to 8.6% in 2011 (Central Statistics Office 2016). There has been a significant rise in life expectancy at birth for males from 58 years in 1986–90 to 67 years in 2011–16, an increase of about 9 years in a 25-year period (1986–90 to 2011–16). For females, the increase has been even higher, at 11 years, from 58 years in 1986–90 to 69 years in 2011–16 (National Institute of Rural Development and Panchayati Raj n.d.). The increase in life expectancy after 60 has been a little higher among women than men (Central Statistics Office 2016). The number of persons aged 60+ years has increased from only 12 million in 1901 to 20 million in 1951, 56.7 million in 1991, 76.6 million in 2001 and 103.8 million in 2011. Growth in the older population by sex is shown in Table 24.1.

A forecast from the National Commission on Population projects that the proportion of older people (60 years and above) is set to more than double from 71 million in 2001 to 173 million in 2026, an increase from 6.9% to 12.4% (Census of India 2011). According to the National Policy on Senior Citizens 2011, the overall population in India will grow by 55%, whereas the population of

TABLE 24.1
Growth of Elderly Population in India (in millions) and Decadal Growth (in %)

Year	Male	Female	Total Population	Decadal Growth (in general population)	Decadal Growth (in older population)
1961	12.4	12.4	24.7	21.6	23.9
1971	15.8	15.8	32.7	24.8	33.7
1981	21.1	21.1	43.2	24.7	33.0
1991	27.3	27.3	56.7	23.9	29.7
2001	37.8	38.9	76.6	21.5	25.2
2011	51.1	52.8	103.8	17.7	35.5

Source: Central Statistics Office 2016.

people in their sixties and above will increase by 326% and those in the age group 80+ by 700% – the fastest-growing group in the years 2000–2050. The growth rates of all age groups are either stable or decreasing, while the growth rate of those over 60 is increasing (Table 24.1). The proportion of the population aged over 60 is set to reach 10% by 2021.

The total urban population is 377.1 million, of which 30.6 million are aged. The total share of the elderly is 8.6%, of which 8.1% is urban; urban aged population growth rose from 4.7% to 8.1% in 2011 (Table 24.2). The elderly working population is 28.5% of the total working population of 41.65%, with a total dependency ratio of 14.2% and urban dependency of 12.4% (Table 24.3). Similar projections by Tiwari et al. suggest that the population of both the working-age and elderly groups is going to increase rapidly by the end of this century (Tiwari et al. 2020, 549). All these statistics give us an idea about the growing old-age population and the pressure it will be exerting upon the country, showing why Indian cities should be designed in a way that is age friendly and sustainable.

From the above tables it is clear that the population of older persons is rising significantly. Tables 24.2 and 24.3 show the trend of increase in urban elderly from 4.7% in 1961 to 8.1% in 2011, along with the rise in the dependency ratio in the cities from 8.7% in 1961 to 12.4% in 2011. This rise in the population and the dependency ratio amongst the older age group poses a serious challenge to the country to design a sustainable and easy-care environment.

TABLE 24.2
Growth of Urban Elderly Population

Year	Urban Growth
1961	4.7
1971	5.0
1981	5.4
1991	5.7
2001	6.7
2011	8.1

Source: Central Statistics Office 2016.

TABLE 24.3
Old-Age Dependency Ratio by Sex

Year	Male	Female	Total	Urban
1961	10.9	10.9	10.9	8.7
1971	11.4	11.6	11.5	8.9
1981	11.8	12.2	12.0	9.2
1991	12.2	12.2	12.2	9.7
2001	12.5	13.8	13.1	10.8
2011	13.6	14.9	14.2	12.4

Source: Central Statistics Office 2016.

24.1.3 Methodology

This study is centered on the United Nations Sustainable Development Goal 11, sustainable cities and communities. It seeks to explore the possibilities of making urban areas elder friendly from a pragmatic philosophical viewpoint (Creswell 2014). The method and techniques used are review-based (traditional review) and narrative techniques, suggesting that the study is qualitative. In the chapter we try to emphasize the need to incorporate the requirements of the older population in the process of policy framing for the sustainable development of Indian cities. In the words of Alley et al., we try to enquire what an 'elder-friendly community' is, which according to them lacks a uniform definition and is "a place where older people are actively involved, valued, and supported with infrastructure and services that effectively accommodate their needs" (Alley et al. 2007, 7). For this we have adopted the traditional literature review method also known as narrative literature review; the needs of old age are identified by analyzing research papers in connection with European nations, as they were the first to witness issues related to aging and have very efficiently supported their older populations (Council of European Development Bank 2014). Taking them as a reference, the common needs of the elderly as appropriate to the Indian context are recognized in relation to the Indian Smart Cities Mission and master plans, which inspect the sustainability of these plans in terms of elder population. The last part of the chapter uses the oral history narration technique (Komolthiti 2016, 68) to portray the perspective of elders in Guwahati city. Seven individuals in Guwahati city were identified using non-probability purposive sampling. Since the aim of this chapter is to perceive sustainable development in the context of elders in Indian cities, individuals aged 60 or more are considered. The diverse selection of individuals is based upon their marital status and type of family. Semi-structured open-ended sets of questions were used for the interview. The idea of this approach is to have in-depth informal conversations with the target group and extract relevant information. Telephone conversations using the native languages of the respondents wherever possible were used for this, Hindi and Assamese being the dominant ones. The interviews were recorded to maintain authenticity and to ensure that important information is not missed out. Personal observations and other important data were recorded in a field notebook. An iterative process was used to minimize errors (Komolthiti 2016, 68). This process helped us extract important excerpts and translate them into English to present them in a form that truly represents the words of the respondents. The identities of individuals have not been revealed and fictitious names are used. Lastly, the narratives are presented in a systematic manner to verify or contradict the findings of the review, and conclusions are drawn.

However, all methodologies are open to criticism, and this work is no different. Lack of good representative samples due to difficulty accessing representatives in the middle of a pandemic, and lockdown in Guwahati city, made the samples somewhat similar in terms of age and income. Being vulnerable to Covid-19, most of the older people restricted their activities; in other cases their families were not very comfortable with live interviews. Most of them did not favor distance interview methods like video conferencing and telephone interviews. This made the interviews representative of a particular class of people, as they were accessed through the researcher's contacts. The approach adopted is very subjective, which is a further criticism of the study. However, an attempt has been made to go deep during the interviews, extracting every possible piece of information from the interviewees. A detailed review of literature helped us minimize errors during the process.

24.2 IDENTIFYING THE NEEDS OF THE AGING POPULATION

On examining the status of the older age group and scrutinizing their needs and problems, it can be said that general concerns with this group relate to health, infrastructure and an improved living environment, as well as community support. Health and aging are very closely connected; the majority of people experience deterioration in health with increasing age. Most older individuals suffer from chronic health problems and issues like dementia; for instance, "more than two of every three people older than age 65 have at least two chronic conditions, and 37 percent have at least four. By age 85, more than half have at least four chronic conditions" (Gleckman 2019, 79). Similarly the need for infrastructural development, especially related to mobility, is a key concern for the older age group (Gant 1997, 207–209). With old age come problems related to accessibility, connectivity and mobility, all of which may be considered as core for the development of the elderly in urban areas.

Hoof et al. (2018) mention that population aging is a very positive yet demanding change for society, more particularly in the developed societies, because urbanization and aging are an indicator of human development, and the interplay between the two gives rise to a variety of situations in the urban environment. The paper "The Challenges of Urban Ageing: Making Cities Age-Friendly in Europe" has identified some of the common needs of the elderly: social involvement, communication and information, citizen involvement and employment, accommodation, transportation and movement, social welfare and health services, outdoor environment and surroundings, respect and social embodiment (Hoof et al. 2018). These can basically be grouped into three main categories as per the WHO categorization of age-friendly cities, i.e., health, participation and security (WHO 2012).

Hoof and Kazak have identified housing needs and age-friendly environments for old age in which the focus is on the construction of large buildings with many members of the older community residing together, and they provide justifications for the positive impact of such practices on the elderly (Hoof and Kazak 2018, 584–585). They also focus on the use of sustainable methods and the role of technology in designing an ideal model for the older generations. Among the most important needs identified by them are efficient entrances and exits to buildings, especially in emergencies, and a community designed with the needs of dementia sufferers in mind.

Locomotor disability is a common issue related to old age, and therefore special needs arise in respect of mobility (Blekesaune and Marit 2018, 2–4). Blekesaune and Marit (2018) also draw attention to housing needs and mention that elderly people's houses are not designed in such a way as to fulfill the needs and demands of increasing age. Thus, for them, the major needs are associated with community, in the form of support or help, efficient transportation, help at home and reform of residential areas to cater to the needs of old age.

Three main housing needs are distinguished by Kavšek and Bogataj. Separating older people from the general population is emphasized, thereby identifying their needs, synthesizing their experience along with active involvement in society with the provision of smaller and structured dwelling units that are cost-effective (Kavšek and Bogataj 2016, 5–6).

Another aspect of planning for the older generation on the grounds of changing demographics, the changing urban environment and its impact on the older generation, with the idea of aging in the place of residence, is provided by Buffel and Phillipson. They view planning for elderly in a unique way, concentrating more on how to plan rather than what to plan, making research and planning policies more rational, recognizing and removing the barriers imposed by the changing nature of cities, involving elders in the planning process and putting them at its core (Buffel and Phillipson 2016, 98–99).

Redefining health care systems and changing existing models for a more inclusive and elder-friendly medical care program is yet another important aspect that needs to be considered. Models and programs such as PACE (Program of All Inclusive Care for the Elderly), CAPABLE (Community Aging in Place, Advancing Better Living for Elders), Medicare Special Needs Plan and Medicaring Communities, as in the United States (Gleckman 2019, 78–84), are examples of preparing and recognizing non-medical as well as medical needs, i.e., care and support.

Ease of living, public transport, accessibility to shops and markets, and safer sidewalks all play a significant role in determining a congenial environment for elders. Gant's paper (1997) is one such effort to recognize and see through the environment that enables elder and disability-friendly features. He takes into consideration efficiency in transportation and pedestrian facilities in the urban areas of Kingston, and also accessibility to markets, along with recognition of noise levels that are disturbing to the city's population (Gant 1997, 207–217).

Sustainable cities and communities are one of the Sustainable Development Goals (SDGs), hence the role of the urban environment as a supporter of different communities is of utmost significance. Incorporation of the concept of the green environment to revitalize urban areas is a recent development in urban planning. In the context of the older population this is of great relevance, with most studies regarding green infrastructure and greenery as an important part of planning for old age. Similarly, Milanovič and Foški focus on the process of urbanization and how it can be made environment friendly. The principal goal of the study is "smart city growth, emphasising the internationalisation of the capital city through urban revitalisation as a city of art, culture and knowledge, a safe and healthy city" (Pichler-Milanovič and Foški 2015, 50–64).

Global Age Friendly Cities: A Guide (WHO 2007) provides a very detailed identification of some of the needs and a checklist for the well-being of the aged population in cities. The checklist takes into consideration age-friendly communications and information, ease of movement, employment and civic involvement, healthy environment, health and welfare, social involvement and

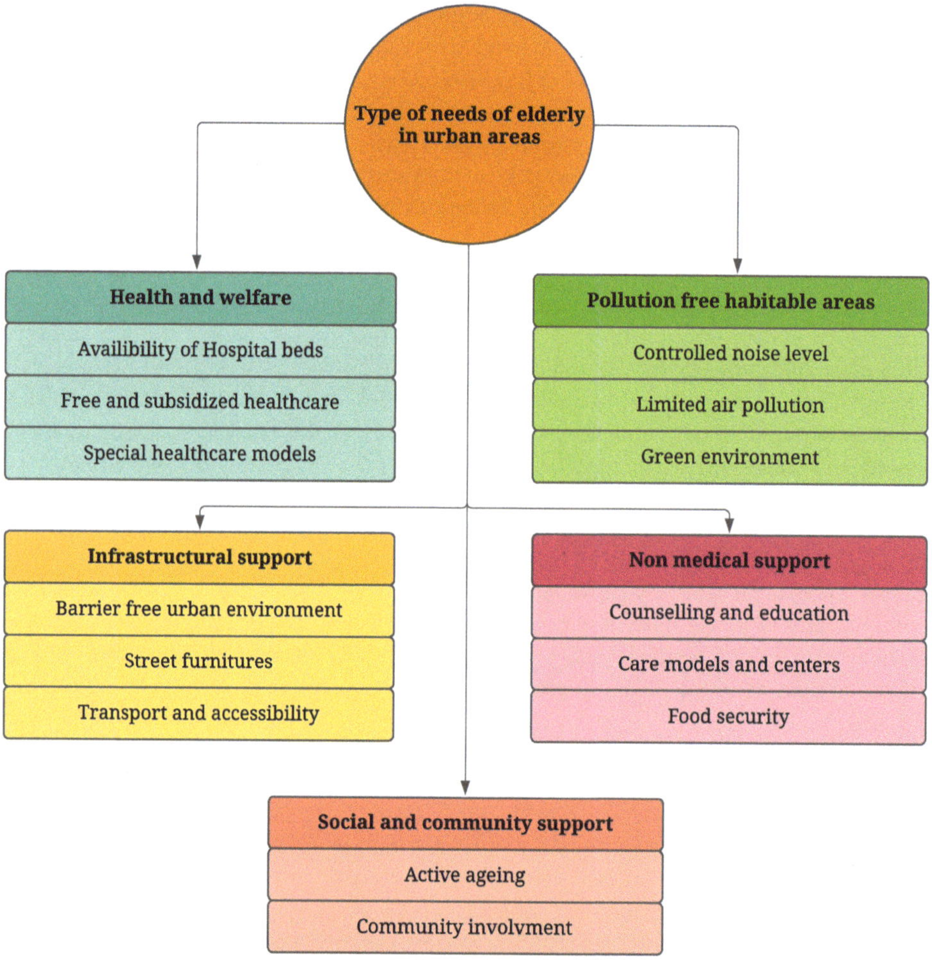

FIGURE 24.1 Needs of the elderly in urban areas. Special healthcare models, e.g., PACE, CAPABLE in USA; care models such as Toyama Prefecture, Minamori Watch-out Network in Japan; and barrier-free environment in Kingston.

(Source: Based on review of literature.)

contribution, and quality accommodation, aiming to serve the purpose of an age-friendly environment by preparing a comprehensive guide for designing cities accordingly.

Kurian and Uchiyama's (2012) study discusses some specific models of care for the elderly in two countries, Japan and the Netherlands. The paper aims to show how these models help make elders empowered, secure and involved in social activities. It presents six models that are very well designed to deal with the diverse needs of elders: Toyama Prefecture, Nagano Prefecture, Minamori Watch Out Network (Japan), The Shelter Housing Zone, The Neighborhood Project and Network of Care and Cure through Laego (Netherlands). The models pursue older persons as the needy, with a focus on delivering care services and making them self-sufficient. Some of the remarkable practices in the models involve encouraging elders to volunteer in physical and social activities, educating them on new technology, keeping a watch on the elders of the community and counseling those in need (Kurian and Uchiyama 2012, 74–81).

From the literatures mentioned above, it can be concluded that the primary needs of elders revolve around health and welfare (free and subsidized health care facilities, number of hospital beds), non-medical support (care centers, food and community support), infrastructural support (barrier-free elder-friendly environment, ease of movement, safe pedestrian routes or sidewalks, seating), accessibility to markets, stores, temples/churches, living conditions (i.e., age-friendly features at home, secure living environment), pollution-free habitable areas, and an environment that supports active aging and aging in place (voluntary involvement in society and community service, employment after retirement, etc.) (Figure 24.1). Attending to all these needs, countries like Japan, the Netherlands, Poland and most other European countries have developed models congenial for elders such as community living, day centers, volunteering agencies to encourage activities among the aged and teach them new technologies, well developed transport networks thoroughly connected to markets, a 'watch-out' volunteering system for older single people or couples, etc. With these ideas of an elder-friendly environment, we now consider Indian city plans and find elements designed for sustainable development of the old-age community in Indian cities.

24.3 INDIA IN THE CONTEXT OF SUSTAINABLE CITIES AND COMMUNITIES

India needs a sustainable development plan for its cities because of the steady growth in its urban areas and continuous growth of population in these areas. The urban population of India has increased significantly, with the current urban population of 31.8% expected to increase to 38.6% by 2036 (Census of India 2011). The expected urban growth is three-quarters (73%) of the total population increase by 2036 and the expected growth in the population during 2011–2036 is about 306 million, of which the urban population is around 234 million. Moreover, currently the urban areas contribute 63% of GDP, which by 2030 is expected to be around 75% (Census of India 2011). This makes the development of urban areas the key concern for planners. In compliance with the SDGs, India is attempting to achieve Goal 11 – sustainable cities and communities – through its various initiatives, schemes and programs (Niti Aayog 2020, 99). With seven targets centering around meeting housing needs, security, efficient transportation, inclusive and integrated planning, resilience to disasters, environmental protection, linking and strengthening national and regional level planning, the aim is to raise the standard of Indian cities to make them more functional for their inhabitants (Niti Aayog 2018, 135). This goal is connected to many other sustainable development goals, such as zero poverty, good health, gender equality, clean drinking water, sanitation, infrastructure and environment protection. Currently the government of India adopts five main indicators to assess progress towards this goal, which are as follows:

i) Houses completed under PMAY as a percentage of net demand assessment for houses.
ii) Urban households living in slums (%)
iii) Wards with 100% door-to-door waste collection (%)
iv) Waste processed (%)

v) Installed sewage treatment capacity as a proportion of sewage generated in urban areas.
(**Niti Aayog** 2019, 150)

According to the 2020 India Voluntary National Review, the overall index score for this goal is 53. An index score of 53 means that just over half the goal has been covered, but with significant sub-national variation (Niti Aayog 2020, 102).

Apart from the indicators mentioned above, the government also aims to reform property rights, transform urban areas and achieve sustainable urbanization. Initiatives like Atal Mission for Rejuvenation and Urban Transformation (AMRUT), Pradhan Mantri Awas Yojana (PMAY), Smart Cities Mission and Ease of Living Index are among the significant initiatives (Niti Aayog 2019; Niti Aayog 2020). Altogether, SDG 11 for India is synonymous with transforming and rejuvenating cities, provision of housing to all, developing full connectivity and transportation, conserving the environment, improving cities for inclusive growth, waste management, and so on (Niti Aayog 2019, 143–144).

An understanding of this goal helps older communities to connect with it. As the name suggests, the goal concerns all inhabitants of a city, especially vulnerable groups, including women, children, older and differently abled people. The rise in the older population indicates a need to include them in city planning. In this context elderly or senior citizens should be considered one of the most integral elements in city planning, as it is always observed that a city that can support its older generation is the one which can efficiently support all of its citizens (WHO 2007, 6). Against this background, we next discuss the Smart Cities Mission, an initiative under SDG Goal 11, and some city master plans in general to understand the type of planning associated with old age in major Indian cities.

24.4 EXPLORING SMART CITIES MISSION AND MASTER PLANS WITH REFERENCE TO OLD AGE

City planning in India these days revolves around smart cities and master plans. Most Indian cities are of older origin, with older versions of plans, policies and sometimes even infrastructure. India is making a concerted effort on urban regeneration and revival (Ministry of Urban Development 2015, 8). The next sections take a thorough look at the Smart Cities Mission and some master plans to see how far they embody the subject of the elderly.

24.4.1 Smart City Mission: India

The definition of smart city varies from one region to another, depending generally upon the available resources, current needs, level of development and willingness to change and reform. In general terms, it is an initiative for making cities capable and effective in supporting their population. Smart cities in simple terms refers to an area that can accomplish the needs of its inhabitants in a way that is functional for all. For comprehensive development, an area should be strong as regards its institutions, infrastructure, physical and social environment; smart cities aim at providing their people with infrastructure, a decent quality of life, a clean and sustainable environment, and smart solutions to problems (Ministry of Urban Development 2015). Currently a total of 100 cities in India have been identified for conversion and establishment as smart cities and projects worth INR 2 trillion that will impact 100 million people have been commenced (Niti Aayog 2020, 103).

The Smart City Mission identifies ten core infrastructural elements:

i. adequate water supply
ii. assured electricity supply
iii. sanitation, including solid waste management

iv. efficient urban mobility and public transport
v. affordable housing, especially for the poor
vi. robust IT connectivity and digitalization,
vii. good governance, especially e-governance and citizen participation
viii. sustainable environment
ix. safety and security of citizens, particularly women, children and the elderly
x. health and education.

(Ministry of Urban Development 2015)

Looking at these ten core elements, the Mission can be regarded as age inclusive. Efficiency in urban mobility and public transport, affordable housing, digital well-being, e-governance, sustainable environment, safety, security and health are of greater significance when issues related to old age need to be addressed. The smart city plan hence very efficiently incorporates the WHO's definition of age-friendly cities in terms of health, participation and security, but the rates and pattern of participation may be different from those in developed parts of the world. Nevertheless, an inclusive society in the Indian context could be the first step towards the concept of active aging, promoting community participation and hence sustainability in urban planning. The key features that show the program's inclusiveness in planning for the old-age population are highlighted in Figure 24.2.

Other important attributes of the Smart Cities Mission include construction of elder-friendly walkways, limiting pollution and congestion, promoting interaction among citizens, making roads pedestrian and cyclist friendly, easy access to important administrative services, preservation of open spaces (green areas, parks, playgrounds), increasing transport connectivity, promotion of transit-oriented development, and online services in different areas or sectors of the city. Hence, it can be seen that the Mission has incorporated some old age-friendly features; however, many things

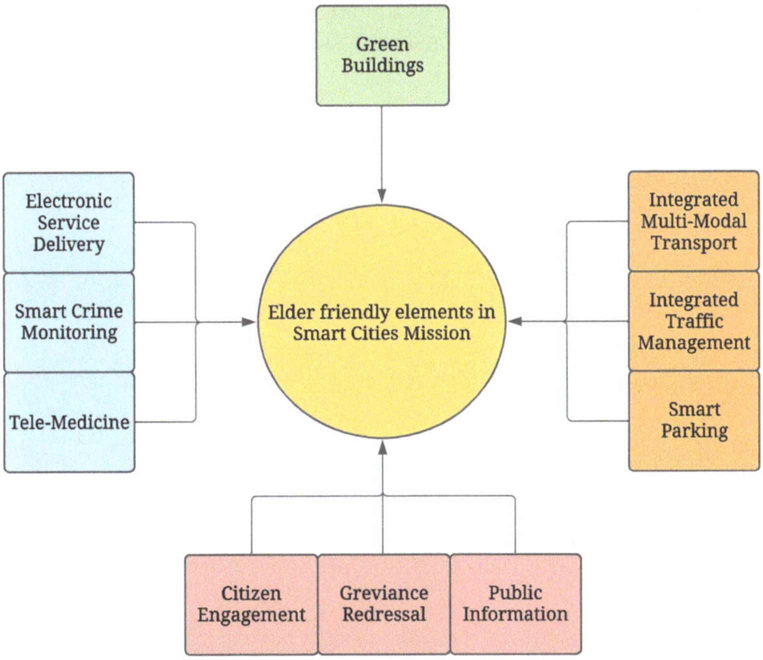

FIGURE 24.2 Elder-friendly elements in Smart Cities Mission.

(Source: Ministry of Urban Development 2015.)

remain unattended to when compared to developed countries. As a developing country, India's core planning concerns are associated with finance, health and security. Issues like empowering the old, improving social interaction, respect and inclusion receive the least concern. There may be different reasons for this. First, being a developing country, India is more concerned with problems like poverty, unemployment, infrastructural development, food security etc., and second, under the existing family pattern, the country still believes in including the elders of the community in important matters, so that although problems do exist in the urban environment, their frequency is much lower than in other parts of the world (Pastithil 2009, 10, 1). Apart from pedestrian-friendly pavements, other issues such as street furniture, signage, road safety, barrier-free environment, ease of movement (including escalators, lifts, public transport etc.) are not clearly mentioned in the Smart Cities Mission. To conclude, the Mission can be regarded as an efficient plan to cater to the needs of the elderly and embraces them as active members of the cities.

24.4.2 City Master Plans

For a more comprehensive insight, we now look at master plans of some major cities. Master plans are very closely connected with the development of Indian cities, being essentially the documentation of the processes involved in the planning and execution of development programs within an area. This section examines whether the reviewed master plans include the concerns of the elderly as a significant element of the urban environment, and if so, what are the issues to be addressed, which issues still remain unevaluated and what planning techniques have been used.

Four master plans are considered, which are representative of megacities, metro cities and million cities (the city classes as per Census of India 2011). Delhi and Mumbai are the oldest cities of India and represent megacities. Bangalore, a newly developed city, is regarded as a metro city. Guwahati, a newly developing city in north-eastern India, is an example of a million plus city.

The general concerns in all these master plans are extension, revitalization and restoration of urban areas as a habitable place. A review of master plans in association with the needs of the elderly

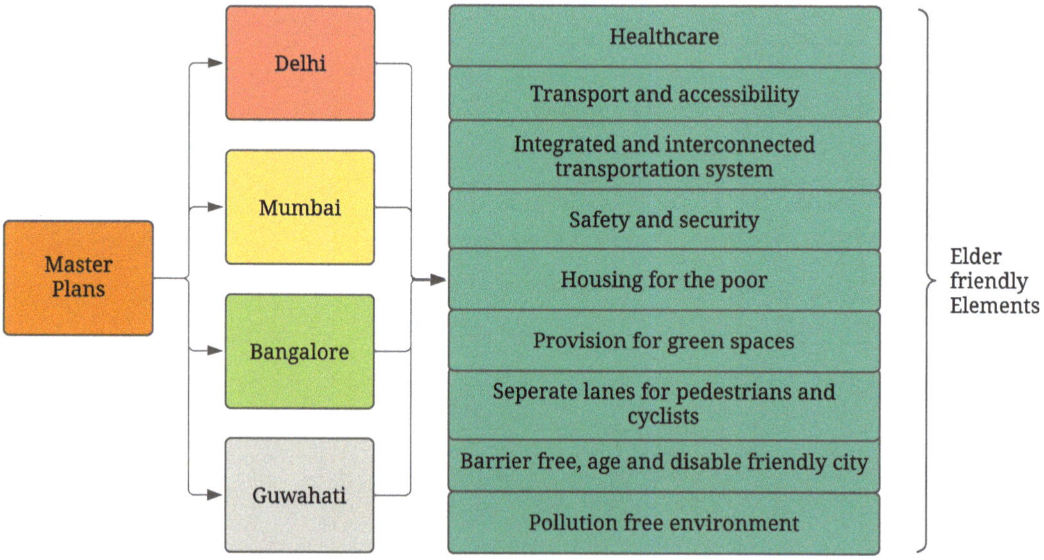

FIGURE 24.3 Elder-friendly elements in master plans.

(Source: Based on reviewed masterplans (Delhi, Mumbai, Bangalore, Guwahati).)

finds that typically, planning is related to health, housing, the provision of green spaces and transportation. Figure 24.3 illustrates elements common to all the plans.

A thorough understanding of the reviewed master plans shows that most of them aim at providing physical infrastructures such as housing, transport, trade and commerce, power etc.; the social infrastructures most dealt with relate to education, green spaces, health, sports/recreational activities and water, etc. This is because of the difference in the level of development in the country. Being a developing country, the problems pertaining to this region are related to meeting basic needs and demands. Complex and complicated needs are still a luxury in the country, so most of our master plans aim to come up with things that can serve basic needs in a city.

Of all the master plans, the Delhi Master Plan is the most detailed and inclusive, because the focus of this plan is to serve the purpose of both physical and social development. The Delhi Master Plan 2021 recognizes the changing demographic structure and increasing old-age population. According to projections by Delhi Development Authority Sub-Group (Master Plan Delhi 2021), the old-age population which in 2011 was 1.42 million (7.8%) will grow to 2.46 million (10.7%) in 2021. Hence, the Delhi Development Authority (DDA) has sought to design an environment that is appropriate for older persons (Delhi Development Authority 2010, 2–3). The main aims of this plan are to provide housing (free or inexpensive) to all, especially the urban poor, to construct and design a transportation system that is fully coordinated and connected, ensure efficient flow of people within or outside the area, to construct separate lanes for cyclist and pedestrians, to increase the number of hospital beds per 1000 population (5 beds/1000), and to provide ample day centers and care homes along with a pollution-free green environment for a healthy life. Other features incorporated in the Delhi Master Plan 2021 are the construction of continuous, unobstructed, slip-proof pavements with appropriate height and width for older and differently abled persons, suitable for all weather conditions and providing direct routes (reducing travel time and distance). The plan also incorporates planting trees alongside pavements and footpaths for shade, lighting and signage to ensure safety, an ample number of suitable crossings, and street furniture such as benches alongside roads. Footbridge construction only if absolutely necessary and to limit pollution in the city, is also regulated. In summary, the Delhi Master Plan is a comprehensive and inclusive plan that tries to solve the everyday problems of elderly people and can act as a model for creating sustainable living and an elder-friendly environment.

Mumbai Master Plan, on the other hand, is more focused on designing a plan that is economically viable for all, probably because of the city's role as the financial capital. According to the statistics, the population aged 60+ has recently declined in this region and the working-age population has increased significantly, which is why employment creation is the core issue in this plan (Mumbai Metropolitan Region Development Authority 2016, 10–18). The master plan aims at providing housing to all because of the high rental values in the city. It also aims to provide some basic necessities in the city that are also beneficial to the older population: reducing traffic congestion, good transportation network, good facilities for pedestrians, accessibility to important places such as markets, pharmacies, hospitals, etc.

The Bangalore Master Plan is more inclined towards making the city environmentally viable and efficient by focusing more on developing techniques to solve pollution and waste disposal problems, which contributes to constructing a habitable environment (Bengaluru Metropolitan Region Development Authority n.d.). Apart from conservation, restoration and reclamation, the Bangalore Master Plan also focuses on improving healthcare facilities in the area to meet the increasing demand. It plans to create a green infrastructure in the city by planting trees and reserving spaces for parks in most areas. Compared to the Delhi Master Plan, this plan is not that inclusive for the older generation and hence lags sustainability in terms of SDG 11.

The Guwahati Master Plan 2025 is inclusive in terms of social and physical elements (Guwahati Metropolitan Development Authority 2009). The focus of the plan is on developing integrated 'intra urban transport'. Good-quality roads with lanes for buses, two wheelers and private cars along with space and separate lanes for pedestrians and cyclists are included in this plan. Protecting and restoring green belts within and around the region is another important goal of the plan. For its older population the plan includes construction of appropriate wide pavements with safety measures such as signage and crossings, escalators installed at major footbridges, the concept of the orange badge in the city enabling vulnerable people to park in close proximity to entrances and exits. Other facilities include the construction of hospitals, an increase in the number of hospital beds, ensuring security by building neighborhood police stations, increasing the number of care homes, day centers and night shelters, etc. The Guwahati Master Plan also has provision for preserving spaces for recreational purposes, ensuring housing for the poor, accessible pharmacies, community health care centers, and markets. To see how efficiently this particular master plan has been drafted and planned in terms of Goal 11, the next section discusses the current perspectives of older people in the city and explores their needs.

24.5 INSIGHTS FROM SENIOR CITIZENS OF GUWAHATI CITY

So far, this study has shown that concern for the elderly is a global issue. In India too, efforts have been made to address their cause, especially in the cities, in order to accomplish SDG 11. We now have an idea of what elder-friendly cities should look like, which is synchronous with the findings of our study. The narratives from older residents of Guwahati city presented here paint a similar picture. On average, most of the problems they experience are related to infrastructure, health and welfare, social and community support, a pollution-free habitable environment, safety and security, and other issues like lack of technological know-how.

Seven representatives from elderly community of Guwahati city were interviewed. Table 24.4 and 24.5 present brief biographical information about the interviewees.

TABLE 24.4
Brief Biographical Information of the Interviewees (1)

Sl. No	Name	Sex	Locality	Age	Marital Status	Family Type	No of Family Members
1.	Pashupati Nath	M	Ambari	79	Widowed	Nuclear	4
2.	Rajendra Sharma	M	Rehabari	71	Married	Nuclear	2
3.	Anil Kr. Choudhury	M	Fatasil	67	Married	Joint	7
4.	Mridul Bhuyan	M	Six Mile	64	Married	Nuclear	1
5.	Jugen Ch. Konwar	M	Chandmari	64	Married	Nuclear	3
6.	Shakib Ali	M	Bamunimaidan	62	Married	Nuclear	2
7.	Neena Prasad	F	Pan Bazzar	61	Widowed	Nuclear	1

TABLE 24.5
Brief Biographical Information of the Interviewees (2)

Sl. No	Name	Source of Income	Language	Caste	Religion	Educational Qualification
1.	Pashupati Nath	Pension	Hindi	General	Hindu	High school
2.	Rajendra Sharma	Business	Gujarati	General	Hindu	High secondary
3.	Anil Kr. Choudhury	Pension	Hindi	General	Hindu	High secondary
4.	Mridul Bhuyan	Pension	Assamese	General	Hindu	Graduate
5.	Jugen Ch. Konwar	Rental Assets	Assamese	General	Hindu	Postgraduate
6.	Shakib Ali	Business	Assamese	General	Islam	Middle school
7.	Neena Prasad	Business	Hindi	General	Hindu	Postgraduate

The Guwahati Master Plan (see Section 24.4.2) has prioritized transportation and connectivity in the city. The main aim of the plan, as already discussed, is protecting and preserving green spaces, developing infrastructure for the vulnerable and improving medical facilities. Guwahati is part of the Smart Cities Mission, and in this context too we have seen provision being made for a better life for city residents, especially for the vulnerable population. Our findings suggest development with regard to the provisions being made for elders in the city. Most of the problems that have arisen are again age specific; people in their early 60s have comparatively lesser problems than those in their late 60s or 70s. Neena Prasad, 61, mentioned

> I have just entered my 60s, people of my age who are physically fit and active generally do not have any issue with public transportation in the city. But I have seen people struggling due to the height of the buses to get in and off them and they are either unfit or older than me. As far as I am concerned, I have no issues with it.

Regarding the infrastructure in the city, they almost all have issues with footpaths, traffic, transportation, unsuitable signage, connectivity to markets, parking etc. Most of them regard transportation as the main problem, and some have highlighted other problems such as accelerating growth and the establishment of commercial activities in residential areas, footpaths obstructed by local vendors, lack of bicycle lanes, unsuitable drainage and waterlogging. Views include the following:

> … there are many problems, a lot of pollution, traffic congestion, connectivity issues etc. Moving from one place to another requires a lot of effort, especially when on public transport. Even if I use a private vehicle, it is very difficult to find a suitable parking spot and I am forced to walk to my destination. This is very problematic as I suffer from a neurological disorder.
>
> Other than the footpaths I don't have any issue. It was constructed about 2 to 3 years back but in front of most of the houses the height is lowered by 1 to 1.5 feet approximately. This creates difficulty in walking as it becomes very inconvenient and risky particularly on rainy days.
>
> (**Pashupati Nath**, 79)

> While I consider public transportation an efficient one but I still think connectivity can be improved. Lack of connectivity in all the places in the city compels persons to use private vehicles. If this could be improved many problems can be solved such as overcrowded markets, lack of walking space, traffic jams, pollution etc.
>
> (**Anil Kumar Choudhury**, 67)

Nevertheless, some others regarded the public transport provision as suitable and highlighted some positive aspects of the city's transportation system along with recent developments.

> I remember my initial days in Guwahati when buses were overloaded and ran on improper roads. But now when I look at the increased number of buses and bus stops, availability of seats, positive passenger behavior, I can state that a lot of work has been done by the government and the city's transportation system has evolved. We have seats reserved for ladies and senior citizens and pink buses are a recent addition.
>
> (**Neena Prasad**, 61)

> Back in 2014 I had major spinal surgery, as a result I was restricted from using any other vehicles than four wheelers. So, I extensively use public transport which I find relatively better than the other places I know. People are often emphatic to me as I use a walking stick which is why I am always offered a seat and an open passage. Also, I consider this act as a matter of respect which our culture has embraced and inherited from years.
>
> (**Anil Kumar Choudhury**, 67)

Satisfaction with social behavior as experienced by Mr Anil Kumar Choudhury contradicts Mr Pashupati Nath's experience. Observation suggests a lack of interaction in the latter compared to others who have considered social behavior satisfactory. Those satisfied by the behavior are mostly very active in society and relatively younger.

Apart from the transportation needs, some others criticize the city's market, as it is sectoral with specialist items in each sector, and emphasize the lack of accessible local markets that can offer everything in one place. Some take the view that elder-friendly elements are totally absent in the city, but as increased provision is being made for older people, the situation is changing.

> Before shifting to Guwahati from Tinsukia, I thought it suitable for living but my assumption was wrong. Both of them are commercial centers, however visiting markets and purchasing things was easier in Tinsukia than in Guwahati. Parking was not an issue, I used to park my car anywhere around and purchased all my stuff from one market. In Guwahati if I have to buy a book I will have to go to Pan Bazaar, for clothes either Fancy Bazar or Ganeshguri which I consider not very appropriate for people of my age. Moreover, the cost of living in Guwahati is higher than Tinsukia.
>
> (**Mridul Bhuyan**, 64)

> I see an overall upliftment in the city's provision for older people. Nowadays visiting any office is easy. Preference is given to older people in the banks, offices or any other public space. There are separate queues and speedy work for us compared to people younger in age.
>
> (**Neena Prasad**, 61)

In terms of medical facilities, the views are relative. Within the state Guwahati's hospitals are among the better equipped with advanced technology but when compared in a wider context there are significant differences of opinion. Since, Guwahati is the capital city it fulfills the needs of people across the state; as Northeast India's largest city it also serves as a medical hub to the neighboring states, which at times results in a shortage of beds for local people, especially those at an economic disadvantage. Apart from this, almost every locality has a pharmacy which is easily accessible even at night. However, the main problem is getting a doctor's appointment, and this needs a smart solution. Some narratives by the elders are:

> Here in Guwahati, doctors are confined to the hospitals and their chambers. Personal visit by a doctor is a rare event in the day hours and becomes almost impossible at night.
>
> (**Anil Kumar Choudhury**, 67)

> Medical facilities in my area are satisfactory. There are pharmacies and hospitals in close proximity to my residence. I know a doctor who resides nearby and at times we can approach him for basic medical help.
>
> (**Mridul Bhuyan**, 64)

However, this contentment with medical facilities is only due to personal acquaintance; it does not imply availability of suitable medical facilities for older people. Those who had access to a doctor in emergency had some personal connection with them.

The findings suggest that security in the city is also relative. There are areas which older people have recognized as safe, whereas there are areas that are not very safe. So, problems like this are area specific. However, compared to other big cities in the country, Guwahati is much more safe and secure, as described by the representative interviewees.

> … there are problems like theft, burgling and other insecurities. For example, I wish to visit the parks nearby but this place lacks suitable ambience for people of my age.
>
> (**Shakib Ali**, 62)

> Guwahati city, when compared to other big cities like Delhi, is a safe place. But there are instances of theft and robbery. My children stay out of the city as they are pursuing higher studies so sometimes,

> I feel vulnerable and insecure. I usually avoid going out of station leaving my wife alone in the house in absence of my children.
>
> (**Jugen Chandra Konwar**, 64)

Social participation and community support is the most relevant for senior citizens in a city. Our study shows that people aged 70 or above are less involved in society and are hence more prone to loneliness in their later years, while the 'young old' (early 60s) are active and more involved in society. This difference shows the need for different types of clubs, day centers, societies or organizations for elders that provide something for everyone.

> Social connection for me is limited. Most of the time I am unaware about the things taking place in my locality as we neighbors barely communicate on a daily basis. Under any instance of emergency, I prefer to seek help from my relatives.
>
> (**Pashupati Nath**, 79)

> I grew up in this locality and have spent my entire life here. So, basically, I am loved and respected by everyone in this area. I enjoy some privileges as I was the former president of a club nearby for 12 consecutive years. Currently I am an active member of the Puja and Bihu committee. Therefore, my interaction and participation in the neighborhood is limitless.
>
> (**Anil Kumar Choudhury**, 67)

One of the significant elements in the Smart Cities Mission is *robust IT connectivity and digitalization*. With this in mind, it is important to have some idea about technological advances, which our findings suggest is not the case with the elders of Guwahati city. Most of them do not use smartphones, or even if they own one their usage is limited. Good technological know-how correlates with good educational qualifications. In this study Jugen Chandra Konwar and Neena Prasad, both postgraduates, considered themselves competent to handle modern devices and software. But the rest of them were confined to basic uses such as making a phone call and messaging.

> ....most of the time it is my son who handles things related to technology. I was a government officer and learnt how to operate smart phones and computers while working. But my usage is basic and confined to making phone calls and playing games.
>
> (**Mridul Bhuyan**, 64)

> I am 64 years old and I feel I am technologically much sounder than others of my age. Surfing the internet and using modern-day gadgets is not a big deal for me. Most of the time I order my medicines online and use different e-commerce platforms to order utilities and other stuff. I even shop online for my children and usually send them gifts when they are away.
>
> (**Jugen Chandra Konwar**, 64)

In connection with green space and pollution-free environment, three significantly different perspectives were observed. Mr Pashupati Nath, 79, attempts to recreate the features of green space on his terrace. He said,

> I am very fond of village life. I wish I could sit in a place with all the village elements (trees, plants, animals, pure air and clear sky). Therefore, I have converted my terrace into a garden with around 200 species of plants and dedicate a significant amount of my time to it, i.e., around 3 to 4 hours a day.

While for Mr Mridul Bhuyan, 64, the solution to this problem was different; for him every locality should have parks and he said,

> I wish I could visit an open space or park regularly. In my area there is no park or open area. They are several kilometers away from my house and therefore not possible for me to visit.

Mrs Neena Prasad, 61, on the other hand opined that

> In the city, I see lack of proper designated places for elders. Had there been one where we could communicate and do different kinds of activities like play games, do exercise or yoga, I would love to be a part of it. At this age most of us feel lonely and such things I believe can even boost our life span.

The above statements were among the notable excerpts from the interview sessions, on the basis of which the following conclusions are drawn.

24.6 CONCLUSION

Sustainable development for the elderly in Indian cities, with reference to smart cities and master plans, is inclusive but has certain limitations. Concerns that are social or emotional in nature are mentioned very little or not at all, whether in the Smart Cities Mission or the master plans. From our interviews with older people in Guwahati city, it appears that it is the social components that are more needed by older people. In fact, most of the infrastructural needs are also related to social needs, such as construction of age-friendly clubs, day centers, care homes, etc. Lack of technical knowledge and awareness is another problem; to solve these, Kurian and Uchiyama's (2012) models from Japan and Netherlands mentioned in Section 24.2 can be helpful. Mr Jugen Chandra Konwar, an elderly resident of Guwahati city, stated that

> Feedback is the key to successful planning. If we replicate the models of successful age-friendly cities while incorporating the basic necessities that a person of my age would require, as well as involve multiple representatives from different age groups, I believe we can get a clear picture of the existing problems.

Therefore, the focus should be on identifying the needs, planning and designing the cities in such a way that they are community friendly and inclusive for all age groups. Only then we can fulfill the objectives of Sustainable Development Goal 11. Despite this, it can be stated that Indian cities have laid foundations for attaining sustainability that will certainly be expanded in the coming years.

REFERENCES

Alley, D., Liebig, P., Pynoos, J., Banerjee, T., Choi, I.H. 2007. "Creating Elder-Friendly Communities." *Journal of Gerontological Social Work* 49, no. 1–2: 1–18. Accessed October 23, 2020. doi: 10.1300/J083v49n01_01.

Bengaluru Metropolitan Region Development Authority. 2031. n.d. *Bangalore Master Plan*. Bangalore, India. Accessed February 22, 2021. http://bmrda.karnataka.gov.in/

Blekesaune, Arild, Marit, Haugen S. 2018. "Ageing in Norwegian Rural and Urban Communities." *European Countryside*, 10 no. 2: 232–246. Accessed July 07, 2020 doi: 10.2478/euco-2018-0014.

Buffel, Tine, Phillipson, Thris. 2016. "Can Global Cities be Age-Friendly Cities? Urban Development and Ageing Population." *Cities* 55: 94–100. Accessed October 23, 2020 doi: 10.1016/j.cities.2016.03.016.

Census of India. 2011. *Population Projections for India and States*. 2019. Report of the Technical Group on Population Projections. Accessed October 07, 2020. censusindia.gov.in

Central Statistics Office. 2016. *Elderly in India-Profile and Programmes*. India: Ministry of Statistics and Programme Implementation, GOI. Accessed February 01, 2021. www.mospi.gov.in

Creswell, John W. 2014. *Research Design: Qualitative, Quantitative, and Mixed Methods Approaches*. 4th ed. Thousand Oaks: SAGE Publication, Inc.

Council of Europe Development Bank. 2014. *Ageing Populations in Europe: Challenges and Opportunities for the CEB*. Accessed October 21, 2020. www.coebank.org

Delhi Development Authority. 2010. *Master Plan for Delhi 2021*. Accessed October 21, 2020. https://dda.org.in/

Gant, Robert. 1997. "Elderly People, Personal Mobility and Local Environment." *Geography* 82, no. 3: 207–217. Accessed December 3, 2020. http://www.jstor.org/stable/40572887

Gleckman, Howard. 2019. "Reimagining the Delivery of Care for Older Adults with Chronic Conditions." *Generations: Journal of the American Society on Aging* 43, no. 1: 78–84. Accessed December 03, 2020. doi: 10.2307/26632567.

Guwahati Metropolitan Development Authority. 2009. *Master Plan for Guwahati Metropolitan Area -2025*. India. Accessed October 21, 2020. https://gdd.assam.gov.in/

Hoof, J Van, Kazak, J.K. 2018. "Urban Ageing." *Indoor and Built Environment* 27, no. 5: 583–586. Accessed October 23, 2020. doi:10.1177/1420326X18768160journals.sagepub.com/home/ibe

Hoof, Joost Van, Kazak, Jan K., Perek-Bialas Jolanta, M., Peek Sebestiaan, T.M. 2018. "The Challenges of Urban Ageing: Making Cities Age-Friendly in Europe." *International Journal of Environmental Research and Public Health* 15, no. 11: 2473. Accessed October 22, 2020. doi: 10.3390/ijerph15112473.

Kavšek, Marta, Bogataj, David. 2016. "Ageing in Place Driving Urban Transformations." *Journal of Universal Excellence* 5, no. 1 (February): 1–12. Accessed July 7, 2020. https://www.fosunm.si/media/pdf/ruo/201651/ruo_035_clanek_kavsek_Cogataj_prejeto_2016_02_04_koncen_za_objavo_doc_1.pdf

Komoltithi, Malisa. 2016. *Leadership Journeys: A Narrative Research Study Exploring Women School Superintendent's Meaning-Making of Leadership Development Experience*. Boston, MA: College of Professional Studies, Northeastern University. June 23, 2021. https://repository.library.northeastern.edu/theses_and_dissertations?utf8=%E2%9C%93&rows=10&q=malisa+komolthiti

Kurian, Rachel, Uchiyama, Chihiro. 2012. "Models of Elderly Care in Japan and The Netherlands: Social Quality Perspectives." *The International Journal of Social Quality* 2, no. 1: 74–88. Accessed February 01, 2021. http://www.jstor.org/stable/23972432

Ministry of Urban Development. 2015. *Smart Cities: Mission Statement and Guidelines*. India: Government of India. Accessed October 10, 2020. https://smartnet.niua.org/content/2dae72ca-e25b-4575-8302-93e8f93b6bf6

Mumbai Metropolitan Region Development Authority. 2016. *Mumbai Metropolitan Regional Plan 2016-2036*. India. Accessed October 22, 2020. mmrda.maharashtra.gov.in

National Institute of Rural Development and Panchayati Raj. n.d. *Policy for the Aged: Opportunities and Challenges*. Accessed January 21, 2021. http://nirdpr.org.in/nird_docs/PolicyfortheAged.pdf

Niti Aayog. 2018. *SDG India Index Baseline Report, 2018*. India: Government of India. Accessed November 4, 2020. http://niti.gov.in

Niti Aayog. 2019. *SDG India Index & Dashboard, 2019-20*. India: Government of India. Accessed July 17, 2021. http://niti.gov.in

Niti Aayog. 2020. *India Voluntary Review 2020 Decade of Action Taking SDG's From Global to Local*. India: Government of India. Accessed July 2, 2021. http://niti.gov.in

Paslithil, A. 2009. "An Aging Indian Population: Issues and Problems." *Proceedings of the Indian History Congress* 70: 1095–1099. Accessed February 28, 2021. http://www.jstor.org/stable/44147753

Pichler-Milanovič, Nataša, Foški, Mojca. 2015. "Green Infrastructure and Urban Revitalisation in Central Europe: Meeting Environmental and Spatial Challenges in the Inner City of Ljubljana, Slovenia." *Urbani Izziv* 26: S50–S64. Accessed December 03, 2020. http://www.jstor.org/stable/24920947

Shah, A.M. 1999. "Changes in the Family and the Elderly." *Economic and Political Weekly* 34, no. 20: 1179–1182. Accessed February 28, 2021. http://www.jstor.org/stable/4407956

Tiwari, A.K., Singh, Brijesh P., Patel, Vaishali. 2020. "Population Projection of India: An Application of Dynamic Demographic Projection Model." *Journal of Critical Reviews* 7, no. 7: 547–555. Accessed October 10, 2020 http://www.jcreview.com/?mno=107364

United Nations. 2020. *The Sustainable Development Goals Report 2020*. Accessed on July 5, 2021. https://unstats.un.org/sdgs/report/2020

World Health Organization. 2007. *Global Age-Friendly Cities: A Guide*. France. Accessed February 03, 2021. https://www.who.int/ageing/publications/Global_age_friendly_cities_Guide_English.pdf

World Health Organization. 2012. *Strategy and Action Plan for Healthy Ageing in Europe, 2012–2020*. Malta, Europe: Regional Committee for Europe Sixty-second session. Accessed October 20, 2020. https://www.euro.who.int/en/health-topics/Life-stages/healthyageing/publications/2012/strategy-and-action-plan-for-healthy-ageing-in-europe,-20122020

Index